Cell Death

Cell Death is a compendium of recent and topical articles from Wiley's landmark *Encyclopedia of Life Sciences* (*ELS*), the leading resource in the life sciences. Spanning the entire spectrum of life sciences, *ELS* features more than 5,400 peer-reviewed articles, which are regularly updated, making it an essential read for life scientists and a valuable resource for teaching. *ELS* is available online at **www.els.net**, in full colour, with new and updated articles added regularly.

Cell Death

Editors

Gerry Melino
MRC Toxicology Unit, University of Leicester, Leicester, UK

David Vaux
La Trobe University, Victoria, Australia

WILEY-
BLACKWELL

Wiley-Blackwell is an imprint of John Wiley & Sons, formed by the merger of Wiley's global Scientific, Technical and Medical business with Blackwell Publishing.

Registered office: John Wiley & Sons Ltd, The Atrium, Southern Gate, Chichester, West Sussex, PO19 8SQ, UK

Other Editorial Offices:
9600 Garsington Road, Oxford, OX4 2DQ, UK
111 River Street, Hoboken, NJ 07030-5774, USA

For details of our global editorial offices, for customer services and for information about how to apply for permission to reuse the copyright material in this book please see our website at www.wiley.com/wiley-blackwell

ISBN: 978-0-470-71573-4

A catalogue record for this book is available from the British Library.

Typeset by MPS Limited, Bangalore, India

Contents

Contributors vii

Preface xi

The Siren's Song: This Death That Makes Life Live 1
Gerry Melino, Richard A Knight and Jean Claude Ameisen

The Origin and Evolution of Programmed Cell Death 13
Jean Claude Ameisen

Cell Death in *C. elegans* 21
Ataman Sendoel and Michael O Hengartner

Caspases and Cell Death 30
Lorraine D Hernandez, Caroline Houde, Maarten Hoek, Brent Butts, Donald W Nicholson and Huseyin Mehmet

The Apoptosome: The Executioner of Mitochondria-mediated Apoptosis 37
Elisabetta Ferraro and Francesco Cecconi

Caspases, Substrates and Sequential Activation 50
John G Walsh and Seamus J Martin

Dismantling the Apoptotic Cell 60
Paula Deming and Sally Kornbluth

The BCL-2 Family Proteins – Key Regulators and Effectors of Apoptosis 69
David L Vaux

BH3-only Proteins 75
Lina Happo, Andreas Strasser and Clare L Scott

Mitochondrial Outer Membrane Permeabilization 90
Melissa J Parsons and Douglas R Green

Mitochondria Fusion and Fission 97
Giovanni Benard, Guihong Peng and Mariusz Karbowski

Death Receptors 110
Peter H Krammer and Inna N Lavrik

Death Receptors at the Molecular Level: Therapeutic Implications 117
Marion MacFarlane

Death Receptor-induced Necroptosis 127
Wim Declercq, Franky Van Herreweghe, Tom Vanden Berghe and Peter Vandenabeele

Inhibitor of Apoptosis (IAP) and BIR-containing Proteins 138
David L Vaux

Structures, Domains and Functions in Cell Death (DD, DED, CARD, PYD) 147
Hao Wu and Yu-Chih Lo

Structure and Function of IAP and Bcl-2 Proteins 156
Mark G Hinds, Peter D Mace and Catherine L Day

Engulfment of Apoptotic Cells and its Physiological Roles 165
Rikinari Hanayama, Masanori Miyanishi, Hiroshi Yamaguchi, Jun Suzuki and Shigekazu Nagata

Autophagy 175
María Isabel Colombo and Hans-Uwe Simon

Autophagy in Nonmammalian Systems 189
Jahda H Hill and Eric H Baehrecke

Apoptosis: Regulatory Genes and Disease 197
James E Vince and John Silke

Caspases in Inflammation and Immunity 212
Philippe M LeBlanc and Maya Saleh

Immunity, Granzymes and Cell Killing 223
Nigel J Waterhouse, Olivia Susanto, Karin A Sedelies and Joseph A Trapani

P53 and Cell Death 230
Kamil Wolyniec, Sue Haupt and Ygal Haupt

Cornification of the Skin: A Non-apoptotic Cell Death Mechanism 240
Eleonora Candi, Richard A Knight and Gerry Melino

Apoptosis: Inherited Disorders 253
Helen C Su and Michael J Lenardo

From Reactive Oxygen and Nitrogen Species to Therapy 262
Scott R McKercher, Tomohiro Nakamura and Stuart A Lipton

Microbial Inhibitors of Apoptosis 272
Georg Häcker

Drug Discovery in Apoptosis 282
Tom O'Brien and Vishva M Dixit

Subject Index 293

Contributors

Jean Claude Ameisen Université Paris-Diderot, Faculté de Médecine Xavier Bichat, Paris, France
The Origin and Evolution of Programmed Cell Death; The Siren's Song: This Death That Makes Life Live

Eric H Baehrecke University of Massachusetts Medical School, Worcester, Massachusetts, USA
Autophagy in Nonmammalian Systems

Giovanni Benard University of Maryland Biotechnology Institute, Baltimore, Maryland, USA
Mitochondria Fusion and Fission

Brent Butts Merck Research Laboratories, Rahway, New Jersey, USA
Caspases and Cell Death

Eleonora Candi University of Tor Vergata, Department of Experimental Medicine, Rome, Italy
Cornification of the Skin: A Non-apoptotic Cell Death Mechanism

Francesco Cecconi Department of Biology, University of Rome Tor Vergata, and IRCCS Fondazione Santa Lucia, Rome, Italy
The Apoptosome: The Executioner of Mitochondria-mediated Apoptosis

María Isabel Colombo IHEM-CONICET- Facultad de Ciencias Médicas, Universidad Nacional de Cuyo, Mendoza, Argentina
Autophagy

Catherine L Day University of Otago, Dunedin, New Zealand
Structure and Function of IAP and Bcl-2 Proteins

Wim Declercq Molecular Signaling and Cell Death Unit, Department for Molecular Biomedical Research, VIB-Ghent University, Ghent, Belgium
Death Receptor-induced Necroptosis

Paula Deming University of Vermont, Burlington, Vermont, USA
Dismantling the Apoptotic Cell

Vishva M Dixit Genentech Inc., South San Francisco, California, USA
Drug Discovery in Apoptosis

Elisabetta Ferraro Department of Biology, University of Rome Tor Vergata, and IRCCS Fondazione Santa Lucia, Rome, Italy
The Apoptosome: The Executioner of Mitochondria-mediated Apoptosis

Douglas R Green St. Jude Children's Research Hospital, Memphis, Tennessee, USA
Mitochondrial Outer Membrane Permeabilization

Georg Häcker Institute for Medical Microbiology, Technische Universität München, Munich, Germany
Microbial Inhibitors of Apoptosis

Rikinari Hanayama Department of Medical Chemistry, Kyoto University Graduate School of Medicine, Kyoto, Japan
Engulfment of Apoptotic Cells and its Physiological Roles

Lina Happo The Walter and Eliza Hall Institute of Medical Research, Melbourne, Victoria, Australia
BH3-only Proteins

Sue Haupt The Peter MacCallum Cancer Centre, Melbourne, Victoria, Australia
P53 and Cell Death

Ygal Haupt Lautenberg Center for General and Tumor Immunology, The Hebrew University Hadassah Medical School, Jerusalem, Israel
P53 and Cell Death

Michael O Hengartner Institute of Molecular Biology, University of Zurich, Zurich, Switzerland
Cell Death in C. elegans

Lorraine D Hernandez Merck Research Laboratories, Rahway, New Jersey, USA
Caspases and Cell Death

Franky Van Herreweghe Molecular Signaling and Cell Death Unit, Department for Molecular Biomedical Research, VIB-Ghent University, Ghent, Belgium
Death Receptor-induced Necroptosis

Jahda H Hill University of Maryland Biotechnology Institute, College Park, Maryland, USA
Autophagy in Nonmammalian Systems

Mark G Hinds Walter and Eliza Hall Institute of Medical Research, Parkville, Victoria, Australia
Structure and Function of IAP and Bcl-2 Proteins

Maarten Hoek Merck Research Laboratories, Rahway, New Jersey, USA
Caspases and Cell Death

Caroline Houde Merck Research Laboratories, Rahway, New Jersey, USA
Caspases and Cell Death

Mariusz Karbowski University of Maryland Biotechnology Institute, Baltimore, Maryland, USA
Mitochondria Fusion and Fission

Richard A Knight MRC Toxicology Unit, University of Leicester, Leicester, UK *and* University College London, London, UK
Cornification of the Skin: A Non-apoptotic Cell Death Mechanism; The Siren's Song: This Death That Makes Life Live

Sally Kornbluth Duke University, Durham, North Carolina, USA
Dismantling the Apoptotic Cell

Peter H Krammer German Cancer Research Center, Heidelberg, Germany
Death Receptors

Inna N Lavrik German Cancer Research Center, Heidelberg, Germany
Death Receptors

Philippe M LeBlanc McGill University, Montreal, Canada
Caspases in Inflammation and Immunity

Michael J Lenardo National Institutes of Health, Bethesda, Maryland, USA
Apoptosis: Inherited Disorders

Stuart A Lipton Burnham Institute for Medical Research, La Jolla, California, USA
From Reactive Oxygen and Nitrogen Species to Therapy

Yu-Chih Lo Weill Cornell Medical College, New York, USA
Structures, Domains and Functions in Cell Death (DD, DED, CARD, PYD)

Peter D Mace University of Otago, Dunedin, New Zealand
Structure and Function of IAP and Bcl-2 Proteins

Marion MacFarlane MRC Toxicology Unit, University of Leicester, Leicester, UK
Death Receptors at the Molecular Level: Therapeutic Implications

Seamus J Martin Department of Genetics, The Smurfit Institute, Trinity College, Dublin, Ireland
Caspases, Substrates and Sequential Activation

Scott R McKercher Burnham Institute for Medical Research, La Jolla, California, USA
From Reactive Oxygen and Nitrogen Species to Therapy

Huseyin Mehmet Merck Research Laboratories, Rahway, New Jersey, USA
Caspases and Cell Death

Gerry Melino MRC Toxicology Unit, University of Leicester, Leicester, UK *and* University of Tor Vergata, Department of Experimental Medicine, Rome, Italy
Cornification of the Skin: A Non-apoptotic Cell Death Mechanism; The Siren's Song: This Death That Makes Life Live

Masanori Miyanishi Department of Medical Chemistry, Kyoto University Graduate School of Medicine, Kyoto, Japan
Engulfment of Apoptotic Cells and its Physiological Roles

Shigekazu Nagata Department of Medical Chemistry, Kyoto University Graduate School of Medicine, Kyoto, Japan
Engulfment of Apoptotic Cells and its Physiological Roles

Tomohiro Nakamura Burnham Institute for Medical Research, La Jolla, California, USA
From Reactive Oxygen and Nitrogen Species to Therapy

Donald W Nicholson Merck Research Laboratories, Rahway, New Jersey, USA
Caspases and Cell Death

Tom O'Brien Genentech Inc., South San Francisco, California, USA
Drug Discovery in Apoptosis

Melissa J Parsons St. Jude Children's Research Hospital, Memphis, Tennessee, USA
Mitochondrial Outer Membrane Permeabilization

Guihong Peng University of Maryland Biotechnology Institute, Baltimore, Maryland, USA
Mitochondria Fusion and Fission

Maya Saleh McGill University, Montreal, Canada
Caspases in Inflammation and Immunity

Clare L Scott The Walter and Eliza Hall Institute of Medical Research, Melbourne, Victoria, Australia
BH3-only Proteins

Karin A Sedelies Apoptosis and Natural Toxicity Laboratory, Peter MacCallum Cancer Centre, Melbourne, Victoria, Australia
Immunity, Granzymes and Cell Killing

Ataman Sendoel Institute of Molecular Biology, University of Zurich, Zurich, Switzerland
Cell Death in C. elegans

John Silke La Trobe University, Melbourne, Victoria, Australia
Apoptosis: Regulatory Genes and Disease

Hans-Uwe Simon Institute of Pharmacology, University of Bern, Bern, Switzerland
Autophagy

Andreas Strasser The Walter and Eliza Hall Institute of Medical Research, Melbourne, Victoria, Australia
BH3-only Proteins

Helen C Su National Institutes of Health, Bethesda, Maryland, USA
Apoptosis: Inherited Disorders

Olivia Susanto Apoptosis and Natural Toxicity Laboratory, Peter MacCallum Cancer Centre, Melbourne, Victoria, Australia
Immunity, Granzymes and Cell Killing

Jun Suzuki Department of Medical Chemistry, Kyoto University Graduate School of Medicine, Kyoto, Japan
Engulfment of Apoptotic Cells and its Physiological Roles

Joseph A Trapani Cancer Cell Death Laboratory, Peter MacCallum Cancer Centre, Melbourne, Victoria, Australia
Immunity, Granzymes and Cell Killing

Tom Vanden Berghe Molecular Signaling and Cell Death Unit, Department for Molecular Biomedical Research, VIB-Ghent University, Ghent, Belgium
Death Receptor-induced Necroptosis

Peter Vandenabeele Molecular Signaling and Cell Death Unit, Department for Molecular Biomedical Research, VIB-Ghent University, Ghent, Belgium
Death Receptor-induced Necroptosis

David L Vaux La Trobe University, Victoria, Australia
Inhibitor of Apoptosis (IAP) and BIR-containing Proteins; The BCL-2 Family Proteins – Key Regulators and Effectors of Apoptosis

James E Vince The University of Lausanne, Epalinges, Switzerland
Apoptosis: Regulatory Genes and Disease

John G Walsh Department of Genetics, The Smurfit Institute, Trinity College, Dublin, Ireland
Caspases, Substrates and Sequential Activation

Nigel J Waterhouse Apoptosis and Natural Toxicity Laboratory, Peter MacCallum Cancer Centre, Melbourne, Victoria, Australia
Immunity, Granzymes and Cell Killing

Kamil Wolyniec The Peter MacCallum Cancer Centre, Melbourne, Victoria, Australia
P53 and Cell Death

Hao Wu Weill Cornell Medical College, New York, USA
Structures, Domains and Functions in Cell Death (DD, DED, CARD, PYD)

Hiroshi Yamaguchi Department of Medical Chemistry, Kyoto University Graduate School of Medicine, Kyoto, Japan
Engulfment of Apoptotic Cells and its Physiological Roles

Preface

Individual cells face three choices: to divide (mitosis), to specialize (differentiate) or to commit suicide (cell death). The balance between these processes ensures that the number of cells in an organism remains essentially in functional equilibrium. While mitosis and differentiation have received detailed attention from cell and molecular biologists for well over a century, physiological cell death has become a major interest only in the last twenty years. Until recent times, most reports of cell death focussed on "accidental" cell death, or "cell killing", where a cell dies because a vital metabolic process necessary for its continued survival is blocked. We now know that in multicellular organisms cell death by suicide is far, far, more common than death of cells because they have been killed.

In retrospect, it is surprising that for a long time biologists never questioned the fate of so many duplicating cells in our body. If we imagine an 80-year-old person in which mitosis proceeded unopposed by any balancing homeostatic death process, he would have around two square km of skin, two tons of bone marrow and lymph nodes, and a gut 16 km long. Indeed, mitosis unchecked by cell death results in neoplastic pathology. Ironically, it was study of just such a neoplasia - follicular lymphoma - that led to identification of the first component of the mechanism for cell suicide. While determining more about the mechanisms for cell death have continued to reveal much about the origins of malignant disease, they have also provided new insights into diseases caused by too much or unregulated cell suicide, such as neurodegenerative diseases.

Searching journal articles in the last twelve months using the terms "cell death" or "apoptosis" yields about 20,000 publications, yet the same search in the year 1987 identifies only 439 publications. The reason for this tremendous growth in interest in cell death research is that many of the molecular mechanisms by which cells kill themselves have been discovered, and abnormalities in the regulation of cell death have been linked to human disease.

Unfortunately, because of the rapid growth of the field, many of the publications on cell death are not totally reliable, have been contradicted, or remain controversial, which can be misleading for students or clinicians new to the area. Nonetheless, several new drugs have been developed based on our current understanding of the molecular mechanisms of death, and many advanced clinical trials look highly promising.

Since cell death has now become translational, and therefore of interest to clinicians, pharmacologists and medical chemists, as well as to basic biologists, it seems an appropriate time to produce this book. In it, we have attempted to clarify the inconsistencies in the literature, in particular by referring to more definitive studies now available using transgenic or gene deleted mice. We have also tried to highlight those as yet unexploited molecular pathways susceptible to therapeutic intervention.

Our thanks go first to our publisher, who stimulated us in this endeavour, and to the scientists who dedicated some of their precious time writing the individual chapters. Our apologies go to our many colleagues who could not be mentioned and properly credited because of time and space limitations. We hope our readers find our efforts insightful and rewarding.

Gerry Melino and David Vaux
Leicester, UK and
Victoria, Australia
February 2010

The Siren's Song: This Death That Makes Life Live

Keynote article

Gerry Melino, *MRC Toxicology Unit, University of Leicester, Leicester, UK*

Richard A Knight, *MRC Toxicology Unit, University of Leicester, Leicester, UK*

Jean Claude Ameisen, *Université Paris-Diderot, Faculté de Médecine, Xavier Bichat, Paris, France*

Individual cells can divide (mitosis), specialize (differentiate) or undergo programmed cell death (apoptosis). The balance between these processes ensures that the number of cells in an organism remains essentially constant. In the past 30 years, the molecular mechanisms of cell death have been identified (caspases, Bcl-2 family, death receptors and apoptosome), with their clinical implications and therapeutic exploitation. Here, we review the entire process from a philosophical and historical viewpoint.

> *Thereby no ship of men ever escapes that comes thither, but the planks of ships and the bodies of men confusedly are tossed by the waves of the sea and the storms of ruinous fire.*

'There is only one serious philosophical problem. It is suicide. It is to judge whether life is or is not worth living'. Thus Albert Camus, following Homer (**Figure 1**), Novalis and Kierkegard puts suicide at the centre of thinking. Since almost 3% of all 650 000 papers published annually in the life sciences are related to apoptosis (cell suicide), it would seem that biology has also entered an existentialist phase.

Individual cells face three choices; to divide (mitosis), to specialize (differentiate) or to commit suicide (apoptosis). The balance between these processes ensures that the number of cells in an organism remains essentially constant (**Figure 2**). However, while mitosis and differentiation have received detailed attention from cell and molecular biologists for well over a century, a morbid fascination with cell death has only recently become a major interest. Indeed, it is surprising that for a long time biologists never questioned the fate of so many duplicating cells in our body. This is particularly true, since if mitosis proceeded unopposed by any balancing homeostatic death process, an 80-year-old person would have *c.*2 Km^2 of skin, 2 tons of bone marrow and lymph nodes, and a gut 16 Km long – the radius of a major world city, including the unpleasant last 500 m (Melino, 2001).

Nevertheless, it took a long time for apoptosis to enter its current position in the limelight. Part of the explanation for this historical imbalance of interest lies in the apoptotic process itself. It is usually 20 times faster than mitosis and sightings of dying cells are rare as they are rapidly absorbed and degraded (phagocytosed) by neighbouring cells. This very speed makes scientific observation and description more difficult. Indeed, *in vivo*, in a given time frame, 20% of mitotic cells are equilibrated by 1% of apoptotic cells, the limit of detection. Hence, in contrast to passive cell death (necrosis) – where leakage and inflammation are distinctive features – apoptotic cells are engulfed and degraded by neighbouring cells without a trace (**Figure 3**). The apoptotic cell and its fragments effectively become surrounded by a dynamically remodelled but impermeable cocoon which prevents extracellular leakage of any potentially harmful intracellular contents, and which would otherwise cause surrounding inflammation and scarring (as observed in toxic death and necrosis). The introduction of the concept of apoptosis/programmed cell death and its molecular interpretation, which is still only partially understood, now allows us to understand how these cellular choices are made and how the entire process evolved (Ameisen, 2002; Koonin and Aravind, 2002). **See also**: Apoptosis: Regulatory Genes and Disease; Engulfment of Apoptotic Cells and its Physiological Roles; The Origin and Evolution of Programmed Cell Death

However, there may be deeper reasons for the late development of interest in cell death. To paraphrase

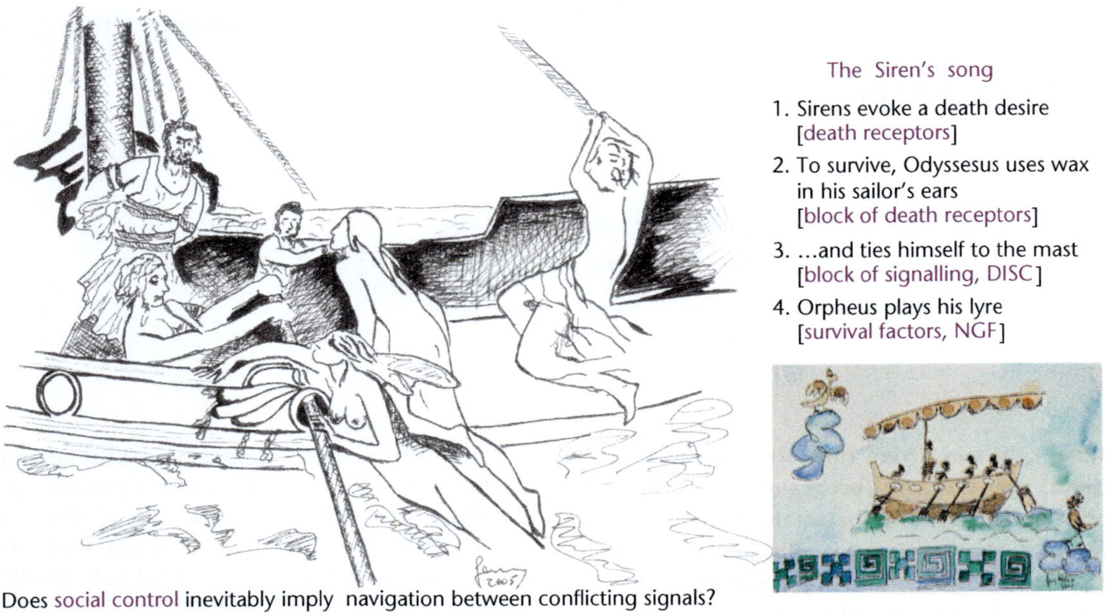

Figure 1 Odysseus is tempted by the Sirens. Homer first describes the death wish of the Siren's song, and the way Odysseus resists to survive. Indeed, Homer describes death (point 1 on the right) and two survival mechanisms (points 2 and 3). Similarly Orpheus (point 4) counteracts death signals by playing survival signals with his song.

Hermann Hesse's views: 'The Orient is pervaded with a totality or religiosity encompassing death, whereas the West is focused on logic and technology which breed divisiveness and which to a degree negate or even ignore death'. Death is therefore marginalized in the Occident, which may therefore account for its late incorporation into scientific consciousness. Although it may have been expected that the study of death would therefore emerge earlier in the East, the lack of development of necessary technology must also be borne in mind. Death (not Being) raises the question of what is, in fact, Being (see **Box 1**). Another German philosopher, Martin Heidegger, asked a related question; what does the verb 'to be' mean? We know that a table and my colleague 'are', but what constitutes their 'areness' as opposed to their absence? What is the Being, and its related modification that we call 'persistence during time'? One possible answer, more applicable to the animate situation, is to regard 'being' as a process, by which we imply at the level of the cell, the individual and even the species, both an identity and a sequence of finite adaptive change over time, and which becomes 'non-being' when the process is ended by fragmentation or death or extinction. In fact, beside technical difficulties, we were not philosophically ready to study 'Death', and in particular 'death from within'.

Thomas Kuhn argued that a scientific revolution begins with the perception of an anomaly. Accordingly, Gunther Stent believed that a scientific discovery starts when a series of implications could not be linked in a logical structure based on current knowledge – a process which can take a long time. Cell death clearly shows such a long gestation (Ameisen, 1999; **Figure 4** and **Figure 5**). Some morphological observations which we would now regard as apoptotic have been made since the middle of the nineteenth century without their biological significance being appreciated until recently. In 1842, Vogt recognized a form of physiological cell death, whereas Flemming, in 1855, used the term chromatolysis to describe the nuclear fragmentation seen during cell death – a characteristic still used, among others, as a hallmark of apoptosis. Other similar descriptions occurred occasionally in the nineteenth and early twentieth centuries. Recently the embryologist Glucksman (1951), the haematologist Bessis (1955) and the biologist Tata (1960) clearly described the morphological phases of apoptosis. In the 1960s, working on insect development, Richard Lockshin recognized the coordinated death of sheets of cells – a process he termed programmed cell death – and which he showed to be energy dependent and to require gene transcription (Lockshin and Williams, 1965). In 1966,

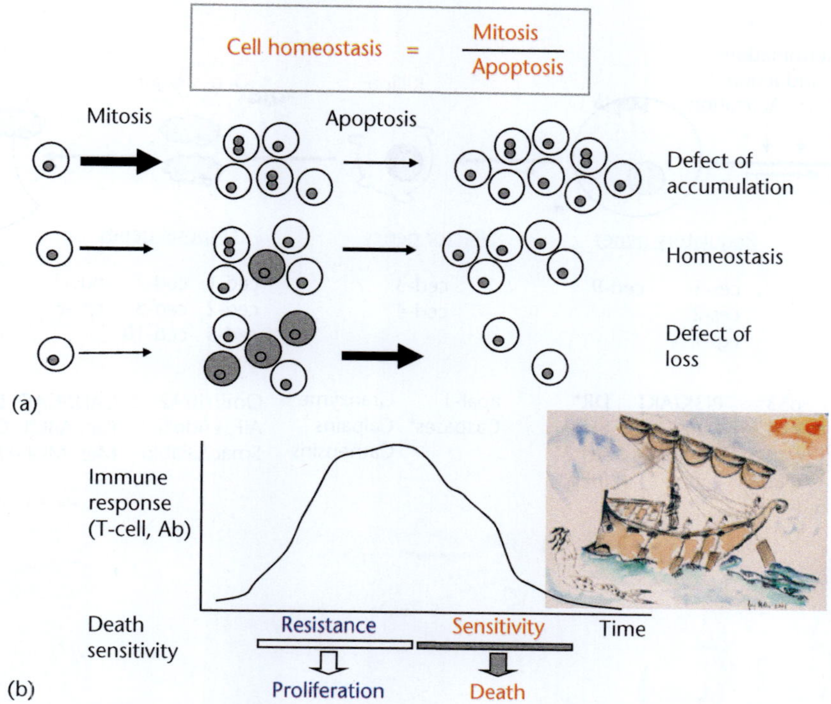

Figure 2 Death and homeostasis. (a) The basic importance of cell death is in counteracting mitosis to regulate homeostasis of cell number in tissues as well as in the entire organism. Consequently, unbalance of mitosis versus apoptosis results in pathologies with accumulation (e.g. cancer) or loss (e.g. neurodegeneration and AIDS) of cell numbers. (b) Physiological events such as immune responses require a tight regulation between death sensitivity and resistance.

John Saunders is able to review 'Death in embryonary systems', showing the role of cell death in moulding the body during development. It was not until the 1970s, when the young Australian pathologist John F Kerr start analysing the morphology of dying cells in histopathology, first alone in Brisbane (1964–1968), then in Edinburgh (1972–1980) where, in collaboration with Andrew H Wyllie and Alistair Currie, the term 'apoptosis' was first used (Kerr *et al.*, 1972). Lockshin and Kerr deserve credit for creating an intellectual framework for all previous observations, moving from scattered observations to interpretation, thus playing a pivotal role in the creation and diffusion of the concept, which has been highly conducive to its development.

It was not until apoptosis moved from the morphological to the mechanistic that it fully acquired scientific credibility and began to provide an intellectual framework for the previous scattered observations (Hengartner, 2000; Krammer, 2000; Meier *et al.*, 2000; Nicholson, 2000; Rich *et al.*, 2000; Savill and Fadok, 2000; Yuan and Yankner, 2000). For this, the credit is largely due to Sydney Brenner, John Sulston and mainly H Robert Horvitz and his collaborators (Michael Hengartner and Junying Yuan) (Metzstein *et al.*, 1998; **Figure 3**). Working with the nematode, *Caenorhabditis elegans*, Sulston first mapped the fate of every cell in the organism, including those that were to commit apoptosis. At first sight, it may seem bioenergetically pointless to generate cells which are then programmed to die. However, the vast majority of cell division is asymmetrical, according, for example, to an anterior–posterior axis of division, generating from a cell A two daughters A and B. If, for example we need two B (e.g. 'neuronal'), a first division generates A–B, and a further division of A, generates A–B, by killing A, we then obtain two B instead of one A and two B (Horvitz and Herskowitz, 1992). This illustrates one of the roles of cell death in the generation of differentiation during development. **See also**: Cell Death in *C. Elegans*

By 1990, Horvitz had shown that apoptosis was determined by several genes, including ced-9 (*The Good*, which blocks apoptosis), ced-3 (*The Bad*, which executes apoptosis) and ced-4 (*The Ugly*, which is an activator of apoptosis) (Avery and Horvitz, 1987;

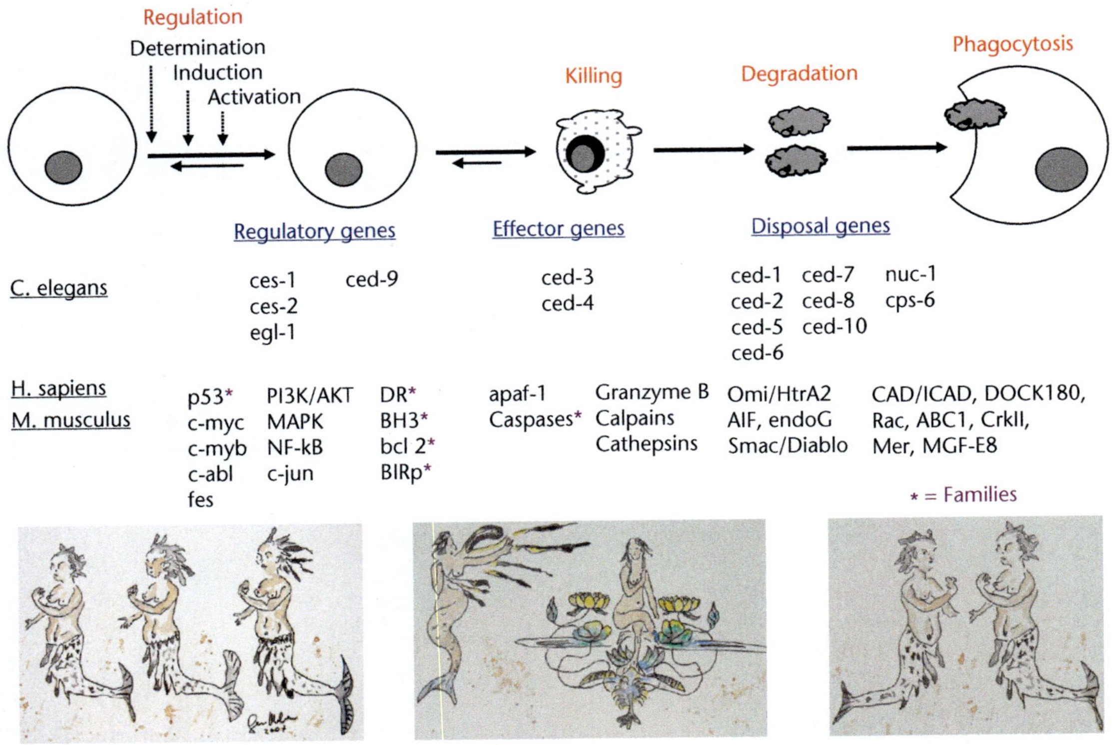

Figure 3 Mechanisms of cell death. Compared to living cells, apoptotic cells show cell shrinkage, smoothness of the cell membrane which remains intact, detachment of the nuclear membrane and condensation of chromatin (with fragmentation of DNA). The dead cell is recognized and phagocytosed by neighbouring cells, thus disappearing from the tissue. The entire process occurs within minutes. The genes involved can be distinguished into regulatory, effector (killing and degradation) and disposal genes, as indicated for the nematode and mammals. (* indicates families of proteins). The basic core mechanism of cell death requires a killer protease (ced-3/caspases) always ready to act, which requires an activator (ced-4/apaf-1) which in turn is repressed by a regulator (ced-9/Bcl-2, related to mitochondria): *ced-9* —| *ced-4* → *ced-3* → death. This core mechanism is activated by an activator (Egl-1/BH-3), and followed by the rapid disposal of the dead corps: Elg-1 —| *ced-9* —| *ced-4* → *ced-3* → death → phagocytosis.

Yuan and Horvitz, 1990; Hengartner *et al.*, 1992). In describing a molecular mechanism, this work provided a new intellectual stimulus. That these molecular developments were crucial is evident from the number of genes and pathways now identified in insects, mammals, as well as other species and their emerging physiological and pathological roles. Crucial, in this process, was the identification of the function of the ced-9 equivalent, Bcl-2 before the sequencing of ced-9 (Vaux *et al.*, 1988). These three main *C. elegans* genes have been highly conserved throughout evolution (Koonin and Aravind, 2002; Ameisen, 2002), such that they, or rather their corresponding gene families, still determine the apoptotic process in mammals. Thus in man, there are 21 *Goods* (the Bcl-2 family), 14 *Bads* (the caspase family), but still (so far) only one *Ugly* (Apaf-1 homologues) (**Figure 3**). **See also**: Caspases and Cell Death; The Apoptosome: The Executioner of Mitochondria-mediated Apoptosis; The Bcl-2 Family Proteins - Key Regulators and Effectors of Apoptosis

Caspases can be the target of viral (as first shown by Lois Miller) or cellular (BIR) inhibitory proteins modulating cell death, and are now being actively exploited for pharmaceutical purposes, for example by Don Nicholson's work (Nicholson *et al.*, 1995). Their molecular effector mechanisms were clarified by the brilliant work of Xiaodong Wang (Liu *et al.*, 1996) and the definition of a novel dedicated cellular organelle, the apoptosome, which activates the *Bads*. These 'master switches' have been highly conserved in evolution so that they, or, again, their equivalent families, still orchestrate apoptosis in mammals (**Figure 3** and **Figure 6**). Like all biological systems, however, death (like life) is not this simple, and apoptosis following mitochondrial damage or death receptor binding may not be inevitable. A number of regulatory sites have

Box 1 Being, Not-being and Death

Being is a fundamental theme in philosophy: the Soul is, in a certain way, the Being [*ens*] (Aristotle): *ens, quod natum est convenire cum omni ente* (soul) (Thomas Aquinas); the *res cogitans* from the *Coito ergo sum* (Descartes). A theme, perhaps, not fully elucidated as yet. Returning to our field, science could be defined as the possibility of interconnections founded on true propositions (Heidegger); and as behaviour/acts of man. Science has the way of being of this Being, defined as Being-in [*Insein*]. Hence, the comprehension of the Being is also the determination of the being of the Being-in. Consequently, science is a way of being of the Being-in, on which the Being relates (being, of the Being-in-the-World in Heidegger In-der-Welt-Sein). Therefore, the fundamental ontology must be found in the existential analysis of the Being-in-the-World, both in an ontic (determined in his being by existence) and ontologic (for its being-determined by existence) sense. Thus, biology, as the science of life, is found on the ontology of the Being-in-the-World. In a modern sense, science results in a phenomenology, from *Φειvoμεvov* (*id est*, manifest, bring to light, apparent) and *Λογoσ* (*id est*, discuss, be true, be false – thus including the possibility that science, as a phenomenology, might result in false assertions). Following Heidegger, the fundamental structural character or mode of the Being-there/here [*Dasein*] is not that of a subject or that of the object, but that of the coherence of the Being-in-the-world [*Insein*], with its related modifications, a concept that we define as time [*weltanshaung, zeitlichkeit, temporalitaet*]; its modalities are the emotional situation (fear, anguish) and the understanding (interpretation, assertion, discussion, curiosity, chat). For Heidegger, the anguish call of the ethical conscience is silence (see also Wittgenstein) and its sense is 'Death' (Being-toward-detah, in Heidegger Sein-zum-Tode).

been described which can interfere with pro-apoptotic signalling, and even some caspase-like molecules have been shown to have antiapoptotic effects. The *Bads* are not always so? This multiplicity of interacting pro- and antiapoptotic factors implies that, as in mitosis, a number of checks and balances are present in the apoptotic programme, and that suicide is carefully considered by the cell as it was by Camus, although a suicidal cell can scarcely be accorded the thoughtful dimension that Camus applied to human suicide. **See also**: Inhibitor of Apoptosis (IAP) and BIR-containing Proteins

There are further complications. How did these molecules and pathways evolve to their current genomic status without having been counter-selected (Ameisen, 1999)? Most likely, the genetic/biochemical subroutines have evolved from other pathways, for example related to mitosis or deoxyribonucleic acid (DNA) damage. It is necessary, moreover, to consider the danger of transferring definitions from one discipline to another, without also bringing confusing implications. Despite the fact that it is suggestive and attractive, the use of the words 'death' or 'suicide' carries implications that are different from those if we would have instead used 'dismantling' or 'disaggregating'. 'Death', for example, implies that there is only one death, that there is nothing after death, and that it is the final event. However, dead cells might 'die' more than once (erythroblasts 'die' when they lose their nuclei and mitochondria to become erythrocytes, and then 'die' again when they are eliminated from circulation; keratinocytes 'die' when differentiated and lose their nuclei and mitochondria, and then 'die' again during desquamation; the same applies to megakaryocytes and platelets). These cells remain active and functional after 'partial death' (e.g. erythrocytes transport oxygen, keratinocytes guarantee the barrier function of the epidermis, platelets provide aggregation and clotting), and death is not the final event (but it precedes differentiation in all earlier examples). Accordingly, if we use the term 'suicide', we bring in, subliminally, anthropological implications derived from the social and philosophical field. We could say that cells commit suicide for the benefit of the organism (altruistic death with social implications). We could also say that the organism kills innocent cells for its own selfish interest (egotistic death). Here, we should consider the definition of 'self' of the cell (I, cell, kill myself for the benefit of the organism), or 'self' of the gene/organism (I, gene/organism, kill the innocent cell for the survival of the genome/organism). But do genes, cells and organisms have a 'self'? **See also**: Cornification of the Skin: a Non-apoptotic Cell Death Mechanism

It is therefore only within the closing years of the last millennium, *The Golden Age*, that apoptosis has been given the scientific interest that its biological significance deserves, providing important research guidelines for the next decade. Recently, differences in the execution of apoptosis are emerging. Possibly, these new emerging facets of the original concept are related to cell-to-cell variation due to the fact that several molecular constituents of the process exists as large redundant molecular families which are differentially expressed in different tissues.

The concept of apoptosis has now reached the level of a fashion, **Figure 4**. With over 19 000 papers annually and three dedicated journals, it may be that over-emphasis is leading to distortion. The concentration on apoptotic death diminishes the physiological and pathological role of necrosis, and fields such as toxicology are desperately searching for a new identity. At the same time, the definition is changing, and acquiring subtle differences. Originally, the term 'programmed cell death' was a definition of a process (genetically and developmentally programmed) whereas 'apoptosis' implied a biochemical character (apoptosis = caspases). Now, the two terms are generally used as synonymous, as opposites to necrosis (passive, nongenetically programmed death). Recently, the term 'cell death' has become more general, including not just apoptosis and necrosis but also other types of death such as: autophagy (Kourtis and Tavernarakis, 2009), caspase-independent cell death, keratinization, Wallerian degeneration and erythrocyte karyorexis. Until definitions are linked to biochemical pathways, in other words, to process, however, these subdivisions remain semantic.

In stating that it is not possible to enter the same river twice, Heraclitus expressed the irreversibility of time. We too, like a flowing river, undergo continuous changes. If the molecules forming our body are continuously changing, what is the meaning of permanence? What controls the changes in the molecules that form our bodies? How do our cells socially interact between themselves to constitute a unified whole? Gradually, the idea emerged that equilibrium and stability of the body is maintained in a dynamic way by signals controlling life and death of single cell. In 1992, Martin Raff developed the concept of 'social control of life and death' (**Figure 7**). This is a concept of enormous power, implying that there are specific survival and death signals, and corresponding receptors on cells. In

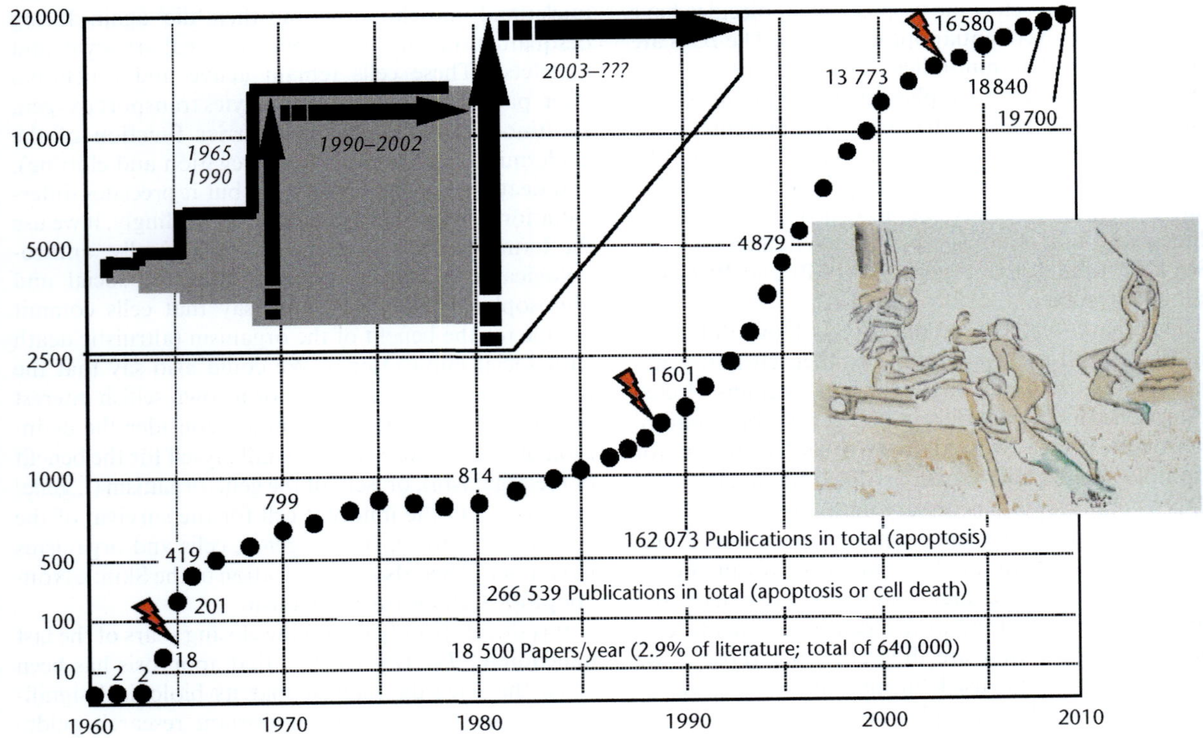

Figure 4 Scientific papers on cell death. A large number of scientific publications have focused on cell death. We might distinguish three phases, from scattered observations before 1965, when the original work in invertebrates and embryology described the phenomenon. From 1990, culminating with the 2002 Nobel Prize, the molecular events were identified. Recently, the detailed mechanisms have been investigated, whereas the clinical relevance, with its potential therapeutic exploitation is being explored. *Inset*, advancement occurs in steps, with pioneering explorative and controversial work, followed by consolidation and refining research.

1965	1. JK Kerr, AH Wyllie, J Currie, RA Lockshin, JR Tata, GT Williams – In the "pre-mechanistic" era, for recognising that cell suicide is a physiological process in most multicellular organisms.
1985	2. S Brenner, J Sulston, HR Horvitz – Determining the genetic pathway of programmed cell death in *C. elegans*, and showing that there are genes involved in cell death but no other process.
1990	3. Y Tsujimoto, CM Croce, ML Cleary, DL Vaux – Cloning of Bcl-2 and identifying it as a regulator of cell death, and showing that inhibition of cell death leads to cancer in humans.
1990	4. S Korsmeyer – Identification of Bax and other pro-apoptotic Bcl-2 family members.
1990	5. HR Horvitz, J Yuan, M Hengartner, G Salvesen, DW Nicholson – Showing that cell death in the worm and apoptosis in mammalian cells are the same, evolutionarily conserved process, and cloning of ced-3 and ced-9, showing that cell death is caspase dependent, and Bcl-2 family members inhibit it upstream.
1990	6. S Nagata, PH Krammer, D Wallach, M Lenardo, DR Green, V Dixit – Identification of components of the Death Receptor triggered apoptosis pathway; identification of related diseases.
1995	7. X Wang, G Kroemer, DR Green, TW Mak, S Lowe, GI Evan, M Karin, M Oren – Role of mitochondria and activation of apaf-1 by cytochrome c. Role of cell death in cancer.
1995	8. J Abrams, H Steller, L Miller, B Hay, RG Korneluk, DL Vaux, X Wang – Identification of insect cell death inhibitors (IAPs and p35) and pro-apoptotic proteins (Reaper, Grim, Hid). Identification of mammalian IAPs and IAP antagonists.
2000	9. Y Ohsumi, DJ Klionsky, B Levine, M Raff, PM Steinert, G Melino – Non-apoptotic cell death, including autophagy, Wallerian Degeneration and skin cornification.
2005	10. SH Rosenberg, T Oltersdorf, SW Fesik - Structure of Bcl-2 family members and development of ABT-263 and ABT-737 in clinical trials.

Figure 5 Who, what, when. One minute history on cell death in ten points: who and what. In addition to preliminary scattered observations that nowadays we would recognize as cell death, we distinguish three arbitrary gross phases of research, definition of the phenomenon (1965–1988), definition of the molecular mechanism (1988–2002), refinement of the molecular pathways with their clinical relevance and therapeutic exploitation, see also (Vaux, 2002). In addition to these 10 points, we would like to stress two major events, (i) the launch of the first dedicated journal in 1994 (Cell Death & Differentiation by Nature-Publishing-Group with G Melino), and (ii) the award of the Nobel Prize for Medicine in 2002 (to S Brenner, J Sulston, HR Horvitz). Years are very indicative, used as quinquennium. Because of the severe space limitation we sincerely apologize to all those colleagues who could not be mentioned here, though we recognize their essential and pivotal contribution to the field.

fact, independently from the *C. elegans* model, Peter Krammer and Shigekazu Nagata identified the first death receptor, CD95, and its ligand CD95L in the early 1990s. Receptors, and their related signals, are of particular importance in multicellular systems, such as the immune system, where signalling is crucial. Such social control of life and death turned out to be of vital importance in complex multicellular systems such as the immune system and the nervous system, where communication between cells is crucial. This implied that the concept of cell death developed late in evolution. Raff's model of apoptosis is linked to multicellularity, and its development therefore follows evolutionary complexity (e.g. immune and neural systems). **See also**: Death Receptors; Death Receptors at the Molecular Level: Therapeutic Implications

To the Sirens first shalt thou come, who bewitch all men, whosoever shall come to them. Whoso draws nigh them unwittingly and hears the sound of the Sirens' voice, never doth he see wife or babes stand by him on his return, nor have they joy at his coming; but the

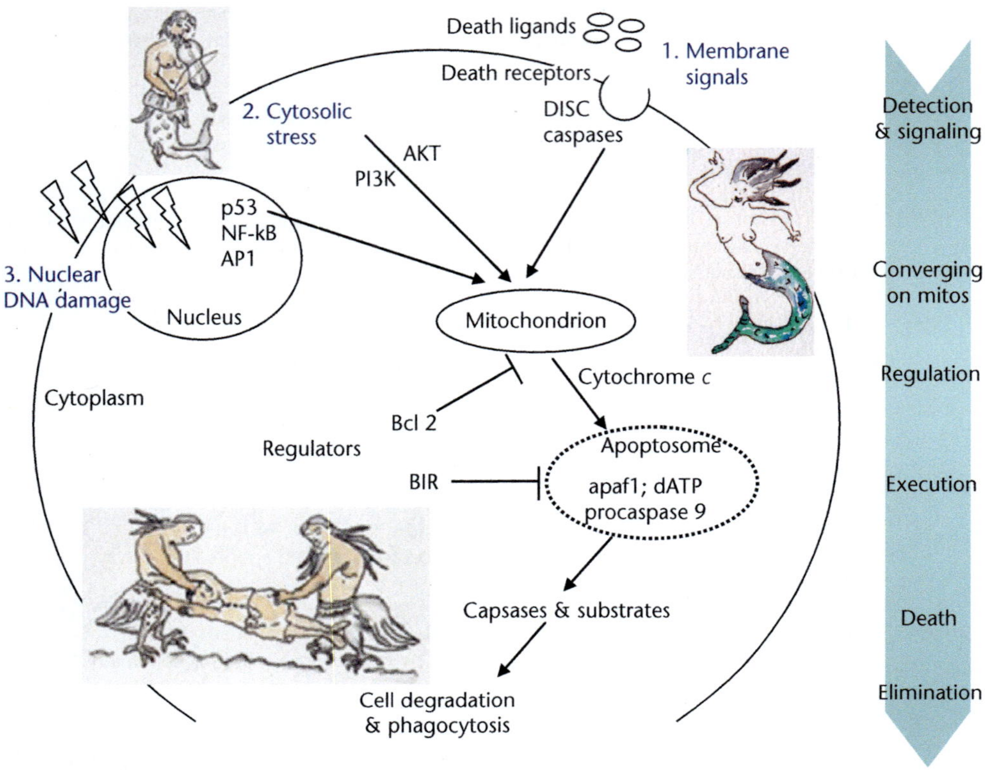

Figure 6 Molecular events of apoptosis. Cell death can be trigged by membrane (1 – death receptor, such as CD95), cytosolic (2 – metabolic stress signals) or nuclear (3 – DNA damage leading to p53 activation) events. Even though several pathways and cross-talk are elicited by individual triggers, not shown, the signals converge on the mitochondrion/apoptosome to activate the downstream effector caspases, which dismantle the cell components. Mitochondria play a regulatory role by releasing activating factors, under the control of the Bcl-2 proteins. The final regulation occurs at the apoptosomal level.

Sirens enchant him with their clear song, sitting in the meadow, and all about is a great heap of bones of men, corrupt in death, and round the bones the skin is wasting. But do thou drive thy ship past, and knead honey-sweet wax, and anoint therewith the ears of thy company, lest any of the rest hear the song; but if thou myself art minded to hear, let them bind thee in the swift ship hand and foot, upright in the mast-stead, and from the mast let rope-ends be tied, that with delight thou mayest hear the voice of the Sirens. And if thou shalt beseech thy company and bid them to loose thee, then let them bind thee with yet more bonds. But when thy friends have driven thy ship past these, I will not tell thee fully which path shall thenceforth be thine, but do thou thyself consider it, and I will speak to thee of either way.

To survive we must resist many death signals, and here the Greek myths return (Ameisen, 1999; Melino, 2001). To resist the persuasive songs of the Sirens (signals) leading to inevitable death, Odysseus was instructed by Circe near Naples to resist the temptation song of the Sirens by blocking hearing (receptors) or by blocking his movements (signalling) with ropes. The poet Orpheus, however, on boarding a ship in the territory of the Sirens (Jason seeking the Golden Fleece in the Argonauts), resisted the songs by singing loudly himself and by playing his lyre. For a mythological view of the Sirenes, see **Box 2**. Thus, he superimposed his life song (antiapoptotic) over the Sirens' death song (pro-apoptotic; see **Figure 1**). Does social control inevitably imply navigating between conflicting signals?

Social control (altruism, cheating, selfishness and combat) is of enormous power in evolution, and signals for this process can be translated to the molecular level. In 1959, the classic work by François Jacob and Jacques Monod demonstrated the existence of 'repressors' through which signals interact to modulate gene

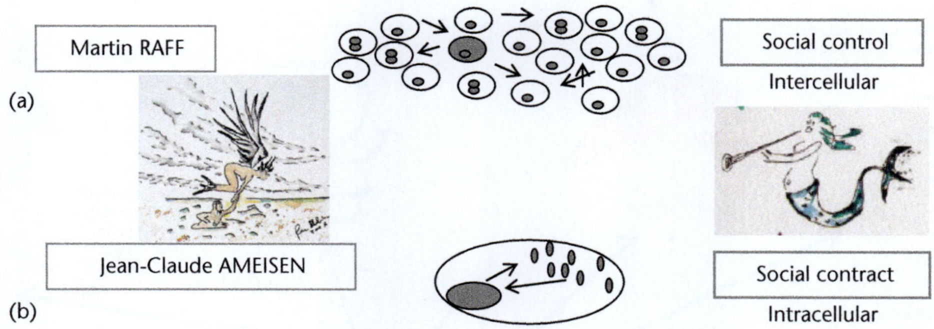

Figure 7 Death as symbiosis or social regulation. Cell death could be conceived as a result of a social cross-regulation between cells that need homoeostatic regulation in multicellular organisms, according to M Raff. Alternatively, it seems implicit in each individual cell to guarantee the ancestral symbiosis between mitochondria and nucleus in eukaryotic cells, according to JC Ameisen.

Box 2 The Sirens

Sirens are the personification of the hot summer dog days, when Sirius (hence their name) lights up and burns high in the sky. Indeed, the great gift that the Sirens gave men is laziness. The Sirens were lazy feathered women aiming at impeding men from working and performing their duties, whereas our mothers and sisters are active. In other words, the opposite of the tenet that men should work and women should dress attractively and go shopping. For Homer, the Sirens evoke a brutal image, against which the pleasant song contrasts in its beauty and purity of spirit. The latter image has been inherited by the Christian cult, as elegantly discussed by George Norman Douglas (1911, *Siren Land*. Pengeum, New York; London (1923 revised)). The pure Siren Parthenope found rest and death in the Gulf of Naples as Saint Lucia. All the Madonnas in Naples (Circe lived near Naples) are Queens of the Sea, like Madonna della Libera, Star of the Sea (Stella Maris), a reincarnation of the antique form of the Sirens, Leucothea, Euploea and Nereid. According to Tertullian, significant Christian figures originate from Mithraism. Mithras, like Christ, is the light of the world; Cybele, like Madonna, is the Grand Mother [Μητηρ] or Magna Mater. Moreover, December 25 was the celebration day of Mithras. Moral regeneration, drinking from the mystic cup, sacramental rituals, consecration of bread and water, confession of sins, flames on the altar, ascetic prayer, rest on Saturday, final judgement, martyrdom, resurrection, hope of immortality, expiation of guilt, baptism, purification of new followers, confirmation, penance, all originated from Mithraism. These, in turn, may be echoes of still older rituals associated with the celebration of the midwinter solstice on December 21/22 at Stone Age sites such as Maes Howe and Stonehenge.

expression. At the molecular level, Changeaux, Monod and Wyman in 1964 created a model of cooperativity with positive and negative homotropic and heterotropic interactions which can finely tune catalytic activity of enzymes, ligand or receptor binding (Wyman and Gill, 1990). Some doubt on the coevolution of apoptosis and multicellularity has been cast by the recognition of a form of cell death in unicellular organisms which is regulated by intercellular signalling and has some morphological and molecular features of multicellular apoptosis. Indeed, ancestral (and modern) unicellular parasites have to protect themselves from the hostile environments of more than one host, and it may be that the pro-survival components of the dialectic that is the apoptotic pathway have a more ancient evolutionary origin than the pro-death signals. It is therefore no surprise that social control extends to unicellular organisms (**Figure 8**). Jean-Claude Ameisen provides examples of repressor models, and in particular toxin-antidote modules, in evolution. In fact, JC Ameisen, P Golstein and JC Reed have described forms of cell death that have been observed in unicellular organisms, including bacteria, the protozoa *Trypanosoma* and *Tetrahymena* and the amoeba *Dictyostelium*. In the bacterium *Escherichia coli* several genes organized as toxin-antidote modules control the equilibrium of life and death. Most are encoded by plasmids, but some by the bacterial chromosome itself. This is the case for the toxin encoded by mazF (which fragments the genome) that is neutralized by the antidote encoded by mazE, which is continuously degraded by a protease, ClpP. It seems therefore, that even unicellular organisms

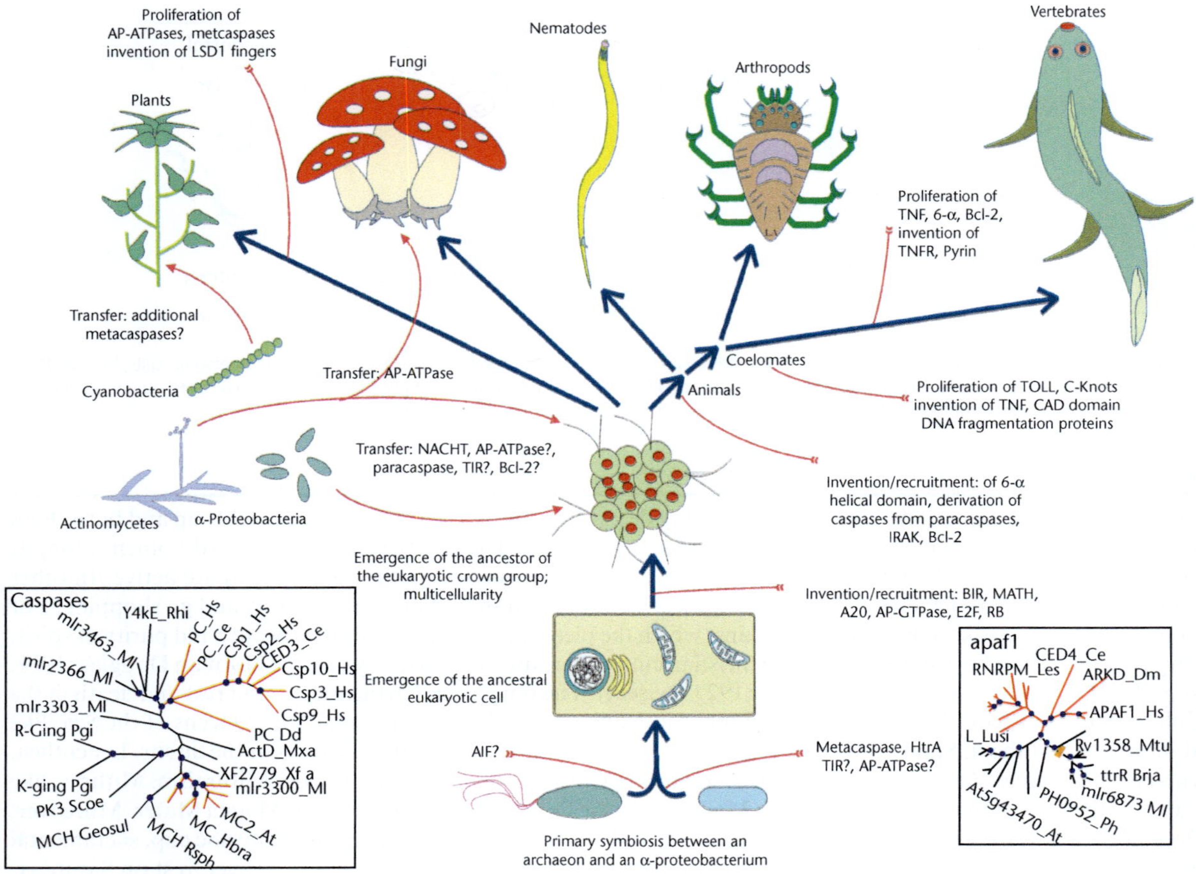

Figure 8 Evolution of cell death. Molecular pathways seem to have evolved from specific biochemical routines and subroutines using specific genetic/proteic modules such as AP-ATPase, Tir, BIR, NACHT and MATH domains which evolved and transferred across evolution. The insets show a simplified evolutionary tree of caspases and apaf-1.

experience life as a continuous inhibition of self-destruction (Ameisen, 2002). As with Sisyphus, the rock he is pushing up the mountain always rolls down before he reaches the summit (The Myth of Sisyphus by Camus); so unicellular organisms may also experience life as a continuous inhibition of death. **See also**: The Origin and Evolution of Programmed Cell Death

Returning to eukaryotic cells, we can now revisit the concept of cell death as a result of intracellular (according to Ameisen) rather than intercellular (according to Raff) signals (Raff, 1992; Ameisen, 2002; **Figure 7**).

It all happened about two billion years ago. Owing to the photosynthetic activity of cyanobacteria, the atmosphere changed from a reducing (hydrogen) to an oxidizing (oxygen) environment. The forms of life changed as a consequence, and evolved. The enormous reactivity of oxygen in its biatomic forms (OO, CO, NO and their redox products) with metals and thiols allowed the development of haem centres which replaced the older iron–sulfur and iron–oxygen centres, allowing more flexible protein structures (e.g. α-helix rather than the more rigid β-structures), leading to more regulation (allostery), and resulting in enormous ranges of electronegativity. At that time interesting symbioses developed, for example between hydrogen-producing and hydrogen-consuming bacteria, maximizing biological performance. Eukaryotic cells may be the result of such a symbiotic liason between bacterial hydrogen-producers (eukaryotic progenitor bacteria) and hydrogen-consumers (mitochondria). This metabolic symbiosis based on hydrogen (or syntrophy), involving an archaebacterium and a eubacterium could thus explain the chimeric-symbiotic origin of the eukaryotic cell. Though some bacteria captured mitochondria giving rise to proto-animal cells, others also captured chloroplasts, becoming proto-vegetal cells. A tight symbiosis implied the presence of essential constituents of one element produced by the other element, and also of toxin-antidote models. However, acquisition of an organelle as potentially dangerous as a respiratory chain necessitates that the dangers are circumscribed. Should this control be compromised, then it is easy to see how mitochondria can kill the cell by exporting cytochrome *c*, Diablo/Smac, Omi/HtrA2 and other yet-to-be-discovered elements, and fragment its genome. Similarly the cell produces the antidote in the form of caspase inhibitors and Bcl-2-like components. Thus, a dialectic is established between producing and consuming elements of the symbiotic partnership. However, the toxic potential of components, such as the electron transport chain, of the consuming partner led to their sequestration within the mitochondria. It would therefore be logical for any mitochondrial damage to lead to the export of signals which kill the cell. The cell in turn protects itself by producing antidotes such as caspase inhibitors and antiapoptotic Bcl family proteins. Life is the equilibrium between these elements and requires the continuous stabilization of such equilibrium, like Sisyphus rolling his stone. Life results from the continuous suppression of death (Ameisen, 2002).

Friends, forasmuch as it is not well that one or two alone should know of the oracles that Circe, the fair goddess, spake unto me, therefore will I declare them, that with foreknowledge we may die, or haply shunning death and destiny escape. First she bade us avoid the sound of the voice of the wondrous Sirens, and their field of flowers, and me only she bade listen to their voices. So bind ye me in a hard bond, that I may abide unmoved in my place, upright in the mast-stead, and from the mast let rope-ends be tied, and if I beseech and bid you to set me free, then do ye straiten me with yet more bonds.

Thus I rehearsed these things one and all, and declared them to my company. Meanwhile our good ship quickly came to the island of the Sirens twain, for a gentle breeze sped her on her way. Then straightway the wind ceased, and lo, there was a windless calm, and some god lulled the waves. Then my company rose up and drew in the ship's sails, and stowed them in the hold of the ship, while they sat at the oars and whitened the water with their polished pine blades. But I with my sharp sword cleft in pieces a great circle of wax, and with my strong hands kneaded it. And soon the wax grew warm, for that my great might constrained it, and the beam of the lord Helios, son of Hyperion. And I anointed therewith the ears of all my men in their order, and in the ship they bound me hand and foot upright in the mast-stead, and from the mast they fastened rope-ends and themselves sat down, and smote the grey sea water with their oars. But when the ship was within the sound of a man's shout from the land, we fleeing swiftly on our way, the Sirens espied the swift ship speeding toward them, and they raised their clear-toned song:

Hither, come hither, renowned Odysseus, great glory of the Achaeans, here stay thy barque, that thou mayest listen to the voice of us twain. For none hath ever driven by this way in his black ship, till he hath heard from our lips the voice sweet as the honeycomb, and hath had joy thereof and gone on his way the wiser. For lo, we know all things, all the travail that in wide Troy-land the Argives and Trojans bare by the gods' designs, yea, and we know all that shall hereafter be upon the fruitful earth.

Understanding apoptosis is understanding this ancient, extremely powerful but dangerous symbiotic liaison. Any change in its equilibrium from inside (DNA damage, metabolic or cell-cycle aberrations) or outside (signals and receptors) will irreversibly activate suicide within minutes. The result is a mitochondria-centred view of life and death (Oberst *et al.*, 2008; **Figure 6**). If, however, the symbiosis occurred later in evolution, in an oxygen-rich environment, this mitochondria-centred view of apoptosis will need rethinking. Will the current primacy of the mitochondrion fade from fashion like Daxx, FLASH and ceramide? Apoptosis, however, is not the sole phenotype of cell suicide. Various pathways of self-destruction seem to coexist in our cells that may have been progressively selected during evolution. And several gene products recruited into these death pathways also seem to participate in the regulation of mitosis and differentiation, blurring the frontiers between 'programmes' of life and death. **See also**: Mitochondria Fusion and Fission; Mitochondrial Outer Membrane Permeabilization; P53 and Cell Death

With the increased prominence of apoptosis in biological science has become a shift in our philosophical attitude to many disease pathologies. Vaux proposed that cancers are not solely disorders of mitosis, but rather a fundamental failure of a pre-neoplastic cell to do the decent thing and commit suicide (Vaux *et al.*, 1988). Similarly, Ameisen proposed that previously rather bewildering diseases such as acquired immunodeficiency syndrome (AIDS; Ameisen and Capron, 1991) and neurodegenerative diseases may result from too much rather than too little programmed cell death (Ameisen, 1999). Several diseases (e.g. cancer and autoimmunity) caused by mutations of the cell death machinery (e.g. CD95, caspases, apaf-1, Bcl-2 and p53) have already been identified. The concept of cell death or apoptosis is therefore pivotal for future research. The complexity and subtlety of the apoptotic process not only allows cells to control their own fate, but also provides pharmacologists and doctors with a new range of therapeutic possibilities to control them. However, the effectiveness and selectivity of these interventions will depend on our capacity to dissect the diverse and subtle interplay that has evolved between the molecular mechanisms that regulate mitosis, differentiation and death. **See also**: Drug Discovery in Apoptosis

References

Ameisen JC (1999) La sculpture du vivant. Le suicide cellulaire ou la mort créatrice. Ed. du Seuil, Paris; Points Seuil 2003.

Ameisen JC (2002) On the origin, evolution, and nature of programmed cell death: a timeline of four billion years. *Cell Death and Differentiation* **9**(4): 367–393.

Ameisen JC and Capron A (1991) Cell dysfunction and depletion in AIDS: the programmed cell death hypothesis. *Immunology Today* **12**(4): 102–105.

Avery L and Horvitz HR (1987) A cell that dies during wild-type *C. elegans* development can function as a neuron in a ced-3 mutant. *Cell* **51**(6): 1071–1078.

Douglas N (1911) *Siren Land.* New York: Penguin. London (1923 rev).

Glucksman A (1951) Cell deaths in normal vertebrate ontogeny. *Biological Reviews Cambridge Philosophical Society* **26**: 59–86.

Hengartner MO (2000) The biochemistry of apoptosis. *Nature* **407**(6805): 770–776.

Hengartner MO, Ellis RE and Horvitz HR (1992) *Caenorhabditis elegans* gene ced-9 protects cells from programmed cell death. *Nature* **356**(6369): 494–499.

Horvitz HR and Herskowitz I (1992) Mechanisms of asymmetric cell division: two Bs or not two Bs, that is the question. *Cell* **68**(2): 237–255.

Kerr JF, Wyllie AH and Currie AR (1972) Apoptosis: a basic biological phenomenon with wide-ranging implications in tissue kinetics. *British Journal of Cancer* **26**(4): 239–257.

Koonin EV and Aravind L (2002) Origin and evolution of eukaryotic apoptosis: the bacterial connection. *Cell Death and Differentiation* **9**(4): 394–404.

Kourtis N and Tavernarakis N (2009) Autophagy and cell death in model organisms. *Cell Death and Differentiation* **16**(1): 21–30.

Krammer PH (2000) CD95's deadly mission in the immune system. *Nature* **407**(6805): 789–795.

Liu X, Kim CN, Yang J, Jemmerson R and Wang X (1996) Induction of apoptotic program in cell-free extracts: requirement for dATP and cytochrome *c*. *Cell* **86**(1): 147–157.

Lockshin RA and Williams CM (1965) Programmed cell death-1. Cytology of degeneration in the intersegmental muscles of the pernyi silkmoth. *Journal of Insect Physiology* **11**: 123–133.

Meier P, Finch A and Evan G (2000) Apoptosis in development. *Nature* **407**(6805): 796–801.

Melino G (2001) The Siren's song. *Nature* **412**: 23.

Metzstein MM, Stanfield GM and Horvitz HR (1998) Genetics of programmed cell death in *C. elegans*: past, present and future. *Trends in Genetics* **14**(10): 410–416.

Nicholson DW (2000) From bench to clinic with apoptosis-based therapeutic agents. *Nature* **407**(6805): 810–816.

Nicholson DW, Ali A, Thornberry NA *et al.* (1995) Identification and inhibition of the ICE/CED-3 protease necessary for mammalian apoptosis. *Nature* **376**(6535): 37–43.

Oberst A, Bender C and Green DR (2008) Living with death: the evolution of the mitochondrial pathway of apoptosis in animals. *Cell Death and Differentiation* **15**(7): 1139–1146.

Raff MC (1992) Social control on cell survival and cell death. *Nature* **356**: 397–400.

Rich T, Allen RL and Wyllie AH (2000) Defying death after DNA damage. *Nature* **407**(6805): 777–783.

Savill J and Fadok V (2000) Corpse clearance defines the meaning of cell death. *Nature* **407**(6805): 784–788.

Vaux DL, Cory S and Adams JM (1988) Bcl-2 gene promotes haemopoietic cell survival and cooperates with c-myc to immortalize pre-B cells. *Nature* **335**(6189): 440–442.

Vaux DL (2002) Apoptosis timeline. *Cell Death and Differentiation* **9**(4): 349–354.

Wyman J and Gill SJ (1990) *Binding and linkage. Functional chemistry of biological macromolecules*. New York: University Books.

Yuan JY and Horvitz HR (1990) The *Caenorhabditis elegans* genes ced-3 and ced-4 act cell autonomously to cause programmed cell death. *Developmental Biology* **138**(1): 33–41.

Yuan J and Yankner BA (2000) Apoptosis in the nervous system. *Nature* **407**(6805): 802–809.

Further Reading

Candi E, Schmidt R and Melino G (2005) The cornified envelope: a model of cell death in the skin. *Nature Reviews Molecular and Cellular Biology* **6**(4): 328–340.

De Laurenzi V and Melino G (2000) Apoptosis. The little devil of death. *Nature* **406**(6792): 135–136.

Melino G, De Laurenzi V and Vousden KH (2002) Friend or foe in tumorigenesis. *Nature Reviews Cancer* **2**(8): 605–615.

The Origin and Evolution of Programmed Cell Death

Jean Claude Ameisen, *Université Paris-Diderot, Faculté de Médecine Xavier Bichat, Paris, France*

Introductory article

Article Contents

- Introduction
- Origins: From the Question 'When' to the Question 'How'
- From Predator/Prey Coevolution to Symbioses: A 'Red Queen' Hypothesis in the Bacteria World
- A Stable Evolutionary Strategy despite High Individual Costs
- The 'Original Sin' Hypothesis: Self-destruction as an Unavoidable Consequence of Life
- *C. elegans* Model: From Paradigm to Paradox
- Walking Around the Evolutionary Bush
- From the Origins of Programmed Cell Death to the Origins of Ageing
- 'There is Grandeur in This View of Life ...'

Programmed cell death and apoptosis have been assumed to emerge with multicellularity, and to depend on specific 'death genes' whose sole effects are execution or repression of cell death. In 1996, I proposed the 'original sin' hypothesis, postulating that the origin of self-destruction is as ancient as the origin of the first cells, and predicting that there are no specific 'death genes'. Rather, an ancestral and unavoidable capacity of effectors of cell survival – of cell metabolism, differentiation, cycling – to induce cell death favoured their continuous selection during evolution for both their 'pro-life' and 'pro-death' activities. Diversification of these effectors was accelerated by their recruitment into host/parasite interactions and symbioses, including the one that gave birth to eukaryote cells. The main prediction of the 'original sin' hypothesis is supported by recent findings showing that effectors of cell death indeed have previously undetected roles in cell survival.

Introduction

> Nothing in biology makes sense except in the light of evolution.
>
> Theodosius Dobzhansky *The American Biology Teacher*, 1973, **35**: 125–129.

Hundred and fifty years ago, Darwin provided an illuminating view of life by proposing that the features of extant organisms could only be understood if their past history of divergence from a common origin – their phylogeny – was taken into account. Darwin attributed a crucial role in evolution to death 'from without', in the form of destructions and 'wars of nature' (Darwin, 1859). But he said nothing about death 'from within'. And he knew almost nothing about the invisible universe of cells, which was just being uncovered. Decades later, Metchnikoff transferred Darwin's concept of 'struggle for existence' within each individual, inside the embryo, as a 'struggle for existence between parts of the animal organisms', and proposed that it was cell death – cell killing by phagocytes – that allowed, during development, the dynamic emergence of 'harmony' from this initial 'disharmony' (Nathan, 2008). In 1951, Glucksman proposed other links between cell death and evolution, by showing that the disappearance of phylogenic vestigial organs during ontogeny resulted from cell death (Glucksman, 1951). And it was the subsequent investigation of various, phylogenetic-distant animals that led to the successive concepts of 'programmed cell death' (PCD), 'cell suicide' and apoptosis, as well as to the discovery of a genetic control of PCD (Ellis and Horvitz, 1986; Lockshin and Zakeri, 2001). Deciphering the core machinery of developmental PCD in the nematode *Caenorhabditis elegans* allowed the identification of evolutionary conserved homologues in vertebrates, arthropods (Meier *et al.*, 2000), as well as sponges and cnidarians (Oberst *et al.*, 2008). And as with most crucial molecular processes, conservation was associated with extensive expansion and diversification. The discovery, in vertebrates, that PCD also plays a central role in adult tissue homeostasis led to the idea that all cells from multicellular animals are intrinsically programmed to self-destruct, and only survive as long as interactions with other cells allow them to suppress this 'default' suicide pathway (Raff, 1992; Vaux, 1993). **See also**: The Siren's Song: This Death that Makes Life Live

PCD also plays an evolutionary conserved role in protection against inner damage caused by genetic alterations and infections. Both host- and microbe-mediated control of PCD are crucial in determining the outcome of most hosts/parasites interactions (Ameisen and Capron, 1991; Clem *et al.*, 1991). The 'Red Queen' metaphor has been proposed as a framework for understanding the selective pressures that drive coevolution of predators and preys (Van Valen, 1973). As Lewis Carroll's Alice has to keep running with the Red Queen just to keep in the same place, the propagation in predators and preys of new weapons, defences and countermeasures allows them just to keep in the same place, namely, to stay alive and reproduce.

Accordingly, if PCD has been an important target of host/parasite arms races, host/parasite coevolution, as well as the lateral gene transfers that they induce, represented a major selective pressure for the diversification of PCD during evolution (Ameisen, 1998).

But when did PCD initially emerge?

Origins: From the Question 'When' to the Question 'How'

The seminal studies of PCD in *C. elegans* (Ellis and Horvitz, 1986; Horvitz, 1999) identified a unique set of four 'death genes' with no other detectable effects than induction or repression of self-destruction. Because of their subsequent diversification, these *C. elegans* death genes were considered as close to their first genetic ancestors assumed to have emerged in the first multicellular animals. PCD, however, was also discovered in plants (Greenberg, 1996), playing an important role in development, bark formation and defences against infections, suggesting that PCD has been crucial for the survival of all multicellular organims. **See also**: Cell Death in *C. Elegans*

The first multicellular organisms emerged around one billion years ago. Were they the first organisms whose cells were endowed with the capacity to self-destruct?

Two reinforcing views led to the conviction that the origin of PCD was concomitant with that of multicellular organisms (Evan, 1994; Vaux *et al.*, 1994). The emergence of 'altruistic' cell suicide was attributed to the selective pressure that applied to multicellularity, because the multicellular body became the evolutionary 'unit of selection', instead of each of its individual cell. In contrast, each cell from any unicellular organism was viewed as an individual carrying an identical probability to give birth to future generations. Accordingly, any mutant gene that might have allowed 'altruistic' cell suicide, would have obligatorily led to the counter-selection of the individual cell expressing such mutant gene.

However, regulated cell death processes were discovered by us and others in unicellular eukaryotes (Ameisen, 1996), and have now been identified in at least 10 species belonging to 5 different branches whose phylogenetic divergence ranges over 1–2 billion years, including slime moulds (Cornillon *et al.*, 1994), kinetoplastids (Ameisen *et al.*, 1995), ciliates (Christensen *et al.*, 1995), dinoflagellates (Vardi *et al.*, 1999) and yeasts (Madeo *et al.*, 1999). Death results from intercellular signalling, allows enforcement of cell differentiation, selection of the fittest cells in a given environment, or the building of transient multicellular bodies made up of dead cell corpses favouring the persistence of long-lived, resistant spores.

PCD also occurs in bacteria, such as *Streptomyces*, and *Myxobacteria* which in adverse environment, form multicellular 'fruiting' bodies of various shapes, including that of a 'tree', in which the 'trunk' and 'branches' made of dead cells support the 'leaves' or 'flowers' made of spores. In *Bacillus subtilis*, such developmental programmes occur in the absence of multicellular body formation. All these seemingly 'altruistic' programmes are triggered by changes in environmental conditions, involve intercellular communication, and are integral parts of the organism life cycle (for a review, see Ameisen, 2002). Thus, PCD seems deeply anchored in all life kingdoms. But how did it emerge?

Because it seemed obvious to consider PCD as an 'altruistic' cell response, the question of the origin of PCD has been equated with the question of the origin of 'altruistic' cell behaviour.

However, I have proposed that there are ways to address the question that are very different from those we had long been accustomed to (Ameisen, 1996, 1998, 2002).

From Predator/Prey Coevolution to Symbioses: A 'Red Queen' Hypothesis in the Bacteria World

Bacteria provide the most fascinating model for addressing the question of the emergence and evolution of PCD in a broad perspective. The genetic and molecular mechanisms participating in the control of cell death in bacteria are very diverse, blurring most of the usual frontiers between unicellular and multicellular behaviours, outside environment and intercellular interactions, death 'from within' and 'from without', 'altruism' and 'selfishness', cooperation and competition, or infections, lateral gene transfers and symbioses. The toxin/antidote modules harboured by numerous infectious mobile genetic elements, such as plasmids and bacteriophage viruses, enforce both the extent and irreversibility of their colonization of bacterial preys by causing the death of uninfected cells. Some of these genetic modules encode paracrine killers which induce death 'from without' by releasing a toxin that kills uninfected or 'cured' neighbour cells, whereas the infected cells are protected by the antidote that they retain. Other modules – the 'addiction modules' – encode a toxin and an antidote that are both retained by the infected cell. The antidote is constantly cleaved by a bacterial protease, coupling the survival of the infected cell to the continuous synthesis of the antidote, and hence to the continuous expression of the toxin/antidote genetic module. If a cell happens to inactivate the plasmid or to escape its segregation during cell division, the 'cured' cell stops producing both the toxin and the antidote. The remaining antidote is cleaved, freeing the remaining long-lived toxin which then executes the cell 'from within' (Yarmolinsky, 1995; Hayes, 2003).

Thus, a vast array of toxin/antidote modules involved in evolutionary arms races between infectious predators and their bacterial preys may have provided the reservoir for the molecular tools (the executioners and their repressors) that allowed the subsequent emergence of regulated, 'altruistic' PCD (Ameisen, 1996, 1998, 2002). In particular, the 'addiction modules' that induce death 'from within' suggest a role for enforced symbiosis – a merging of heterogeneous genetic entities into a new identity – in the emergence of regulated self-destruction; 'self', in this case referring to a symbiont, a new community made up by the bacteria and the plasmid. And an additional similarity between this 'addictive' symbiotic self-destruction process and the process we call PCD is that they both result from a 'default' pathway.

Accordingly, I proposed a model in which successive steps of symbiotic events – between bacteria and their addiction modules of plasmid origin, and later between eukaryote cells and their mitochondria endosymbionts of bacterial origin – might have accounted for the continuous selection, radiation and evolution of regulated PCD throughout life kingdoms (Ameisen, 1996, 1998, 2002).

Such an evolutionary scenario extended the concept of 'social control' of cell survival and cell death beyond that exerted at the level of the colony, by considering each cell itself as an evolving 'society' in which competition and cooperation between heterogeneous genomes, compartments and organelles will influence the cell fate in terms of life and death (Ameisen, 1998, 2002).

A Stable Evolutionary Strategy despite High Individual Costs

The blurring of frontiers between killing and self-destruction, cooperation and competition can also be observed in bacteria in situations that do not involve plasmid infection. Most bacterial species have a multicellular way of life, involving intercellular communication in response to changing environments, through the release of density-dependent 'quorum factors' (Kaiser, 1996), which control multiple gene expression and induce various collective behaviours such as luminescence, biofilm formation or differentiation into surviving or dying cells, respectively.

These developmental processes often involve a succession of steps of symmetry breaking in the colony. For example, in *B. subtilis* a decrease in nutrients will stochastically induce in some cells – the future survivor cells – the expression of a differentiation factor, the sporulation factor SpoA (González-Pastor *et al.*, 2003). SpoA acts as both a sword and an armour, as an executioner 'from without' for cells that do not express it, and as a protector 'from within' for cells that have synthesized it. Cells that have not expressed SpoA die, providing new nutrients to the cells that have expressed it. Once the population is entirely composed of survivors, if the environment continues to be detrimental, another step of reciprocal differentiation will further break symmetry. Each cell initiates an incomplete process of asymmetric division, and a criss-cross exchange of transcription factors through the intercellular membrane that links the big mother cell and the small daughter cell will induce differentiation of the daughter cell into a spore, and death of the mother cell (Losick and Stragier, 1992). These successive processes of symmetry breaking and intercellular signalling – coupling survival of a part of the colony with death of another part – can be viewed either as examples of 'murders' by which some cells survive by killing their neighbours or rather as examples of socially regulated 'altruistic' self-destruction allowing the survival of a part of the colony at the expense of the dismissal of another part.

Other forms of regulated cell death processes closely resembling cell suicide exist in various bacteria species. They involve 'addiction modules' that reside inside the bacterial chromosome in the absence of any plasmid (Aizenman *et al.*, 1996; Hayes, 2003; Engelberg-Kulka *et al.*, 2006), suggesting an initial lateral gene transfer of the addiction module, either from plasmid to bacterial chromosome or the other way around. Whatever the case, the 'default' death pathway resulting from a repression of the expression of these chromosomal 'addiction modules' occurs in response to environmental stress, and leads to the cleavage of the remaining antidote, the freeing of the remaining long-lived toxin and the induction of death 'from within'. In *Escherichia coli*, the repression of the *mazE*/*mazF* addiction module in response to nutrient shortage, phage infection or deoxyribonucleic acid (DNA) damage, and the resulting 'death from within' induced by the MazF toxin, is under intercellular 'social control': it requires intercellular signalling mediated by the extranuclear death factor (EDF) 'quorum factor', whose release depends on bacterial density (Kolodkin-Gal *et al.*, 2007). Thus, in the face of environmental aggressions that will cause death 'from without', the premature death 'from within' of a part of the colony favours the survival of another part, that will benefit from feeding on the self-destructing neighbour cells.

Interestingly, such adverse environmental conditions can also trigger a process of chromosomal DNA rearrangement and mutations operating 'from within' through the induction of an SOS-stress response (Bjedov *et al.*, 2003). Hence, it is all the more striking that despite the existence of such potent mechanisms of genetic diversification, self-destruction escape mutants do not rapidly emerge and overtake the whole colony. Death-escaping 'cheater' mutants, biasing differentiation towards spores, have indeed been identified in some *Myxobacteria* species, but their persistence depends on the presence of 'ready-to-die' neighbours, implying the existence of constraints limiting the spread of such escape mutants (Velicer *et al.*, 2000). The emergence and evolution of regulated cell death processes, including self-destruction, might only represent a particular and extreme example of several other cooperation processes in various bacterial species, suggesting that cooperation may be under strong selection pressures and may have represented a somehow stable evolutionary strategy despite its high individual costs (Rainey and Rainey, 2003).

The 'Original Sin' Hypothesis: Self-destruction as an Unavoidable Consequence of Life

I believed that the role of host/parasite interactions, lateral gene transfer and symbioses outlined earlier only represented subsequent developments on a more ancient theme (Ameisen, 1996). I proposed that an ancestral pleiotropy of the molecular tools allowing self-destruction – a multifunctional involvement in both 'pro-life' and 'pro-death' activities – has been critical for the emergence and persistent selection of the self-destruction processes (Ameisen, 1996). In such an evolutionary context, the selective advantages of the 'pro-death' activity of such tools at the level of the colony, in terms of improved survival of some members of the colony at the expense of the premature dismissal of others in inhospitable environments, would have been strongly reinforced by the selective advantages of their 'pro-life' activity at the level of each cell from the colony, in terms of improved individual survival as long as self-destruction is not induced. Such a pleiotropy, or multifunctionality not only provided an explanation for the propagation of PCD during evolution, but also for its very evolutionary origin, in the framework of a model that I have termed 'the original sin' hypothesis (Ameisen, 1996, 1998, 2002).

Briefly, the hypothesis postulated that most molecular effectors of vital functions, such as metabolism, differentiation or cell cycle, will induce stochastic self-destruction in any cell when their activity is not regulated by other cell survival effectors that act as partial antagonists. In such a view, the potential executioners and repressors of PCD were already present from the onset in the first cells – and thus in the last universal common ancestor (LUCA) of all extant species – among most effectors of various vital functions (Ameisen, 1998, 2002).

According to this hypothesis, the capacity to self-destruct can be viewed as an 'original sin' of the earliest cells, an ancestral consequence of their very capacity to self-organize, survive and reproduce. As discussed earlier (Ameisen, 1998, 2002), this view predicted that as long as the executionary tools which become progressively selected (e.g. in host/pathogen conflicts) for their killing properties retain at least some of their vital functions, such persistent pleiotropy will strongly favour further positive selection. This view also linked the evolution of mechanisms which control death 'from within' to those which control genetic diversification 'from within'. Finally, this view made two testable predictions (i) in any species, there should be more than a single PCD pathway and (ii) more importantly, there should be no effector involved in the execution of cell death that does not also participate in some vital function (Ameisen, 1996, 1998, 2002).

But could such a view be reconciled with the implications of the *C. elegans* paradigm of PCD?

C. elegans Model: From Paradigm to Paradox

Two of the most wide ranging conceptual implications of the studies of PCD in *C. elegans* were derived from the exploration of genetic mutants with Ced-3 or Ced-4 loss-of-function, and with both Ced-3 or Ced-4 and Ced-9 loss-of-function. These studies implied that (i) there was only one molecular pathway of PCD in *C. elegans* and (ii) the executioners of PCD (Ced-3 and Ced-4), as well as the repressor of PCD (Ced-9) had no other possible function than the execution and repression of cell death, respectively (Ellis and Horvitz, 1986; Ellis *et al.*, 1991; Horvitz, 1999). This led to the concept of the existence of specific, death genes that emerged and became selected during evolution for their sole capacity to induce or repress self-destruction.

In a somehow paradoxical manner, however, cell death seemed for a long time not to play any significant role in *C. elegans*, raising the very question of how such 'death genes' had been conserved while they made no apparent contribution to the organism's fitness. During a long period, this puzzling problem was rarely raised, probably because the explicative power of the paradigm by far outweighed this cryptic paradox. However, a series of recent findings indicates that multifunctionality of the products of 'death genes' is indeed a common feature in phylogenetically widely diverging branches of the evolutionary bush, ranging from bacteria to vertebrates, and including *C. elegans* itself (Ameisen, 2004). **See also**: Cell Death in *C. Elegans*

Walking Around the Evolutionary Bush

In mammals, a series of findings progressively revealed that (i) several PCD pathways coexist in parallel and (ii) gene products involved in the execution or repression of PCD also have important vital functions, such as metabolism, differentiation or cell cycle. Examples not only included all upstream inducers of PCD, such as the tumour necrosis factor (TNF) superfamily of death ligands and receptors, the initiator caspases, the intramitochondrial protein cytochrome *c*, the p53 tumour suppressor and pleiotropic survival and self-destruction processes such as autophagy. But they also involved crucial downstream repressors of cell death, such as Bcl-2, and more importantly, crucial inducers and executioners of self-destruction that were until recently considered as lacking any other activity, such as apoptosis-activating factor 1 (Apaf-1), the executioner caspases, the intramitochondrial proteins endonuclease G, Omi/HtrA2 and apoptosis-inducing factor (AIF), the Bad and Bid proapoptotic BH3-only members of the Bcl-2/Bax family and last, but not least, the two essential proapoptotic Bcl-2 family actors of mitochondria-mediated death, Bax and Bak (for a review, see Ameisen, 2004; Galluzzi *et al.*, 2008; and for Bax and Bak, see Karbowski *et al.*, 2006; Jones *et al.*, 2007; Sheridan *et al.*, 2008). **See also**: Apoptosis: Regulatory Genes and Disease

Pleiotropy of molecular executioners of PCD is not a particular feature of mammalian complexity. Findings in bacteria have also suggested a previously undetected pleiotropy of toxins, assumed to have no other effect than death induction (Pedersen *et al.*, 2003). Very recent work in *E. coli* indicated that on repression of the bacterial chromosome-encoded *mazE*/*mazF* 'addiction module', the MazF toxin exerts a bifunctional effect, inducing death in 90% of the colony, and allowing survival of the remaining 10% (Amitai *et al.*, 2009). Briefly, MazF, when freed of its antidote, induces a rapid and sharp inhibition of the synthesis of most proteins, while increasing the synthesis of a small number of short proteins, that include both 'pro-death' and 'pro-life' effectors. The latter appear to allow survival only in the presence of factors released by dying bacteria. Thus, the MazF toxin acts as an essential survival factor in a minority of cells once it has induced death in a significant proportion of the colony, allowing a coupling of death with the survival of at least some members of the colony.

Concerning *C. elegans* itself, recent findings also revealed that there may be more avenues towards death than the single Egl1/Ced-9/Ced-4/Ced-3 pathway (Derry *et al.*, 2001). More importantly, concerning pleiotropy, the *C. elegans* p53-like 1 (CEP-1) homologue plays a crucial role in meiosis, independent of its pro-death function, and is required for whole animal survival in stressful environmental conditions (Derry *et al.*, 2001). Ced-3 and Ced-4 play a role in the whole body fitness of *C. elegans* by allowing resistance to infections, an effect that seems not directly related to their 'pro-death' activity (Liu *et al.*, 2006). Ced-9 and Egl-1 regulate mitochondrial dynamics independently of any involvement in cell death (Delivani *et al.*, 2006), and Ced-4 is involved in DNA repair in adult germ cells (Zermati *et al.*, 2007). Thus, *C. elegans* may end up being part of the pleiotropy model, rather than an exception.

And, for us, and for our health, this ancient evolutionary conserved pleiotropy might render the manipulation of cell death for therapeutic purposes more complex than once expected.

From the Origins of Programmed Cell Death to the Origins of Ageing

As PCD, ageing seems to result from the involvement of pleiotropic effectors involved in vital functions (Williams, 1957; for a review, see Ameisen, 1999, 2004). And as PCD, ageing was long assumed to have emerged in multicellular organisms. However, obligate ageing is also a feature of at least three unicellular organisms, yeasts (Jazwinski, 1996) and two bacterial species, including *E. coli* (Stewart *et al.*, 2005).

In the yeast *Saccharomyces cerevisiae*, a mother cell gives birth to approximately 20 daughter cells, and then becomes sterile and dies. Thus, the apparent eternal youth and fecundity of a yeast colony, in fact, results from endless successive generations of short-lived cells. Ageing in yeast results in part from an asymmetric distribution of some molecular components (such as damaged proteins and circular ribosomal DNA minicircles) whose accumulation in the mother cell precipitates sterility and death, whereas their initial lack at birth in daughter cells endows them with youth and fecundity (Shcheprova *et al.*, 2008). The premature dismissal of the mother cells is one of the basic mechanisms that may allow the generation of the paradoxical molecular phenomenon that we call youth (Sheldrake, 1974), which endows cells, which are ever older in terms of their genealogical age, to begin their existence with the same life expectancy and fecundity that their youngest, long-gone ancestors. I believe that this process of symmetry breaking – which resembles PCD in that it segregates between survival and death – also operates in most, if not all, unicellular organisms, and that the apparently symmetric process of cell division usually masks subtle intercellular segregation mechanisms that allow the propagation of life (Ameisen, 1999, 2004).

'There is Grandeur in This View of Life ...'

Random variations at each generation resulting from 'descent with modification', and death 'from without' resulting from the confrontation with the environment were the main features of the Darwinian theory of evolution.

In such a conceptual framework, PCD and ageing may represent two particular instances in which the selection of a regulated enforcement of premature death 'from within' may have provided enhanced fitness and survival in the face of inhospitable outside environments and of inner damages caused by metabolism. In parallel, the emergence and selection of mechanisms allowing the regulated induction of genetic diversification 'from within' has endowed cells – and organisms – with a capacity to change identities in the face of ever changing, deleterious environments. An ancestral, pleiotropic, pro-survival activity of at least some of the functional domains of the molecular effectors controlling death, ageing and genetic diversification 'from within' might have favoured their initial selection, their constant availability for de novo selection, their radiation through lateral gene transfers and symbioses, and their progressive propagation and diversification in most – if not all – species during evolution (Ameisen, 1999, 2004).

Hundred and fifty years ago, Charles Darwin concluded *The Origin of Species* by stating, 'Thus, from the war of nature, from famine and death, the most exalted object which we are capable of conceiving, namely the production of higher animals, directly follows. There is grandeur in this view of life ...'

Part of this grandeur might also reverberate in the view that regulated processes of premature death, operating from within, may have paradoxically favoured life's persistence in the face of the inescapable threat of destruction inflicted by its outer and inner environment. And it remains to be assessed to what extent such an ancient, blind and evermore complex game with death – with its own dismissal – may have played a crucial role in the long journey of life through time, and in its continuous capacity to produce, in Darwin's words, "endless forms most beautiful and most wonderful."

References

Aizenman E, Engelberg-Kulka H and Glaser G (1996) An *E. coli* chromosomal "addiction module" regulated by guanosine [corrected] 3′,5′-bispyrophosphate: a model for programmed bacterial cell death. *Proceedings of the National Academy of Sciences of the USA* **93**: 6059–6063.

Ameisen JC (1996) The origin of programmed cell death. *Science* **272**: 1278–1279.

Ameisen JC (1998) The evolutionary origin and role of programmed cell death in single celled organisms: a new view of executioners, mitochondria, host-pathogen interactions, and the role of death in the process of natural selection. In: Lockshin R, Zakeri Z and Tilly J (eds) *When Cells Die*, pp. 3–56. New York: Wiley-Liss, Inc.

Ameisen JC (1999) La Sculpture du Vivant. Le suicide cellulaire ou la mort créatrice. Paris, Editions du Seuil, (Points Seuil, 2003).

Ameisen JC (2002) On the origin, evolution, and nature of programmed cell death: a timeline of four billion years. *Cell Death and Differentiation* **9**: 367–393.

Ameisen JC (2004) Looking for death at the core of life in the light of evolution. *Cell Death and Differentiation* **11**: 4–10.

Ameisen JC and Capron A (1991) Cell dysfunction and depletion in AIDS: the programmed cell death hypothesis. *Immunology Today* **12**: 102–105.

Ameisen JC, Idziorek T, Billaut-Mulot O *et al.* (1995) Apoptosis in a unicellular eukaryote (*Trypanosoma cruzi*): implications for the evolutionary origin and role of programmed cell death in the control of cell proliferation, differentiation and survival. *Cell Death and Differentiation* **2**: 285–300.

Amitai S, Kolodkin-Gal I, Hananya-Meltabashi M, Sacher A and Engelberg-Kulka H (2009) *E. coli* MazF leads to the simultaneous selective synthesis of both "death proteins" and "survival proteins". *PLoS Genetics* **5**(3): e1000390; Epub March 13.

Bjedov I, Tenaillon O, Gérard B *et al.* (2003) Stress-induced mutagenesis in bacteria. *Science* **300**: 1404–1409.

Christensen ST, Wheatley DN, Rasmussen MI and Rasmussen L (1995) Mechanisms controlling death, survival and proliferation in a model unicellular eukaryote *Tetrahymena thermophila*. *Cell Death and Differentiation* **2**: 301–308.

Clem RJ, Fechheimer M and Miller LK (1991) Prevention of apoptosis by a baculovirus gene during infection of insect cells. *Science* **254**: 1388–1390.

Cornillon S, Foa C, Davoust J *et al.* (1994) Programmed cell death in *Dictyostelium*. *Journal of Cell Science* **107**: 2691–2704.

Darwin C (1859) *On the origin of species by means of natural selection*. London: John Murray. The Complete work of Charles Darwin online: http://darwin-online.org.uk/.

Delivani P, Adrain C, Taylor RC, Duriez PJ and Martin SJ (2006) Role for CED-9 and Egl-1 as regulators of mitochondrial fission and fusion dynamics. *Molecular Cell* **21**: 761–773.

Derry WB, Putzke AP and Rothman JH (2001) *Caenorhabditis elegans* p53: role in apoptosis, meiosis, and stress resistance. *Science* **294**: 591–595.

Ellis HM and Horvitz HR (1986) Genetic control of programmed cell death in the nematode *C. elegans*. *Cell* **44**: 817–829.

Ellis RE, Yuan J and Horvitz HR (1991) Mechanisms and functions of cell death. *Annual Review of Cell Biology* **7**: 663–698.

Engelberg-Kulka H, Amitai S, Kolodkin-Gal I and Hazan R (2006) Bacterial programmed cell death and multicellular behavior in bacteria. *PLoS Genetics* **2**(10): e135.

Evan GA (1994) Old cells never die, they just apoptose. *Trends in Cell Biology* **4**: 191–192.

Galluzzi L, Joza N, Tasdemir E *et al.* (2008) No death without life: vital functions of apoptotic effectors. *Cell Death and Differentiation* **15**: 1113–1123.

Glucksman A (1951) Cell deaths in normal vertebrate ontogeny. *Biological Reviews of the Cambridge Philosophical Society* **26**: 59–86.

González-Pastor JE, Hobbs EC and Losick R (2003) Cannibalism by sporulating bacteria. *Science* **301**: 510–513.

Greenberg JT (1996) Programmed cell death: a way of life for plants. *Proceedings of the National Academy of Sciences of the USA* **93**: 12094–12097.

Hayes F (2003) Toxins-antitoxins: plasmid maintenance, programmed cell death, and cell cycle arrest. *Science* **301**: 1496–1499.

Horvitz HR (1999) Genetic control of programmed cell death in the nematode *Caenorhabditis elegans*. *Cancer Research* **59**(suppl.): 1701–1706.

Jazwinski SM (1996) Longevity, genes, and aging. *Science* **273**: 54–59.

Jones RG, Bui T, White C *et al.* (2007) The proapoptotic factors Bax and Bak regulate T cell proliferation through control of endoplasmic reticulum Ca(2+) homeostasis. *Immunity* **27**: 268–280.

Kaiser D (1996) Bacteria also vote. *Science* **272**: 1598–1599.

Karbowski M, Norris KL, Cleland MM, Jeong SY and Youle RJ (2006) Role of Bax and Bak in mitochondrial morphogenesis. *Nature* **443**: 658–662.

Kolodkin-Gal I, Hazan R, Gaathon A, Carmeli S and Engelberg-Kulka H (2007) A linear pentapeptide is a quorum-sensing factor required for mazEF-mediated cell death in *E. coli*. *Science* **318**: 652–655.

Liu WH, Lin YL, Wang JP *et al.* (2006) Restriction of *vaccinia virus* replication by a ced-3 and ced-4-dependent pathway in *Caenorhabditis elegans*. *Proceedings of the National Academy of Sciences of the USA* **103**: 4174–4179.

Lockshin RA and Zakeri Z (2001) Timeline. Programmed cell death and apoptosis: origins of the theory. *Nature Reviews. Molecular Cell Biology* **2**: 545–550.

Losick R and Stragier P (1992) Crisscross regulation of cell-type-specific gene expression during development in *B. subtilis*. *Nature* **355**: 601–604.

Madeo F, Fröhlich E, Ligr M *et al.* (1999) Oxygen stress: a regulator of apoptosis in yeast. *Journal of Cell Biology* **145**: 757–767.

Meier P, Finch A and Evan G (2000) Apoptosis in development. *Nature* **407**: 796–801.

Nathan C (2008) Metchnikoff's legacy in 2008. *Nature Immunology* **9**: 695–698.

Oberst A, Bender C and Green DR (2008) Living with death: the evolution of the mitochondrial pathway of apoptosis in animals. *Cell Death and Differentiation* **15**: 1139–1146.

Pedersen K, Zavialov AV, Pavlov MY *et al.* (2003) The bacterial toxin RelE displays codon-specific cleavage of mRNAs in the ribosomal A site. *Cell* **112**: 131–140.

Raff MC (1992) Social controls on cell survival and cell death. *Nature* **356**: 397–400.

Rainey PB and Rainey K (2003) Evolution of cooperation and conflict in experimental bacterial populations. *Nature* **425**: 72–74.

Shcheprova Z, Baldi S, Frei SB, Gonnet G and Barral Y (2008) A mechanism for asymmetric segregation of age during yeast budding. *Nature* **454**: 728–734.

Sheldrake AR (1974) The ageing, growth and death of cells. *Nature* **250**: 381–385.

Sheridan C, Delivani P, Cullen SP and Martin SJ (2008) Bax- or Bak-induced mitochondrial fission can be uncoupled from cytochrome *c* release. *Molecular Cell* **31**: 570–585.

Stewart EJ, Madden R, Paul G and Taddei F (2005) Aging and death in an organism that reproduces by morphologically symmetric division. *PLoS Biology* **3**: e45. Epub February 1.

Van Valen L (1973) A new evolutionary law. *Evolutionary Theory* **1**: 1–30.

Vardi A, Berman-Frank I, Rozenberg T *et al.* (1999) Programmed cell death of the dinoflagellate *Peridinium gatunense* is mediated by CO(2) limitation and oxidative stress. *Current Biology* **9**: 1061–1064.

Vaux DL (1993) Towards an understanding of the molecular mechanisms of physiological cell death. *Proceedings of the National Academy of Sciences of the USA* **90**: 786–789.

Vaux DL, Haeker G and Strasser A (1994) An evolutionary perspective on apoptosis. *Cell* **76**: 777–779.

Velicer GJ, Kroos L and Lenski RE (2000) Developmental cheating in the social bacterium *Myxococcus xanthus*. *Nature* **404**: 598–601.

Williams GC (1957) Pleiotropy, natural selection and the evolution of senescence. *Evolution* **11**: 398–411.

Yarmolinsky MB (1995) Programmed cell death in bacterial populations. *Science* **267**: 836–837.

Zermati Y, Mouhamad S, Stergiou L *et al.* (2007) Non-apoptotic role for Apaf-1 in the DNA damage checkpoint. *Molecular Cell* **28**: 624–637.

Further Reading

Albert B, Johnson A, Lewis J *et al.* (2007) *Molecular Biology of the Cell*. New York: Garland Science.

Ameisen JC (2008) Dans la Lumière et les Ombres. Darwin et la bouleversement du monde. Paris, Fayard/Sevil.

Doolittle WF and Bapteste E (2007) Pattern pluralism and the tree of life hypothesis. *Proceedings of the National Academy of Sciences of the USA* **104**: 2043–2049.

Gould SJ (2002) *The Stucture of Evolutionary Theory*. Cambridge, MA: The Belknap Press of Harvard University Press.

Kaufmann S (1993) *The Origins of Order. Self-organization and Selection in Evolution*. New York: Oxford University Press.

Kirschner M, Gerhart J and Mitchison T (2000) Molecular 'vitalism'. *Cell* **100**: 79–88.

Koonin EV and Aravind L (2002) Origin and evolution of eukaryotic apoptosis: the bacterial connection. *Cell Death and Differentiation* **9**: 394–404.

Lewin B (2007) *Genes IX*. Boston: Jones and Bartlett Publishers.

Margulis L and Sagan D (1997) *Microcosmos. Four Billion Years of Microbial Evolution*. Berkeley: University of California Press.(First published by Summit Books 1986).

Maynard Smith J and Szathmary E (1997) *The Major Transitions in Evolution*. New York: Oxford University Press. (First published 1995).

Cell Death in *C. elegans*

Advanced article

Ataman Sendoel, *Institute of Molecular Biology, University of Zurich, Zurich, Switzerland*

Michael O Hengartner, *Institute of Molecular Biology, University of Zurich, Zurich, Switzerland*

Article Contents

- *ced-3, ced-4, ced-9* and *egl-1* are Conserved Genes Essential for Programmed Cell Death in *C. elegans*
- Regulation of Cell Death during Somatic Development
- Engulfment of Dying Cells
- Germline Apoptosis
- DNA Damage-induced Germline Apoptosis
- Conclusion

Programmed cell death plays a central role in the development of most multicellular animals. During the development of *Caenorhabditis elegans*, a total of 1090 cells are generated, 131 of which are destined to die. Genetic studies focusing on the control of the fate of these 131 cells revealed an evolutionary conserved set of genes essential for all programmed cell deaths in *C. elegans*. In a cell undergoing apoptosis, the BH3-only domain protein EGL-1 binds to the CED-9–CED-4 complex on the outer mitochondrial membrane resulting in the release of CED-4, which in turn activates the effector caspase CED-3. These at the time pioneering findings established *C. elegans* as a prime model system to study apoptosis, a system that still today provides a stage for new inspiring science, such as studies on *C. elegans* apoptotic cell clearance and on deoxyribonucleic acid (DNA) damage-induced apoptosis.

'What is life? Life – that means to continuously repel something that wants to die', wrote Nietzsche in 1892 (Nietzsche, 1892). Approximately 100 years later, scientists began to molecularly dissect the programmed cell death pathway and to understand the importance of cell death for a living organism, a story in which the tiny nematode *Caenorhabditis elegans* played a leading part. **See also**: Death Receptors at the Molecular Level: Therapeutic Implications; The Siren's Song: This Death that Makes Life Live

The analysis of programmed cell death in the worm was preceded and enabled by the elucidation of the complete *C. elegans* cell lineage (Sulston *et al.*, 1983a). Knowledge of the complete *C. elegans* cell lineage laid the groundwork to study mechanisms responsible for various aspects of the development of this animal (Sulston and Horvitz, 1981). The development of *C. elegans* is largely invariant and gives rise to animals with exactly the same number of cells and to individuals with near identical anatomy. Knowledge of the complete *C. elegans* lineage was used to identify genes that control cell lineage by analysing mutants at the single cell level for abnormalities in cell number (Sulston and Horvitz, 1981). One particular aspect of the cell lineage seemed arcane: of the 1090 cells that are generated during hermaphrodite development, only 959 are found in the adult; the remaining 131 cells are systematically eliminated, as part of the normal developmental programme (Sulston *et al.*, 1983a). Because the cell death pattern is highly reproducible – the same 131 cells die, each of them at a characteristic developmental time – this cell elimination process is known as programmed cell death (**Figure 1**).

There are two waves of developmental cell death: 113 of the 131 die during embryogenesis, mostly between 250 and 450 min after fertilization, whereas a second smaller wave is observed during the second larval stage (Sulston *et al.*, 1983b). These dead cells are easily detectable with Nomarski (DIC) optics and have been extensively characterized by both light and electron microscopy. The morphological changes include loss of texture and an increase in refractivity of the dying cell, culminating in a stage where the corpses are readily detectable by DIC as highly refractile disks (Sulston and Horvitz, 1977; **Figure 2**). The ultrastructural analysis of these cells undergoing programmed cell death showed features such as condensation of the cytoplasm and compaction of nuclear chromatin – characteristics highly reminiscent of mammalian apoptosis (Kerr *et al.*, 1972). This similarity on the ultrastructural level between mammalian apoptosis and programmed cell death in *C. elegans* led to the hypothesis that similar molecular mechanisms might underlie both processes.

Cells that die during *C. elegans* development derive from diverse parts in the cell lineage, and include many different cell types, including neurons, neuron-associated cells, muscle cells and hypodermal cells (Sulston

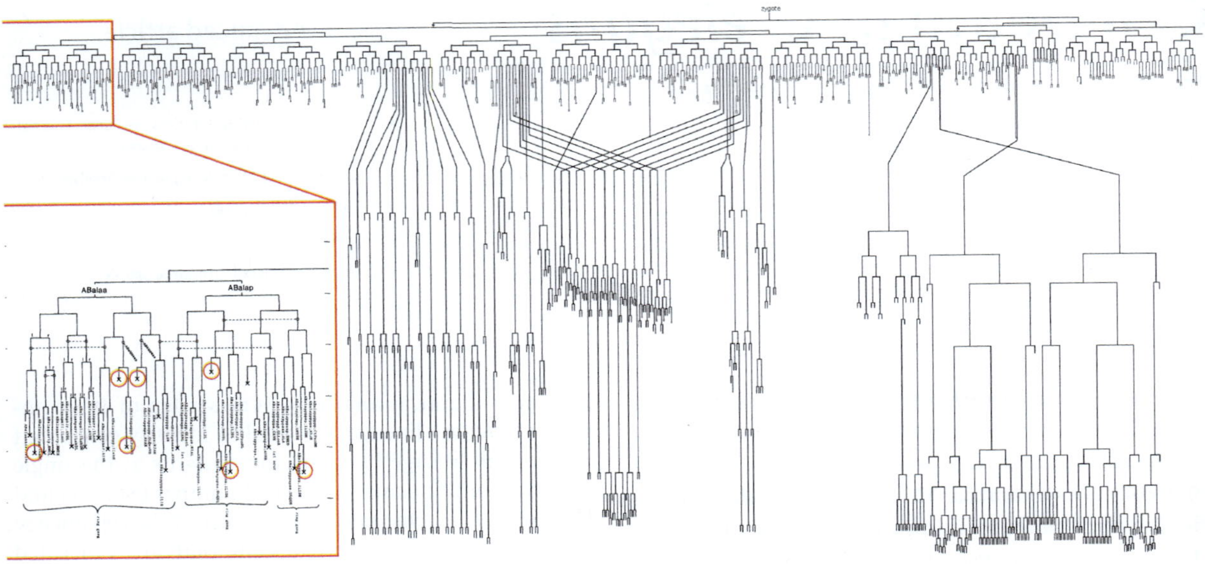

Figure 1 The cell lineage of *C. elegans*. The complete *C. elegans* cell lineage (pattern of cell divisions from the zygote to the adult hermaphrodite). In the magnified panel, cells that are doomed to die are highlighted with a red circle. Adapted from Sulston *et al.* (1983a, b).

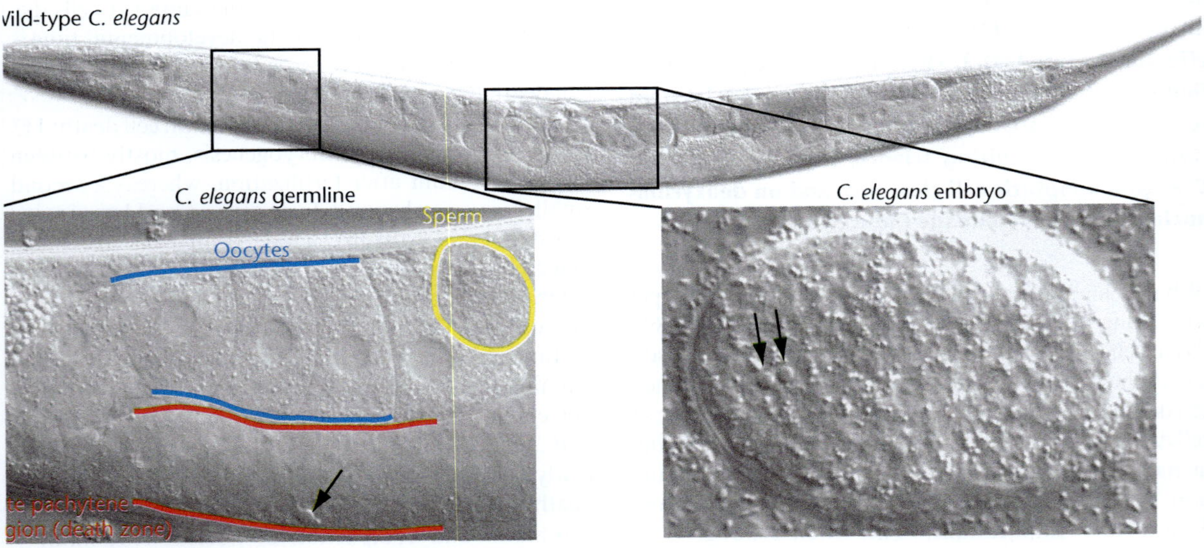

Figure 2 Developmental and germline apoptosis in *C. elegans*. Developmental and germline apoptosis in *C. elegans*, as visualized by Nomarski (DIC) optics. A first major wave of developmental cell death is observed in embryos 250–450 min after fertilization where 113 cells die (right side). At the second larval stage, another 18 somatic cells undergo programmed cell death. The third wave of apoptosis occurs in the pachytene region of the adult gonad (left side), where approximately half of the germ cells die by apoptosis. Arrows indicate apoptotic cell corpses.

et al., 1983a). The reproducibility of cell death during *C. elegans* development suggested that programmed cell death could be in fact seen simply as another type cell fate, albeit a rather terminal one. An important corollary of the hypothesis that a cell dies as a consequence of its cell fate is that programmed cell death ought to be controlled by specific genes. Subsequent genetic screens for abnormal cell death found indeed such genes. In the next sections, we will briefly discuss some of the major classes of cell death genes identified in *C. elegans*.

ced-3, ced-4, ced-9 and *egl-1* are Conserved Genes Essential for Programmed Cell Death in *C. elegans*

Mutations that completely inactivate the genes *ced-3* and *ced-4* (cell death abnormal) prevent all programmed cell death in *C. elegans* (except for some cells in the male tail), indicating that CED-3 and CED-4 have pro-apoptotic properties (Ellis and Horvitz, 1986; Yuan and Horvitz, 1990). CED-3 encodes a protease of the caspase family and was the first identified caspase with a role in apoptosis. Though hundreds of caspase substrates have been identified in mammals, the number of known CED-3 targets is still very limited (Xue and Horvitz, 1997; Chan *et al.*, 2000; Taylor *et al.*, 2007). The function of CED-4 – homologous to the mammalian apoptotic adaptor protein (APAF1) – is to promote autocatalytic cleavage of the precursor form of CED-3 (Yuan and Horvitz, 1992). In addition, CED-4 produces a second transcript which is much less abundant and codes for a slightly larger form, CED-4L, which seems to have opposite effects: overexpression of CED-4L prevents cell death (Yuan and Horvitz, 1992; Shaham and Horvitz, 1996). The molecular mechanism of action of CED-4L is still unresolved. An attractive possibility is that CED-4L acts as a poison subunit, and can interfere with apoptosis formation or function (see later).

Interestingly, animals mutant in *ced-3* or *ced-4* have now additionally 131 fully differentiated cells (14% more cells); its nervous system increases from 302 to 407 cells. Despite this dramatic increase in cell numbers, lack of apoptosis does not result in any major abnormalities at the morphological or behavioural level. Several aspects such as life span, locomotion or feeding are indistinguishable between wild-type and *ced-3* mutant animals (Ellis and Horvitz, 1986).

What happens to the cells that fail to die in *ced-3* mutants? In many cases these 'undead' cells adopt a cell fate similar to that of their sister cells, or of cells at homologous positions in other lineages (Ellis and Horvitz, 1986). However, differentiation of undead cells appears to be generally less efficient and more variable than that of normal cells. Nevertheless, several observations suggest that undead cells are healthy and capable, at least under certain conditions, of normal function. For example, some undead cells can functionally substitute for a sister cell that is eliminated by laser ablation. In one extreme case, the undead cell P12.pp can even usurp the fate normally assigned to its sister, forcing the latter to take over a different fate (Avery and Horvitz, 1987).

CED-9 acts as a negative regulator of programmed cell death in *C. elegans* (Hengartner *et al.*, 1992). Loss-of-function alleles of *ced-9* result in extensive cell death of cells that normally survive, due to inappropriate activation of programmed cell death. The cells that additionally die as a result of the *ced-9* mutation are different in different animals, the pattern of the additional cells dying is thus not fixed. In contrast, a gain-of-function allele of *ced-9* leads to the survival of cells that are normally doomed to die. Mutations in either *ced-3* or *ced-4* can fully prevent ectopic programmed cell death caused by *ced-9* loss-of-function indicating that *ced-3* and *ced-4* genetically act downstream of *ced-9*. CED-9 is homologous to the human proto-oncogene Bcl-2 (B-cell lymphoma), a similarity that was jointly responsible at that time to establish *C. elegans* cell death as major topic in apoptosis research.

The ability of CED-9 to inhibit apoptosis is regulated by the pro-apoptotic BH3 domain protein EGL-1 (EGg Laying defective), which acts genetically as a negative regulator of CED-9. As is the case with *ced-3* and *ced-4*, loss of *egl-1* function results in the loss of all developmental apoptosis (**Figure 3**). By contrast, overexpression of *egl-1* in embryos leads to ectopic cell death (Conradt and Horvitz, 1998).

At the molecular level, programmed cell death is controlled through a cascade of protein–protein interactions between these four core apoptotic proteins (Conradt and Horvitz, 1998; del Peso *et al.*, 2000; Chen *et al.*, 2000). In cells that should survive, CED-9 sequesters CED-4 in an inactive complex on the outer mitochondrial membrane. In cells fated to die, EGL-1 binds through its BH3 domain to CED-9, inducing a conformational change in CED-9 that untethers CED-4 from CED-9 and allows translocation of CED-4 to the perinuclear region, where it activates apoptosis (Wu *et al.*, 1997a, b; Spector *et al.*, 1997). CED-4 activates apoptosis by forming the *C. elegans* apoptosome: a CED-4 tetramer into which several proCED-3 molecules are recruited. The high local concentration of proCED-3 molecules is thought to then allow processing of proCED-3 into the active protease (Yang *et al.*, 1998), which then cleaves apoptotic substrates and kills the cell (**Figure 4**).

Although the programmed cell death pathway is highly conserved at the genetic level between *C. elegans* and mammals, there are several important distinctions at

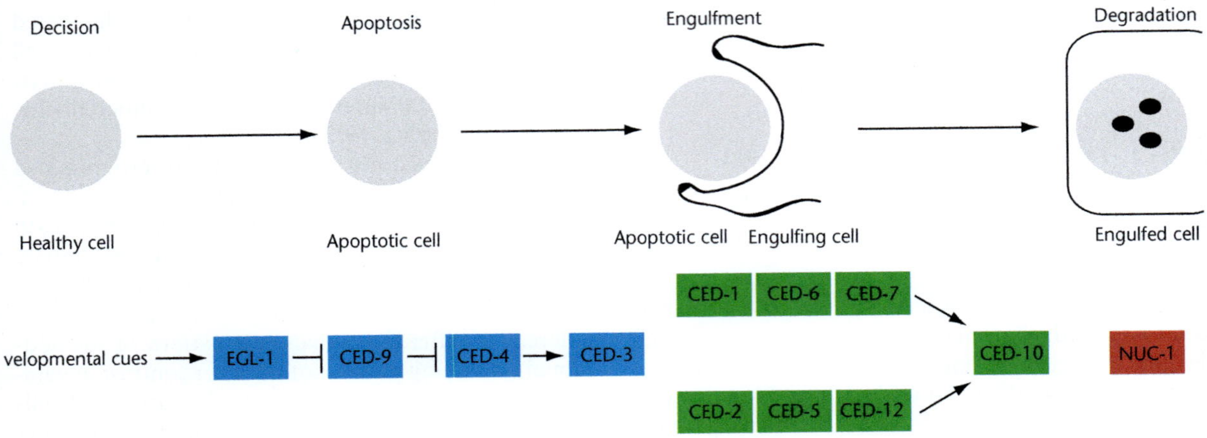

Figure 3 A conserved genetic pathway for developmental programmed cell death in *C. elegans*. Programmed cell death includes three distinct steps: the decision to die, engulfment of the apoptotic cell and degradation of the engulfed cell. The core apoptotic machinery consists of *egl-1*, *ced-9*, *ced-4* and *ced-3* and is activated by tissue-specific developmental cues. Apoptotic cells are then engulfed by neighbouring cells due to activation of two partially redundant pathways – *ced-1*, *ced-6*, *ced-7* and *ced-2*, *ced-5*, *ced-12*, which converge at the Rac1 homologue *ced-10*. Degradation of engulfed cells depend on different genes such as *nuc-1*, an endonuclease that was the first identified cell death mutant in *C. elegans* (Ellis *et al.*, 1991).

the biochemical level. Mitochondria are central in mammals for the intrinsic apoptotic pathway (Green *et al.*, 1998). To induce apoptosis, pro-apoptotic Bcl-2 family members must bind to the outer mitochondrial membrane and compete with pro-survival Bcl-2 family members for mitochondrial permeabilization. As a result of mitochondrial permeabilization (which occurs via mechanisms that are still under debate), the electron carrier cytochrome *c* is released into the cytosol, where it associates with Apaf-1 (apoptosis-activating factor-1) and the procaspase-9 to form the mammalian apoptosome (Li *et al.*, 1997). By contrast, several observations suggest that cytochrome *c* has no active role in programmed cell death in *C. elegans*. First, CED-4 lacks the WD40 repeat which is responsible in Apaf-1 for cytochrome *c* binding. Second, *in vitro* purified proteins of the core apoptotic machinery are sufficient to activate CED-3 (Hu *et al.*, 1998; Yan *et al.*, 2005). **See also**: Apoptosis: Regulatory Genes and Disease; The Apoptosome: The Executioner of Mitochondria-mediated Apoptosis

Regulation of Cell Death during Somatic Development

How is the highly reproducible pattern of developmental apoptosis determined? Elegant studies led to the conclusion that much of developmental apoptosis in *C. elegans* is regulated at the level of *egl-1* transcription: cells fated to die express *egl-1*; cells that survive do not. Whether a cell expresses *egl-1* in turn depends on the specific constellation of transcriptional activators and repressors present within the cell. For some cells, such as the NSM (neuro-secretory motor neuron) sisters, detailed transcriptional regulatory networks have been identified. The interested reader can find more information on these networks in several recent reviews (Lettre and Hengartner, 2006; Conradt and Xue, 2005).

Engulfment of Dying Cells

C. elegans does not have any professional motile macrophages. Rather, most cells are able to recognize and engulf dying cells. During embryogenesis, apoptotic cells are engulfed simply by a neighbour, whereas post-embryonically, mainly epithelial cells are engaged to take over this function. Engulfment of cells in *C. elegans* is a highly efficient process: dying cells are usually completely engulfed within an hour of the first morphological evidence of apoptosis. Mutations in genes that function in engulfment result in a 'persistent cell corpse' phenotype, where the dead cells remain unengulfed in the worm for a long time.

Two partially redundant pathways control engulfment in *C. elegans* (**Figure 3**). The first pathway contains the candidate receptor CED-1, the adaptor protein CED-6 and the ABC transporter CED-7 – homologues of MEGF10 (multiple EGF-like-domains)/LRP1 (low

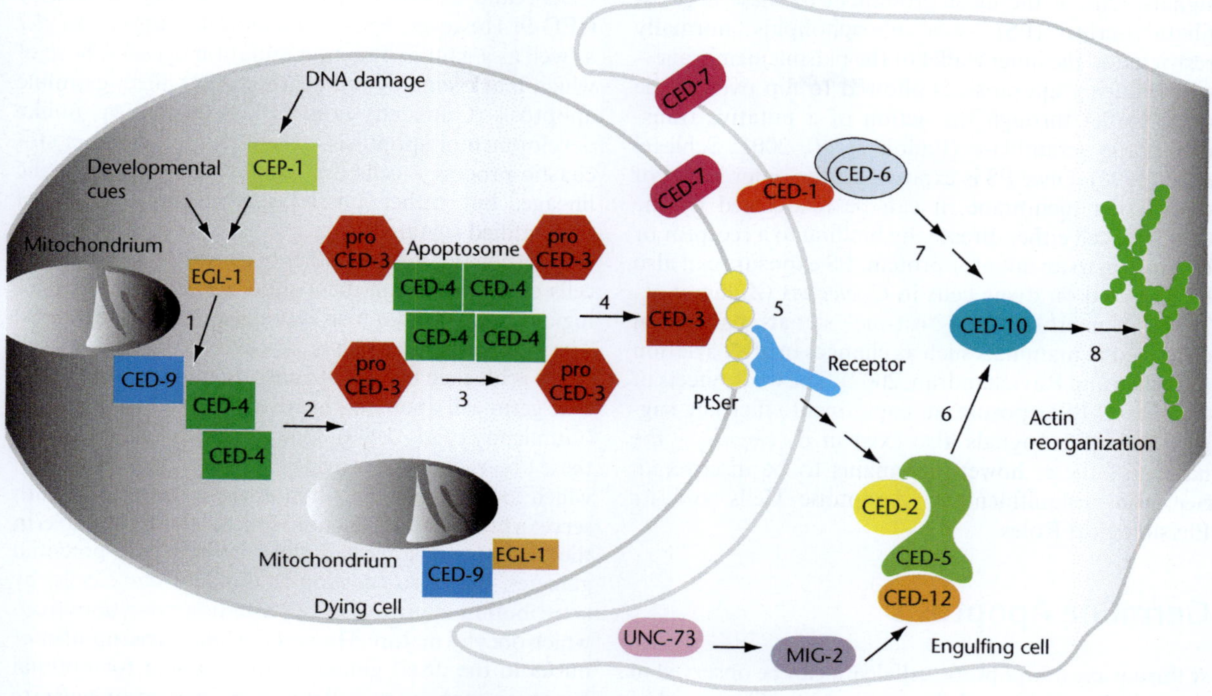

Figure 4 Molecular model of apoptosis induction and cell corpse removal in *C. elegans*. Apoptosis is initiated by the BH3-only domain protein EGL-1, which binds to the CED-9–CED-4 complex on the outer mitochondrial membrane (1). Upon EGL-1 binding, CED-4 dimer is released from CED-9 and recruits proCED-3 molecules to build the so-called apoptosome (2). CED-3 activation (4) occurs by proteolytic cleavage (3). Removal of the apoptotic cell is triggered by receptor-mediated recognition of 'eat-me' signals such as phosphatidylserine (5). Intracellular signalling that drives engulfment of the dying cell involves two partially redundant pathways consisting of *unc-73, mig-2, ced2, ced-5, ced-12* (6) and *ced-1, ced-6, ced-7* (7). These two signalling pathways may converge at the Rac1 homologue CED-10 to regulate actin reorganization (8).

density lipoprotein receptor-related protein), GULP (PTB domain-containing engulfment adapter protein) and ABCA1 (ABC transporter A family member). CED-1 and CED-6 act together in the engulfing cell: CED-1 likely acts as an apoptotic cell receptor, clustering around apoptotic cells and signalling recognition of apoptotic corpses via the adaptor protein CED-6. By contrast, genetic experiments suggest that CED-7 activity is required in both the engulfing and the dying cell (Kinchen and Ravichandran, 2007).

The second engulfment pathway consists of the bipartite RacGEF CED-5/CED-12, and its upstream regulators CED-2, MIG-2 (abnormal cell MIGration), and UNC-73 (UNCoordinated; mammalian homologues: Dock 180/Elmo, and CrkII, RhoG and Trio, respectively; **Figure 3**).

The two pathways converge onto the Rac1 homologue CED-10, which in turn mediates actin rearrangements necessary for corpse internalization (Ellis *et al.*, 1991). Rac functions as a molecular switch, cycling between inactive guanosine diphosphate (GDP)-bound and active guanosine triphosphate (GTP)-bound conformations on the basis of activating GEFs (guanine nucleotide exchange factors) and hydrolysis-promoting GAPs (GTPase-activating proteins). CED-5 and CED-12 act as a bipartite GEF for CED-10 during engulfment, whereas UNC-73 is the GEF for MIG-2. By contrast, the GAPs involved in cell corpse engulfment remain to be identified (Ellis *et al.*, 1991; Kinchen and Ravichandran, 2007).

The evolutionary conservation of engulfment genes between worms and mammals suggests that the process of cell corpse clearance is also quite ancient, and likely arose soon after – or perhaps even before – the invention of apoptosis. Consistent with the latter alternative, the *C. elegans* engulfment machinery can recognize and remove not only apoptotic, but also necrotic cells (Chung *et al.*, 2000).

How are dying cells in *C. elegans* recognized as such? In mammals, apoptotic cells expose numerous 'eat-me' signals. One of the most prominent of these is phosphatidylserine (PS) – a phospholipids normally restricted to the inner leaflet of the plasma membrane – which during apoptosis is allowed to flip over to the outer leaflet through the action of a putative transmembrane scramblase (Fadok *et al.*, 2001; Schlegel *et al.*, 1996). Once PS is exposed on the outer leaflet of the plasma membrane, it can be recognized by the engulfing cell either directly by binding to a receptor or by binding to an adaptor protein. PS exposure can also be observed on dying cells in *C. elegans* (Züllig *et al.*, 2007). Numerous other 'eat-me' signals have been described in mammals such as changes in glycosylation (Kinchen and Ravichandran, 2007). The weak effects of changes in PS exposure on engulfment efficiency suggest that other signals also exist in *C. elegans* – the nature of these, however, remains to be discovered. **See also**: Engulfment of Apoptotic Cells and its Physiological Roles

Germline Apoptosis

A third wave of apoptotic cell death can be observed in the germline of the adult hermaphrodite. The gonad of the *C. elegans* adult hermaphrodite consists of two symmetrical, U-shaped tubes, which are linked to a common uterus. The distal end of each gonadal tube is capped by a somatic distal tip cell (DTC). The LAG-2 (Lin-12 And Glp-1 phenotype) growth factor, expressed by the DTC, maintains the stem cell potential of the germ cells and promotes their mitotic proliferation. As germ cells move outside the zone of influence of the DTC, they stop dividing, enter meiosis and progress into the pachytene stage of meiosis I. Near the bend of the gonadal tube, cells exit pachytene in response to activation of the ras/MAPK (mitogen-activated protein kinase) pathway and begin to enlarge in size. In the proximal arm, germ cells progress to diakinesis of prophase I and increase in size to form oocytes. A special feature of the germline is that most germ cell nuclei are only partially enclosed by the plasma membrane, and are connected with each other via a large common cytoplasm, the rachis (**Figure 2**; Gartner *et al.*, 2008).

Approximately half of all presumptive oocytes undergo apoptosis shortly before germ cells exit the pachytene of prophase I. Because these deaths occur under normal growth conditions, they are referred to as 'physiological germ cell apoptosis'. Germ cell and developmental apoptosis share many morphological features (**Figure 2**). Furthermore, both depend on CED-3 and CED-4 function, and both are blocked by CED-9. However, loss-of-function mutations in *egl-1* as well as a gain-of-function mutation in *ced-9*, both of which block somatic cell death, do not affect germline apoptosis (Gumienny *et al.*, 1999). Moreover, unlike developmental apoptosis, germline apoptosis is a stochastic process – cells die not according to a specific lineage, but rather possibly in response to as-yet unidentified signals.

Why does *C. elegans* bother to generate so many germ cells only to let them die a short period later? Physiological germline apoptosis is not essential for oogenesis. The number of offspring is, however, reduced when cell death is blocked by loss-of-function of *ced-3*, suggesting that germ cell apoptosis is advantageous for oogenesis (Gumienny *et al.*, 1999). One possible explanation for these observations is provided by the 'nurse cell model', which postulates that physiological germ cell death serves a homeostatic function. Indeed, the many nuclei in the distal part of the germline are not only potential gametes, they also function as de facto nurse cells, by contributing to growth of the common syncytium from which oocytes mature. However, while a large number of nuclei in the distal gonad arm is required for optimal oogenesis, only a few cell nuclei are needed to generate the mature oocytes found in the proximal gonad. Physiological germ cell apoptosis thus could be used to cull the supernumerary nuclei. Consistent with this hypothesis, germ cell apoptosis is restricted to the female germline, as spermatogenesis in *C. elegans* does not require any nurse cells.

DNA Damage-induced Germline Apoptosis

The cells of an organism are under the constant threat of genotoxic stress, be it as a consequence of replication, irradiation, reactive oxygen species or hydrolysis. Therefore, multiple protective mechanisms must exist to maintain the integrity of the genome. Genotoxic stress activates in metazoans a series of pathways that result in deoxyribonucleic acid (DNA) repair, a transient cell cycle arrest or – if the damage is significant enough – in apoptotic cell death.

Several of these DNA damage responses can readily be studied in *C. elegans*. For example, exposure of adult hermaphrodites to ionizing radiation (IR) results in proliferation arrest of the mitotic germ stem cells, and a dramatic increase in germ cell apoptosis (Gartner *et al.*, 2000). Interestingly, the hermaphrodite germline seems to be unique in this respect: neither somatic cells nor the male germline possess the ability to undergo apoptosis following genotoxic stress.

DNA damage-induced apoptosis is also dependent on the core apoptotic machinery: germline apoptosis following IR is completely blocked in *ced-3* and *ced-4* mutant animals whereas *ced-9* mutants are hypersensitive to IR (Gartner *et al.*, 2000). In contrast to physiological germline apoptosis, however, the BH-3-only domain protein EGL-1 is required for apoptotic cell death following IR. Moreover, DNA damage-induced apoptosis is blocked by *ced-9 (n1950)* gain-of-function mutation, an allele which does not affect physiological germ cell death (Gumienny *et al.*, 1999; Gartner *et al.*, 2000).

Genetic studies identified many mutants with specific defects in DNA damage-induced apoptosis. Most of these genes code for highly conserved DNA damage response proteins. For example, HUS-1 (human HUS1 related) and MRT-2 (MoRTal germline) are part of the conserved 9-1-1 proliferating cell nuclear antigen (PCNA)-like complex associating with DNA double-strand breaks, whereas RAD-5 (RADiation sensitivity abnormal/yeast RAD-related)/CLK-2 (CLocK (biological timing) abnormality) is a functionally conserved protein implicated in S-phase regulation (Hofmann *et al.*, 2002; Ahmed and Hodgkin, 2000; Ahmed *et al.*, 2001). The *C. elegans* homologues of mammalian ATM (ataxia telangiectasia mutated) and ATR (ataxia telangiectasia and Rad3-related protein) kinases (*atm-1* and *atl-1*) as well as of the checkpoint kinases CHK-1 and CHK-2 (*chk-1* and *chk-2*) are also part of various DNA damage response pathways (Stergiou and Hengartner, 2004).

The tumour suppressor p53 plays a central role in DNA damage response pathways in mammals. Its *C. elegans* homologue, *cep-1* (*C. elegans* p53-like), promotes DNA damage-induced apoptosis but is dispensable for physiological germ cell as well as for somatic cell death (Derry *et al.*, 2001). As is the case in mammals, p53 promotes apoptosis through transcriptional upregulation of BH3 domain proteins (EGL-1 and CED-13) (Schumacher *et al.*, 2005a, b; Oda *et al.*, 2000). Why this transcriptional response is restricted to the female germline is not yet known.

Interestingly, although *cep-1* is seen as the p53 homologue in terms of function, its sequence is more closely related to the p53 family member p63 (Suh *et al.*, 2006). Moreover, p63 has been reported to be constitutively expressed in female germ cells during meiotic arrest and is essential in a process of DNA damage-induced oocyte death not involving p53. It is thus possible that the original function of the p53 family was to control apoptosis in response to DNA damage in the germline, which the somatic functions of p53 being developed only at a later time point.

How is CEP-1 (C. Elegans P53-like protein) in the *C. elegans* germline regulated? One major regulation occurs, surprisingly, at the messenger ribonucleic acid (mRNA) level: the KH domain protein GLD-1 (defective in germ line development) has been shown to directly bind to the *cep-1* 3′-UTR and mediate its translational repression in the early meiotic stages of oogenesis (Schumacher *et al.*, 2005a, b). This inhibition is likely important to prevent CEP-1 from activating apoptosis in response to the DNA double-strand breaks introduced during meiotic recombination. Several other negative regulators have been proposed: *akt-1* and *akt-2* – kinases implicated in insulin signalling – have been shown to negatively regulate DNA damage-induced apoptosis in the *C. elegans* germline via regulating CEP-1 activity (Quevedo *et al.*, 2007). Furthermore, the IASPP (Inhibitor of ASPP protein) homologue in *C. elegans*, APE-1 (APoptosis Enhancer), has been shown to bind to p53 and could thus act as a CEP-1 regulator. Consistent with this observation, depletion of APE-1 function resulted in increased CEP-1-mediated apoptosis (Bergamaschi *et al.*, 2003). Nevertheless, the set of identified CEP-1 regulators is clearly not complete. Indeed, to a large extent, negative regulation of mammalian p53 relies on MDM2 (murine double minute)-mediated nuclear export and proteasomal degradation (Zhang and Xiong, 2001). However, there are no clear homologues of MDM2, nor its negative regulator p14ARF (alternate reading frame) in the *C. elegans* genome. It will be thus interesting to elucidate the whole regulatory network of CEP-1, to see how this key protein is controlled in the worm, and to ultimately understand more profoundly the evolution of this branch of the DNA damage response pathway. **See also**: P53 and Cell Death

Conclusion

The field of apoptosis has clearly demonstrated that a stringent reductionist view of a biological problem can efficiently reveal the framework of the solution, and that such a framework often also holds true for more complicated organisms. ‘Everything should be made as simple as possible, but not simpler’, said Einstein. Despite its relative simplicity, *C. elegans* is still sophisticated enough that it will allow several more generations of biologists to study its apoptotic pathways before we fully understand them. The tiny worm still holds big secrets.

References

Ahmed S, Alpi A, Hengartner MO and Gartner A (2001) *C. elegans* RAD-5/CLK-2 defines a new DNA damage checkpoint protein. *Current Biology* **11**(24): 1934–1944.

Ahmed S and Hodgkin J (2000) MRT-2 checkpoint protein is required for germline immortality and telomere replication in *C. elegans*. *Nature* **403**(6766): 159–164. doi:10.1038/35003120.

Avery L and Horvitz HR (1987) A cell that dies during wild-type *C. elegans* development can function as a neuron in a ced-3 mutant. *Cell* **51**(6): 1071–1078.

Bergamaschi D, Samuels Y, O'Neil NJ *et al.* (2003) iASPP oncoprotein is a key inhibitor of p53 conserved from worm to human. *Nature Genetics* **33**(2): 162–167. doi:10.1038/ng1070.

Chan SL, Yee KS, Tan KM and Yu VC (2000) The *Caenorhabditis elegans* sex determination protein FEM-1 is a CED-3 substrate that associates with CED-4 and mediates apoptosis in mammalian cells. *Journal of Biological Chemistry* **275**(24): 17925–17928. doi:10.1074/jbc.C000146200.

Chen F, Hersh BM, Conradt B *et al.* (2000) Translocation of *C. elegans* CED-4 to nuclear membranes during programmed cell death. *Science* **287**(5457): 1485–1489.

Chung S, Gumienny TL, Hengartner MO and Driscoll M (2000) A common set of engulfment genes mediates removal of both apoptotic and necrotic cell corpses in *C. elegans*. *Nature Cell Biology* **2**(12): 931–937. doi:10.1038/35046585.

Conradt B and Horvitz HR (1998) The *C. elegans* protein EGL-1 is required for programmed cell death and interacts with the Bcl-2-like protein CED-9. *Cell* **93**(4): 519–529.

Conradt B and Xue D (2005) Programmed cell death. *WormBook: The Online Review of C. elegans Biology* 1–13. doi:10.1895/wormbook.1.32.1.

Derry WB, Putzke AP and Rothman JH (2001) *Caenorhabditis elegans* p53: role in apoptosis, meiosis, and stress resistance. *Science* **294**(5542): 591–595. doi:10.1126/science.1065486.

Ellis HM and Horvitz HR (1986) Genetic control of programmed cell death in the nematode *C. elegans*. *Cell* **44**(6): 817–829.

Ellis RE, Jacobson DM and Horvitz HR (1991) Genes required for the engulfment of cell corpses during programmed cell death in *Caenorhabditis elegans*. *Genetics* **129**(1): 79–94.

Fadok VA, de Cathelineau A, Daleke DL, Henson PM and Bratton DL (2001) Loss of phospholipid asymmetry and surface exposure of phosphatidylserine is required for phagocytosis of apoptotic cells by macrophages and fibroblasts. *Journal of Biological Chemistry* **276**(2): 1071–1077. doi:10.1074/jbc.M003649200.

Gartner A, Boag PR and Blackwell TK (2008) Germline survival and apoptosis. *WormBook: The Online Review of C. elegans Biology* 1–20. doi:10.1895/wormbook.1.145.1.

Gartner A, Milstein S, Ahmed S, Hodgkin J and Hengartner MO (2000) A conserved checkpoint pathway mediates DNA damage – induced apoptosis and cell cycle arrest in *C. elegans*. *Molecular Cell* **5**(3): 435–443.

Green Z, Reed JC and Douglas R (1998) Mitochondria and apoptosis. *Science* **281**(5381): 1309–1312. doi:10.1126/science.281.5381.1309.

Gumienny TL, Lambie E, Hartwieg E, Horvitz HR and Hengartner MO (1999) Genetic control of programmed cell death in the *Caenorhabditis elegans* hermaphrodite germline. *Development* **126**(5): 1011–1022.

Hengartner MO, Ellis RE and Horvitz HR (1992) *Caenorhabditis elegans* gene ced-9 protects cells from programmed cell death. *Nature* **356**(6369): 494–499. doi:10.1038/356494a0.

Hofmann ER, Milstein S, Boulton SJ *et al.* (2002) *Caenorhabditis elegans* HUS-1 is a DNA damage checkpoint protein required for genome stability and EGL-1-mediated apoptosis. *Current Biology* **12**(22): 1908–1918.

Hu Y, Ding L, Spencer DM and Núñez G (1998) WD-40 repeat region regulates Apaf-1 self-association and procaspase-9 activation. *Journal of Biological Chemistry* **273**(50): 33489–33494.

Kerr JF, Wyllie AH and Currie AR (1972) Apoptosis: a basic biological phenomenon with wide-ranging implications in tissue kinetics. *British Journal of Cancer* **26**(4): 239–257.

Kinchen JM and Ravichandran KS (2007) Journey to the grave: signaling events regulating removal of apoptotic cells. *Journal of Cell Science* **120**(13): 2143–2149. doi:10.1242/jcs.03463.

Lettre G and Hengartner MO (2006) Developmental apoptosis in *C. elegans*: a complex CEDnario. *Nature Reviews. Molecular Cell Biology* **7**(2): 97–108. doi:10.1038/nrm1836.

Li P, Nijhawan D, Budihardjo I *et al.* (1997) Cytochrome *c* and dATP-dependent formation of Apaf-1/caspase-9 complex initiates an apoptotic protease cascade. *Cell* **91**(4): 479–489.

Nietzsche F (1892) *Die fröhliche Wissenschaft "la gaya scienza"*. Ernst Schmeitzner Verlag.

Oda E, Ohki R, Murasawa H *et al.* (2000) Noxa, a BH3-only member of the Bcl-2 family and candidate mediator of p53-induced apoptosis. *Science* **288**(5468): 1053–1058.

del Peso L, Gonzalez VM, Inohara N, Ellis RE and Núñez G (2000) Disruption of the CED-9-CED-4 complex by EGL-1 is a critical step for programmed cell death in *Caenorhabditis elegans*. *Journal of Biological Chemistry* **275**(35): 27205–27211. doi:10.1074/jbc.M000858200.

Quevedo C, Kaplan DR and Brent Derry W (2007) AKT-1 regulates DNA-damage-induced germline apoptosis in

C. elegans. *Current Biology* **17**(3): 286–292. doi:10.1016/j.cub.2006.12.038.

Schlegel RA, Callahan M, Krahling S, Pradhan D and Williamson P (1996) Mechanisms for recognition and phagocytosis of apoptotic lymphocytes by macrophages. *Advances in Experimental Medicine and Biology* **406**: 21–28.

Schumacher B, Hanazawa M, Lee M-H *et al.* (2005b) Translational repression of *C. elegans* p53 by GLD-1 regulates DNA damage-induced apoptosis. *Cell* **120**(3): 357–368. doi:10.1016/j.cell.2004.12.009.

Schumacher B, Schertel C, Wittenburg N *et al.* (2005a) *C. elegans* ced-13 can promote apoptosis and is induced in response to DNA damage. *Cell Death and Differentiation* **12**(2): 153–161. doi:10.1038/sj.cdd.4401539.

Shaham S and Horvitz HR (1996) An alternatively spliced *C. elegans* ced-4 RNA encodes a novel cell death inhibitor. *Cell* **86**(2): 201–208.

Spector MS, Desnoyers S, Hoeppner DJ and Hengartner MO (1997) Interaction between the *C. elegans* cell-death regulators CED-9 and CED-4. *Nature* **385**(6617): 653–656. doi:10.1038/385653a0.

Stergiou L and Hengartner MO (2004) Death and more: DNA damage response pathways in the nematode *C. elegans*. *Cell Death and Differentiation* **11**(1): 21–28. doi:10.1038/sj.cdd.4401340.

Suh E-K, Yang A, Kettenbach A *et al.* (2006) p63 protects the female germ line during meiotic arrest. *Nature* **444**(7119): 624–628. doi:10.1038/nature05337.

Sulston JE and Horvitz HR (1977) Post-embryonic cell lineages of the nematode, *Caenorhabditis elegans*. *Developmental Biology* **56**(1): 110–156.

Sulston JE and Horvitz HR (1981) Abnormal cell lineages in mutants of the nematode *Caenorhabditis elegans*. *Developmental Biology* **82**(1): 41–55.

Sulston JE, Schierenberg E, White JG and Thomson JN (1983a) The embryonic cell lineage of the nematode *Caenorhabditis elegans*. *Developmental Biology* **100**(1): 64–119.

Sulston JE, Schierenberg E, White JG and Thomson JN (1983b) The embryonic cell lineage of the nematode *Caenorhabditis elegans*. *Developmental Biology* **100**(1): 64–119.

Taylor RC, Brumatti G, Ito S *et al.* (2007) Establishing a blueprint for CED-3-dependent killing through identification of multiple substrates for this protease. *Journal of Biological Chemistry* **282**(20): 15011–15021. doi:10.1074/jbc.M611051200.

Wu D, Wallen HD, Inohara N and Nuñez G (1997b) Interaction and regulation of the *Caenorhabditis elegans* death protease CED-3 by CED-4 and CED-9. *Journal of Biological Chemistry* **272**(34): 21449–21454.

Wu D, Wallen HD and Nuñez G (1997a) Interaction and regulation of subcellular localization of CED-4 by CED-9. *Science* **275**(5303): 1126–1129.

Xue D and Horvitz HR (1997) *Caenorhabditis elegans* CED-9 protein is a bifunctional cell-death inhibitor. *Nature* **390**(6657): 305–308. doi:10.1038/36889.

Yan N, Chai J, Lee ES *et al.* (2005) Structure of the CED-4-CED-9 complex provides insights into programmed cell death in *Caenorhabditis elegans*. *Nature* **437**(7060): 831–837. doi:10.1038/nature04002.

Yang X, Chang HY and Baltimore D (1998) Essential role of CED-4 oligomerization in CED-3 activation and apoptosis. *Science* **281**(5381): 1355–1357.

Yuan J and Horvitz HR (1992) The *Caenorhabditis elegans* cell death gene ced-4 encodes a novel protein and is expressed during the period of extensive programmed cell death. *Development* **116**(2): 309–320.

Yuan JY and Horvitz HR (1990) The *Caenorhabditis elegans* genes ced-3 and ced-4 act cell autonomously to cause programmed cell death. *Developmental Biology* **138**(1): 33–41.

Zhang Y and Xiong Y (2001) Control of p53 ubiquitination and nuclear export by MDM2 and ARF. *Cell Growth & Differentiation: The Molecular Biology Journal of the American Association for Cancer Research* **12**(4): 175–186.

Züllig S, Neukomm LJ, Jovanovic M *et al.* (2007) Aminophospholipid translocase TAT-1 promotes phosphatidylserine exposure during *C. elegans* apoptosis. *Current Biology* **17**(11): 994–999. doi:10.1016/j.cub.2007.05.024.

Further Reading

Mangahas PM and Zhou Z (2005) Clearance of apoptotic cells in *Caenorhabditis elegans*. *Seminars in Cell & Developmental Biology* **16**(2): 295–306. doi:10.1016/j.semcdb.2004.12.005.

Caspases and Cell Death

Advanced article

Lorraine D Hernandez, *Merck Research Laboratories, Rahway, New Jersey, USA*
Caroline Houde, *Merck Research Laboratories, Rahway, New Jersey, USA*
Maarten Hoek, *Merck Research Laboratories, Rahway, New Jersey, USA*
Brent Butts, *Merck Research Laboratories, Rahway, New Jersey, USA*
Donald W Nicholson, *Merck Research Laboratories, Rahway, New Jersey, USA*
Huseyin Mehmet, *Merck Research Laboratories, Rahway, New Jersey, USA*

Article Contents

- History and Classification
- Structure and Active Site
- Substrate Recognition and Mechanism of Action
- Sequence of Action
- Substrates during Apoptosis
- Inhibitors
- Caspase Knockouts
- Pathological Implications
- Therapeutic Outlook

Caspases (Cysteinyl ASPartate-specific proteASE) are cysteine proteases involved in cell death. The caspase family is comprised of 12 proteins in humans, seven in *Drosophila melanogaster* and a single protein in *Caenorhabditis elegans'*. All caspases consist of three structural domains: a prodomain, a large subunit and a small subunit. The catalytically active enzymes are formed either by proteolytic cleavage of the subunits, or through a proximity-induced activation process involving the prodomain. The cleavage site of caspase substrates is indicated by the positions P4, P3, P2 and P1. P1 is always aspartate, while amino acids in the other positions are highly variable. By cleaving over 1000 substrates in the cell, caspase activation results in the complete dismantling and ultimate death of the cell. Caspase activity can be regulated at a pharmacological level and has become a therapeutic target for the treatment of many diseases.

History and Classification

The first caspase activity was described in 1989 by Sleath and Schmidt as a proteolytic event required for the activation of interleukin-1β. In 1993, the group led by Bob Horvitz discovered the cell death gene *ced-3* encodes a protease belonging to the same family. In 1999, the first human genetic disease caused by mutation of a caspase (type 10) was identified by Michael Lenardo (Wang *et al.*, 1999). Phylogenetic analyses, and also studies of substrate specificity, indicate that caspases can be subdivided into three different groups. **Figure 1a** reports the present classification of the human caspases. **See also**: Caspases, Substrates and Sequential Activation

Group I includes caspases 1, 4, 5 and 12, which are involved in inflammation. The prototype of this group is caspase 1, previously known as interleukin-1β conversion enzyme, or ICE (Black *et al.*, 1989; Kostura *et al.*, 1989; Yuan *et al.*, 1993). Its enzymatic activity was identified in 1989, but the enzyme was only purified and cloned in 1992, a year after Lois Miller identified the first caspase inhibitor, the viral protein p35. Caspase 1, or ICE, controls the maturation of IL-1β actively participating in the inflammatory response. Site P4 of the substrate is preferentially a hydrophobic residue. Caspase 12 is absent in most humans with less than 2% of humans of African descent expressing the protein.

Group II includes caspases 3, 6 and 7, shown by a star in **Figure 1a**. The prototype of this group is caspase 3, the homologue of ced-3 in *Caenorhabditis elegans*. Group II caspases are directly involved in apoptosis in the terminal effector phases. Group II caspases have a very small prodomain, indicating a simple regulation of their enzymatic activation. These caspases are in general activated directly by another caspase in a cascade of enzymatic amplification. Site P4 of the substrate is preferentially an Asp (D) residue. Site P3 of the substrate is preferentially a Glu (E) residue. Therefore this subfamily preferentially cleaves substrates with the sequence DExD. Caspase 6 however has a VEHD preference.

Group III consists of caspases 2, 8, 9 and 10. These caspases have a very large prodomain, and therefore a complex mechanism of activation. In fact, these caspases need molecular adaptors to be enzymatically activated, in contrast to Group II caspases. Caspases 8 and 10 are part of the signal transduction mechanism of apoptosis receptors such as CD95; in this case, the adaptor molecule is FADD. Caspase 9 is activated in the apoptosome, and the necessary adaptor molecule is Apaf-1 (apoptosis-activating factor-1). Site P4 of the

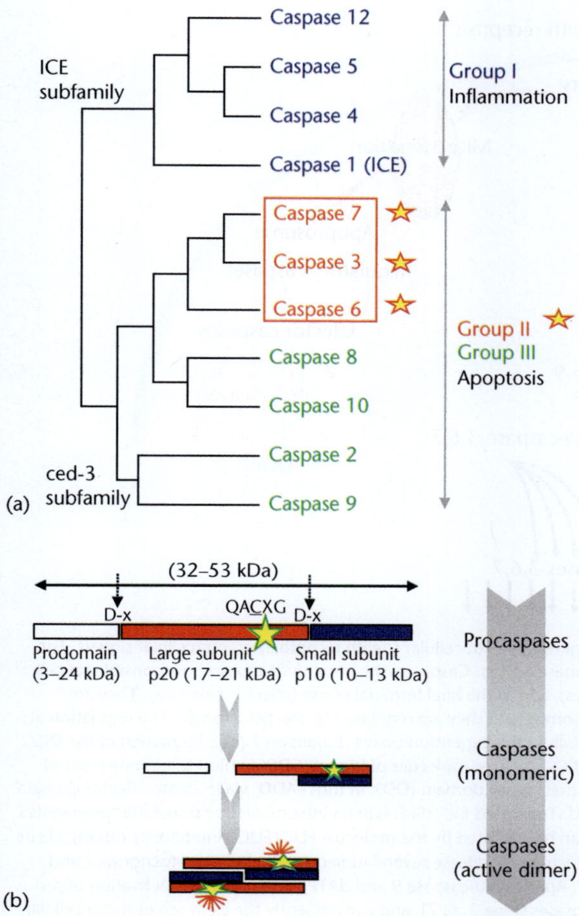

Figure 1 The human caspases family. (a) Classification of the human caspases. According to the phylogenic tree, caspases are subdivided into two major subfamilies, ICE and ced-3. The former are involved in inflammation, and the latter in apoptosis. The caspases with a very short prodomains (< 30 aa) are boxed (type 3, 6, 7); all the other enzymes have long prodromains (> 10 kDa) subject to complex regulation. The enzymes can be divided according to their proteolytic specificity into group I, involved in cytokine maturation, group II, involved in the final effector phase of apoptosis (indicated by a star) and group III, involved in the upstream regulation of apoptosis. At least two gene clusters have been identified, consistent with some caspases arising from tandem gene duplication. Caspase 13 is an error of sequencing; caspase 12 is mutated into a nonfunctional enzyme in up to 90% of the human population. Caspase 14 has no apparent role in apoptosis or inflammation preventing its classification. The scheme has been modified from Nicholson (1999). (b) Structural organization and activation of caspases. The caspases are produced as pro-enzymes (32–53 kDa), including a prodomain (3–24 kDa), the large subunit (17–21 kDa) containing the active site, and the small subunit (10–13 kDa). To become enzymatically active, these three components must be proteolytically cleaved (D-x site), a phenomenon that is regulated by the prodomain itself; this allows the assembly of the large and small subunit, and dimerization into a functional enzyme.

substrate is preferentially an aliphatic residue. Caspase 2 can be activated in the PIDDosome complex, so named because of the adapter molecule *p*53-**i**nduced protein with a *d*eath *d*omain. However, while these caspases have a long prodomain they can also be cleaved and activated by other caspases, allowing them to be both 'regulator' or 'effector' caspases, depending on the context of activation (see later).

The last caspase family member identified is caspase 14. It has a role in keratinocyte differentiation, but apparently none in apoptosis or inflammation, preventing its classification in any of the previous groups. **See also**: Cornification of the Skin: A Non-apoptotic Cell Death Mechanism

Structure and Active Site

As already described, caspases consist of three structural domains: a prodomain, a large subunit (also named as p20) and a small subunit (called p10). The enzyme is composed of the small and large subunits (p20/p10, approximately 30 kDa). The active cysteine is in the p20 subunit and the active site can be classified into four different subsites S4–S1 based on available crystal structure data. The S1 subsite, which attacks the carboxyl group of the polypeptide side chain in P1 (the site of cleavage is Asp – D), is composed of p20 as well as p10 subunits. Similarly, the site for recognition and positioning of the substrate (S4–S1) is furnished by both p20 and p10, even though the residues mostly responsible for specificity (S4) are contained in p10.

Caspases function enzymatically as dimers of identical molecules bound to each other. **Figure 1b** shows a very schematic view of dimer formation. Each monomer (p20/p10) is formed by a compact cylinder dominated by six β sheets and five α helices distributed on the opposite site of the plane formed by the β sheets. The dimer (p20/p10)2 contains two (p20/p10) units aligned head to tail, which position their respective active sites on opposite sides of the molecule. Although there are two active sites, their cooperation has never been observed. The orientation of the molecule indicates a specific mechanism of activation, well regulated with security systems (a 'safety catch').

For Group III caspases, which contain a long prodomain, enzymatic activation in the initial phases of apoptosis occurs through interaction with adaptor proteins and does not require cleavage and reorganization of the subunits. Instead, clustering of several intact caspase 8 or caspase 9 molecules in complexes such as DISC (death initiation signalling complex) and

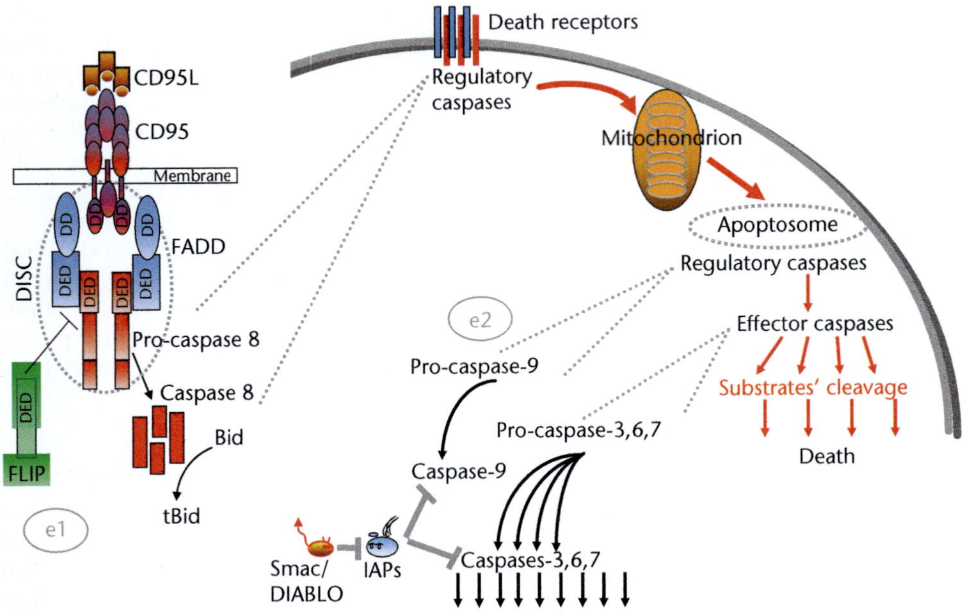

Figure 2 Caspases and cell death. In mammalian cells, apoptosis can be triggered by extracellular (death receptors) or intracellular signals. The signal converges to the mitochondrion/apoptosome where the final effector phase occurs. Caspases, the proteolytic enzymes responsible for cell death, are involved both in the upstream regulatory phase (regulatory caspases), and in the final terminal phase (effector caspases). They are activated by an adaptor molecule (FADD in the DISC, or Apaf-1 in the apoptosome), and they are regulated by the Bcl-2 family. The regulation at the level of the death receptor, and of the apoptosome is shown in greater details in the expansion boxes. *Expansion 1 (e1)*: Formation of the DISC, following the activation of a death receptor such as CD95 (also called Fas or APO-1). Three molecues of ligand (CD95L) bind three molecules of receptor (CD95), allowing the recruitment of the adaptor molecule FADD via their death domain (DD). In turn FADD, via its death effector domain (DED), recruits and activates caspase 8, which cleaves the specific substrate Bid. Truncated Bid (tBid) is in fact the molecular signal that propagates the death signal to the mitochondria and the apoptosome. This mechanism can be inhibited by the molecule FLIP (FLICE-inhibitory protein), via its DED. *Expansion 2 (e2)*: When activated by apoptotic signals (e.g. by tBid), mitochondria release several molecules, including cytochorme *c* and DIABLO/Smac. The former goes to the apoptosome (formed by cytochrome *c*, Apaf-1, pro-caspase 9 and dATP), and allows the activation of pro-caspase 9. This in turns activates several molecules of downstream effector capsases (type 3, 6, 7), and consequently the cleavage of many cellular substrates results in cell death. Still at its terminal stage, cell death can be inhibited by IAPs, whereas Smac/DIABLO can remove the protection by IAPs. The cascade of proteolytic amplification created by caspase 9 (apical regulatory caspase) and caspases 3, 6, and 7 (downstream effector caspases) is extremely powerful. DISC, death initiation signalling complex and IAP, inhibitors of apoptosis proteins.

the apoptosome, respectively, is sufficient to trigger their activity (**Figure 2**). Recently, caspase 2 has also been shown to undergo this 'proximity-induced activation' by clustering in the PIDDosome to become activated.

Substrate Recognition and Mechanism of Action

The substrate recognition site of the caspase is indicated as S4–S1, at which the substrate is bound at residues P4–P1. This recognition guarantees an absolute specificity. The S4–S1 substrate recognition site varies in each caspase, even though S1 is absolutely constant and highly tight. The deep and basic S1 pocket is responsible for the absolute specificity to bind Asp (D) as the substrate residue P1. In fact, Asp is physically constrained and held by a hydrogen bond with Arg179, Gln283 and Arg341 (the numbers refer to caspase 3). The dimensions of the S1 subsite are reduced to a minimum, and are therefore responsible for the lack of tolerance for substitutions at P1. Bond sites P2 and P3 (subsites S2 and S3) are more tolerant, less specific and more heterogeneous in the various caspases. Therefore the substrate (or the inhibitor) positions are itself within the S1–S4 site, creating interactions as hydrogen bonds with Ser339 and Arg341 (conserved in practically all of the caspases). The P4 (S4)-binding site is the major determinant of specificity.

Caspases are cysteine proteases, possessing a catalytic triad which includes a cysteine (active site) in the vicinity of a histidine. The third component of the classical catalytic triad typical of cysteine proteases (typically a serine) is not constant in caspases, and its function can be differently substituted by different groups in different components of the family. For example, if we consider caspase 3, subsequent to substrate binding, the catalysis can take place with the catalytic diad Cys285 and His237. The third element of the cysteine protease triad is in fact substituted by an 'oxyanion hole'. This is a pocket formed by amide nitrogens from Gly238 and Cys285 (conserved in all caspases), where hydrogen bonds the carbonyl oxygen of the P1 residue of the substrate during catalysis.

Sequence of Action

Although the regulatory caspases of the death receptor (8 and 10) or of the apoptosome (9) or of the PIDDosome (2) participate in signal transmission, the effector caspases (type 3, 6 and 7) determine the death of the cell. Effector caspases are diverse and are not activated in parallel, but in a sequential manner, resulting in an amplification cascade of proteolytic cleavage (vaguely similar to the cascade of proteolytic activation and amplification of the complement system or blood coagulation). These cascades have the effect of augmenting the signal.

In the apoptosome, the first caspase to be activated is caspase 9. This in turn activates caspases 3 and 7. Caspase 3 is also able to proteolytically activate caspase 9, therefore creating a reciprocating loop of potentiation. Caspases 3 and 7 proteolytically activate caspases 6 and 2 (Inoue *et al.*, 2009). At this point, caspases 3, 7, 6 and 2 are highly activated and work to proteolytically cleave their intracellular substrates and determine cell death itself. A diagram of this order of activation is depicted in **Figure 3**. Depending on the stimulus, caspase 2 can also be activated by the PIDDosome and function as a 'regulator' caspase. Caspase 6 can also proteolytically activate the apical regulatory caspases 8 and 10, creating another potentiation loop, which further enhances the death receptor signals.

Schematically we can define the molecular phases of apoptosis in the following manner (see **Figure 3**): (1) *Initiation*, activation of death receptors (caspases 8 and 10), (2) *Determination*, activation of the apoptosome (caspase 9), (3) *Amplification*, increase in the type and number of active caspases (caspases 3, 7, 6 and 2) and apical reactivation loop (caspases 8 and 10), (4) *Demolition*, proteolytic cleavage of over 700 different substrates which determine the death of the cell.

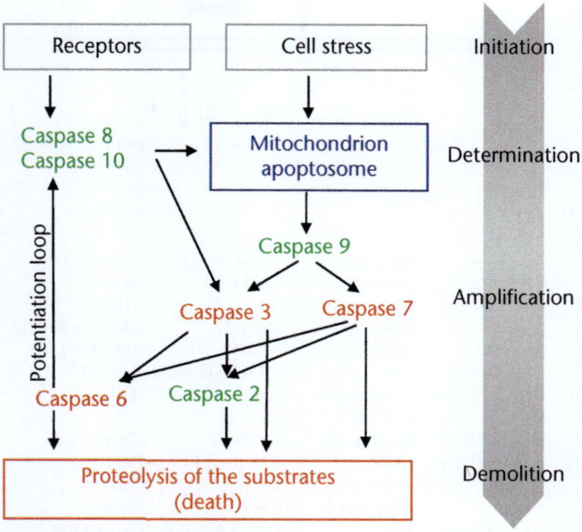

Figure 3 Sequential activation of caspases. In general, caspases with a long prodomain are involved in the upstream regulation and activation of the apoptotic pathway; they require a tight regulation and activation, and cleave very specific substrates. Caspases with a short prodomain are involved in the amplification of the effector cascade; thus they have a very simple, fast and direct activation, and they also have many substrates, whose cleavage and inactivation finally kills the cell. Caspase 6 seems to be able to reactivate the upstream regulatory caspases, creating a feedback forward potentiation loop that strongly enhances the amplification of the death signal. Colour code is in keeping with **Figure 1**.

Substrates during Apoptosis

Experimental studies indicate that over 1000 different polypeptides can be generated during apoptosis. A comprehensive database of caspase substrates listing more than 700 substrates reported in the literature (CASBAH) has been compiled by Luthi and Martin (http://bioinf.gen.tcd.ie/casbah/). Recent high-throughput efforts have almost doubled the number of caspase substrates and have yielded a less-biased dataset of physiological caspase substrates. The relative lack of overlap between the previously reported substrates and those defined by unbiased approaches indicate that many more caspase substrates remain to be discovered. **Figure 4b** indicates some of the principal substrates. The proteolytic cleavage of these substrates clearly renders the cell incapable of performing its proper functions.

Inhibitors

For practical reasons we distinguish pharmacological (**Table 1A**) and natural (**Table 1B**) inhibitors of caspases.

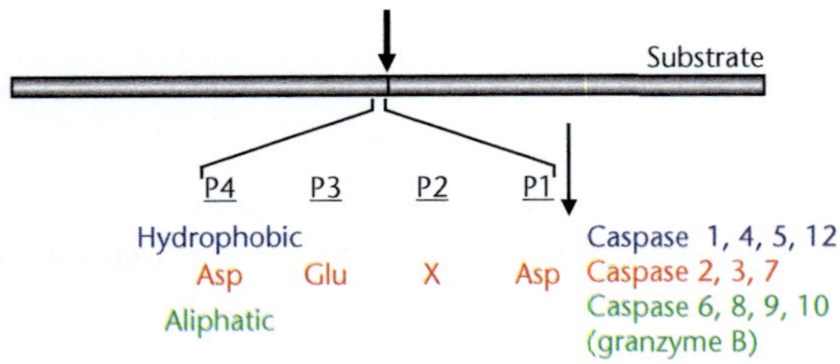

(a)

(b)

DxxD site	DxxD site	Non-DxxD site	Unknown site
DNA-PKs	MEKK-1	STAT1	wee1
PARP	Mst1	Sp-1	CDC27
Rad51	PRK2	SRP p72	SAF-A/hnRNP-U
Acinus	PP2A	NF-kB	hnRNP-A1
CAD/DFF45	FAK	PITSLRE kinases	Ras GAP
DNA-RTC140	Fodrine (all)	PAK-2	Raf1
Rb	Gelsolin	p59 fyn	Akt 1
mdm2/HDM	Keratin-18	CaMK-IV	Cbl e Cbl-b
p21-waf1/cip	LAP2	p28 Bap31	PKN
NuMA	Nup153	Actin	Catenin-β, -γ
ATM	Rabaptin-5	Gas2	Kinectin
U1-70KsnRNP	APC	Lamin A and B	Caplastatin
hnRNP-C1/C2	Hsp90	Bcl-Xl	Ataxin-3
SREB	UbqCE NEDD4	Bid	AMPA receptors
IkB-alfa	Bcl2	APP-beta	p27 kip1
D4-GDI	Presenilin2	Pro-IL-16	MCM3
cPLA2	Huntigtin	Pro-caspases	
PKC-δ, -τ, -ζ	SBMA-AR		
Atrofin-1	p53, p63, p73		

Figure 4 Caspase proteolytic specificity. (a) Substrate recognition. The caspases recognize a tetrapeptide motif corresponding to the four residues P4-P3-P2-P1. Though the position at P3 and P1 seems to be obligatory, the position P4 allows the classification of three subfamilies; see also **Figure 1**. This property has facilitated the identification of group-specific inhibitors. Colour code is in keeping with **Figure 1**. (b) Substrates. This list of substrates is incomplete; a more comprehensive list of approximately 700 substrates, with their role in cell death is reported in http://bioinf.gen.tcd.ie/casbah/. The caspase substrates can have a single cleavage site (e.g. DQTD in Gelsonin), nested multiple sites (e.g. DEVDGVD in PARP), redundant clustered sites (e.g. DSLD-(X13)-DEED-(X16)-DLND-(X32)-DGTD in Huntigtin) or distal multiple sites (e.g. DEPD and DAVD in ICAD).

The understanding of the molecular mechanism of catalysis and of the properties of the P4–P1 sites of the substrate were the basis for designating pharmacological inhibitors, both of polypeptide type and of low molecular weight compounds. The electrophyllic groups which interact irreversibly with the catalytic Cys are aldehydes, nitriles and ketones, of which the latter are much more stable *in vivo* and therefore are most easily applied to pharmaceutics. Caspase inhibitors are all constituted according to the following scheme: (tetrapeptide)-C)-CH2-X. Where X signifies a –Cl or –F (fluoro- or chloro-methyl ketone, abbreviated fmk), –N2 (diazomethylketone) or –oCOR (acyloxymethylketone). The peptide portion, stabilized, determines the specific selectivity of the various caspases. Thus, zVAD-fmk (*N*-benzyloxycarbonyl-Val-Ala-Asp(O-Me) fluoromethyl ketone) is a nonspecific inhibitor of almost all of the caspases. DEVD-fmk, however, is specifically selective for effector caspases such as caspases 3, 6 and 7. **Table 1** gives an idea of the specific inhibitors for different caspases.

In 1991, Lois Miller first identified a natural inhibitor of apoptosis, before even the discovery of caspases. The protein p35 produced by baculovirus (*Autographa californica*) is a potent and selective inhibitor of caspases. Subsequently, numerous other viral caspase inhibitors were identified, such as cytokine response modifier A (CrmA). Inhibition of caspases by these specific proteins is one of the important pathways among which viruses regulate apoptosis.

Numerous cellular proteins exist with similar actions to these viral caspase inhibitors. They are called IAPs (inhibitors of apoptosis proteins). **Table 1B** shows the different viral and mammalian IAPs with their relative inhibitory effects on each caspase. All of the proteins which inhibit caspases, whether viral or cellular, are characterized by structural domains called BIR (baculovirus IAP repeat). Consequently all IAP proteins are also called BIR proteins (BIRP). However, the contrary does not hold: there are BIRP which do not show IAP activity. This fact is due to the prokaryotic origin of BIRP which are multifunctional. Many BIRP in humans (e.g. survivin) are passenger proteins in the mitotic spindle and are implicated in cell division. **See also**: Inhibitor of Apoptosis (IAP) and BIR-containing Proteins

Caspase Knockouts

The creation of transgenic mice lacking specific caspases has been accomplished for most caspases. **Table 2** displays the phenotypes for these mice. Because caspases play a determining role in apoptosis, sometimes specifically, sometimes redundantly, the phenotypes are extremely diverse, as indicated by the table. The role of caspases in the receptor mechanism is indicated by the caspase 8 knockout. The similarity of the phenotypes of the knockouts of caspases 3, 9 and Apaf-1 indicate the presence of a very important mechanism in the apoptosome by which the amplification of the caspase cascade is initiated and regulated. **See also**: The Apoptosome: The Executioner of Mitochondria-mediated Apoptosis

Table 1 Inhibitors of caspases

Panel A: Peptide inhibitors

	zVAD-fmk	aDEVD-cho	aYVAD-cho
Caspase 1	2	15	1
Caspase 2	2400	1700	>10 000
Caspase 3	40	1	>10 000
Caspase 4	130	130	360
Caspase 5	5	200	160
Caspase 6	100	30	>10 000
Caspase 7	40	2	>10 000
Caspase 8	2	1	350
Caspase 9	4	60	1000
Caspase 10	–	10	400

Panel B: Natural inhibitors

Caspase inhibitor	Source	Target caspase(s)	Strength of inhibition (where known)
CrmA	Cowpox virus	Caspase 1	K_i 4–10 pM
	Caspase 8		K_i < 340 pM
	Caspase 10		K_i 4–17 nM
P35	Baculovirus	Broad spectrum, for example, inhibits caspases 1, 3, 6, 8, 7 and 10	K_is 0.1–9 nM
OpIAP	Baculovirus	Does not directly inhibit caspase activity	
XIAP	Mammalian	Caspase 3	IC50 0.1–2 nM
		Caspase 7	IC50 1–10 nM
		Caspase 9	IC50 10 nM
cIAP1	Mammalian	Does not directly inhibit caspases at physiological concentrations	
cIAP2	Mammalian	Does not directly inhibit caspases at physiological concentrations	
ILP-2	Human	Does not directly inhibit caspases at physiological concentrations	
ML-IAP	Mammalian	Does not directly inhibit caspases at physiological concentrations	
Survivin	Mammalian	Does not directly inhibit caspases at physiological concentrations	
NAIP	Mammalian	Does not directly inhibit caspases at physiological concentrations	
BRUCE	Mammalian	Does not directly inhibit caspases at physiological concentrations	

Notes: BRUCE, BIR repeat-containing ubiquitin-conjugating enzyme; cIAP, cellular inhibitor of apoptosis; ILP-2, inhibitor of apoptosis protein (IAP)-like protein-2; ML-IAP, melanoma inhibitor of apoptosis protein; NAIP, neuronal apoptosis inhibitory protein; OpIAP, Orgyia pseudotsugata inhibitor of apoptosis; XIAP, X-linked inhibitor of apoptosis.
Source: Modified from Ekert *et al.* (1999) and Callus and Vaux (2007), with permission from Nature Publishing Group. See also Nicholson (2000). Inhibition constant (K_i, nM) shown is $t_{1/2}$ at 1 μM in seconds. NI, non inhibitory and Inib, inhibitory.

Pathological Implications

The execution of apoptosis is dependent on the activation of caspases, and as a result, caspases play a determining role in all pathologies of excessive apoptosis. For example, caspases play a crucial role in the *in vivo* activation of death described in myocardial infarction and in cerebral ischaemia, conditions in

Table 2 Caspase knockouts mice

	Development	Apoptotic phenotype
Caspase 1	Normal	None, defect in inflammation only
Caspase 2	Normal	Germinal cells
Caspase 3	Lethal PN	Neural
Caspase 6	Normal	None
Caspase 7	Normal	Mild in mouse embryonic fibroblast
Caspases 3 and 7	Lethal PN	Neural, cardiac and thymocytes
Caspase 8	Lethal E	Death receptors (CD95, TNF and DR3)
Caspase 9	Lethal E	Neural
Caspase 11[a]	Normal	Thymocytes and neural
Caspase 12	Normal	None, enhanced inflammation only

Notes: Often, the severity of the phenotype depends on the genetic background of the animals, as for example described in caspase 3. E, embryonically lethal; PN, perinatally lethal and TNF, tumour necrosis factor.
[a]Caspase 11 in the mouse is the orthologue of human caspases 4 and 5. Caspase 10 is absent in mice.

which cell death occurs both by apoptosis as well as by necrosis.

Caspases can also play an aberrant role in regulatory processes determining the major propensity or vulnerability of cells to lethal insults. Examples are the possible role played by caspases in triplet diseases such as Huntington disease (HD) and in neurodegeneration like in Alzheimer disease. In HD, low level of intracellular activation of caspase 3 liberates an *N*-terminal fragment of the protein huntingtin (four DxxD sites in a cluster), containing the Glu expansion, facilitating a pathological aggregation. In turn, this aggregate sensitizes more caspase 3, forming a vicious cycle of activation and aggregation of poly Glu. Cleavage of huntingtin by caspase 6 is also required for neuronal dysfunction and degeneration in HD. The cell, thus sensitized, facilitates the activation of caspase 8 (from the CD95 receptor, or directly by an effector caspase, or even stimulated indirectly by the poly Glu aggregate), determining the death of the cell. This sensitization can occur over an extended period of time (years).

In Alzheimer disease, caspase 3 alters the normal processing of the amyloid β precursor protein (APP), removing the *C*-terminal. APP, deprived of the intramolecular reinternalization signal, proceeds down a route of degradation that results in the formation of the amyloid β (Aβ) peptide, which can in turn favour the activation of apoptosis by means not yet clearly understood. In addition, caspase 3 has been implicated in the cleavage of tau leading to the formation of pathological tau filaments. In either case, caspase 3 seems to augment the baseline levels (threshold?) of activation of cell death, rendering the cell more sensitive to death, and participating in the pathogenetic mechanisms of these important illnesses. **See also**: Drug Discovery in Apoptosis

Therapeutic Outlook

Selective peptide inhibitors of different caspases (see **Table 1A**) have been successfully developed, even though they have problems penetrating the cell membrane, exhibit electrophilic promiscuity and low stability in solution, making them more susceptible to attack by biological nucleophiles such as cathepsin. Though the current molecules are poorly adapted to clinical use, they have provided the research community with useful analytical tools. Several companies are currently in advanced stages of development for new classes of low molecular weight nonpeptide inhibitors. These show a much higher potency and specificity, opening new therapeutic perspectives. Nonpeptide inhibitors such as the 'isatins' have potencies up to 1000 times greater than the peptide class of zVAD-fmk, with substantial improvements *in vivo*. Though these studies are still in the early stages, the work holds promise for future therapeutic application.

References

Black RA, Kronheim SR and Sleath PR (1989) Activation of interleukin 1β by a co-induced protease. *FEBS Letters* **247**: 386–390.

Callus BA and Vaux DL (2007) Caspase inhibitors: viral, cellular and chemical. *Cell Death and Differentiation* **14**(1): 73–78.

Ekert PG, Silke J and Vaux DL (1999) Caspase inhibitors: viral, cellular and chemical. *Cell Death and Differentiation* **6**(11): 1081–1086.

Inoue S, Browne G, Melino G and Cohen GM (2009) Ordering of caspases in cells undergoing apoptosis by the intrinsic pathway. *Cell Death and Differentiation* **16**(7): 1053–1061.

Kostura MJ, Tocci MJ, Limjuco G *et al.* (1989) Identification of a monocyte specific pre- interleukin 1β convertase activity. *Proceedings of the National Academy of Sciences of the USA* **86**: 5227–5231.

Nicholson DW (1999) Caspase structure, proteolytic substrates, and function during apoptotic cell death. *Cell Death and Differentiation* **6**(11): 1028–1042.

Nicholson DW (2000) From bench to clinic with apoptosis-based therapeutic agents. *Nature* **407**(6805): 810–816.

Wang J, Zheng L, Lobito A *et al.* (1999) Inherited human caspase 10 mutations underlie defective lymphocyte and dendritic cell apoptosis in autoimmune lymphoproliferative syndrome type II. *Cell* **98**: 47–58.

Yuan J, Shaham S, Ledoux S, Ellis HM and Horvitz HR (1993) The *C. elegans* cell death genes ced-3 encodes a protein similar to mammalian interleukin 1β-converting enzyme. *Cell* **75**: 641–652.

Further Reading

Fischer U, Laenicke RU and Schultze-Ostoff K (2003) Many cuts to ruin: a comprehensive update of caspase substrates. *Cell Death and Differentiation* **10**: 76–100.

Franchi L, Eigenbrod T, Muñoz-Planillo R and Nuñez G (2009) The inflammasome: a caspase-1-activation platform that regulates immune responses and disease pathogenesis. *Nature Immunology* **10**(3): 241–247.

Fuentes-Prior P and Salvesen GS (2004) The protein structures that shape caspase activity, specificity, activation and inhibition. *Biochemical Journal* **384**: 201–232.

Hotchkiss RS and Nicholson DW (2006) Apoptosis and caspases regulate death and inflammation in sepsis. *Nature Review of Immunology* **6**(11): 813–822.

Lüthi AU and Martin SJ (2007) The CASBAH: a searchable database of caspase substrates. *Cell Death and Differentiation* **14**: 641–650.

Melino G (2001) The Siren's song (concept: apoptosis). *Nature* **412**: 23.

Shi Y (2004) Caspase activation: revisiting the induced proximity model. *Cell* **117**(7): 855–888.

Slee EA, Adrian C and Martin SJ (1999) Serial killers: ordering caspase activation events in apoptosis. *Cell Death and Differentiation* **6**: 1067–1074.

Zheng TS, Hunot S, Kuida K and Flavell RA (1999) Caspase knockouts: matter of life and death. *Cell Death and Differentiation* **6**: 1043–1053.

The Apoptosome: The Executioner of Mitochondria-mediated Apoptosis

Elisabetta Ferraro, *Department of Biology University of Rome Tor Vergata, and IRCCS Fondazione Santa Lucia, Rome, Italy*

Francesco Cecconi, *Department of Biology University of Rome Tor Vergata, and IRCCS Fondazione Santa Lucia, Rome, Italy*

Advanced article

Article Contents

- Introduction
- The Apoptosome Structure: Its Components and Its Assembly
- Modulation of the Apoptosome Formation
- Apoptosome-like Complexes in Evolution. Is the Mitochondrial Pathway of Apoptosis Conserved?
- The Role of the Apoptosome in Mammalian Development
- Final Remarks

The apoptosome is a molecular complex of two major components – the adapter protein apoptotic protease activating factor 1 (Apaf1) and the protease caspase-9. These are assembled during apoptosis upon Apaf1 interaction with cytochrome *c*, which is released from the intermembrane space of mitochondria under precise cell death stimuli. Apoptosome assembly triggers effector caspase activation, which – in turn – drives cell demise. Several proteins bind and regulate apoptosome. This complex is found in vertebrates, even though similar functions are played by ortholog proteins through slightly different mechanisms also in

lower eukaryotes. Given the involvement of apoptosome dysfunctions in human diseases, this complex is also a relevant molecular target in biomedicine.

Introduction

Stimuli that trigger apoptosis in mammalian cells engage the cell death machinery in a variety of ways which can be summarized into two major categories of apoptotic pathways: the extrinsic or receptor-mediated pathway and the intrinsic or mitochondria-mediated pathway. The extrinsic and the intrinsic pathways of apoptosis are often interconnected, depending on the cell type and on the apoptotic stimulus. In general, signals generated by most of the death-promoting agents all converge, at some point, on mitochondria. The intrinsic pathway is characterized by the release of important proapoptotic factors from the intermembrane space of mitochondria into the cytosol and which promote the assembly of a multiproteic complex called *apoptosome*. The formation of the apoptosome leads to the activation of effector caspases which execute cell death through cleavage of multiple protein substrates. Here we discuss the molecular structure and the cellular functions of this complex. **See also**: Apoptosis: Regulatory Genes and Disease; Caspases and Cell Death; Caspases, Substrates and Sequential Activation; Mitochondrial Outer Membrane Permeabilization

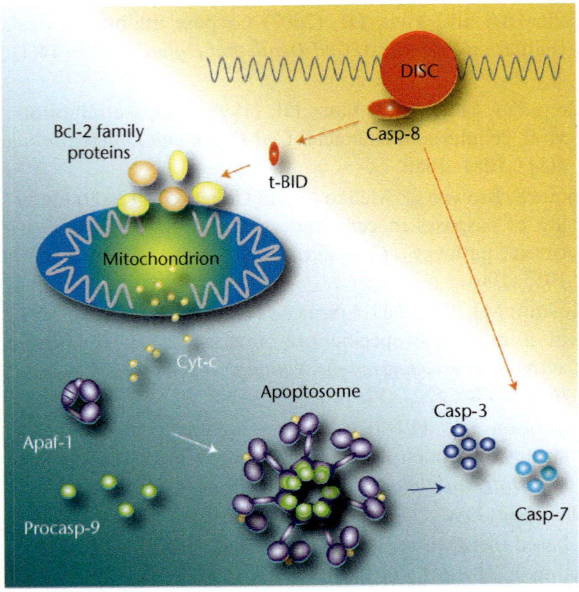

Figure 1 The intrinsic (or mitochondria-mediated) pathway of apoptosis and its correlation with the extrinsic pathway. The intrinsic pathway of apoptosis is highlighted in blue: cytochrome *c* release from mitochondria is modulated by proteins belonging to the Bcl-2 family. In the cytosol, cytochrome *c* binds Apaf1 and induces apoptosome assembly and activation of caspase-9 which, in turn, activates caspase-3 and caspase-7. Effector caspases are responsible for the executioner phase of apoptosis. The extrinsic pathway of apoptosis is highlighted in orange: the assembly of a complex called DISC (death-inducing signalling complex) promotes activation of caspase-8 which can cleave BID and trigger cytochrome *c* release from mitochondria and apoptosome formation or can directly cleave effector caspases.

The Apoptosome Structure: Its Components and Its Assembly

The most important proapoptotic factor released from mitochondria during the intrinsic pathway of apoptosis is the cytochrome *c* (Liu *et al*., 1996). Once in the cytosol, the cytochrome *c* binds to the apoptotic protease activating factor (Apaf1) and, in the presence of 2′-deoxyadenosine triphosphate (dATP) (or, at higher concentration, ATP), promotes the oligomerization of Apaf1 in a caspase-activating complex known as the apoptosome (Zou *et al*., 1997). The initiator procaspase-9 is then recruited to the complex by binding to Apaf1 to form a holoenzyme complex. Subsequently, active caspase-9 cleaves and activates the effector caspase-3 and 7 responsible for the execution phase of apoptosis (**Figure 1**).

Apaf1

The core of the apoptosome is the adapter protein Apaf1. This large protein (approximately 140 kD depending on the isoforms) is the mammalian orthologue of *Caenorhabditis elegans* death-promoting protein-4 (CED-4), an essential protein involved in programmed cell death in the nematode *C. elegans* (see later). Three regions can be identified in Apaf1: the *N*-terminal caspase-recruitment domain (CARD, residues 1–90), the central nucleotide-binding and oligomerization region (nucleotide-binding, Apaf1/Rgene/CED-4; NB-ARC, 128-586) which shares a high degree of homology with CED-4 and, third, a *C*-terminal WD40 region (613–1248) (**Figure 2a**).

The CARD domain interacts with the CARD domain of the procaspase-9 upon apoptosome formation. The NB-ARC region is at the centre of Apaf1-oligomerization and consists of a nucleotide-binding and oligomerization domain (NOD) and a superhelical domain (HD2). The NOD region is an adenosine triphosphatase

(ATPase) domain (belonging to the AAA+ family of ATPases) organized into a nucleotide α/β binding domain (NBD) with conserved Walker boxes A (P-loop 155–161) and B (239–243) required for dATP/ATP and Mg^{2+} binding and a short α-helical domain (HD1). HD1 is followed by a winged-helix domain (WHD). WHD are domains usually found in deoxyribonucleic acid (DNA)-binding proteins such as transcription factors (Yu *et al.*, 2005; Riedl and Salvesen, 2007) (**Figure 2b–d**). The *C*-terminal WD40 region of Apaf1 contains 13 WD-40 repeats organized into two circular β-propeller-shaped domains. The first domain contains seven WD40 repeats and the second six WD40 repeats. WD-40 repeats are sequences about 40 aminoacids long which often terminate in Trp-Asp (WD); they are usually involved in protein–protein interactions.

Six different Apaf1 isoforms have been identified in *Homo sapiens* (Apaf1XL, L, S, M, Xs and ALT). They result from differential splicing and differ by insertion of an extra 43 aminoacids long WD40 repeat and/or the insertion of 11 aminoacids after the CARD domain. The form containing 13 WD-40 repeats and the 11 aminoacid insertion (Apaf1XL) is 1248 aminoacids long. It is considered the 'canonical' sequence since it is the most stable and the one functionally active (**Figure 2a–d**). Two isoforms corresponding to human Apaf1XL and Apaf1L have been found in mouse.

Apoptosome assembly

Normally, Apaf1 resides in the cytosol in a compact autoinhibited shape. Electron microscopy images suggest that, in the inactive conformation, the WD40 domain of Apaf1 folds back onto the rest of the protein therefore inhibiting Apaf1 oligomerization and procaspase-9 interaction (**Figure 2b–d**). In fact, removal of the WD40 domain results in constitutive binding and activation of caspase-9 (Bao and Shi, 2007). Once released into the cytosol, cytochrome *c* binds to the WD40 domain of Apaf1 and displaces it from the rest of the protein. However, this is not enough. In fact, although the lock is released, the CARD and the NB-ARC regions of Apaf1 remain in their autoinhibited conformation, as indicated by the crystal structure of WD40-deleted Apaf1 in which the bound ADP acts as an organizing centre for four adjoining domains locking Apaf1 in an inactive conformation; the CARD domain is not accessible and cannot recruit caspase-9 and the ATPase domain is blocked and is unable to form oligomers (**Figure 2b–d**) (Riedl *et al.*, 2005). Cytochrome *c* binding to WD40 and WD40 displacement needs to be followed by binding and hydrolysis of ATP or dATP (referred to as (d)ATP) which allows Apaf1 to assume a more extended conformation in which CARD and ATPase domains are no longer blocked (Acehan *et al.*, 2002).

Since Apaf1 in its monomeric and autoinhibited form binds dATP to its ATPase domain, it has been recently proposed that cytochrome *c* binding to Apaf1 allows the hydrolysis of the already bound dATP (2′-deoxyadenosine diphosphate) to dADP. dADP is subsequently replaced by exogenous dATP; Apaf1 is, indeed, a *bona fide* ATPase; dATP hydrolysis and subsequent exchange of dADP by dATP are both necessary steps for inducing conformational changes that are essential for the assembly of a functional apoptosome rather than nonfunctional aggregates. Based on this hypothesis, ATP hydrolysis would only happen in one reaction cycle; after the exchange, dATP bound to Apaf1 remains unhydrolysed during apoptosone formation. However, the kinetics of apoptosome assembly and the precise role of the (d)ATP and ATPase activity of Apaf1 need to be further characterized (Bao and Shi, 2007).

The three-dimensional structure of the apoptosome was determined by cryo-microscopy in 2002 from Akey's lab (Acehan *et al.*, 2002). In the presence of cytochrome *c* and (d)ATP the apoptosome has a wheel-like structure with a seven-fold symmetry in which seven molecules of Apaf1 oligomerize through their CARD and NOD domains. The central ring of the apoptosome is formed by the conjugation of seven CARD and NOD domains of Apaf1; the seven spikes extending from the central hub are made up of the HD2 and WD40 domains. The WD40 domains are organized into two β propellers which form two lobes (**Figure 3**). Cytochrome *c* binds to Apaf1 between the two WD40 lobes. Procaspase-9 is recruited to the apopotosme through its CARD domain which binds Apaf1 CARD domain at the central hub of this structure (Acehan *et al.*, 2002). The high local concentration of procaspase-9 on the hub may facilitate the recruitment of monomers from solution to form dimers. Each dimer of caspase-9 contains only one active catalytic site, likely the one of the monomer bound to Apaf1 through the CARD domain (Renatus *et al.*, 2001; Acehan *et al.*, 2002). It has been suggested that apoptosomes with bound procaspase-9 are prone to form dimers in which the top surface of each apoptosome is oriented towards the dimer interface (Acehan *et al.*, 2002). Caspase-9 and caspase-3 were found within the native apoptosome whereas caspase-6 and caspase-7 were not (Hill *et al.*, 2004). The exact placement of the CARD domain and the NB-ARC region in the apoptosome can be explained by two models described in **Figure 3** (Riedl and Salvesen, 2007).

Caspase-9

Procaspase-9 is an initiator caspase. It exists as an inactive monomer in the cytosol of nonapoptotic cells.

The human procaspase-9 is ubiquitously expressed. It is a 416 aa protein characterized by a long *N*-terminal CARD prodomain (1–92) a large catalytic subunit (140–289), a small subunit (331–416) and two linker regions

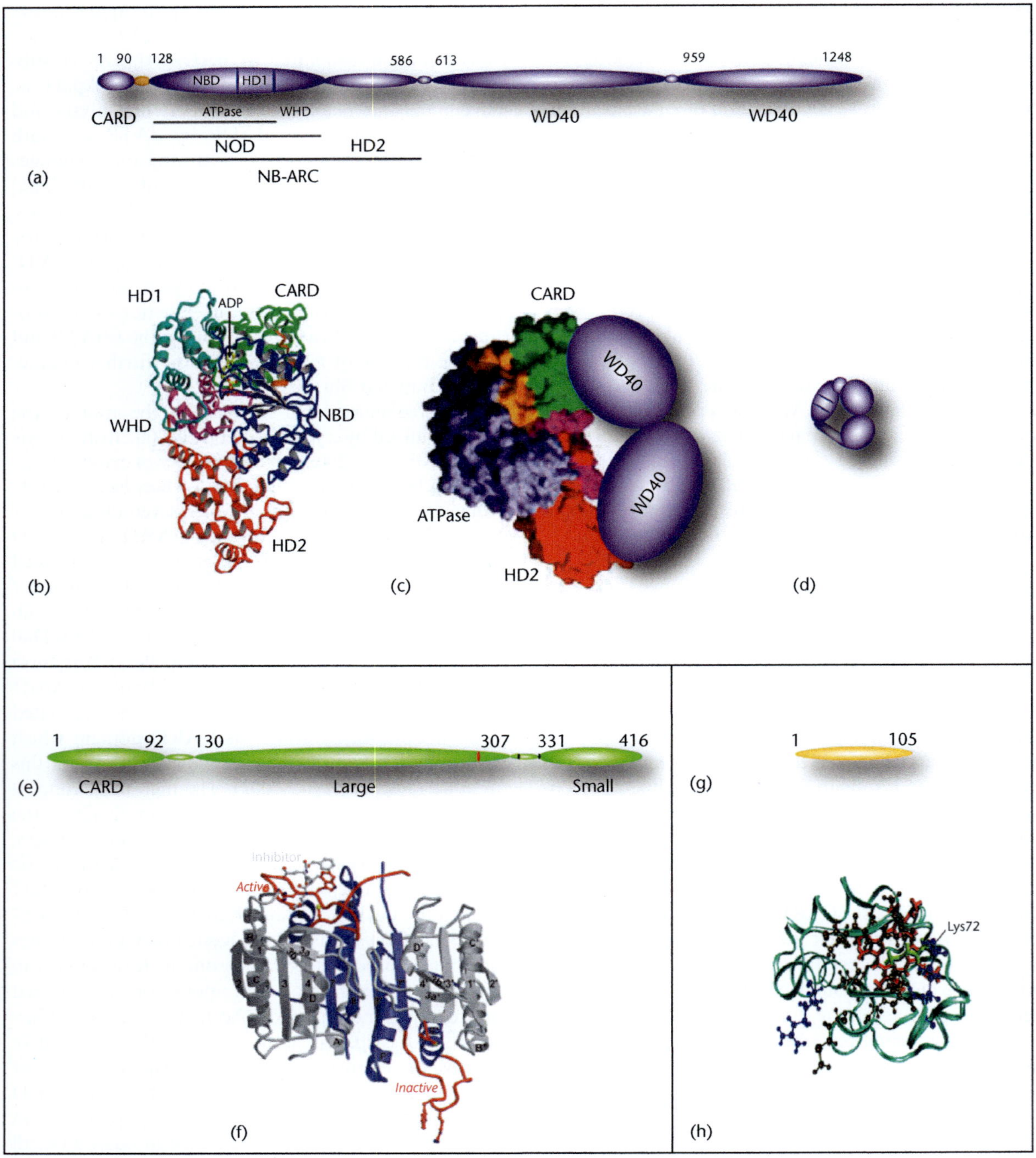

between these domains (**Figure 2e, f**). Two isoforms of procaspase-9 have been identified in human (caspase-9L and caspase-9S). These isoforms are generated by alternative splicing and display opposing functions. Caspase-9L (considered the canonical sequence) mediates cell death whereas caspase9-S lacks the large catalytic subunit and functions as a caspase-9 dominant-negative inhibitor of the apoptosome by directly binding the Apafl CARD domain and interfering with recruitment of procaspase-9L. In mouse two isoforms of caspase-9 were also identified but one is not functional and does not act as a dominant-negative, whereas in rat a caspase-9 dominant-negative isoform has been identified. **See also**: Caspases, Substrates and Sequential Activation

Although effector caspases are mainly activated upon proteolytic cleavage by apical caspases, the initiator caspases become active upon binding through their long prodomains to large molecular complexes assembled around adapter proteins. This binding allows an increased local concentration and oligomerization of the procaspases which is followed by their autoproteolytic cleavage in the loop between the large and the small subunits. However, the interchain autocatalytic cleavage (at Asp315) appears not to be the critical step in procaspase-9 activation. Different models have been proposed to explain caspase-9 activation. The 'induced proximity' model states that, upon recruitment into the apoptosome, procaspase-9 monomers are brought into close proximity of one another and hence are able to autoprocess themselves (Salvesen and Dixit, 1999). Another hypothesis ('proximity-driven dimerization model') is that, upon apoptosome binding, monomers of inactive procaspase-9 are brought into close proximity at a high concentration. This induces their dimerization which is sufficient for caspase-9 activation (the interdomain loop is long and this feature may allow the zymogen to adopt an active conformation) whereas autoproteolysis (which do happen within the apoptosome) merely stabilizes the dimers (Boatright *et al.*, 2003; Pop *et al.*, 2006). The 'induced conformation model' is based on the observation that caspase-9 has a much higher level of catalytic activity within the apoptosome than when free in solution. This suggests that a conformational change must occur in the active site of the apoptosome-bound caspase-9. This conformational change might occur even without monomer-to-dimer transition, for example, by binding the surface necessary for homo-dimerization and stabilizing the active conformation; or by inducing caspase-9 dimerization and activation not only by increasing caspase-9 local concentration but also through additional interactions between the apoptosome and caspase-9; or by assembling the dimeric caspase-9 in a higher order complex (interaction between homodimers: homo-tertramers). It has been hypothesized that additional interactions between Apafl and caspase-9 different from those between their CARD domains might be required for caspase-9 activation within the apoptosome (Bao and Shi, 2007).

Cytochrome c

Human cytochrome *c* is a 105 aminoacids protein encoded by a nuclear gene and synthetized into the

Figure 2 Structure of the main apoptosome components: Apaf1, caspase-9 and cytochrome *c*. (a) Schematic representation of Apaf1 domain organization. Human Apaf1 can be divided into three functional regions. A CARD domain, a NB-ARC region and a WD-40 region. The NB-ARC region is formed by a NOD region and a superhelical domain HD2. The NOD region contains the AAA+ ATPase region, in which the α/βNBD domain and an HD1 domain can be distinguished. The HD1 domain is followed by a WDH domain. The bridging helix between the CARD and the NBD domain is drawn in orange. The WD40 region is organized into two domains: The first one contains seven WD40 repeats, whereas the C-terminal region contains six WD-40 repeats. (b) Ribbon diagram of the structure of WD-40-deleted Apaf1 (1–591) bound to ADP. The CARD domain (green) packs against the α/β NBD domain (blue), the short helical domain HD1 (cyan) and the winged-helix domain WHD (magenta). ADP binds the hinge region between the α/β domain and HD1 but is also coordinated by a few critical residues from the WHD domain which, by sensing the bound nucleotide, seems to be a key element for the regulation of the autoinhibited state of Apaf1. In this autoinhibited conformation, the CARD domain is not accessible for recruiting procaspase-9; blocking the ATPase domain does not allow the oligomerization of Apaf1 (Protein Data Bank accession code 1Z6T) (excerpt from Riedl *et al.*, 2005). (c) Model for the full-length autoinhibited Apaf1 conformation. The same structure depicted in **Figure 1b** is here seen in another orientation; its probable association with the WD-40 domain interacting with the *N*-terminal region is visible (excerpt from Riedl *et al.*, 2005). (d) Simplified scheme of **Figure 2c** (see also **Figure 1**). (e) Schematic representation of caspase-9 domain organization. Human procaspase-9 contains a CARD domain, followed by a large subunit and a small subunit separated by linker regions. The catalytic cysteine residue is Cys287 (red mark). Two black marks indicate Asp315 (the interchain autocleavage between the large and the small subunit) and the caspase-3 cleavage site Asp330, respectively. (f) Schematic representation of dimeric CARD-deleted caspase-9. Caspase-9 is composed of two domains; the large subunit of each domain is coloured in grey and the small subunit is coloured in blue. The bound inhibitor molecule is depicted in grey. Only the catalytic side in the left domain is active and, therefore, able to bind the inhibitor. In the right domain, the same residues are transposed from their catalytic conformation into a novel structure incapable of catalysis (Protein Data Bank accession code 1JXQ) (excerpt from Renatus *et al.*, 2001). (g) Human cytochrome *c* is a 105 aa long protein. (h) Schematic representation of holocytochrome. Lys 72 is enlighted, since it is essential for the stability of the interaction between cytochrome *c* and Apaf1. The heme group is colored in red (Protein Data Bank accession code 1AKK) (excerpt from Kalanxhi and Wallace, 2007).

cytosol (apo-cytochrome *c*). From the cytosol it is carried inside the intermembrane space of mitochondria where the enzyme haeme-lyase catalyses the incorporation of the heme group (holo-cytocrome *c*) (**Figure 2g, h**). Heme binding induces a conformational change of the holo-cytocrome *c* which keeps it inside the intermembrane space within the cristae. Inside mitochondria, cytochrome *c* operates as part of the respiratory chain by shuttling electrons through its heme group between Complex III (cytochrome *c* reductase) and Complex IV (cytochrome *c* oxidase).

During apoptosis cytochrome *c* is thought to be released from mitochondria in two phases: mobilization and then translocation through the mitochondrial outer membrane. Cytochrome *c* is a soluble component of the respiratory chain. It is anchored to the inner membrane thanks to its association with the phospholipid cardiolipin. It has been proposed that since cytochrome *c* is positively charged it electrostatically binds the anionic cardiolipin. Alternatively one acyl chain of cardiolipin might insert itself into the hydrophobic channel present in cytochrome *c* (Kalanxhi and Wallace, 2007). There are two isoforms of cytochrome *c*: the testicular and the somatic form, both able to promote apoptosis (Ow *et al.*, 2008).

The apo-cytochrome *c* is not able to activate Apaf1, so signifying that the heme group is essential for its apoptotic function. In this way the cell is protected from cytosolic cytochrome *c* during cytochrome *c* synthesis and translocation into mitochondria (Ow *et al.*, 2008). It has been reported that the redox capacity of cytochrome *c* is not necessary for apoptosome assembly; however, more recently, it has been shown that the oxidized form of cytochrome *c* (Fe^{3+}) can induce apoptosome activation, whereas the reduced form (Fe^{2+}) cannot. Cytochrome *c* seems to form a stable interaction with the apoptosome although previous observations suggest that cytochrome *c* may dissociate from the assembled apoptosome (Hill *et al.*, 2004; Yu *et al.*, 2005).

Modulation of the Apoptosome Formation

Apoptosome formation and activity are tightly regulated at several levels through physiological mechanisms, ensuring that the apoptosome complex is only fully assembled and functional when the cell is irrevocably committed to die.

A first level of regulation is the control of cytochrome *c* release from mitochondria. In addition to cytochrome *c*, other proapoptotic factors are located inside the mitochondrial intermembrane space. They are released on mitochondrial outer membrane permeabilization (MOMP). MOMP occurs through an as-yet-unidentified mechanism and is mainly regulated by proteins belonging to the Bcl-2 (B-cell lymphoma protein-2) family as reviewed by Ow and coworkers (Ow *et al.*, 2008). **See also**: Mitochondria Fusion and Fission; Mitochondrial Outer Membrane Permeabilization

Second mitochondria-derived activator of caspases/direct inhibitor of apoptosis protein (IAP)-binding protein with low pI (*Smac*/*DIABLO*) and high-temperature-requirement protein A2 (*HtrA2*/*Omi*) are another two important regulators of apoptosome activity released from mitochondria as well as cytochrome *c*. Once in the cytosol, they accelerate apoptosis by binding and inhibiting the inhibitors of apoptosis protein (*IAPs*) which are inhibitors of caspases. It has been proposed that IAPs serve as a mechanism to prevent accidental cell death by blocking caspase activity. When MOMP occurs, cytochrome *c* is released, apoptosome is formed and Smac/DIABLO and Omi/HtrA2 are also released. Smac and Omi bind to the baculovirus IAP repeat (BIR) domains of IAPs via *N*-terminal IAP-binding motifs (IBMs), therefore inhibiting IAPs interaction with caspases which are now free and active. Among mammalian IAPs, XIAP (X-linked inhibitor of apoptosis) is the strongest inhibitor of caspase-3, -7 and -9 *in vitro*. XIAP binds caspase-9 and inhibits its dimerization whereas it inhibits dimeric processed caspase-3 and -7 by binding and blocking their active site. Moreover, IAPs act as ubiquitin E3 ligases thereby promoting autoubiquitination, ubiquitination and degradation of protein to which they bind, for example, caspases and IAPs inhibitors. XIAP can indeed induce caspase-9 and SMAC ubiquitination and degradation (Vaux and Silke, 2005). XIAP was found to be part of the apoptosome where it might be responsible for the stable recruitment of active caspase-3 to the complex (Hill *et al.*, 2004). **See also**: Inhibitor of Apoptosis (IAP) and BIR-containing Proteins

Upon apoptosome activation, active caspase-3 is able to process caspase-9 at Asp330, presumably as part of a feedback amplification loop. It appears that cleavage by caspase-3 does not activate caspase-9, but enhances apoptosis by alleviating XIAP inhibition of caspase-9. Some authors report Apaf1 cleavage by caspase-3 at different sites. However, the functional meaning of this cleavage is not clear yet.

Apoptosome activity is inhibited through phosphorylation of procaspase-9 at different sites. The protein kinase PKB-Akt phosphorylates the serine 196, so inhibiting human caspase-9 activation. However, this phosphorylation site is absent in other mammals.

Kinase ERK directly phosphorylates caspase-9 at threonine 125 whereas protein kinase PKCzeta phosphorylates serine 144 of human caspase-9, thus inhibiting its activation. Furthermore the protein kinase PKA signalling alters the formation of the apoptosome by preventing binding between caspase-9 and Apaf1 (Schafer and Kornbluth, 2006).

Both caspase-9 and Apaf1 are regulated at transcriptional and translational level. As already seen, six splicing isoforms of Apaf1 have been identified. However, their physiological meaning needs to be clearly defined whereas the short isoform of caspase-9 has an inhibitory role on the apoptosme assembly. Apaf1 is transcriptionally regulated by E2F and p53; the hypermethylation of the Apaf1 gene promoter causes silencing of Apaf1 expression (Moroni *et al.*, 2001). Furthermore, a new regulatory mechanism of Apaf1 expression has recently been proposed; E2F1 could have a differential capacity to induce Apaf1 transcription based on the association of the Apaf1 promoter with active chromatin in developing neurons and repressed chromatin in mature neurons (Wright *et al.*, 2007).

It has been hypothesized that cytochrome *c* concentration in the cytosol has to reach a threshold (which might vary significantly depending on the levels of other regulators such as IAPS) to induce apoptosome formation and this would represent an additional level of regulation of the apoptosome assembly.

Differential localization of procaspase-9 (mitochondria and nucleus) and nitrosylaton of the mitochondrial pool have been reported. Moreover, association of Apaf1 with discrete domains in the plasma membrane may constitute a novel mechanism of apoptosome inhibition (Sun *et al.*, 2005). It has also been shown that cytochrome *c* is nitrosylated on its heme iron in mitochondria during apoptosis, this being presumed to be a proapoptotic modification (Ow *et al.*, 2008).

The apoptosome function has been found to be regulated by many modulators although their role is still being debated. Apaf1-interacting protein (*APIP*) binds to the CARD domain of Apaf1 in competition with caspase-9, negatively regulating the apotosome activation. For its part, *Aven* seems to inhibit apoptosis by binding Apaf1 and impairing its oligomerization, even though it has been identified as a Bcl-x_L interactor. Tumour-up-regulated CARD-containing antagonist of caspase nine (*TUCAN/CARDINAL*) is a CARD-containing protein thus able to bind and inhibit caspase-9. *NAC* binds Apaf1 through a CARD–CARD interaction whereas hepatocellular carcinoma antigen 66 (*HCA66*) interacts with the Apaf1 NBD domain. Both proteins positively promote apoptosome activity. It has been suggested that acetylcholinesterase (*AChE*) might promote the interaction between cytochrome *c* and Apaf1. Interactions of AChE with caveolin-1 and subsequently with cytochrome *c* appear to be indispensable for apoptosome formation (Schafer and Kornbluth, 2006; Piddubnyak *et al.*, 2007).

Although the association between the Apaf1 homologue CED-4 and CED-9 (the homologue of the anti-apoptotic *Bcl-2 family* members) might suggest it, the interaction between Apaf1 and Bcl-2 proteins is controversial. However, more recently, a Bcl-x_L binding region inside the ATPase domain of Apaf1 has been identified.

Heat shock proteins (*Hsps*) are also modulators of the apoptosome activity. They are induced by a variety of cellular stresses and work protecting the cell. It has been reported that *Hsp70* inhibits apoptosome formation by binding to Apaf1 and preventing procaspase-9 recruitment. *Hsp90* inhibits cytochrome *c* mediated oligomerization of Apaf1. Nonphosphorylatable Hsp90 binds more strongly to Apaf1 and inhibits its oligomerization. *Hsp27* sequesters cytochrome *c* inhibiting cytochrome binding to Apaf1. Hsp27 also binds caspase-3 inhibiting its proteolytic activation by caspase-9.

Putative HLA-DR-associated protein-1 (*PHAPI*) also known as mapmodulin and prothymosin-α (*ProT*) have been identified as strong apoptosome regulators although they do not directly bind to apoptosome components (Jiang *et al.*, 2003). ProT inhibits apoptosome assembly acting as an oncoprotein whereas PHAPI facilitates caspase-9 activation and enhances the catalytic activity of caspase-3 in the apoptosome complex probably through stabilization of caspase-3 dimers. In fact, it acts as tumour suppressor.

Apoptosome-like Complexes in Evolution. Is the Mitochondrial Pathway of Apoptosis Conserved?

Apoptosis has been deeply studied in three model organisms: in the nematode *Caenorhabditis elegans*, in the arthropode *Drosophila melanogaster* and in some mammals (vertebrates) (Horvitz, 2003; Hay and Guo, 2006). Apoptosis is a well-conserved form of cell death; it is indeed carried out through the activation of caspases. In addition, a protein homologous to Apaf1 and responsible for caspase activation is present in all organisms. Most of the players in *C. elegans* apoptosis have homologues in *D. melanogaster* and mammals. However, the way in which they interact is different in many aspects. In particular the role of mitochondria

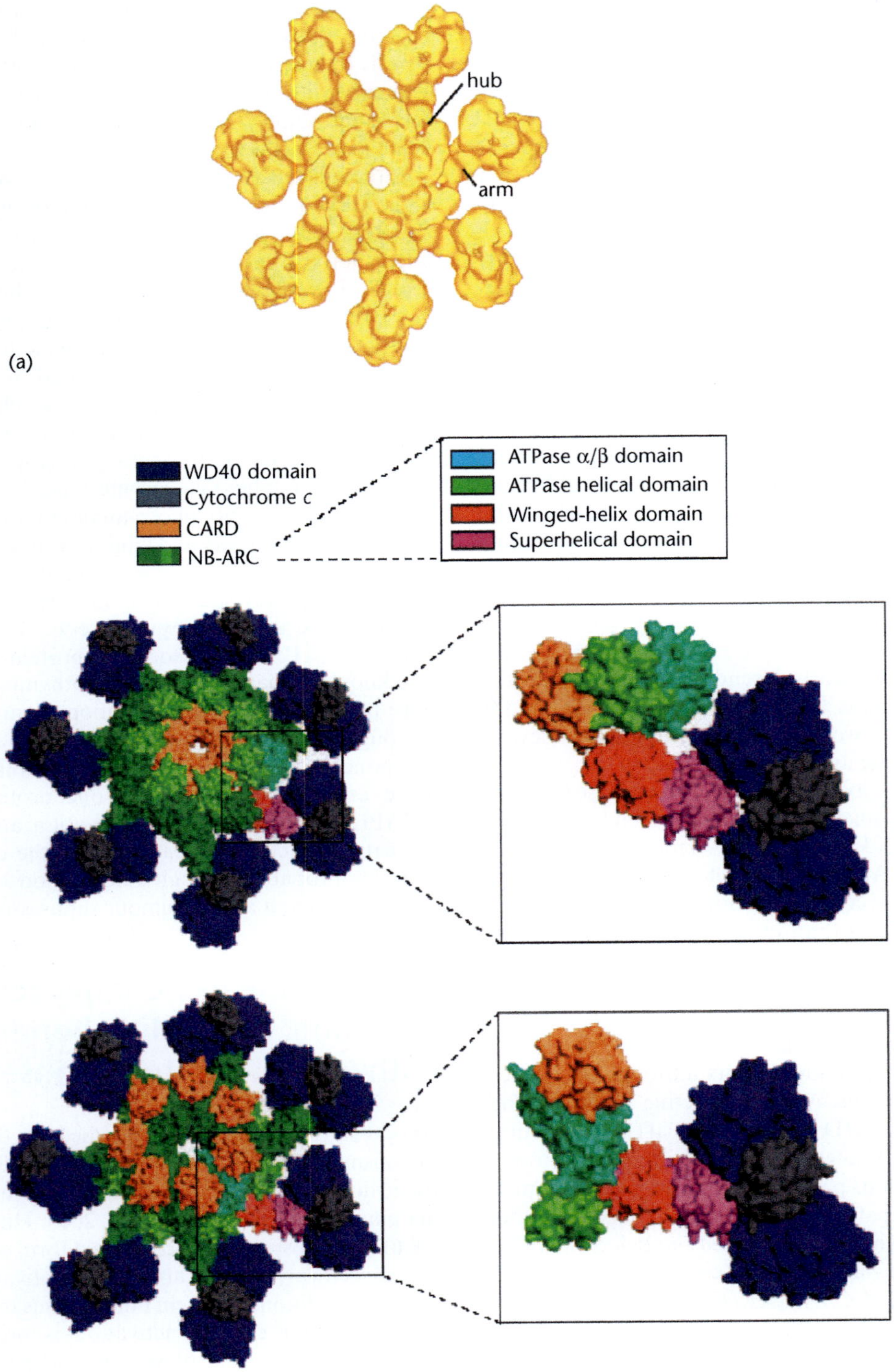
hub
arm
WD40 domain
Cytochrome c
CARD
NB-ARC
ATPase α/β domain
ATPase helical domain
Winged-helix domain
Superhelical domain

(a)

(b)

and cytochrome *c* in apoptosome formation and caspase activation is highly controversial in organisms other than mammals.

Caenorhabditis elegans

In *C. elegans*, CED-9 (an homologue of Bcl-2) is normally anchored to the mitochondria and associated with a symmetric dimer of the Apaf1 homologue, CED-4. CED-9 only interacts with one of the two molecules of CED-4. Apoptotic induction causes the upregulation of the BH3-only protein EGL-1 which binds to CED-9 and inhibits its interaction with CED-4. CED-4 dimers are therefore free to further dimerize and to form a tetrameric apoptosome-like structure which, in turn, binds and activates the CARD-bearing caspase CED-3 which will cause cell demise (**Figure 4a**) (Lettre and Hengartner, 2006). **See also**: Cell Death in *C. Elegans*

Mitochondria in *C. elegans* apoptosis seems to be merely a platform for the CED-9-CED-4 complex and to have no functional roles. In this organism, in fact, CED-9 binds mitochondria as well as Bcl-2. However, whereas Bcl-2 acts inhibiting MOMP, in worms neither MOMP nor cytochrome *c* are necessary for caspase activation. Furthermore CED-4 does not possess the WD-40 domain necessary for cytochrome *c* binding. In mammals, Bcl-2 proteins have also been implicated in mitochondrial fusion and fission and mitochondrial remodelling has been associated with MOMP. Therefore, localization of CED-9 to mitochondria and the role of Bcl-2 in mitochondria dynamics both suggest that mitochondria might have a role in *C. elegans* apoptosis. Indeed, it appears that mitochondria dynamics do play a role in nematode cell death since EGL-1 causes mitochondrial fission whereas CED-9 antagonizes this function and upregulation of the mediating fission protein Drp1 induces apoptosis. However, mitochondria remodeling in *C. elegans* was not found to be associated with MOMP and cytochrome *c* release. Indeed, its precise role in *C. elegans* apoptosis remains unclear (Oberst *et al.*, 2008).

Drosophila melanogaster

As regards the fruit fly *D. melanogaster*, the homologue of Apaf1 in *Drosophila* is the WD repeats containing protein Dark. The initiator caspase Dronc interacts with Dark through its *N*-terminal CARD domain. In the presence of dATP Dark and Dronc form an octameric apoptosome whose structure has been revealed by cryo-microscopy and which appears not to contain the cytochrone *c* (Yu *et al.*, 2006). Upon Dark interaction, Dronc is activated and cleaves the executioner caspases Drice and Dcp-1 (homologue to mammalian caspase-3) (**Figure 4b**). However, in contrast to caspase-9, the prodomain of Dronc is cleaved off during its activation. Moreover, the autocatalytic cleavage of Dronc induces its stable dimerization and strongly enhances its catalytic activity.

Both CED-4 and Dark contain an *N*-terminal CARD domain followed by an NB-ARC region with an AAA+ type ATPase domain as well as Apaf1. However, they do not seem to have ATPase activity. The crystal structure of the CED-4-CED-9 complex shows that ATP is bound to the active site and that Dark needs dATP for apoptosome formation. The ATPase activity of CED-4 and Dark however, is questionable and, presumably, dispensable. Moreover, Dark appears not to need cytochrome *c* to activate Dronc whereas a key role in regulating *Drosophila* apoptosis is played by DIAP1, which is homologous to the mammalian IAPs proteins. DIAP1 binds processed Dronc and inhibits Dronc–Dark interaction. DIAP1 can also bind Dronc and favour its degradation. Apoptosis is induced by upregulation of DIAP1 antagonists (Grim, HID and Reaper). DIAP1 antagonists cause DIAP1 autoubiquitination and degradation, thereby allowing Dronc activation (**Figure 4b**) (Vaux and Silke, 2005). This is in contrast with what is observed in mammals, where SMAC cannot ubiquitinate XIAP. DIAP1 is also able to autoubiquitination and to Grim, HID and Reaper ubiquitination and degradation.

Therefore, unlike mammals, in fly, DIAP1 and its antagonists play a central role in apoptosis whereas cytochrome *c* appears to be dispensable for caspase activation. Moreover, while mammalian IAPs antagonists (Smac/DIABLO and Omi/HtrA2) are released from mitochondria, in flies they are upregulated. Therefore it seems that mitochondria do not play any role in apoptosis. *Drosophila* possesses two cytochrome *c* genes (*cyt-c-d* and *cyt-c-p*). The importance of cytochrome *c* in *D. melanogaster* apoptosis is strongly

Figure 3 Shape and structure of the human apoptosome. (a) Top view of the 12.8 Å resolution 3D structure of the wheel-shaped apoptosome. The central hub and the seven arms are revealed (excerpt from Yu *et al.*, 2005). (b) Two models for domain arrangement in the NB-ARC region. Top: In the *bridge model* the winged helix bridges two neighbouring ATPase domains (cyan and green) to form the oligomer. In this way an inner CARD (orange) ring appears to be surrounded by an outer ring formed by the ATPase and the WHD (red) domains. Bottom: In the *AAA+-like model* the ATPase domains form by themselves the oligomer by binding back to back onto each other. In this model the CARD domains form a looser ring on the top of the NOD region (excerpt from Riedl and Salvesen, 2007). The colour code is not the same as in **Figure 2**.

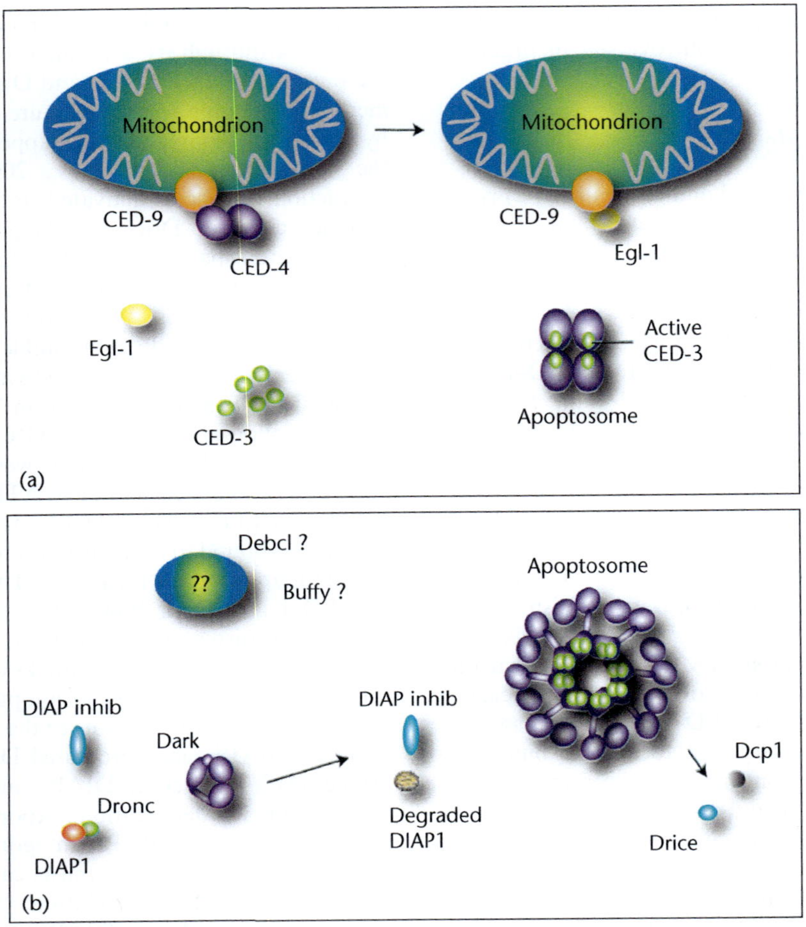

Figure 4 Apoptosome formation in *C. elegans* and in *D. melanogaster*. (a) In *C. elegans*, in nonapoptotic cells, dimers of CED-4 asymmetrically interact with CED-9 which is bound to mitochondria. Apoptotic induction promotes EGL-1 upregulation. Egl-1 binds CED-9 and displaces it from CED-4, which is thus free to oligomerize as a tetrameric complex in the presence of ATP. On oligomerization, CED-4 activates CED-3 which will perform the final steps of apoptosis. (b) In *Drosophila*, normally Dronc is inhibited by DIAP1 and apoptotic induction upregulates Dronc, Dark and Diap-1 inhibitors (Diap-1 inhib: Reaper, HID amd Grim). Diap-1 inhibitors bind and induce DIAP1 degradation thus releasing Dronc inhibition which, in the presence of dATP, can bind oligomerized Dark. In the apoptosome Dronc is activated and, in turn, cleaves and activates the effector caspases Drice and Dcp1. Roles of mitochondria and Buffy and Debcl are unclear in this system.

debated and most data suggest that cytochrome *c* is not essential for apoptosis (Yu *et al.*, 2006). However, one hypothesis posits that even though there is no cytochrome *c* release, cytochrome *c* undergoes a conformational change which presents an epitope and that the apoptosome is formed at the mitochondrial surface. The conformational change might be due to alterations in mitochondrial morphology and is caspase-dependent (Oberst *et al.*, 2008).

Buffy (antiapoptotic) and Debcl (proapoptotic) are Bcl-2 homologous proteins in *Drosophila*. Unlike *C. elegans* and mammals they do not have a pivotal role in apoptosis. This has led scientists to think that *Drosophila* might have had an atypical evolution in which the mitochondrial pathway has disappeared or the role of mitochondria in apoptosis is not conserved among phyla as previously thought. Nonetheless, recently it has been found that, during apoptosis, Repear and HID cause caspase-dependent mitochondrial fragmentation and cytochrome *c* release (Abdelwahid *et al.*, 2007). This role is distinct from their role as DIAP1 antagonists and apoptotic activators. It has also been found that Drp1 mediates mitochondrial fragmentation and that, although cytochrome *c* is not necessary for apoptosis, Drp1 knockout induced a reduction of cell death. It might be that mitochondria disruption is not the cause of caspase activation, but is a way to ensure and enhance apoptosis. Mitochondria fragmentation might have an essential (even though not fully understood) role in apoptosis. Indeed in all organisms such fragmentation is linked to apoptotic factors; in nematode and mammals it is related to Bcl-2 proteins. In flies (where Bcl-2 proteins have a less important role in apoptosis) this role in mitochondria fragmentation is assumed by HID and Repear (Oberst *et al.*, 2008).

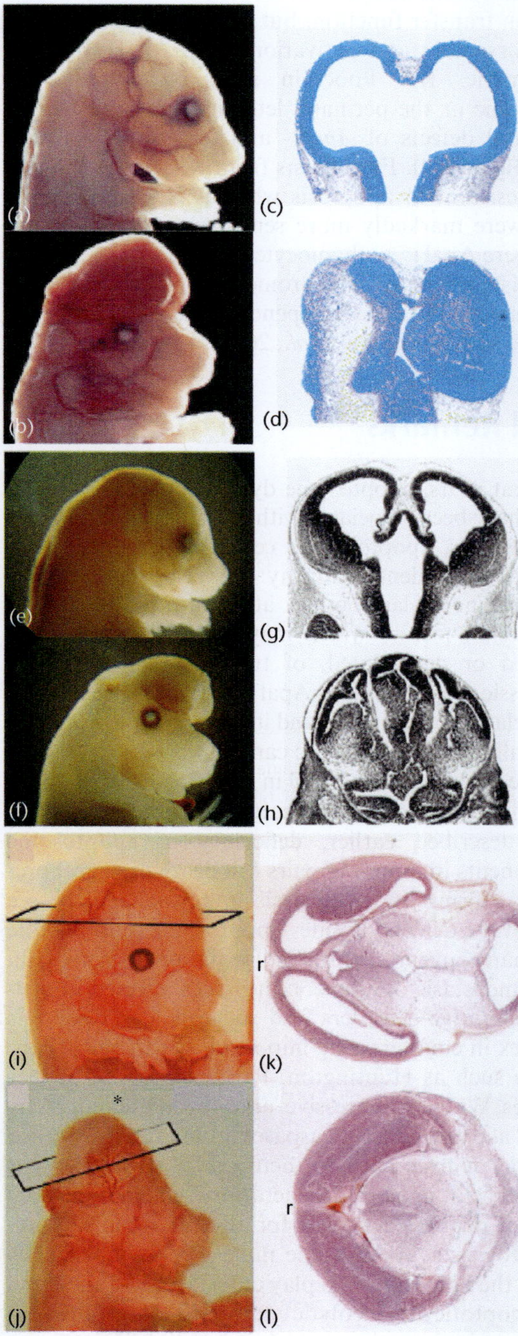

Figure 5 Phenotypes related to Apoptosome components' gene targeting. Gross and histological analysis of the neural embryonic phenotype of caspase-9 (a–d), Apaf1 knockout (e–h) and cytochrome c knockout/in (i–l) mice (excerpt from Kuida *et al.*, 1998; Cecconi *et al.*, 2008 and Hao *et al.*, 2005, respectively). Whole mounts of e16.5 caspase $9^{+/-}$ and $^{-/-}$ embryos (a, b), e16.5 Apaf1 wt and $^{-/-}$ embryos (e, f) and KA/+ control and KA/KA mutant e14.5 embryos (i, j) often show brain overgrowth and protrusion (indicated by an asterisk). Transverse sections of e10.5 caspase-9 mutant and corresponding wild-type brains stained with toluidine-blue (c, d). Transverse sections of e12.5 Apaf1 mutant and corresponding wild-type brains (g, h). Coronal sections of e14.5 KA/KA mutant embryos and corresponding KA/+ brains (k, l). In most cases, overexpansion of the neuroephitelium causes ventricular occlusion. C, caudal; r, rostral.

The Role of the Apoptosome in Mammalian Development

Ablation of apoptosome components causes embryonic or perinatal death. This means that apoptosis mediated by the apoptosome plays a central role in embryogenesis.

Caspase-9 knockout

Two papers have reported the knockout of caspase-9 in mice (Hakem *et al.*, 1998; Kuida *et al.*, 1998). These studies show that Caspase-9 deficiency is, in most cases, prenatally lethal. Caspase-9 mice exhibit prominent brain malformations with expansion and protrusion of the brain and cranial tissues from the skull and defects in inner ear development caused by defective apoptosis in the rapidly proliferative neuroepithelium (neural progenitor cells, NPCs). At stage e10.5 a defect in neural tube closure of the hindbrain is observed. Exencephaly appears at stage e16.5 (**Figure 5a–d**). No apparent histological abnormalities were observed in various other tissues. However, caspase-9 thymocytes show defective apoptosis upon treatment with a subset of stimuli.

Caspase-9 knockout $^{-/-}$ mice display a strikingly similar phenotype to that of caspase-3 knockout mice: however, the phenotype of caspase-9 knockout mice is more severe than that of caspase-3 knockout mice. This suggests that caspases other than caspase-3, such as caspase-7, also function downstream of caspase-9 during brain development.

Apaf1 knockout

Apaf1-deficient embryos die after embryonic stage e16.5 and before birth. The disruption of *Apaf1* in mice leads to an abnormal phenotype of many organs, severe craniofacial malformations, brain overgrowth, persistence of interdigital webs, dramatic alterations of the lens, retina and eye vascular system and abnormal inner ear development (Cecconi *et al.*, 1998; Yoshida *et al.*, 1998). Thus, *Apaf1* knockout does not precisely mimic *caspase 3* and *caspase 9* knockouts, which exhibit a predominantly neuronal phenotype; the number of developmental alterations observed in Apaf1-deficient embryos are actually more widespread, being distributed all over the organism. As for other organs affected, the caspase-9 deficient phenotype of the inner ear appears slightly milder than the *Apaf1* phenotype. At stage 12.5 Apaf1-deficient embryos exhibit marked hyperplasia of the diencephalon and midbrain embryonic ventricular zone where NPCs are confined. The telencephalic vesicles are abnormally folded and reduced in size because of the supernumerary cells that often occlude the intraventricular space. Exencephaly is visible at e16.5 due to defect of skull and facial midline cleft (**Figure 5e–h**).

The importance of Apaf1, caspase-3 and caspase-9 during neurogenesis lies especially in the regulation of cell death of neuronal progenitors cells. In the developing brain of these mutants apoptosis was found to be reduced. The NPCs which should undergo apoptosis continue to proliferate and undergo differentiation, providing supernumerary cells which generate protruding forebrain masses. Inhibition of cell death in the neural tube also often leads to the generation of spina bifida.

The spontaneous autosomal recessive mutation affecting neural tube development (*fog, forebrain overgrowth*) has been shown to map in the *Apaf1* chromosomal *locus*, although the corresponding mutation was uncharacterized. The analysis of *Apaf1/fog* double heterozygous embryos, which exhibit defects consistent with those reported for *fog* and *Apaf1* single homozygotes, has strongly supported the hypothesis that *fog* is an *Apaf1* hypomorphic allele.

Cytochrome c deficiency

Cytochrome *c* knockout mice die in utero at embryonic stage 10.5 because of a deficit in oxydative phosphorylation. For this reason 'knock-in' mice in which the normal allele is replaced by a mutant cytochrome *c* (Lys72Ala or KA allele) were generated (Hao *et al.*, 2005). Lys72 is essential for the stability of the interaction between cytochrome *c* and Apaf1; the Lys72Ala mutant form of cytochrome *c* retains normal electron transfer function, but fails to induce apoptosome assembly and activation since Apaf1 remains monomeric. KA knock-in mice recapitulate the embryonic or the perinatal lethality and brain developmental defects of *Apaf1*- and *caspase-9*-knockout mice (**Figure 5i–l**). Fibroblasts from KA knock-in mice were resistant to apoptosis whereas KA/KA thymocytes were markedly more sensitive to death stimuli than were Apaf1$^{-/-}$ thymocytes. These studies suggest the existence of a cytochrome *c*- and apoptosome-independent but Apaf1-dependent mechanism (s) for caspase activation (Ow *et al.*, 2008).

Final Remarks

In recent years, apoptosome dysfunction and dysregulation have been associated with several human diseases. In particular, apoptosome's central role in the mitochondria-dependent pathway of cell death places this multimolecular complex at the top list of tumour suppression factors which may be responsible, when mutated or deregulated, of tumorigenesis or cancer progression. For example, Apaf1 promoter silencing has been related to melanoma and its dysregulation seems to be a hallmark of pancreatic cancer, whereas a decrease caspase-9 activity is involved in a number of lung, ovarian and testicular cancers (Schafer and Kornbluth, 2006).

As described earlier, deficiency of apoptosome components in mouse impairs cell death in neural precursors during embryogenesis and results in severe neural tube defects. Besides apoptosome putative role in human congenital brain malformation, its activity in the homeostasis of adult brain tissues affected in neurodegenerative disorders also suggests a role for this complex in counteracting important pathological conditions such as Huntington, Parkinson or Alzheimer diseases. When it is excessive, apoptosome activity may indeed activate effector caspases and induce unwanted neuronal demise. The issue here is whether cell death is at all relevant in neurodegenerations, whether the final death of neurons is critical for affected brain physiology. However, apoptosome molecules may also turn out, in the near future, to play crucial roles in sublethal (nonapoptotic) proteolytic activities during neurogenesis or in synaptic transmission.

For these reasons, a full comprehension of structure, activity and fine-tuning of the apoptosome lays down a new frontier for biomedical research.

References

Abdelwahid E, Yokokura T, Krieser RJ *et al.* (2007) Mitochondrial disruption in *Drosophila* apoptosis. *Developmental Cell* **12**: 793–806.

Acehan D, Jiang X, Morgan DG *et al.* (2002) Three-dimensional structure of the apoptosome: implications for assembly, procaspase-9 binding, and activation. *Molecular Cell* **9**: 423–432.

Bao Q and Shi Y (2007) Apoptosome: a platform for the activation of initiator caspases. *Cell Death and Differentiation* **14**: 56–65.

Boatright KM, Renatus M, Scott FL *et al.* (2003) A unified model for apical caspase activation. *Molecular Cell* **11**: 529–541.

Cecconi F, Alvarez-Bolado G, Meyer BI, Roth KA and Gruss P (1998) Apafl (CED-4 homolog) regulates programmed cell death in mammalian development. *Cell* **94**: 727–737.

Cecconi F, Piacentini M and Fimia GM (2008) The involvement of cell death and survival in neural tube defects: a distinct role for apoptosis and autophagy? *Cell Death and Differentiation* **15**: 1170–1177.

Hakem R, Hakem A, Duncan GS *et al.* (1998) Differential requirement for caspase 9 in apoptotic pathways *in vivo*. *Cell* **94**: 339–352.

Hao Z, Duncan GS, Chang CC *et al.* (2005) Specific ablation of the apoptotic functions of cytochrome *c* reveals a differential requirement for cytochrome *c* and Apaf-1 in apoptosis. *Cell* **121**: 579–591.

Hay BA and Guo M (2006) Caspase-dependent cell death in *Drosophila*. *Annual Review of Cell and Developmental Biology* **22**: 623–650.

Hill MM, Adrain C, Duriez PJ, Creagh EM and Martin SJ (2004) Analysis of the composition, assembly kinetics and activity of native Apaf-1 apoptosomes. *EMBO Journal* **23**: 2134–2145.

Horvitz HR (2003) Worms, life, and death (Nobel lecture). *Chembiochem* **4**: 697–711.

Jiang X, Kim HE, Shu H *et al.* (2003) Distinctive roles of PHAP proteins and prothymosin-alpha in a death regulatory pathway. *Science* **299**: 223–226.

Kalanxhi E and Wallace CJ (2007) Cytochrome *c* impaled: investigation of the extended lipid anchorage of a soluble protein to mitochondrial membrane models. *Biochemical Journal* **407**: 179–187.

Kuida K, Haydar TF, Kuan CY *et al.* (1998) Reduced apoptosis and cytochrome *c*-mediated caspase activation in mice lacking caspase 9. *Cell* **94**: 325–337.

Lettre G and Hengartner MO (2006) Developmental apoptosis in *C. elegans*: a complex CEDnario. *Nature Reviews. Molecular Cell Biology* **7**: 97–108.

Liu X, Kim CN, Yang J, Jemmerson R and Wang X (1996) Induction of apoptotic program in cell-free extracts: requirement for dATP and cytochrome *c*. *Cell* **86**: 147–157.

Moroni MC, Hickman ES, Lazzerini Denchi E *et al.* (2001) Apaf-1 is a transcriptional target for E2F and p53. *Nature Cell Biology* **3**: 552–558.

Oberst A, Bender C and Green DR (2008) Living with death: the evolution of the mitochondrial pathway of apoptosis in animals. *Cell Death and Differentiation* **15**: 1139–1146.

Ow YP, Green DR, Hao Z and Mak TW (2008) Cytochrome *c*: functions beyond respiration. *Nature Reviews. Molecular Cell Biology* **9**: 532–542.

Piddubnyak V, Rigou P, Michel L *et al.* (2007) Positive regulation of apoptosis by HCA66, a new Apaf-1 interacting protein, and its putative role in the physiopathology of NF1 microdeletion syndrome patients. *Cell Death and Differentiation* **14**: 1222–1233.

Pop C, Timmer J, Sperandio S and Salvesen GS (2006) The apoptosome activates caspase-9 by dimerization. *Molecular Cell* **22**: 269–275.

Renatus M, Stennicke HR, Scott FL, Liddington RC and Salvesen GS (2001) Dimer formation drives the activation of the cell death protease caspase 9. *Proceedings of the National Academy of Sciences of the USA* **98**: 14250–14255.

Riedl SJ, Li W, Chao Y, Schwarzenbacher R and Shi Y (2005) Structure of the apoptotic protease-activating factor 1 bound to ADP. *Nature* **434**: 926–933.

Riedl SJ and Salvesen GS (2007) The apoptosome: signalling platform of cell death. *Nature Reviews. Molecular Cell Biology* **8**: 405–413.

Salvesen GS and Dixit VM (1999) Caspase activation: the induced-proximity model. *Proceedings of the National Academy of Sciences of the USA* **96**: 10964–10967.

Schafer ZT and Kornbluth S (2006) The apoptosome: physiological, developmental, and pathological modes of regulation. *Developmental Cell* **10**: 549–561.

Sun Y, Orrenius S, Pervaiz S and Fadeel B (2005) Plasma membrane sequestration of apoptotic protease-activating factor-1 in human B-lymphoma cells: a novel mechanism of chemoresistance. *Blood* **105**: 4070–4077.

Vaux DL and Silke J (2005) IAPs – the ubiquitin connection. *Cell Death and Differentiation* **12**: 1205–1207.

Wright KM, Smith MI, Farrag L and Deshmukh M (2007) Chromatin modification of Apaf-1 restricts the apoptotic pathway in mature neurons. *Journal of Cell Biology* **179**: 825–832.

Yoshida H, Kong YY, Yoshida R *et al.* (1998) Apafl is required for mitochondrial pathways of apoptosis and brain development. *Cell* **94**: 739–750.

Yu X, Acehan D, Menetret JF *et al.* (2005) A structure of the human apoptosome at 12.8A resolution provides insights into this cell death platform. *Structure* **13**: 1725–1735.

Yu X, Wang L, Acehan D, Wang X and Akey CW (2006) Three-dimensional structure of a double apoptosome formed by the *Drosophila* Apaf-1 related killer. *Journal of Molecular Biology* **355**: 577–589.

Zou H, Henzel WJ, Liu X, Lutschg A and Wang X (1997) Apaf-1, a human protein homologous to *C. elegans* CED-4, participates in cytochrome *c*-dependent activation of caspase-3. *Cell* **90**: 405–413.

Further Reading

Chipuk JE and Green DR (2008) How do BCL-2 proteins induce mitochondrial outer membrane permeabilization? *Trends in Cell Biology* **18**: 157–164.

Fadeel B, Ottosson A and Pervaiz S (2008) Big wheel keeps on turning: apoptosome regulation and its role in chemoresistance. *Cell Death and Differentiation* **15**: 443–452.

Ferraro E, Corvaro M and Cecconi F (2003) Physiological and pathological roles of Apaf1 and the apoptosome. *Journal of Cellular and Molecular Medicine* **7**: 21–34.

Garrido C, Brunet M, Didelot C *et al.* (2006) Heat shock proteins 27 and 70: anti-apoptotic proteins with tumorigenic properties. *Cell Cycle* **5**: 2592–2601.

Pellegrini L and Scorrano L (2007) A cut short to death: Parl and Opa1 in the regulation of mitochondrial morphology and apoptosis. *Cell Death and Differentiation* **14**: 1275–1284.

Steller H (2008) Regulation of apoptosis in *Drosophila*. *Cell Death and Differentiation* **15**: 1132–1138.

Caspases, Substrates and Sequential Activation

John G Walsh, *Department of Genetics, The Smurfit Institute, Trinity College, Dublin, Ireland*

Seamus J Martin, *Department of Genetics, The Smurfit Institute, Trinity College, Dublin, Ireland*

Advanced article

Article Contents

- Introduction
- Caspases Classification and Structure
- Pathways That Lead to Caspase Activation
- Demolition Phase of Apoptosis
- Immune Clearance and Immune Tolerance of Apoptotic Cells
- Conclusions
- Acknowledgements

Superfluous, aged or damaged cells are eliminated from tissues via a controlled cell death process, called apoptosis. Apoptosis is coordinated by a group of cysteine aspartic acid-specific proteases (caspases) that become specifically activated within cells destined to die. Some of these proteases act as initiators, their role being to interface with signalling events and initiate the proteolytic cascade, whereas others act as executioner enzymes and carry out the internal dismantling of the cell that results in death. During the terminal phase of apoptosis the executioner caspases (caspases-3 and -7) simultaneously cleave hundreds of protein substrates to terminate cell viability and produce the characteristic apoptotic phenotype. This large-scale proteolysis also dismembers the cell into discrete fragments that are recognized and removed by scavenging phagocytes. Recent evidence also suggests that the actions of executioner caspases may disable molecules that are capable of initiating or exacerbating immune responses if released into the extracellular space.

Introduction

The fate of individual cells in a multicellular organism is carefully controlled to ensure that unwarranted population expansions or contractions do not occur. It is now well established that cell death, just like cell division, is a highly regulated or 'programmed' process. Programmed cell death is critical for the removal of superfluous cells during embryonic development and tissue remodelling as well as in the resolution of immune responses.

Central to this form of cell death are a group of cysteine aspartic acid-specific proteases (caspases) that coordinate this process from within the cell destined to die. Some of these proteases (i.e. caspases-8 and -9) are classed as initiators, their role being to interface with signalling events and initiate the cell death process. Other members of this protease family, such as caspases-3, -6 and -7, are termed executioner caspases, as they execute the terminal events in apoptosis that usher in the death of the cell. During the terminal phase of apoptosis these executioner caspases carry out a large-scale proteolytic demolition that shuts down the physiology of the cell and dismantles it into discrete fragments that can be removed by scavenging phagocytes. **See also**: Apoptosis: Regulatory Genes and Disease

Caspases Classification and Structure

Caspases are conserved across the animal lineages. Homologues have been identified in the worm *Caenorhabditis elegans* (i.e. Ced-3), the fly *Drosophila melanogaster* (i.e. Dronc and *Drosophila* interleukin-1β converting enzyme), as well as in mammals. In humans, the caspase family of proteases contains 12 members (**Figure 1**). Overall, 15 mammalian caspases have been reported to date. Mammalian caspases are broadly divided into two major groups based on their role in two distinct events namely macrophage activation (inflammatory caspases) and programmed cell death (apoptotic caspases) (Logue and Martin, 2008). The inflammatory caspases (i.e. caspases-1, -4 and -5) are involved in the maturation of cytokines released from macrophages such as interleukin-1β and interleukin-18. To date, the panel of known inflammatory caspase substrates is essentially limited to these two cytokines although it is likely that more substrates will be identified in time. Certain scenarios leading to caspase-1 activation, namely the sensing of cytoplasmic deoxyribonucleic acid (DNA) by the AIM2 (absent in melanoma 2) inflamasome, are also associated with a loss of cell viability (Teresa Fernandes-Alnemri *et al.*, 2009). However, AIM2/caspase-1-dependent cell death is a necrosis-like phenomenon that has been termed pyroptosis. The pro-apoptotic members of the caspase family include caspases-3, -6 -7, -8 and -9. Compared to the inflammatory caspases, the characterization of the apoptotic caspase cascade is much more complete and has been found to involve hundreds of potential caspase substrates (Lüthi and Martin, 2007). **See also**: Caspases and Cell Death; Dismantling the Apoptotic Cell

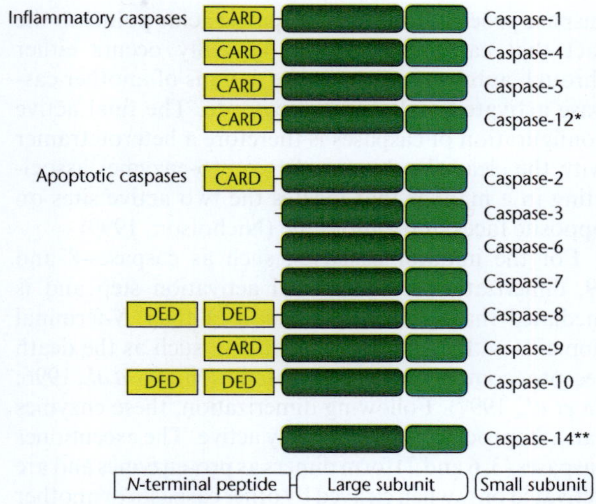

Figure 1 Members of the human caspase family of proteases. All caspases are characterized by an *N*-terminal domain followed by a large and small subunit. *Full length caspase-12 is only expressed in a subset of individuals of African decent. **Caspase-14 is a relatively uncharacterized member implicated in the differentiation of skin cells.

Structurally, all caspases are composed of an *N*-terminal pro-domain followed by a large and small subunit (**Figure 1**). The length of the *N*-terminal pro-domain is an important determinant of caspase function and its position in the caspase activation cascade. Caspases with long *N*-terminal domains are often referred to as 'initiators', whereas those with short *N*-terminal domains are termed 'executioners'. Among the initiator caspases there are two related yet distinct *N*-terminal domain types known as the caspase recruitment domain (CARD) and the death effector domain (DED). Through these domains initiator caspases are able to interface with upstream signalling platforms and undergo activation. Executioner caspases do not possess these domains; therefore, their activation in response to a cell death stimulus must be mediated through an initiator caspase. **See also**: Structure, Domains and functions in Cell Death (DD, DED, CARD, PYD)

All caspases are produced as largely inactive pro-enzymes and require proteolytic processing as well as dimerization to become fully active (Logue and Martin, 2008; Taylor *et al.*, 2008). Processing of the pro-enzyme occurs at an aspartic acid located between the large and small subunits. Caspases, as their name suggests, have a strict requirement for aspartic acid in the P1 position of their substrate cleavage sites. The fact that internal caspase processing sites are aspartic acids points to the fact that caspase activation typically occurs either through autoprocessing, or the actions of another caspase activated earlier in the cascade. The final active configuration of caspases is therefore a heterotetramer with the cleaved subunits of two pro-enzymes associating in a manner that locates the two active sites on opposite faces of the molecule (Nicholson, 1999).

For the initiator caspases, such as caspases-8 and -9, dimerization is the crucial activation step and is mediated through the association of their *N*-terminal domains with an activating platform such as the death receptor complex or the apoptosome (Muzio *et al.*, 1996; Li *et al.*, 1997). Following dimerization, these enzymes can autoprocess to become fully active. The executioner caspases (3, 6 and 7) form dimers as pro-enzymes and are activated only when cleaved by other caspases or another class of proteases known as the granzymes (**Figure 2**; Slee *et al.*, 2001; Adrain *et al.*, 2005). **See also**: Caspases and Cell Death

Pathways That Lead to Caspase Activation

There are three major routes leading to caspase activation and apoptotic cell death. These have been termed the extrinsic, intrinsic and granzyme B pathways. In all three scenarios, activation or introduction of an apical protease (initiator caspase or granzyme B) is the first step in initiating the caspase activation cascade. The range of substrates cleaved by these initiators is relatively narrow, but, importantly, it includes the executioner caspases that once activated can go on to mediate hundreds of proteolytic events within the cell. In this way the caspase cascade is amplified from tightly regulated events occurring at a specific location in the cell to a cell-wide proteolytic maelstrom which impacts on all aspects of cellular physiology. **See also**: Apoptosis: Regulatory Genes and Disease

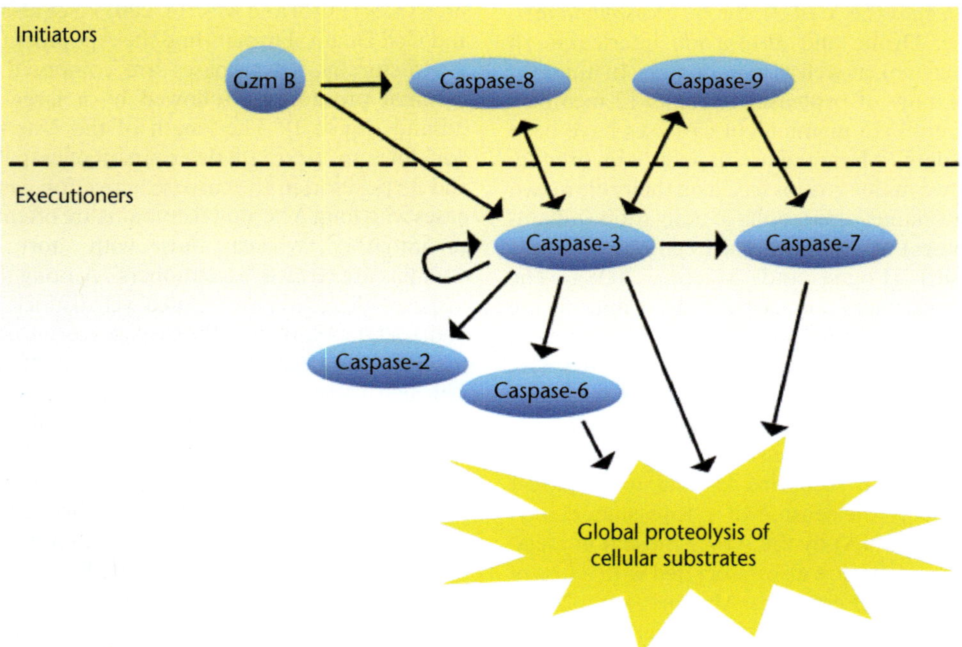

Figure 2 Sequential activation of caspases during apoptosis. Initiator caspases such as caspases-8 and -9 are activated by pro-apoptotic signals. These enzymes in turn activate the executioner enzymes such as caspases-3 and -7. Caspase-3 is particularly important in propagating the cascade through activation of other executioner enzymes as well through positive feed back onto the initiators. The cytotoxic lymphocyte protease granzyme B (GzmB), although not a caspase, has a proteolytic specificity similar to the caspases and can directly process and activate caspases-3, -7 and -8 on delivery into target cells.

Extrinsic pathway

The extrinsic pathway is so called because the instruction to die is provided by another cell, such as a cytotoxic T lymphocyte (CTL), through extracellular signalling in the form of death receptor engagement. The binding of receptors such as Fas, TRAIL (tumour necrosis factor (TNF)-related apoptosis-inducing ligand) receptor or TNF receptor expressed on the target cell to corresponding ligands expressed on cytotoxic cells leads to the recruitment of an activation platform known as the death receptor complex (**Figure 3a**). This complex is composed of adapters such as FADD (Fas-associated death domain) as well as the initiator enzyme caspase-8 (Muzio *et al.*, 1996). Caspase-8 associates with this complex through its DED. Active caspase-8 subsequently processes caspases-3 and -7. These executioner caspases in turn carry out the apoptotic dismantling of the targeted cell. Additionally, caspase-8 may also cleave and activate a pro-apoptotic protein known as BID (Luo *et al.*, 1998). **See also**: Death Receptors

Granzyme B pathway

As an alternative to the engagement of death receptors on the surface of target cells, cytotoxic natural killer (NK) cells and CTLs may kill their targets through the delivery of lytic granules across the immunological synapse (**Figure 3b**; Cullen and Martin, 2008). Contained within these granules are two key constituent proteins, perforin and granzyme B. Perforin is thought to permeabilize the plasma membrane of the target cell, possibly through the formation of channels (Keckler, 2007) and this facilitates the entry into the cell of granzyme B, a proteolytic enzyme that can initiate

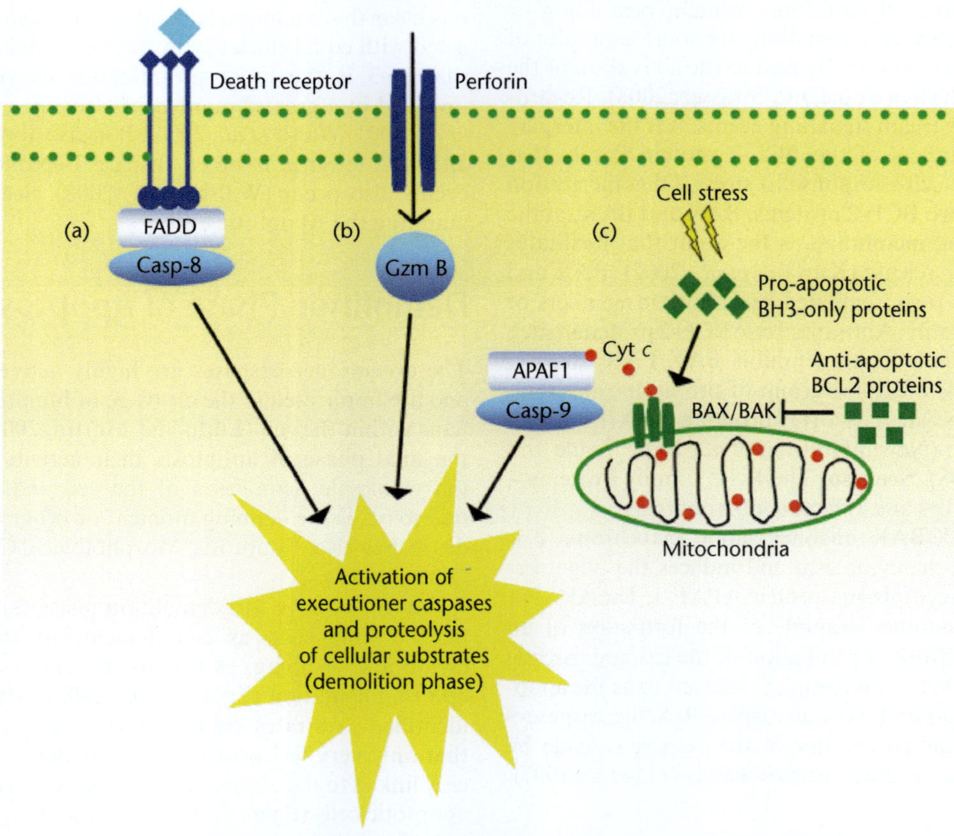

Figure 3 Pathways to caspase activation and cell death. Pro-apoptotic signals are routed through three major pathways. (a) Extrinsic (death receptors), (b) granzyme B (lytic granules derived from cytotoxic lymphocytes) or (c) intrinsic (release of cytochrome *c* [Cyt *c*] from the mitochondrial intermembrane space). All pathways lead to the activation of a protease cascade involving members of the caspase family of proteases.

apoptosis. This is achieved in part through the ability of granzyme B to cleave and activate the executioner caspases-3 and -7 (Adrain *et al.*, 2005). In humans, granzyme B-mediated processing of BID and the subsequent activation of the intrinsic pathway is also thought to be a key event (Cullen *et al.*, 2007). **See also**: Immunity, Granzymes and Cell Killing

Intrinsic pathway

The intrinsic pathway is centred on events occurring at mitochondrial membranes. Specifically, it is the release of cytochrome *c* from mitochondria that triggers the activation of the caspase cascade and essentially represents a point of no return for the dying cell (**Figure 3c**). As the 'intrinsic' label suggests, many triggers of the intrinsic pathway involve cellular stresses that generate death-inducing conditions. DNA damage, transcriptional/translational inhibition, protein misfolding or loss of growth factor signalling are some examples of events that can potentially lead to the activation of the intrinsic pathway (Youle and Strasser, 2008). Regardless of the upstream signalling events, it is the interplay between members of the BCL-2 protein family that regulates the cell's response to stress. Oligomerization of either of two BCL-2 proteins, BAX and BAK, at the mitochondrial membrane, is the event that facilitates cytochrome *c* release (Kuwana *et al.*, 2002). BAX and BAK are, in turn, regulated by yet other members of the BCL-2 family. Anti-apoptotic BCL-2 proteins (such as BCL-2, BCL-X, etc.) inhibit BAX/BAK channel opening while a diverse group of pro-apoptotic BH3-only proteins (such as BID) promote BAX/BAK oligomerization (Newm Kuwana *et al.*, 2002; Youle and Strasser, 2008). **See also**: The Bcl-2 Family Proteins – Key Regulators and Effectors of Apoptosis

Upon BAX/BAK oligomerization cytochrome *c* is released into the cytoplasm and induces the oligomerization of the cytoplasmic protein APAF-1. The APAF-1 oligomer is another example of the formation of an activating platform for initiation of the caspase cascade (Hill *et al.*, 2004). This complex, referred to as the apoptosome, recruits and activates caspase-9. Active caspase-9 can now act as an initiator of the caspase cascade by cleaving and activating caspases-3 and -7 (Li *et al.*, 1997).

Effectors of demolition

As already described, the common endpoint of all three cell death pathways is the activation of the executioner caspases-3, -6 and -7. Among these enzymes, caspases-3 has the unique ability to proteolytically process several other caspases both upstream (caspases-8 and -9) and downstream (caspases-3, -6 and -7) of its own position within the pathway (Walsh *et al.*, 2008).

The ability of caspase-3 to activate other caspases points to an important question regarding the executioner caspases. Namely, what is the relative contribution that each enzyme makes to the processing of other cellular substrates during the final phase of apoptosis? Caspase-6 may have a specific role in the cleavage of certain nuclear lamin proteins but outside of this small window little else is known (Slee *et al.*, 2001). The closely related orthologues caspases-3 and -7 have often been considered to be functionally redundant. This view has been supported by screening studies using synthetic small peptide substrates as well as mouse knockout data (Thornberry *et al.*, 1997; Lakhani *et al.*, 2006). However, on closer inspection there is a clear evidence of significant functional divergence. Although it is clear that an important subset of substrates is targeted with equal efficiency by both enzymes, in general caspase-3 is the more promiscuous enzyme and is involved in a greater number of cleavage events (Slee *et al.*, 2001; Walsh *et al.*, 2008). It is also apparent that, although fewer in number, caspase-7-specific cleavage events also occur (Walsh *et al.*, 2008). **See also**: Dismantling the Apoptotic Cell

Demolition Phase of Apoptosis

The executioner caspases are highly active proteases and are implicated in the cleavage of hundreds of proteins within the cell (Lüthi and Martin, 2007). During the final phase of apoptosis their activity results in the wholesale demolition of the cell and, in a very real sense, is the defining moment of programmed cell death. **See also**: Apoptosis: Morphological Criteria and Other Assays

We can observe this demolition phase through hallmark events such as cell detachment and plasma membrane blebbing, as well as nuclear condensation and fragmentation (Taylor *et al.*, 2008). Although the identified substrates are numerous, it is a curious fact that only very few proteolytic events have been explicitly linked to the observed alterations occurring within apoptotic cells (**Figure 4**). It would seem that many caspase-dependent cleavage events represent a general shutdown of the cellular machinery 'one of a thousand cuts', whereas still others are innocent bystanders caught up in the proteolytic process (Taylor *et al.*, 2008). **See also**: Autophagy

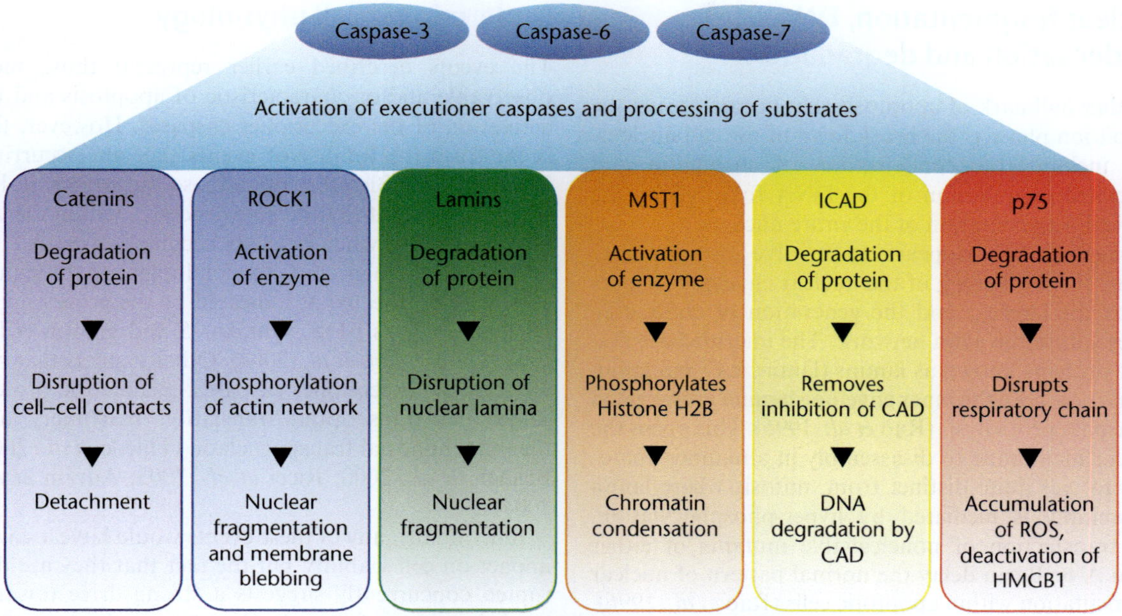

Figure 4 Consequences of substrate processing by the executioner caspases during the demolition phase of apoptosis. Six representative caspases substrates are shown, but it is important to note that hundreds of proteins are cleaved by the executioner caspases during the demolition phase of apoptosis.

Detachment, rounding and blebbing

For the vast majority of cells, cytochrome *c* release from mitochondria represents an irreparable loss of viability. The late occurring events of programmed cell death mediated by the executioner caspases are primarily directed at preparing the cell for efficient removal by phagocytes. As such, cells that are part of a connected tissue layer must be extricated from their position anchored to neighbouring cells and the extracellular matrix. A number of proteins involved in cell–cell connectivity have been identified as caspase substrates. One such group of proteins subject to caspase proteolysis are the β- and γ-catenins. These proteins mediate the interaction between cadherins and the cytoskeleton, and the observation that catenins are cleaved suggests a break in cadherin based cell–cell contacts (Brancolini *et al.*, 1997). More directly, cadherins themselves are also cleaved by caspases and therefore implicated in cell detachment (Steinhusen *et al.*, 2001). The association between the cell and the extra cellular matrix may be disrupted by cleavage of the proteins focal adhesion kinase and $P130^{Cas}$ (Levkau *et al.*, 1998).

In addition to breaking links with their neighbours, what is clear when observing apoptotic cells is that such cells undergo significant rearrangements of the cytoskeletal network. Not only do apoptotic cells round and retract but the plasma membrane also undergoes dramatic blebbing (extrusion of the membrane leading to the release of apoptotic bodies). Microtubules, microfilaments and their associates are affected by caspase-mediated proteolysis. The list of cytoskeletal substrates includes proteins such as actin, myosin, tubulin, gelsolin, fodrin and vimentin (Communal *et al.*, 2002; Gerner *et al.*, 2000; Walsh *et al.*, 2008). However, for many of these reported substrates, the link between proteolysis and any functional outcome has been more often inferred than expressly demonstrated. An exception is the Rho-associated kinase, ROCK1. Caspase proteolysis of ROCK1 removes its *C*-terminal autoinhibitory domain yielding a constitutively active form of ROCK1. Active ROCK1 can then go on to phoshorylate myosin light chains. The phosphorylation of myosin light chains induces an increase in the actin–myosin force and therefore an increase in contractility leading to membrane blebbing (Coleman *et al.*, 2001). Interestingly, although this event requires caspase-dependent processing of ROCK1, it is remarkably independent of other caspase-mediated proteolysis, as cells transfected with active ROCK1 undergo blebbing in the absence of apoptosis (Coleman *et al.*, 2001).

Nuclear fragmentation, DNA condensation and degradation

Another hallmark of apoptosis that occurs during the demolition phase is the breakdown of the cell nucleus. This includes the condensation of chromatin and associated degradation of the DNA followed by the physical fragmentation of the entire nucleus.

Fragmentation appears to involve two separate events: the weakening of the nuclear envelope (in particular the lamina) and the generation of contractile forces within the actin network. The intermediate filament proteins known as lamins (lamin A/C and lamin B) that compose the inner envelope lamina are targeted by caspase proteolysis (Rao *et al.*, 1996). This opens the nuclear membrane to disassembly in a manner analogous to, yet quite distinct from, mitosis where lamin disassembly is mediated by hyper-phosphorylation. The transfection of noncleavable mutants of either lamin A or B can delay the normal pattern of nuclear fragmentation within apoptotic cells (Rao *et al.*, 1996).

However, weakened membrane integrity does not cause the nucleus to passively fall apart. To fragment and for those fragments to disperse, as is observed during apoptosis, the nucleus must be physically torn asunder. The forces required to achieve this are cytoskeletal and are again mediated by the ROCK1 phosphorylation of myosin light chain. This affects the actin mesh surrounding the nucleus and promotes actin–myosin-based contraction.

Before the fragmentation of the nucleus, the chromatin/DNA inside are subject to condensation and degradation. In fact the laddering of DNA during apoptosis was an early marker for this form of cell death. The condensation of chromatin appears to be primarily due to the caspase-dependent activation of another kinase, Mst1 (Cheung *et al.*, 2003). Activated Mst1 enters the nucleus and phosphorylates the histone protein H2B. The phosphorylation of histone proteins is similarly involved in chromosome condensation during cell division but the specific event of H2B phosphorylation appears to be unique to apoptotic cells (Cheung *et al.*, 2003).

Following chromatin condensation, the DNA is degraded by a nuclease known as CAD (caspase activated dnase). This was identified as an active constituent of apoptotic or caspase-3-treated cell lysates (Enari *et al.*, 1998). $CAD^{-/-}$ cells fail to undergo oligonucleosomal fragmentation although cell death otherwise proceeds (Samejima *et al.*, 2001). The mechanism of CAD activation is through caspase cleavage of the inhibitor of CAD (ICAD; Enari *et al.*, 1998).

Undermining cell physiology

The events described earlier represent those most observable and/or characteristic of apoptosis and the activation of the executioner caspases. However, this is far from the totality of events that are occurring. Through identification of caspase substrates it has become apparent that diverse processes within the cell are undermined when caspases become activated.

Both the Golgi and the endoplasmic reticulum appear to be affected as indicated by the processing of proteins such as p115, syntaxin 5 and giantin (Chiu *et al.*, 2002; Lowe *et al.*, 2004). Other systems that run afoul of the executioner caspases, include the proteosome, the transcription/translation machinery and the mitochondrial transport chain (Thiede *et al.*, 2001; Bushell *et al.*, 2000; Ricci *et al.*, 2003; Adrain *et al.*, 2004).

Individually, any of these events would have a severe impact on cell viability but the fact that they are disrupted concurrently suggests a strong drive towards total shutdown of the cell. However, in the context of the severe cytoskeletal disruption that is mediated by substrates such as ROCK1, some events could be viewed as turning the lights out in an already crumbling building. Perhaps, the widespread targeting of these pathways not only ensures that there is no escape from the actions of the caspases but also serves to dampen down the overall levels of biochemical activity within the cell and therefore enhance its inert status.

Immune Clearance and Immune Tolerance of Apoptotic Cells

Although a clear understanding of the significance of most caspase-dependent cleavage events has yet to be achieved, the major endpoint must be the presentation of apoptotic bodies to the surveying phagocytes responsible for cleaning up the cellular debris. As stated earlier, to the immune system, the difference between necrotic and apoptotic cells is one of night and day. Though necrotic cell death is characterized by the release of self-derived danger signals and inflammation, apoptosis is largely inert or even refractory to an inflammatory response (Birge and Ucker, 2008). Therefore, if cytochrome *c* release alone is sufficient to kill a cell, much of the caspase-dependent processing of substrates is likely to be geared towards getting an apoptotic cell ready to meet the immune system on good terms.

The detachment and fragmentation into apoptotic bodies is a key event that aids in the removal of apoptotic cells by phagocytes; however, to achieve this apoptotic cells must be labelled as targets. A number of changes/candidate molecules expressed on the membrane of apoptotic cells have been proposed to be involved including phosphatidylserine and oxidized-low-density lipoprotein (Martin *et al.*, 1996; Chang *et al.*, 1999; Greenberg *et al.*, 2006). Along with these membrane changes apoptotic cells also release chemoattractive factors such as lysophosphatidylcholine that allow phagocytes to home in on their location. Lysophosphatidylcholine release is caspase-3-dependent, requiring a cleavage-mediated activation of phospholipase A (Lauber *et al.*, 2003). Additionaly, a very recent report has made the claim that release of a chemokine (CX3CL1/fractalkine) from apoptotic B cells specifically attracts macrophages (Truman *et al.*, 2008). Interestingly, this release was also apparently caspase dependent and involved the cleavage of a membrane-bound form of the chemokine from the cell surface. However, whether the caspase involvement in this processing event is direct or indirect remains unclear.

Another important part of being parsed and packaged for removal is the potential antigenicity of the apoptotic cell. Danger signals released from necrotic cells are not typically single purpose immune signals such as cytokines but rather 'homeostatic' cellular constituents that serve the internal cellular physiology of normal healthy cells. These molecules only come into contact with the immune system when they are released from the insulating cocoon of the cell by the deregulated rupture of the plasma membrane during necrosis. Examples of such signals include HMGB1, Hsp70 and DNA–protein complexes such as nucleosomes (Todryk *et al.*, 1999; Napirei *et al.*, 2000; Scaffidi *et al.*, 2002). Because apoptotic cells maintain plasma integrity before their ingestion by phagocytes, this has been cited as the primary reason behind their lack of immunogenicity (Birge and Ucker, 2008). Although membrane integrity is likely to have a significant effect on the immune system's tolerance of apoptotic cells, it does not appear to be the whole story. It is clearly observable in cell culture that apoptotic cells that cannot be taken up by phagocytes eventually do lose their membrane integrity and undergo secondary necrosis. If the plasma membrane was the only barrier to potential danger signals, then cells that have undergone secondary necrosis should activate the immune system in a similar manner to primary necrotic cells. In fact, studies using bone marrow-derived mouse macrophages have found that exposure to late-stage secondary necrotic cells does not stimulate a comparable response to that of primary necrotic cells (Patel *et al.*, 2006). This suggests that apoptotic cells are deactivated from the inside out, a scenario that seems to almost insist on a role for caspases.

An example of such a mechanism has recently been reported for the danger signal HMGB1. Kazama *et al.* (2008) demonstrated that caspase-3-mediated proteolysis of the mitochondrial protein p75 leads to an accumulation of reactive oxygen species within the apoptotic cell. These oxidizing species were sufficient to deactivate HMGB1, which interestingly was found to be able to escape from early apoptotic cells (Kazama *et al.*, 2008). This seems to provide a clear example of how caspase-mediated changes in the internal environment of apoptotic cells could reduce their inflammatory potential. Another event that may be of significant consequence in relation to maintaining immune tolerance is the fragmentation of the nucleus. Naked DNA, as well as DNA–protein complexes, have been reported as a major source of autoimmune stimulation in diseases such as Lupus (Napirei *et al.*, 2000). DNAse $II^{-/-}$ mouse macrophages that are unable to digest such nulceosome fragments from ingested apoptotic bodies promote a deregulated immune pathology. This pathology is enhanced in mice also deficient in CAD and therefore apoptosis-associated nuclear fragmentation (Kawane *et al.*, 2003). This suggests that this hallmark effect on the nucleus observed in dying cells may function to reduce its antigenicity.

Evidence for the direct processing of a danger signal by executioner caspases is yet to be reported in the literature. However, because of the large number of caspase substrates, it is not surprising that associated or related proteins have been identified as substrates. Examples include the members of the Hsp90 complex and the ribonucloproteins. However, at this time, these associations only provide a source of speculation, and definitive evidence for direct caspase-dependent inactivation of alarmins remains to be found.

The identification and validation of any proposed danger signal must overcome a number of hurdles, not the least of which is a detectable presence in pathophysiological conditions (Kono and Rock, 2008). The critical requirement that apoptosis avoids activating the immune response suggest that the study of programmed cell death, particularly during the demolition phase, may help to identify new danger signals and to provide corroborating evidence of the importance of those already identified.

Conclusions

In excess of 600 proteins have been found to be cleaved by caspases during apoptosis in mammals and this number is still growing. Owing to this almost embarrassing profusion of caspase substrates, the challenge now is to identify the subset of these substrate proteins that have real functional significance for apoptosis. Although it is possible that all substrate proteins that are cleaved by caspases during apoptosis contribute to the demise of the cell, it does seems rather unlikely that all caspase substrates play equally important roles in this process. It seems rather more probable that many proteins are simply caught up in the proteolytic mayhem that breaks out within a cell during apoptosis and become cleaved by caspases – by accident rather than design – during the terminal phase of cell death.

Unfortunately, there is no easy way to identify functionally important caspase substrates from the proteolytic noise that inevitably accompanies a process where several proteases become activated within a short time-frame. One way of approaching the question of whether a caspase substrate is potentially more relevant than others is to ask whether this protein is also cleaved by caspases in other species. However, this approach has not been adopted in many cases and has led to generalizations concerning the significance of some substrates for apoptosis that are unlikely to be borne out on more detailed analysis. Though it is certainly possible that some caspase substrates may be cleaved in a species-specific manner, it seems implausible that the same functional endpoints are achieved through proteolysis of a completely different array of proteins in different species. This view is buttressed by observations that many of the same morphological endpoints are seen in apoptotic cells from organisms as divergent as flies and man. Thus, it is probable that a conserved cohort of caspase substrates are cleaved by worm, fly and mammalian caspases and that these represent the legitimate targets that ensure controlled cell destruction, as well as the morphological and functional hallmarks of apoptosis.

Notwithstanding some of the unresolved issues discussed earlier, a combination of genetic and biochemical evidence has established beyond any doubt that the caspase proteases are the key effector enzymes in apoptosis. On activation in response to diverse forms of cellular stress, the actions of the caspases commit the cell to a rapid and controlled death that minimizes damage to neighbouring cells and prevents activation of the immune system.

Acknowledgements

We thank Science Foundation Ireland for their ongoing support of work in our laboratory. Work in the Martin Laboratory is supported by grants from the European Union Marie Curie Research Training Network (Apoptrain), Science Foundation Ireland (SRCG20336), The Wellcome Trust (082749), The Health Research Board of Ireland and Cancer Research Ireland. JGW is supported by a postgraduate scholarship from The Irish Research Council for Science and Engineering Technologies (IRCSET).

References

Adrain C, Creagh EM, Cullen SP and Martin SJ (2004) Caspase-dependent inactivation of proteasome function during programmed cell death in *Drosophila* and man. *Journal of Biological Chemistry* **279**(35): 36923–36930.

Adrain C, Murphy BM and Martin SJ (2005) Molecular ordering of the caspase activation cascade initiated by the cytotoxic T lymphocyte/natural killer (CTL/NK) protease granzyme B. *Journal of Biological Chemistry* **280**: 4663–4673.

Birge RB and Ucker DS (2008) Innate apoptotic immunity: the calming touch of death. *Cell Death and Differentiation* **15**(7): 1096–1102.

Brancolini C, Lazarevic D, Rodriguez J and Schneider C (1997) Dismantling cell-cell contacts during apoptosis is coupled to a caspase-dependent proteolytic cleavage of beta-catenin. *Journal of Cell Biology* **139**(3): 759–771.

Bushell M, Wood W, Clemens MJ and Morley SJ (2000) Changes in integrity and association of eukaryotic protein synthesis initiation factors during apoptosis. *European Journal of Biochemistry* **267**(4): 1083–1091.

Chang MK, Bergmark C, Laurila A *et al.* (1999) Monoclonal antibodies against oxidized low-density lipoprotein bind to apoptotic cells and inhibit their phagocytosis by elicited macrophages: evidence that oxidation-specific epitopes mediate macrophage recognition. *Proceedings of the National Academy of Sciences of the USA* **96**(11): 6353–6358.

Cheung WL, Ajiro K, Samejima K *et al.* (2003) Apoptotic phosphorylation of histone H2B is mediated by mammalian sterile twenty kinase. *Cell* **113**(4): 507–517.

Chiu R, Novikov L, Mukherjee S and Shields D (2002) A caspase cleavage fragment of p115 induces fragmentation of the Golgi apparatus and apoptosis. *Journal of Cell Biology* **159**(4): 637–648.

Coleman ML, Sahai EA, Yeo M *et al.* (2001) Membrane blebbing during apoptosis results from caspase-mediated activation of ROCK I. *Nature Cell Biology* **3**(4): 339–345.

Communal C, Sumandea M, de Tombe P *et al.* (2002) Functional consequences of caspase activation in cardiac myocytes. *Proceedings of the National Academy of Sciences of the USA* **99**(9): 6252–6256.

Cullen SP, Adrain C, Lüthi AU, Duriez PJ and Martin SJ (2007) Human and murine granzyme B exhibit divergent substrate preferences. *Journal of Cell Biology* **176**(4): 435–444.

Cullen SP and Martin SJ (2008) Mechanisms of granule-dependent killing. *Cell Death and Differentiation* **15**(2): 251–262.

Enari M, Sakahira H, Yokoyama H *et al.* (1998) A caspase-activated DNase that degrades DNA during apoptosis, and its inhibitor ICAD. *Nature* **391**(6662): 43–50.

Fernandes-Alnemri T, Yu J, Datta P, Wu J and Alnemri ES (2009) AIM2 activates the inflammasome and cell death in response to cytoplasmic DNA. *Nature* **258**: 509–513.

Gerner C, Frohwein U, Gotzmann J *et al.* (2000) The Fas-induced apoptosis analyzed by high throughput proteome analysis. *Journal of Biological Chemistry* **275**(50): 39018–39026.

Greenberg ME, Sun M, Zhang R *et al.* (2006) Oxidized phosphatidylserine-CD36 interactions play an essential role in macrophage-dependent phagocytosis of apoptotic cells. *Journal of Experimental Medicine* **203**(12): 2613–2625.

Hill MM, Adrian C, Duriez PJ, Creagh EM and Martin SJ (2004) Analysis of the composition, assembly kinetics and activity of native Apaf-1 apoptosomes. *EMBO Journal* **23**: 2134–2145.

Kawane K, Fukuyama H, Yoshida H *et al.* (2003) Impaired thymic development in mouse embryos deficient in apoptotic DNA degradation. *Nature Immunology* **4**(2): 138–144.

Kazama H, Ricci JE, Herndon JM *et al.* (2008) Induction of immunological tolerance by apoptotic cells requires caspase-dependent oxidation of high-mobility group box-1 protein. *Immunity* **29**(1): 21–32.

Keckler MS (2007) Dodging the CTL response: viral evasion of Fas and granzyme induced apoptosis. *Frontiers in Bioscience* **12**: 725–732.

Kono H and Rock KL (2008) How dying cells alert the immune system to danger. *Nature Reviews. Immunology* **8**(4): 279–289.

Kuwana T, Mackey MR, Perkins G *et al.* (2002) Bid, Bax, and lipids cooperate to form supramolecular openings in the outer mitochondrial membrane. *Cell* **111**(3): 331–342.

Lakhani SA, Masud A, Kuida K *et al.* (2006) Caspases 3 and 7: key mediators of mitochondrial events of apoptosis. *Science* **311**: 847–851.

Lauber K, Bohn E, Kröber SM *et al.* (2003) Apoptotic cells induce migration of phagocytes via caspase-3-mediated release of a lipid attraction signal. *Cell* **113**(6): 717–730.

Levkau B, Herren B, Koyama H, Ross R and Raines EW (1998) Caspase-mediated cleavage of focal adhesion kinase pp125FAK and disassembly of focal adhesions in human endothelial cell apoptosis. *Journal of Experimental Medicine* **187**(4): 579–586.

Li P, Nijhawan D, Budihardjo I *et al.* (1997) Cytochrome *c* and dATP-dependent formation of Apaf-1/caspase-9 complex initiates an apoptotic protease cascade. *Cell* **91**(4): 479–489.

Logue SE and Martin SJ (2008) Caspase activation cascades in apoptosis. *Biochemical Society Transactions* **36**(part 1): 1–9.

Lowe M, Lane JD, Woodman PG and Allan VJ (2004) Caspase-mediated cleavage of syntaxin 5 and giantin accompanies inhibition of secretory traffic during apoptosis. *Journal of Cell Science* **117**(part 7): 1139–1150.

Luo X, Budihardjo I, Zou H, Slaughter C and Wang X (1998) Bid, a Bcl2 interacting protein, mediates cytochrome *c* release from mitochondria in response to activation of cell surface death receptors. *Cell* **94**(4): 481–490.

Lüthi AU and Martin SJ (2007) The CASBAH: a searchable database of caspase substrates. *Cell Death and Differentiation* **14**: 641–650.

Martin SJ, Finucane DM, Amarante-Mendes GP, O'Brien GA and Green DR (1996) Phosphatidylserine externalization during CD95-induced apoptosis of cells and cytoplasts requires ICE/CED-3 protease activity. *Journal of Biological Chemistry* **271**(46): 28753–28756.

Muzio M, Chinnaiyan AM, Kischkel FC *et al.* (1996) FLICE, a novel FADD-homologous ICE/CED-3-like protease, is recruited to the CD95 (Fas/APO-1) death-inducing signaling complex. *Cell* **85**(6): 817–827.

Napirei M, Karsunky H, Zevnik B *et al.* (2000) Features of systemic lupus erythematosus in Dnase1-deficient mice. *Nature Genetics* **25**(2): 177–181.

Nicholson DW (1999) Caspase structure, proteolytic substrates, and function during apoptotic cell death. *Cell Death and Differentiation* **6**(11): 1028–1042.

Patel VA, Longacre A, Hsiao K *et al.* (2006) Apoptotic cells, at all stages of the death process, trigger characteristic signaling events that are divergent from and dominant over those triggered by necrotic cells: implications for the delayed clearance model of autoimmunity. *Journal of Biological Chemistry* **281**(8): 4663–4670.

Rao L, Perez D and White E (1996) Lamin proteolysis facilitates nuclear events during apoptosis. *Journal of Cell Biology* **135**(6 part 1): 1441–1455.

Ricci JE, Gottlieb RA and Green DR (2003) Caspase-mediated loss of mitochondrial function and generation of reactive oxygen species during apoptosis. *Journal of Cell Biology* **160**(1): 65–75.

Samejima K, Tone S and Earnshaw WC (2001) CAD/DFF40 nuclease is dispensable for high molecular weight DNA cleavage and stage I chromatin condensation in apoptosis. *Journal of Biological Chemistry* **276**(48): 45427–45432.

Scaffidi P, Misteli T and Bianchi ME (2002) Release of chromatin protein HMGB1 by necrotic cells triggers inflammation. *Nature* **418**(6894): 191–195.

Slee EA, Adrian C and Martin SJ (2001) Executioner caspase-3, -6, and -7 perform distinct, non-redundant roles during the demolition phase of apoptosis. *Journal of Biological Chemistry* **276**: 7320–7326.

Steinhusen U, Weiske J, Badock V *et al.* (2001) Cleavage and shedding of E-cadherin after induction of apoptosis. *Journal of Biological Chemistry* **276**(7): 4972–4980.

Taylor RC, Cullen SP and Martin SJ (2008) Apoptosis: controlled demolition at the cellular level. *Nature Reviews of Molecular Cell Biology* **9**(3): 231–241.

Thiede B, Dimmler C, Siejak F and Rudel T (2001) Predominant identification of RNA-binding proteins in Fas-induced apoptosis by proteome analysis. *Journal of Biological Chemistry* **276**(28): 26044–26050.

Thornberry NA, Rano TA, Peterson EP *et al.* (1997) A combinatorial approach defines specificities of members of the caspase family and granzyme B. Functional relationships established for key mediators of apoptosis. *Journal of Biological Chemistry* **272**: 17907–17911.

Todryk S, Melcher AA, Hardwick N *et al.* (1999) Heat shock protein 70 induced during tumor cell killing induces Th1 cytokines and targets immature dendritic cell precursors to enhance antigen uptake. *Journal of Immunology* **163**(3): 1398–1408.

Truman LA, Ford CA, Pasikowska M *et al.* (2008) CX3CL1/fractalkine is released from apoptotic lymphocytes to stimulate macrophage chemotaxis. *Blood* **112**(13): 5026–5036.

Walsh JG, Cullen SP, Sheridan C *et al.* (2008) Executioner caspase-3 and caspase-7 are functionally distinct proteases. *Proceedings of the National Academy of Sciences of the USA* **105**(35): 12815–12819.

Youle RJ and Strasser A (2008) The BCL-2 protein family: opposing activities that mediate cell death. *Nature Review of Molecular and Cellular Biology* **9**(1): 47–59.

Further Reading

Adrain C, Brumatti G and Martin SJ (2006) Apoptosomes: protease activation platforms to die from. *Trends in Biochemical Science* **31**(5): 243–247.

Ashkenazi A and Dixit VM (1998) Death receptors: signaling and modulation. *Science* **281**(5381): 1305–1308.

Chipuk JE and Green DR (2008) How do BCL-2 proteins induce mitochondrial outer membrane permeabilization? *Trends in Cell Biology* **18**(4): 157–164.

Cohen GM (1997) Caspases: the executioners of apoptosis. *Biochemical Journal* **326**: 1–16.

Dix MM, Simon GM and Cravatt BF (2008) Global mapping of the topography and magnitude of proteolytic events in apoptosis. *Cell* **134**(4): 679–691.

Green DR, Ferguson T, Zitvogel L and Kroemer G (2009) Immunogenic and tolerogenic cell death. *Nature Review of Immunology* **9**(5): 353–363.

Mahrus S, Trinidad JC, Barkan DT *et al.* (2008) Global sequencing of proteolytic cleavage sites in apoptosis by specific labelling of protein *N* termini. *Cell* **134**(5): 866–876.

Martin SJ and Green DR (1995) Protease activation during apoptosis: death by a thousand cuts? *Cell* **82**(3): 349–352.

Slee EA, Harte MT, Kluck RM *et al.* (1999) Ordering the cytochrome *c*-initiated caspase cascade: hierarchical activation of caspases-2, -3, -6, -7, -8, and -10 in a caspase-9-dependent manner. *Journal of Biological Chemistry* **144**: 281–292.

Dismantling the Apoptotic Cell

Paula Deming, *University of Vermont, Burlington, Vermont, USA*

Sally Kornbluth, *Duke University, Durham, North Carolina, USA*

Advanced article

Article Contents

- Introduction
- Caspases: Regulators of the Apoptotic Process
- Concluding Remarks

Apoptosis is a programme of cell death that results in dramatic morphological and biochemical changes in the dying cell due to the systematic dismantling of cellular architecture and functional pathways. The proteins that execute the apoptotic programme are a group of proteases termed caspases (cysteine-dependent aspartate-specific protease). Caspases proteolytically cleave a host of cellular substrates at aspartate residues, which may render them either functionally inactive or confer novel activities that help to promote cellular demise. Substrates targeted by caspases during the apoptotic programme include proteins involved in maintaining various aspects of cytoskeletal and organelle architecture as well as

proteins that function in signalling networks critical for cell function. Following the execution phase of apoptosis, the cellular corpse is packaged in an orderly fashion into membrane-bound apoptotic bodies that are sensed by phagocytes, which neatly engulf the dead cell without eliciting an immune response.

Introduction

Apoptosis, also known as programmed cell death, is the process of eliminating unwanted or damaged cells through the orderly dismantling of cellular architecture and modulation of key biochemical signalling modules. Cellular destruction, which is accomplished without ever exposing the extracellular environment to the intracellular contents of the dying cell, is largely executed by a group of proteases known as caspases (cysteine-dependent aspartate-specific protease), that functionally modify numerous endogenous proteins through proteolytic cleavage at specific aspartate residues. Functional alteration of these caspase substrates not only promotes the physical breakdown of core structural components but also modifies the cellular environment to facilitate cellular execution. Critical viability pathways are inactivated whereas those that amplify the apoptotic signal are turned on. The cellular remnants are then neatly packaged into membrane-bound apoptotic bodies that are subsequently engulfed by phagocytes. The process of apoptosis allows for removal of the dead cell in the absence of an immune response which is crucial during the development of tissues and organisms as well as to the maintenance of normal physiological processes. **See also**: Apoptosis: Regulatory Genes and Disease; The Siren's Song: This Death that Makes Life Live

Caspases: Regulators of the Apoptotic Process

Caspases are a group of highly conserved cysteine-dependent proteases that uniquely cleave protein substrates after aspartate residues. Of the 14 mammalian caspases, 7 of them are known to play a role in apoptosis. The distinct phenotypes of knockout mice have revealed important information regarding the crucial role of the apoptotic caspases during development and the apoptotic process and have highlighted the fact that individual caspases play distinct roles in organismal development and in carrying out the apoptotic programme (**Table 1**). Induction of apoptosis involves a signalling cascade with sequential activation of initiator caspases (caspases 8, 10, 9 and 2) and effector caspases (caspases 3, 7 and 6). The caspases are produced as catalytically inactive zymogens or proenzymes containing a prodomain, a large and a small subunit (**Figure 1**). Initiator and effector caspases can be distinguished based on their structure as well as their position within the apoptotic signalling cascade. Initiator, or apical, caspases have a long prodomain that contains either a death effector domain (DED) or caspase recruitment domain (CARD). The activation of initiator caspases requires binding to specific oligomeric adaptor protein complexes such as the 'apoptosome' involved in activation of caspase 9, the 'PIDDosome' that activates caspase 2, and the 'death-inducing signalling complex', or DISC, involved in activation of caspase 8. Recruitment of the initiator caspases to these specific cellular platforms results in a conformational change of the initiator caspase which then renders the caspase catalytically active. Once active, these initiators function to activate the effector caspases through proteolytic cleavage. Cleavage of effector caspases at distinct aspartate residues results in the restructuring of the large and small subunit, subsequent dimerization and

Table 1 Apoptotic caspase knockouts

Caspase	Developmental effect	Apoptotic defect
Initiator		
Caspase 2	Excessive germ cells	Germ cell specific
Caspase 9	Embryonic lethal, neural	Mitochondrial pathway
Caspase 8	Embryonic lethal	Death receptor pathway
Effector		
Caspase 3	Perinatal lethality[a], neural	Lack of or delayed morphological changes
Caspase 7	Normal	
Caspase 6	Normal	

[a]Perinatal lethality occurs in select genetic backgrounds.

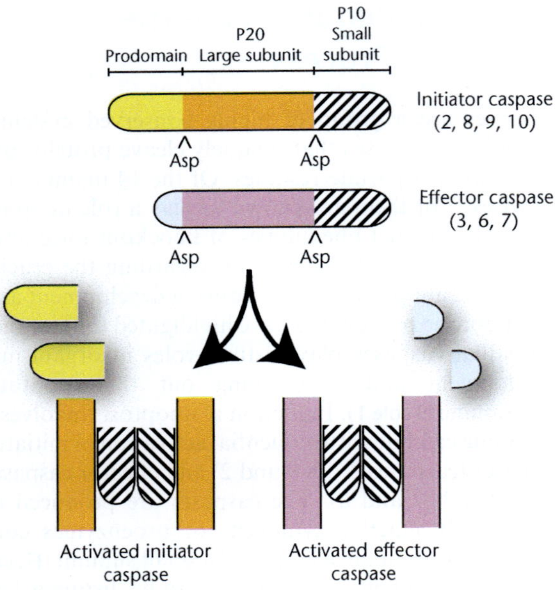

Figure 1 Structure and activation of apoptotic caspases. All caspases are produced as catalytically inactive zymogens or proenzymes containing a prodomain, a large (p20) and a small subunit (p10). The prodomain of the initiator (also known as apical) caspases (2, 8, 9 and 10) is much longer than that of effector caspases 3, 6 and 7. For the initiator caspases, binding to adaptor proteins results in the cleavage of the prodomain and subsequent rearrangement of the large and small subunits to produce a catalytically active heterodimer. Once active, initiator caspases proteolytically cleave effector caspases at distinct aspartate residues, to release the prodomain, large and small subunits. This then allows for the rearrangement of the large and small subunits to form the active heterodimer.

activation. The active effector caspases then proteolytically degrade a host of intracellular proteins to carry out the cell death programme. Although there is redundancy between the effector caspases with regard to their cellular substrates, caspase 3 is the predominant protease that drives demolition of the cell. Indeed caspase 7 deficiency does not appear to influence development or apoptosis induced by a number of stimuli whereas caspase 3 deficiency is perinatally lethal in some genetic backgrounds, and mouse embryo fibroblasts from caspase 3-deficient mice display a prominent defect in the phenotypic morphological changes associated with apoptosis (**Table 1**). This signalling cascade is not strictly unidirectional as amplification of the apoptotic signal is mediated in part through positive feedback loops whereby activated effector caspases (e.g. caspase 3) can proteolytically cleave initiator caspases (e.g. caspase 9) to further enhance their activity. Given the nature of this amplification cascade (akin to the coagulation protease cascade), once a critical threshold of caspase activity is reached, a massive amount of caspase activity is quickly generated thereby committing the cell to death. **See also**: Caspases and Cell Death; Caspases, Substrates and Sequential Activation

Activation of the apoptotic caspases

In response to a diverse array of signals, caspase activation is mediated by two evolutionarily conserved pathways, the intrinsic and extrinsic death pathways. For signals that arise from within the cell, (e.g. deoxyribonucleic acid (DNA) damage) pro-apoptotic members of the Bcl-2 protein family (Bax and Bak) are stimulated to permeablize the mitochondrial outer membrane, resulting in the release of the respiratory chain protein, cytochrome *c* into the cytoplasm. The presence of cytoplasmic cytochrome *c* drives the oligomerization of an adaptor protein, Apaf-1, which serves as the platform for caspase 9 activation. Following formation of the active Apaf-1 (apoptosis-activating factor 1)/caspase 9-containing apoptosome, effector caspases are cleaved by caspase 9 and cellular demolition is set in motion (**Figure 2**). Extracellular ligands that activate death receptors trigger apoptosis through an extrinsic pathway. The formation of the DISC complex serves as the platform for the recruitment and activation of the initiator caspases 8 and/or 10. Once active, depending on the apoptotic stimulus, these initiator caspases can either directly activate effector caspases or proceed through the mitochondrial pathway by cleaving the Bcl-2 family protein, Bid (which contains a single Bcl-2 homology region known as the BH3 domain). Truncated Bid (tBid) then induces mitochondrial permeabilization through activation of Bax and Bak which in turn activates caspase 9 and downstream effector caspases (**Figure 2**). Regardless of the apoptotic signal and the pathway by which apoptosis proceeds (intrinsic or extrinsic death pathway) the end goal is the same, that is, to activate the effector caspases which in turn will dismantle the cell through proteolytic cleavage of intracellular targets.

Effect of caspase-mediated cleavage on target substrates – gain- or loss-of-function

As a large portion of the proteome (more than 400 different proteins) are known to be cleaved by effector caspases during apoptosis, it is difficult to ascertain whether cleavage of a particular protein results in a functional effect on the apoptotic process or is merely a consequence of the massive proteolysis occurring in the cell at that time (Dix *et al.*, 2008; Fischer *et al.*, 2003; Johnson and Kornbluth, 2008; Luthi and Martin, 2007; Mahrus *et al.*, 2008). Caspase-mediated proteolysis may either functionally inactivate or activate substrates as cleavage occurs at sites both between and within functional domains. The resultant cleavage products may be unstable protein fragments that are rapidly degraded or stable effector proteins that have enhanced activity and in some cases acquire novel functions (**Figure 3**). Gain-of-function substrates typically act to positively regulate the apoptotic process. For example, caspase 3-mediated cleavage of the Rho-associated coiled coil kinase 1, ROCK1, induces the release of an autoinhibitory domain and renders the serine/threonine kinase constitutively active (Coleman *et al.*, 2001). Proteolytic activation of ROCK1 promotes actin- and myosin-dependent contractility, resulting in cellular contraction and membrane blebbing, two characteristic morphological features of apoptosis (Coleman *et al.*, 2001; Shi and Wei, 2007). Although modification of some caspase substrates results in functional activation, there are far more known instances where substrate cleavage renders the target protein inactive. As detailed below, such is the case for numerous proteins that are crucial for maintaining cell structure and viability.

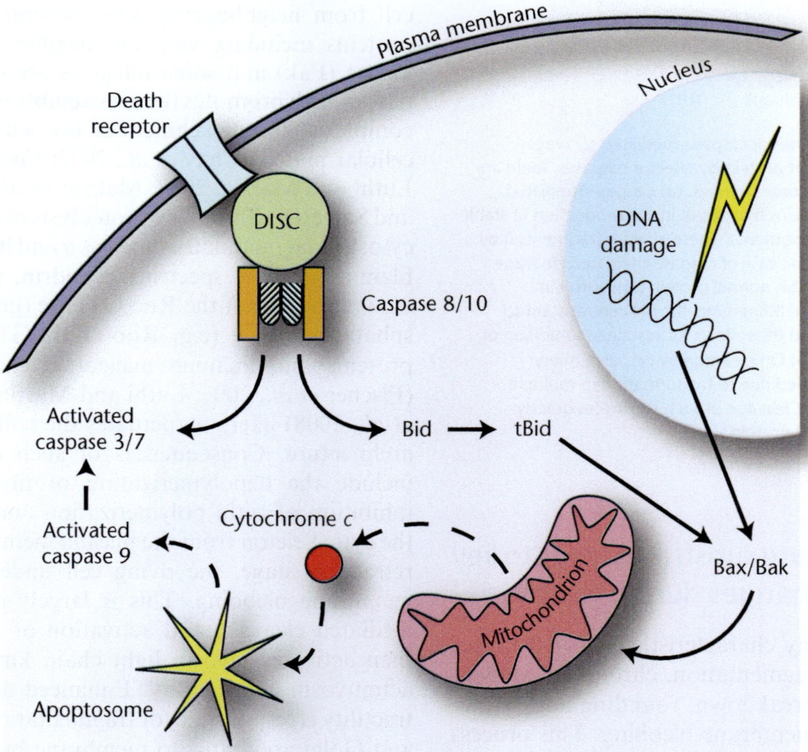

Figure 2 Activation of caspases through intrinsic and extrinsic death pathways. Caspase activation may proceed through either an external (extrinsic) or internal (intrinsic) death pathway, dependent on the stimulus. The intrinsic pathway proceeds through the mitochondria via activation of the pro-apoptotic Bcl-2 family members (Bax and Bak) which induce mitochondrial permeabilization and the release of cytochrome *c*. Cytoplasmic cytochrome *c* drives the formation of the apoptosome and activation of initiator caspase 9. Once active, caspase 9 functions to proteolytically cleave and activate the effector caspases 3 and 7, which then go on to cleave numerous intracellular proteins to dismantle the cell. The binding of extracellular ligands to death receptors triggers the formation of a DISC complex resulting in the activation of initiator caspases 8 and/or 10. These caspases can either directly activate caspase 3 or signal through the mitochondria via a Bid-dependent cleavage event. Truncated Bid (tBid) then activates Bax/Bak which induce mitochondrial permeabilization, the release of cytochrome *c*, formation of the apoptosome and the activation of caspases 9 and 3.

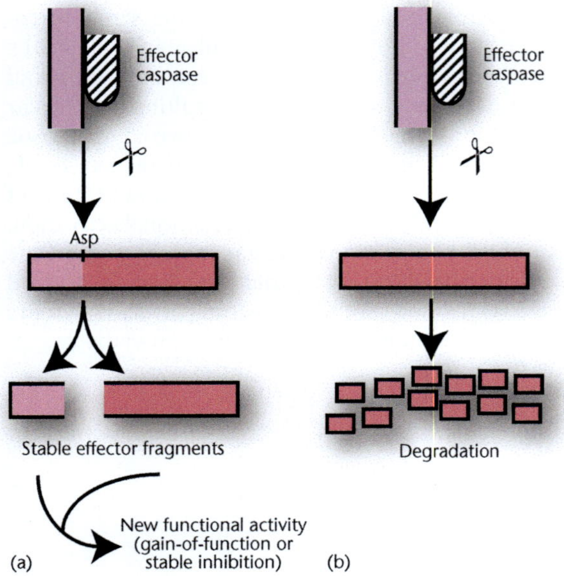

Figure 3 Functional outcomes of caspase-mediated cleavage. Following cleavage of a target protein by effector caspases, there are several probable functional consequences. (a) Caspase-mediated cleavage of intracellular proteins may result in the production of stable, functionally active effector fragments. These effector fragments may become newly active (as in the case of caspase-mediated cleavage of ROCK1) or may act to inhibit normal protein function (e.g. caspase-mediated cleavage of IKB renders the protein resistant to proteosomal degradation and thus allows for sustained inhibition of NFκB, see text for details). (b) On cleavage by caspases many substrates are quickly degraded due to the formation of multiple unstable protein fragments. Cleavage of such substrates quickly depletes the cell of the target protein.

Cleavage of caspase substrates leads to key morphological changes during apoptosis

Apoptotic cells display characteristic cellular changes that include DNA fragmentation, chromatin condensation, cytoskeletal breakdown, rounding and retraction of the cell and membrane blebbing. This process culminates in the packaging of the cellular remnants into membrane-bound apoptotic bodies that are then sensed by phagocytes and subsequently engulfed in an immunologically silent manner. Although over 400 intracellular substrates for the effector caspases have been identified, only a subset have known functions in bringing about the morphological and biochemical changes that occur during cell programmed cell death. The execution stage of apoptosis can be simplified into three morphological stages, all of which are dependent on caspase activity: cell release or retraction, membrane blebbing and condensation (**Figure 4**).

Cytoskeletal events and membrane blebbing

During the release stage, loss of cell–cell contact, cell adhesion and collapse of the cytoskeleton cause the dying cell to round up and become 'released' from its surroundings. Caspases orchestrate these events by altering the cytoskeleton via cleavage of assorted regulators of the actin microfilament system. For instance, cleavage-induced inactivation of proteins integral in establishing and maintaining cell–cell contacts, such as cadherins, β catenin, desmocollin-3, desmoglein-1 and 3, desmoplakin and plakophilin (Dix *et al.*, 2008; Fischer *et al.*, 2003; Luthi and Martin, 2007; Mahrus *et al.*, 2008), mediates retraction of the dying cell from neighbouring cells. Several focal adhesion proteins including vinculin, paxillin, focal adhesion kinase (Fak) and some integrins are targeted by caspases which promotes the disassembly of focal adhesion complexes and detachment of the cell from the extracellular matrix (Chay *et al.*, 2002; Fischer *et al.*, 2003; Luthi and Martin, 2007; Mahrus *et al.*, 2008; Timmer and Salvesen, 2007). The proteolysis of numerous other cytoskeletal modulators such as α and β actin, α actinin, filamin, gelsolin, spectrins, α fodrin, troponin T and core regulators of the Rho GTPase (guanosine triphosphatase) family (e.g. Rho-GDI, GTPase-activating proteins and guanine nucleotide exchange factors) (Fischer *et al.*, 2003; Luthi and Martin, 2007; Mahrus *et al.*, 2008) likely perpetuates the collapse of cellular architecture. Consequences of such cleavage events include the depolymerization of filamentous actin, inhibition of actin polymerization and uncoupling of the cytoskeleton from the plasma membrane. After the retraction stage, the dying cell undergoes extensive membrane blebbing. This is largely due to caspase-mediated cleavage and activation of ROCK1, which then activates myosin light chain kinase to increase actomyosin contractility. Enhanced actomyosin contractility is responsible for fragmentation of the nucleus and Golgi apparatus to membrane buds that are dispersed throughout the cell body, and blebbing of the plasma membrane.

Nuclear events

The hallmark nuclear changes associated with apoptosis are chromatin condensation and internucleosomal DNA fragmentation. Although chromosome condensation is an identifying feature of apoptotic cells, the cellular factors that mediate this process have only

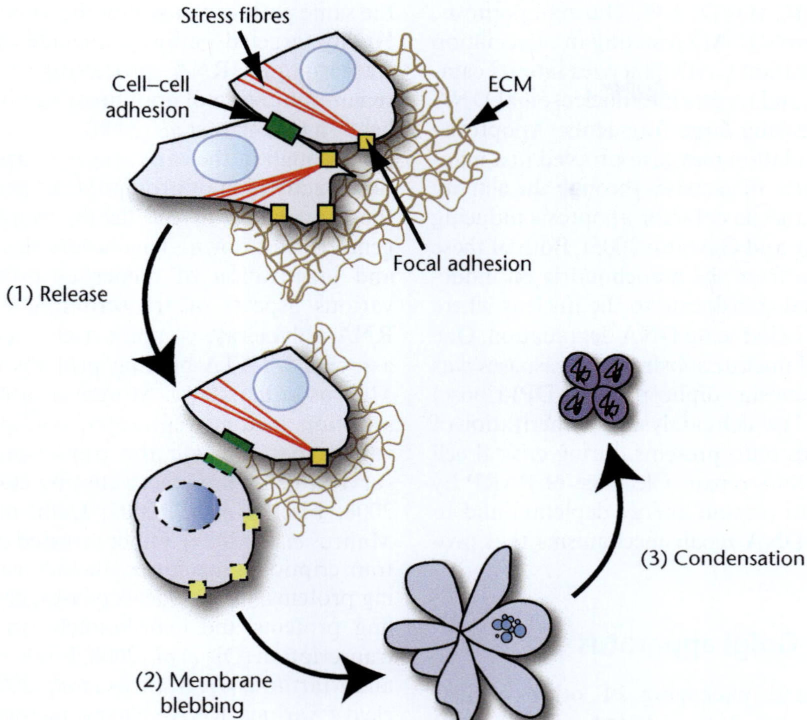

Figure 4 Morphological stages of apoptosis. Caspase-mediated cleavage of proteins essential for the stability and maintenance of cell–cell contacts, focal adhesions and the actin cytoskeleton results in the loss of cell–cell contacts, the detachment of cells from the substratum and collapse of the cytoskeletal architecture. The cell morphologically appears retracted, as it is 'released' from attachment to surrounding cells and matrix. Caspase-dependent activation of ROCK1 promotes actomyosin contractility, subsequent fragmentation of the nucleus and Golgi apparatus and membrane blebbing. Lastly, the broken down cellular remnants are tightly packaged and condensed into apoptotic bodies.

partly been identified. One protein that when cleaved by caspases promotes chromosome condensation is the nuclear apoptotic chromatin condensation inducer (Acinus) (Sahara *et al.*, 1999). The role of Acinus during apoptosis may reach beyond chromosome condensation as it is part of the apoptosis- and splicing-associated protein (ASAP) complex which regulates ribonucleic acid (RNA) processing during apoptosis (Schwerk *et al.*, 2003). Moreover in some apoptotic cells Acinus participates in the process of DNA degradation either by regulating the activity of apoptotic nucleases or their access to DNA (Joselin *et al.*, 2006). In addition to the chromatin condensing, DNA becomes dissociated from the matrix via degradation of nuclear matrix/scaffold attachment regions (M/SARs) and the matrix itself is broken down. In this regard, caspase-mediated cleavage of scaffolding attachment factor (SAF-A) results in the detachment of DNA from structural sites within the nuclear matrix and destabilization of the chromatin (Gohring *et al.*, 1997). The nuclear envelope is broken down through cleavage of multiple nuclear lamina proteins (e.g. A- and B-type lamins, Lap-2 (lamin-associated protein 2)) (Buendia *et al.*, 1999). Moreover, nuclear pore complexes become clustered and consequently deregulated via targeted proteolysis of several nucleoporins (e.g. Nup 153, Nup 214 and Nup 50) and associated nuclear pore proteins (e.g. Tpr, nuclear pore-associated filament protein, RanBP1) (Fischer *et al.*, 2003; Luthi and Martin, 2007; Mahrus *et al.*, 2008). This collapse of the nuclear architecture coupled with increased actomyosin-contractility (discussed above) promotes the nuclear fragmentation typically observed in apoptotic cells.

In addition to structural changes in the nuclear architecture, DNA is characteristically degraded by nucleases that are activated in response to apoptotic stimuli. The best studied of these is caspase-activated DNase (CAD)/DNA fragmentation factor 40 (DFF40), which, in healthy, nonapoptotic cells, is inhibited through the tight binding of the inhibitory chaperone inhibitor of CAD (ICAD)/DFF45. During apoptosis, active caspase 3 cleaves ICAD resulting in dissociation from CAD. On liberation CAD dimerizes into its catalytically active form and targets internucleosomal DNA for degradation releasing large fragments. Apoptosis-induced DNA degradation may also proceed in a manner that is independent of caspases through the activity of endonuclease G and its cofactor apoptosis-inducing factor (AIF) (Widlak and Garrard, 2005). Both of these proteins are released from the mitochondria on induction of apoptosis and translocate to the nucleus where they can co-operate in inducing DNA degradation. One of the first identified nuclear substrates of caspases was the nuclear poly(adenosine diphosphate (ADP)-ribose) polymerase (PARP-1) which catalyses polymerization of ADP ribose moieties onto proteins during critical cell processes such as DNA repair. Cleavage of PARP by caspases functions to prevent energy depletion and to shut down cellular DNA repair mechanisms that promote survival.

Changes in the Golgi apparatus

The disassembly and packaging of other cellular organelles also occurs during apoptosis. Like the nucleus, the Golgi apparatus undergoes a dramatic morphological change whereby it is disassembled, fragmented into small vesiculo-tubular elements, dispersed in the cell body and eventually incorporated into apoptotic bodies. These changes are dependent on caspase-mediated proteolysis of several key Golgi proteins. For example, cleavage of golgin 160 and GRASP65, proteins that function to maintain structural support and stack Golgi cisternae, respectively, destabilizes the overall organization of the Golgi apparatus (Lane *et al.*, 2002; Mukherjee *et al.*, 2007). Cleavage of p115, a protein that both facilitates membrane fusion by tethering vesicles to the Golgi apparatus and provides structural support, not only compromises Golgi structure but alters p115 such that it translocates to the nucleus where it enhances the apoptotic process (Mukherjee and Shields, 2009).

Caspases alter the transcriptional and translational machinery

In addition to the physical breakdown and packaging of cellular components, caspases systematically dismantle signalling networks critical for cell viability. Specific biochemical pathways are efficiently shut down through the cleavage by caspases of multiple components along the same pathway or within the same protein complex. Highly targeted pathways include those that regulate transcription, RNA processing, translation, DNA repair, inhibition of apoptosis, survival and DNA replication (Mahrus *et al.*, 2008).

Although in the early stages of apoptosis transcriptional activation of pro-apoptotic genes is required for the progression to cell death, transcription of many genes is shut down. This occurs through the cleavage and inactivation of numerous proteins involved in various aspects of transcription. For example, the RNA polymerase complex itself is cleaved along with associated TATA-binding proteins (Dix *et al.*, 2008; Mahrus *et al.*, 2008). Moreover, a multitude of transcription factors, repressors, co-repressors and activators that regulate the transcription of numerous survival genes are inactivated by caspases (Dix *et al.*, 2008; Fischer *et al.*, 2003; Luthi and Martin, 2007; Mahrus *et al.*, 2008). Other targeted components of the transcriptional machinery include various RNA-binding proteins, histone deacetylases, chromatin remodelling proteins and exoribonucleases which terminate transcription (Dix *et al.*, 2008; Fischer *et al.*, 2003; Luthi and Martin, 2007; Mahrus *et al.*, 2008). Caspases also cleave various RNA-splicing factors and heteroribonuclear proteins which may alter the ability of the cell to process pre-mRNAs (messenger RNA) (Dix *et al.*, 2008; Luthi and Martin, 2007; Mahrus *et al.*, 2008; Schwerk and Schulze-Osthoff, 2005). Just as the effects of caspase activity on transcriptional processes are abundant, the synthesis of a vast majority of mRNAs is also markedly reduced during apoptosis due to a global inhibition of translation. This inhibition results from the caspase-dependent cleavage of multiple translation initiation factors (e.g. eIF4B, eIF3β, p35, eIF2α and proteins of the eIF4G family) and elongation factors (e.g. EFIβ)· Moreover, caspases cleave and activate dsRNA (double stranded RNA)-activated protein kinase (PKR), which also leads to inhibition of eukaryotic initiation factor eIF2α (Luthi and Martin, 2007; Timmer and Salvesen, 2007). In contrast, a small subset of mRNAs containing IRES (internal ribosome entry sequence) elements within their 5′-untranslated region (UTR) are not subject to global inhibition and are instead translated efficiently. It is in this manner that *de novo* synthesis of pro-apoptotic proteins containing so-called death IRESes (e.g. c-Myc, Apaf-1 and death-associated protein 5, DAP5, also named p97 and NAT1, *N*-acetyl transferase 1) occurs (Marash and Kimchi, 2005). Caspases can influence the translation of such death IRES-containing mRNAs via cleavage of DAP5, which produces an effector fragment that serves to stimulate IRES-dependent translation and thereby enhance the apoptotic programme (Henis-Korenblit *et al.*, 2002; Marash and Kimchi, 2005).

Survival pathways are targeted during apoptosis

Several biochemical pathways involved in protecting the cell are inactivated by caspases during apoptosis. A classic example is the signalling pathway involving the pro-survival transcription factor, nuclear factor-κB (NFκB). Although, the NFκB p65 subunit is still able to bind to DNA following cleavage by caspases, it behaves as a dominant negative protein as it is no longer able to transactivate target survival genes (Fischer *et al.*, 2003). Moreover, cleavage at the *N*-terminus of the NFκB inhibitor, IκB, renders the inhibitor resistant to degradation by the proteasome leading to an overall stabilization of NFκB inhibition. Several growth factors required for the survival of cells are also cleaved and inactivated by effector caspases (e.g. epidermal growth factor, EGF). The apoptotic signal is amplified as a result of caspase activity directed towards both anti- and pro-apoptotic members of the Bcl-2 protein family. The apoptotic inhibitors Bcl-2 and Bcl-x(L), for example, are cleaved by caspases, which not only renders them unable to inhibit apoptosis but converts them to pro-apoptotic proteins. Caspase-induced proteolysis of Bid activates its pro-apoptotic function and allows for the apoptotic signal to proceed through the mitochondria to activate downstream effector caspases. Proteins that regulate the Bcl-2 family of proteins are also targeted by caspases. This is the case for the survival protein kinase, Akt, which normally functions to inhibit the activity of components of the intrinsic death machinery such as the pro-apoptotic protein Bad. Cleavage-induced degradation of Akt relieves these inhibitory events and therefore promotes the apoptotic signal (Widmann *et al.*, 1998; Xu *et al.*, 2002).

The apoptotic cell calls for its own disposal

An important aspect of programmed cell death is the swift removal of the dying cell which prevents leakage of the cellular contents into the extracellular milieu and avoids an immune reaction. Cells undergoing programmed cell death orchestrate their own disposal by sending signals that call in phagocytic cells and trigger engulfment. One such signal is the chemoattractant lipid lysophosphatidylcholine, LPC. Caspase 3-mediated cleavage and activation of calcium-independent phospholipase A (iPLA) results in the hydrolysis of plasma membrane bound phosphatidylcholine, generating arachadonic acid and LPC (Lauber *et al.*, 2003). Release of LPC into the extracellular environment serves to recruit professional phagocytes such as monocytes and macrophages. It should be noted that in some circumstances, cells neighbouring the apoptotic bodies are capable of engulfment which would make it unnecessary for professional phagocytes to migrate to clear the corpses. Regardless of which cell type carries out engulfment, the process is triggered on recognition of 'eat me' signals displayed on the surface of apoptotic cells. The best characterized of these 'eat me' signals is phosphatidylserine (PS). During apoptotic cell death, PS moves from the inner leaflet of the plasma membrane to the outer leaflet and serves as a major recognition signal for phagocytes. Other 'eat me' signals implicated in the induction of phagocytosis include changes in the surface charge of glycoproteins and glycolipids, an oxidized low-density lipoprotein (LDL)-like moiety, the expression of a particular intercellular adhesion molecule (ICAM3) and the binding of opsonizing proteins (e.g. thrombospondin and the complement protein Cq1) to the surface of the apoptotic cell. Phagocytic cells can interact with apoptotic corpses indirectly through bridging proteins that bind both to molecules such as PS displayed on the dying cell surface and to various membrane receptors (e.g. scavenger receptors) on the phagocytes. Additionally, some cells may express membrane receptors that are able to directly interact with PS displayed on the corpse surface (Miyanishi *et al.*, 2007; Park *et al.*, 2007). Apoptotic cells not only display 'eat me signals', but lose the expression of viability markers such as CD31 which normally function to suppress engulfment. The engulfment of apoptotic bodies is a unique process that results in the disposal of the dead cell in an immunologically silent manner. This is reinforced by the subsequent release of anti-inflammatory cytokines by the phagocytic cell. **See also**: Engulfment of Apoptotic Cells and its Physiological Roles

Caspase-independent cell death

Alternate mechanisms of regulated cellular suicide that occur independent of caspases also exist and include autophagy (also known as type II cell death), necroptosis and PARP-1-mediated necrotic death. In addition to serve as backup mechanisms to mediate cell death when the traditional apoptotic machinery is compromised, these caspase-independent processes may be important for cell killing under circumstances of pathogen infection, bioenergetic crisis and DNA

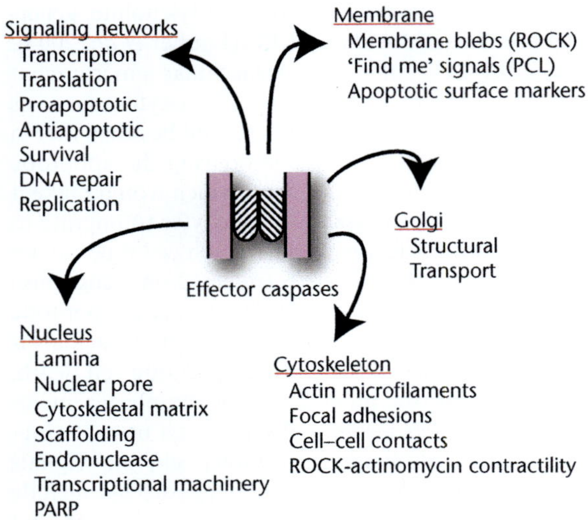

Figure 5 Caspases affect cell death by targeting major networks important for cell architecture and viability. Effector caspases dismantle the cell by cleaving key cellular components that function to maintain cell structure and viability. Target proteins include those involved in: (1) signalling networks that regulate cellular processes such as transcription, translation, apoptosis, replication, (2) nuclear structure and function, (3) membrane structure and integrity, (4) cytoskeletal architecture and (5) Golgi structure and function.

damage. **See also**: Cornification of The Skin: A Non-Apoptotic Cell Death Mechanism

Concluding Remarks

In summary, apoptosis is a caspase-mediated process that proceeds in a tightly controlled, energy-dependent manner. An abundance of intracellular proteins are proteolytically cleaved during this process in what might seem like intracellular mayhem. However, a number of key components of cellular architecture and signalling networks (**Figure 5**) are known to be systematically targeted by caspases. This allows for dismantling of the cell in an orderly fashion and its efficient, specialized clearance by phagocytes, all in the absence of an immune response.

References

Buendia B, Santa-Maria A and Courvalin JC (1999) Caspase-dependent proteolysis of integral and peripheral proteins of nuclear membranes and nuclear pore complex proteins during apoptosis. *Journal of Cell Science* **112**(part 11): 1743–1753.

Chay KO, Park SS and Mushinski JF (2002) Linkage of caspase-mediated degradation of paxillin to apoptosis in Ba/F3 murine pro-B lymphocytes. *Journal of Biological Chemistry* **277**: 14521–14529.

Coleman ML, Sahai EA, Yeo M *et al.* (2001) Membrane blebbing during apoptosis results from caspase-mediated activation of ROCK I. *Nature Cell Biology* **3**: 339–345.

Dix MM, Simon GM and Cravatt BF (2008) Global mapping of the topography and magnitude of proteolytic events in apoptosis. *Cell* **134**: 679–691.

Fischer U, Janicke RU and Schulze-Osthoff K (2003) Many cuts to ruin: a comprehensive update of caspase substrates. *Cell Death and Differentiation* **10**: 76–100.

Gohring F, Schwab BL, Nicotera P, Leist M and Fackelmayer FO (1997) The novel SAR-binding domain of scaffold attachment factor A (SAF-A) is a target in apoptotic nuclear breakdown. *EMBO Journal* **16**: 7361–7371.

Henis-Korenblit S, Shani G, Sines T *et al.* (2002) The caspase-cleaved DAP5 protein supports internal ribosome entry site-mediated translation of death proteins. *Proceedings of the National Academy of Sciences of the USA* **99**: 5400–5405.

Johnson CE and Kornbluth S (2008) Caspase cleavage is not for everyone. *Cell* **134**: 720–721.

Joselin AP, Schulze-Osthoff K and Schwerk C (2006) Loss of Acinus inhibits oligonucleosomal DNA fragmentation but not chromatin condensation during apoptosis. *Journal of Biological Chemistry* **281**: 12475–12484.

Lane JD, Lucocq J, Pryde J *et al.* (2002) Caspase-mediated cleavage of the stacking protein GRASP65 is required for Golgi fragmentation during apoptosis. *Journal of Cell Biology* **156**: 495–509.

Lauber K, Bohn E, Krober SM *et al.* (2003) Apoptotic cells induce migration of phagocytes via caspase-3-mediated release of a lipid attraction signal. *Cell* **113**: 717–730.

Luthi AU and Martin SJ (2007) The CASBAH: a searchable database of caspase substrates. *Cell Death and Differentiation* **14**: 641–650.

Mahrus S, Trinidad JC, Barkan DT *et al.* (2008) Global sequencing of proteolytic cleavage sites in apoptosis by specific labeling of protein N termini. *Cell* **134**: 866–876.

Marash L and Kimchi A (2005) DAP5 and IRES-mediated translation during programmed cell death. *Cell Death and Differentiation* **12**: 554–562.

Miyanishi M, Tada K, Koike M *et al.* (2007) Identification of Tim4 as a phosphatidylserine receptor. *Nature* **450**: 435–439.

Mukherjee S, Chiu R, Leung SM and Shields D (2007) Fragmentation of the Golgi apparatus: an early apoptotic event independent of the cytoskeleton. *Traffic* **8**: 369–378.

Mukherjee S and Shields D (2009) Nuclear import is required for the pro-apoptotic function of the Golgi protein p115. *Journal of Biological Chemistry* **284**: 1709–1717.

Park D, Tosello-Trampont AC, Elliott MR *et al.* (2007) BAI1 is an engulfment receptor for apoptotic cells upstream of the ELMO/Dock180/Rac module. *Nature* **450**: 430–434.

Sahara S, Aoto M, Eguchi Y *et al.* (1999) Acinus is a caspase-3-activated protein required for apoptotic chromatin condensation. *Nature* **401**: 168–173.

Schwerk C, Prasad J, Degenhardt K *et al.* (2003) ASAP, a novel protein complex involved in RNA processing and apoptosis. *Molecular and Cellular Biology* **23**: 2981–2990.

Schwerk C and Schulze-Osthoff K (2005) Regulation of apoptosis by alternative pre-mRNA splicing. *Molecular Cell* **19**: 1–13.

Shi J and Wei L (2007) Rho kinase in the regulation of cell death and survival. *Archivum Immunologiae et Therapiae Experimentalis* **55**: 61–75.

Timmer JC and Salvesen GS (2007) Caspase substrates. *Cell Death and Differentiation* **14**: 66–72.

Widlak P and Garrard WT (2005) Discovery, regulation, and action of the major apoptotic nucleases DFF40/CAD and endonuclease G. *Journal of Cell Biochemistry* **94**: 1078–1087.

Widmann C, Gibson S and Johnson GL (1998) Caspase-dependent cleavage of signaling proteins during apoptosis. A turn-off mechanism for anti-apoptotic signals. *Journal of Biological Chemistry* **273**: 7141–7147.

Xu J, Liu D and Songyang Z (2002) The role of Asp-462 in regulating Akt activity. *Journal of Biological Chemistry* **277**: 35561–35566.

Further Reading

Croft DR, Coleman ML, Li S *et al.* (2005) Actin-myosin-based contraction is responsible for apoptotic nuclear disintegration. *Journal of Cell Biology* **168**: 245–255.

Degterev A and Yuan J (2008) Expansion and evolution of cell death programmes. *Nature Reviews. Molecular Cell Biology* **9**: 378–390.

Li J and Yuan J (2008) Caspases in apoptosis and beyond. *Oncogene* **27**: 6194–6206.

Martelli AM, Zweyer M, Ochs RL *et al.* (2001) Nuclear apoptotic changes: an overview. *Journal of Cell Biochemistry* **82**: 634–646.

Nagata S, Nagase H, Kawane K, Mukae N and Fukuyama H (2003) Degradation of chromosomal DNA during apoptosis. *Cell Death and Differentiation* **10**: 108–116.

Ranger AM, Malynn BA and Korsmeyer SJ (2001) Mouse models of cell death. *Nature Genetics* **28**: 113–118.

Ravichandran KS and Lorenz U (2007) Engulfment of apoptotic cells: signals for a good meal. *Nature Reviews. Immunology* **7**: 964–974.

The BCL-2 Family Proteins – Key Regulators and Effectors of Apoptosis

David L Vaux, *La Trobe University, Victoria, Australia*

Advanced article

Article Contents

- Introduction
- Structure
- Binding between Family Members
- Evolution of BCL-2 Family Proteins
- Integration and Regulation of Apoptotic Signals
- Roles *In Vivo*
- Apoptosis and Cancer
- BCL-2 Antagonist Drugs

BCL-2 was the first cloned component of the mechanism for apoptosis – the process by which metazoan cells commit suicide – to be recognized. Mammals are now known to carry genes for a large number of BCL-2 like proteins, some of which, like BCL-2 itself, inhibit apoptosis, and others that promote or are required for apoptosis. Direct interactions between BCL-2 family members are essential for the proper regulation and implementation of apoptosis during development and for homoeostasis. Abnormalities to the regulation of cell death, such as those caused by mutations to genes for BCL-2 family members prevent apoptosis occurring when it should, and can lead to diseases including cancer. Understanding the roles of BCL-2 family members, their structures and how they interact, has allowed the development of novel agents for the treatment of cancer that are now in clinical trials.

Introduction

BCL-2 was identified as the product of a gene translocated onto the immunoglobulin heavy chain locus in the *t*(14;18) translocations typical of follicular lymphoma (Tsujimoto *et al.*, 1984). The association of translocations activating BCL-2 with lymphoma suggested that it was an oncogene. However, when this was tested by overexpressing BCL-2 in cytokine-dependent cell lines in tissue culture, it did not behave like other oncogenes known at the time. Rather than acting as they did to stimulate cell proliferation, allow cells to grow in the absence of growth factors or enable the cells to form tumours in mice, BCL-2 allowed the cells to survive in a G_0 state when cytokine was removed. Instead of undergoing apoptosis in the absence of growth factors, cells transfected with BCL-2 survived in a quiescent state, but, importantly, were able to grow and proliferate once more when cytokine was restored (Vaux *et al.*, 1988).

These experiments showed that inhibition of physiological cell death could lead to cancer, and BCL-2 thus became the first example of a new kind of oncogene – an inhibitor of apoptosis – the first cloned component of the mechanism for cell death to be identified in any organism.

For several years, BCL-2 was the only component of the mechanism for apoptosis that had been cloned. In the nematode *Caenorhabditis elegans*, a pathway specific for developmentally programmed cell death had been deduced from classical genetic experiments (Ellis and Horvitz, 1986). It initially remained unclear whether these findings had any relevance to mammalian biology or apoptosis. However, expression of human BCL-2 in the worm was able to greatly reduce programmed cell death during nematode development (Vaux *et al.*, 1992). This meant that human BCL-2 was able to impact on components of the worm's cell death machinery, and showed that programmed cell death and apoptosis were the same, evolutionarily conserved process. It also implies that BCL-2 would function to inhibit a process requiring homologues of the worm's killer genes, *ced*-3 and *ced*-4. The subsequent cloning of the nematode cell death (*ced*) genes confirmed that the mechanisms were conserved, because CED-9 had similar sequence to BCL-2 (Hengartner and Horvitz, 1994). Furthermore, cloning of *ced-3* revealed that it encoded a cysteine protease, suggesting that apoptosis in mammalian cells was a BCL-2 controllable process that involved activation of proteases. **See also**: Apoptosis: Regulatory Genes and Disease; Cell Death in *C. Elegans*

Since that time, a large family of BCL-2-like proteins have been identified, either by sequence similarity or by the ability to bind in coimmunoprecipitations, yeast 2-hybrid experiments or using phage display. Some BCL-2 family members, such as BCL-2 itself, inhibit cell death, whereas others promote it.

Structure

Members of the BCL-2 family are recognized by presence of one or more BCL-2 homology (BH) domains (**Figure 1**). All cell death pathways that can be controlled by BCL-2, or similar antiapoptotic family members Mcl-1, BCL-x, BCL-w and A1 (Bfl-1), require the presence of either of the proapoptotic members BAX or BAK (Lindsten *et al.*, 2000). (Bok/Mtd is another proapoptotic BCL-2 family member like BAX and BAK, but its expression is largely restricted to the

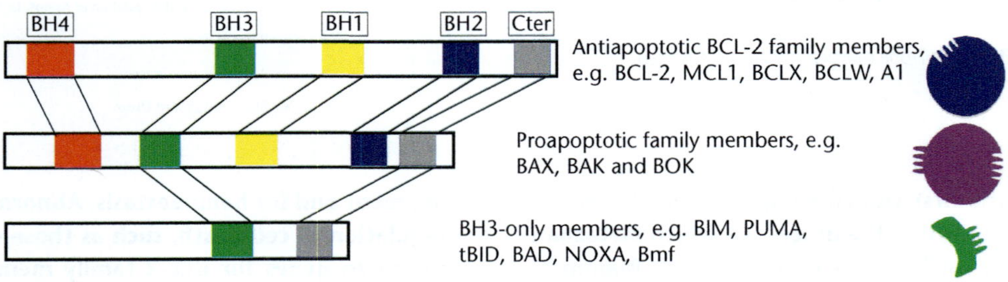

Figure 1 There are three classes of BCL-2 family proteins: Antiapoptotic family members (top, blue symbol); BAX, BAK and Bok (middle, purple symbol) and BH3-only proteins (top, green symbol). All BCL-2 family members bear a number of BCL-2 homology (BH) motifs. Although all have a BH3 motif (green), antiapoptotic BCL-2 family members and the proapoptotic proteins BAX, BAK and Bok, all bear a BH1 (yellow), a BH2 (blue), a BH3 (green) and a BH4 (red) motif. Most BCL-2 family members also have a hydrophobic region at the carboxy-terminus (grey) that can act as a transmembrane domain. When antiapoptotic BCL-2 family members and the proapoptotic multiple BH motif containing proteins fold, the BH3, BH1 and BH2 motifs form a hydrophobic pocket (depicted as indentations in the diagrams on the right) that is capable of binding to the BH3 α helices (depicted as fingers) of other BCL-2 family members.

ovaries (Inohara *et al.*, 1998).) Although BAX and BAK are proapoptotic, they adopt a very similar fold to BCL-2, and bear the BCL-2 homology (BH) domains BH1, BH2 and BH3. Although BAX and BAK are often said not to bear a BH4 domain, comparison of this region (α helix 1) in the 3D structures indicates no substantial difference between the pro and anti-apoptotic family members, so in **Figure 1** BAX and BAK are also shown to have a BH4 domain. **See also**: Structure and Function of IAP and Bcl-2 Proteins

The third class of BCL-2 family member proteins are the 'BH3-only' proteins, which, like BAX and BAK, promote apoptosis. Members of this class bear only the BH3 domain. Although BID is classed as a BH3-only protein, when its 3D structure was determined it was found to have overall similarity to BCL-2 that was not just limited to the BH3 domain (Chou *et al.*, 1999; McDonnell *et al.*, 1999). **See also**: BH3-Only Proteins

Both the antiapoptotic BCL-2 family members and BAX and BAK are almost entirely composed of α helices (Hinds and Day, 2005). The surface bears a shallow hydrophobic groove that can bind the BH3 helix of other family members. This groove is made up from parts of BH1, BH2 and BH3 domains. For example, in crystal complexes a peptide corresponding to the BH3 domain of BAK was found to lie in hydrophobic groove of BCL-x (Sattler *et al.*, 1997) (**Figure 2**).

The *C*-termini of the antiapoptotic proteins, as well as BAX and BAK, contain hydrophobic residues that allow binding to membranes within the cell, such as the outer membrane of the mitochondria. The hydrophobic *C*-terminus of BAX folds over the protein, and occupies the hydrophobic groove in healthy cells, but unfolds enabling it to interact with membranes, and allowing other proteins access to the groove (Suzuki *et al.*, 2000).

The BH3-only proteins (with the exception of BID) appear to be largely unstructured, but the BH3 domain itself is α helical when bound to an antiapoptotic family member such as BCL-x. Although the sequence of BID is very different to BCL-2, it nevertheless folds into a protein with similar tertiary structure. BID can be cleaved by caspase 8 in cells undergoing apoptosis, and the truncated form, tBID, retains the BH3 domain via which it can bind to other BCL-2 family members to promote apoptosis. **See also**: BH3-Only Proteins

Binding between Family Members

The presence of either BAX or BAK is required for mammalian cells to undergo apoptosis by the mechanism controlled by the antiapoptotic BCL-2 family proteins (Lindsten *et al.*, 2000).

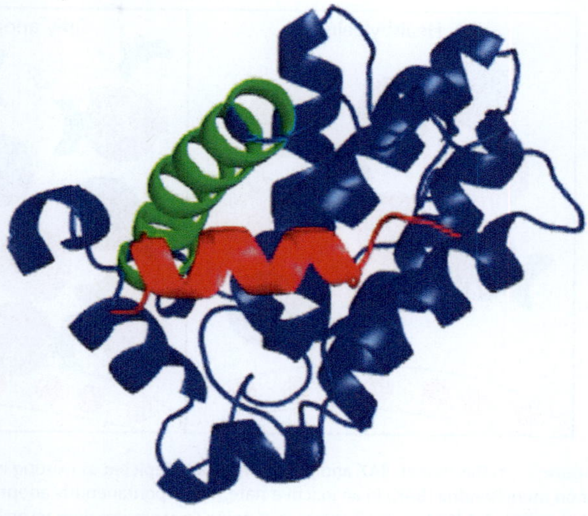

Figure 2 Complex of Bcl-x (blue) with the BH3 helix peptide from BAK (red). The BAX BH3 peptide fits into a hydrophobic groove on the surface of Bcl-x. The BH3 of Bcl-x itself is coloured green. Note that Bcl-x, like all multidomain BCL-2 family members, is almost entirely composed of α helices. BCL-2 antagonist compounds were designed to occupy the same groove as the BAK peptide shown here. Structure from 1bxl drawn using Polyview-3D http://polyview.cchmc.org/.

Although BAK binds most strongly to MCL1 and BCLX, BAX binds with similar strength to BCLX, BCL-2, BCLW and MCL1. BAX and BAK appear to bind to BH3-only member proteins much more weakly, but can interact directly with BIM, PUMA and tBID.

The BH3-only proteins BAD, Bik and Bmf bind most strongly to BCL-2, BCLW and BCLX; NOXA prefers MCL1 and A1; whereas BIM, tBID and PUMA can bind to all of the antiapoptotic BCL-2 family members (van Delft and Huang, 2006).

These interactions suggest that antiapoptotic family members keep cells alive in two ways: first by binding to BAX and BAK to keep them in an inactive state, and second by binding to and sequestering BH3-only proteins (**Figure 3**). On an apoptotic stimulus, the anti-apoptotic proteins are degraded, or levels of active BH3-only proteins are increased. This can relieve inhibition of BAX and BAK, and can free the BH3-only proteins such as BIM, tBID and PUMA that can promote BAX/BAK activation. Either way, BAX and BAK associate with the outer mitochondrial membranes, adopt an active conformation and oligomerize, such that the integrity of the outer mitochondrial membrane is breached, and cytochrome *c* is released

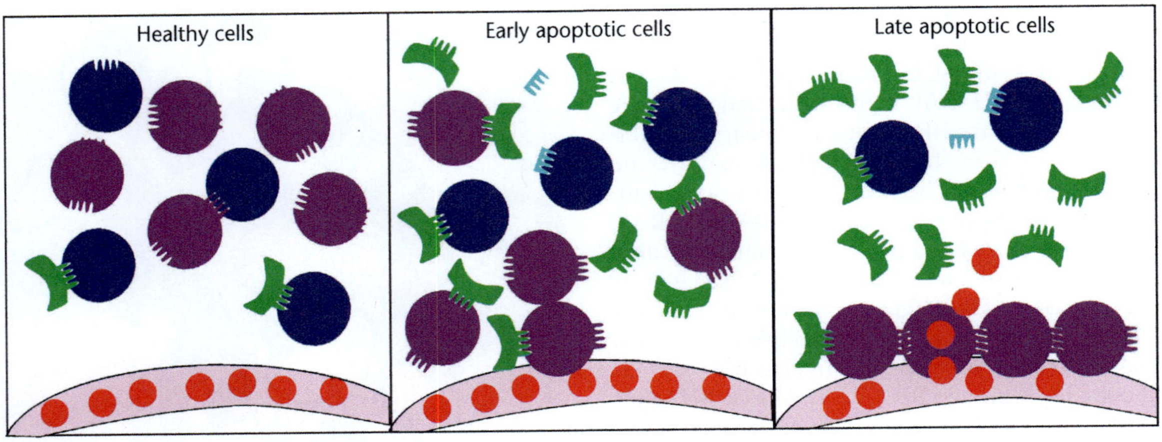

Figure 3 In this model, BAX and BAK (purple) are depicted as existing in 3 states. In healthy cells, they exist as monomers in the cytoplasm (BAX) or on mitochondria (BAK) in an inactive state. They spontaneously adopt a 'ready' conformation (BH3 'fingers' exposed), which allows them to bind to BCL-2, MCL1 or BCLX (blue), but doing so promotes their reversion into the inactive state (BH3 not exposed). BH3-only proteins (green) can bind to BCL-2, MCL1 and BCLX, preventing them from inactivating BAX and BAK. Some BH3-onlys, such as BIM, PUMA and tBID can also interact with BAX and BAK on the mitochondria (pink) to increase the rate at which they adopt the active conformation (middle panel) that somehow makes the outer membrane permeable, so that cytochrome c (red) is released (right panel). Artificial BCL-2 antagonists, such as ABT-263 (light blue) can act like BH3-only proteins by binding to antiapoptotic BCL-2 family proteins. According to this model, apoptosis can be triggered by increasing the concentration of BH3-only proteins, adding an artificial BCL-2 antagonist, or depleting the cells of antiapoptotic BCL-2 family members. Although BAX and BAK can spontaneously activate to allow mitochondrial membrane permeability, this might be accelerated by the presence of certain BH3-only proteins on the mitochondrial outer membrane. In healthy (non-apoptotic) cells, only small amounts of antiapoptotic BCL-2 members need to be bound to BAX and BAK at any one time.

into the cytoplasm. Although BAX or BAK alone can oligomerize to form pores in artificial membrane vesicles (Lovell *et al.*, 2008), some researchers believe the pores in mitochondrial membranes of intact cells are composed of or also contain other proteins.

Evolution of BCL-2 Family Proteins

The similarity of the 3D structure of BAX and BAK, the antiapoptotic BCL-2 family members and BID (Hinds and Day, 2005), indicate that they are all derived from the same ancestral gene. It is possible that a BAX-like gene evolved first, and encoded a protein that was capable of oligomerizing on membranes due to possession of both an exposed α helix (i.e. the BH3) as well as a hydrophobic pocket capable of binding to it (**Figure 3**). Duplication of this gene, and subsequent mutation, led to generation of proteins (e.g. BCL-2) that had similar overall structure but were antiapoptotic because they lost the ability to expose a BH3 domain in a way that could be bound by another protein, but they could still bind to the BH3 domain of BAX. Other duplications of the ancestral BAX gene led to the evolution of the BH3-only proteins, which could not cause apoptosis on their own, but could promote oligomerization of BAX by binding their BH3 domains into the hydrophobic core of BAX, favouring its active conformation. These BH3-only proteins retained the ability to bind to antiapoptotic family members, effectively neutralizing them and could also bind to BAX and BAK, stabilizing them in the form capable of oligomerizing.

Integration and Regulation of Apoptotic Signals

Activation of BAX and BAK, and hence apoptosis, are chiefly controlled in two ways, either by activation of BH3-only proteins, or by decreasing the abundance of antiapoptotic BCL-2 family members. Some BH3-only proteins, such as PUMA and NOXA, exist at low levels in healthy cells, but can be transcriptionally upregulated, for example, by p53 following deoxyribonucleic acid (DNA) damage. Other BH3-only proteins are regulated proteolytically (e.g. BID) or by phosphorylation (e.g. BAD). Increasing the levels of active BH3-only proteins allows them to antagonize the antiapoptotic family members, liberating BAX and BAK, or by activating them on the mitochondrial outer membrane.

Expression of the antiapoptotic BCL-2 family protein MCL1 can be induced by cytokines, providing one mechanism by which growth factors can promote cell survival (Kozopas *et al.*, 1993). MCL1 has a short half-life (approximately 2 h), and is constantly being degraded by the proteasome, so reducing its rate of transcription causes its levels to drop quickly, resulting in apoptosis. Fine-tuning of the levels of MCL1 is achieved by phosphorylation, which might stop it binding to proapoptotic BCL-2 family members, or influence how it is ubiquitylated (Zhong *et al.*, 2005). Furthermore, the half-life of MCL1 can be increased or decreased by the levels of the competing MCL1 ubiquitin ligases and deubiquitylases.

Expression of A1 and BCLX are regulated at least in part by nuclear factor-κB (NFκB) transcription factors (Zong *et al.*, 1999) which are, in turn, controlled by receptors such as TNF receptor superfamily members and components of the system for innate immunity.

Expression of several members of the BCL-2 family are regulated by micro ribonucleic acids (microRNAs) (reviewed in Garzon and Croce, 2008). For example, miR122 targets BCLW, miR-29b downregulates MCL1, and miR15 and miR16 can induce apoptosis by targeting BCL-2. However, miR17 is reported to promote apoptosis by increasing levels of BIM, whereas miR125b inhibits apoptosis by down-regulating BAK.

Roles *In Vivo*

BCL-2 was identified because its gene is activated by translocations in follicular lymphoma. This suggested that inhibition of apoptosis could promote cancer, a notion confirmed by generation of BCL-2 transgenic mice, which develop a similar, slowly progressing disease (McDonnell *et al.*, 1989). Although transgenic experiments such as this have confirmed the role of BCL-2 family members as oncogenes (the antiapoptotic family members) or tumour-suppressor genes (the proapoptotic family members), gene deletion studies have revealed more about the requirement for these proteins in physiological circumstances.

Although deletion of the genes for either BAX or BAK alone results in only a subtle phenotype, deletion of the genes for both has dramatic (and surprising) consequences. Whereas most BAX:BAK double knock out (DKO) embryos die before birth, some survive and the occasional mouse has survived into adulthood, with relatively normal appearance other than persistence of interdigital webbing and other minor abnormalities (Lindsten *et al.*, 2000). Because cells derived from BAX:BAK DKO embryos are completely resistant to all apoptotic stimuli that act by the pathway that can be inhibited by BCL-2, this means that cell death necessary for morphological development can still occur even though there is no apoptosis. This suggests that apoptosis by the BAX/BAK-dependent mechanism is not essential for development, and cells that normally die in this way become quiescent and eventually succumb for some other reason, such as metabolic failure. Apoptosis therefore seems to be needed for rapid removal of cells, and the suppression of tumor development, rather than being essential for cell suicide per se.

BCL-2-gene-deleted mice usually survive for several weeks after birth, but die due to kidney failure resulting from abnormal tubule development (Veis *et al.*, 1993). These animals also turn grey from premature death of melanocytes, and are lymphopenic due to apoptosis of lymphoid cells. This suggests that the earliest essential requirements for BCL-2 are to maintain survival of lymphocytes, melanocyte stem cells and a type of cell needed for proper kidney development. Interestingly, BCL-2 KO mice that are also heterozygous or mutant for BIM do not develop kidney disease, do not become grey as rapidly and have relatively normal numbers of lymphocytes (Bouillet *et al.*, 2001). This shows that BIM is the BH3-only protein responsible for death of these cells.

Deletion of BCLX is embryonic lethal. Although much development appears normal, foetuses die at about E14, and the most severe abnormalities are in the developing brain (Motoyama *et al.*, 1995). Interestingly, mice with a point mutation in BCL-x had reduced numbers of platelets, indicating BCL-x is a major determinant of platelet lifespan (Mason *et al.*, 2007). MCL1 knockout mouse embryos succumb even earlier, before implantation possibly around E4 (Rinkenberger *et al.*, 2000).

A1 can be induced by cytokines such as granulocyte–macrophage colony stimulating factor (GM-CSF), tumour necrosis factor α (TNFα) and interleukin-1β (IL-1β)· It inhibits neutrophil apoptosis during inflammatory responses, and deletion of the genes for one of the four A1 members in the mouse (A1a) leads to a reduction in circulating neutrophil numbers and impaired neutrophil survival during inflammatory responses (Hamasaki *et al.*, 1998).

There are also a number of viruses that carry BCL-2 like genes, including adenovirus, *Epstein–Barr virus*, Karposi-associated herpes virus. Presumably, these viruses use the BCL-2 like protein to delay defensive apoptosis of the host cell (White, 2006).

Apoptosis and Cancer

Cancers grow when the rate of cell proliferation exceeds the rate of apoptosis. Inhibition of cell death pathways can therefore contribute to cancer growth, but treatments that increase the rate of apoptosis in cancers could be therapeutic.

The association of translocations that activate BCL-2 with follicular lymphoma, and the demonstration that transgenic overexpression of BCL-2 could cause cancer or accelerate onset of cancer caused by oncogenes such as c-myc, showed that blocking apoptotic pathways could lead to cancer in experimental models, but was also a common cause of cancer in humans.

Several further BCL-2 family members have been implicated in causing cancer in humans, but they might also play roles in cancer progression and resistance to treatment.

Although some mutations directly affect cell death genes, such as mutation of BAX in colorectal carcinomas (Yagi *et al.*, 1998), natural selection operates on genetically unstable cancer cells to favour those with higher levels of antiapoptotic proteins, or lower levels of proapoptotic proteins. Such cells would also have greater resistance to radiation and chemotherapy. Thus, even mutations to genes that act upstream of the cell death effector pathways, such as p53, might act by allowing survival of a cancer cell, even if no members of the BCL-2 family were mutated.

BCL-2 Antagonist Drugs

The finding that peptides corresponding to the BH3 domain of proapoptotic family members could bind to antiapoptotic BCL-2 family members and induce apoptosis prompted the design of synthetic BH3-mimetic compounds to use for chemotherapy as BCL-2 antagonists. To date, the most successful of these has been ABT-263 produced by Abbott Inc. This compound triggers apoptosis in some cell types, but acts specifically, as it does not kill BAX:BAK double knock out cells. Phase II clinical trials are currently underway with ABT-263 (Tse *et al.*, 2008). If proven efficacious, these drugs will vindicate the global scientific effort investigating the mechanisms of apoptosis over the last quarter century.

See also: Drug Discovery in Apoptosis

References

Bouillet P, Cory S, Zhang L, Strasser A and Adams J (2001) Degenerative disorders caused by Bcl-2 deficiency prevented by loss of its BH3-only antagonist Bim. *Developmental Cell* **1**: 645–653.

Chou JJ, Li HL, Salvesen GS, Yuan JY and Wagner G (1999) Solution structure of BID, an intracellular amplifier of apoptotic signaling. *Cell* **96**: 615–624.

van Delft MF and Huang DC (2006) How the Bcl-2 family of proteins interact to regulate apoptosis. *Cell Research* **16**: 203–213.

Ellis HM and Horvitz HR (1986) Genetic control of programmed cell death in the nematode *C. elegans*. *Cell* **44**: 817–829.

Garzon R and Croce CM (2008) MicroRNAs in normal and malignant hematopoiesis. *Current Opinion in Hematology* **15**: 352–358.

Hamasaki A, Sendo F, Nakayama K *et al.* (1998) Accelerated neutrophil apoptosis in mice lacking A1-a, a subtype of the bcl-2-related A1 gene. *Journal of Experimental Medicine* **188**: 1985–1992.

Hengartner MO and Horvitz HR (1994) *C. elegans* cell survival gene ced-9 encodes a functional homolog of the mammalian proto-oncogene bcl-2. *Cell* **76**: 665–676.

Hinds MG and Day CL (2005) Regulation of apoptosis: uncovering the binding determinants. *Current Opinion in Structural Biology* **15**: 690–699.

Inohara N, Ekhterae D, Garcia I *et al.* (1998) Mtd, a novel bcl-2 family member activates apoptosis in the absence of heterodimerization with bcl-2 and bcl-x-l. *Journal of Biological Chemistry* **273**: 8705–8710.

Kozopas KM, Yang T, Buchan HL, Zhou P and Craig RW (1993) Mcl1, a gene expressed in programmed myeloid cell differentiation, has sequence similarity to bcl2. *Proceedings of the National Academy of Sciences of the USA* **90**: 3516–3520.

Lindsten T, Ross AJ, King A *et al.* (2000) The combined functions of proapoptotic Bcl-2 family members Bak and Bax are essential for normal development of multiple tissues. *Molecular Cell* **6**: 1389–1399.

Lovell JF, Billen LP, Bindner S *et al.* (2008) Membrane binding by tBid initiates an ordered series of events culminating in membrane permeabilization by Bax. *Cell* **135**: 1074–1084.

Mason KD, Carpinelli MR, Fletcher JI *et al.* (2007) Programmed anuclear cell death delimits platelet life span. *Cell* **128**: 1173–1186.

McDonnell TJ, Deane N, Platt FM *et al.* (1989) bcl-2-immunoglobulin transgenic mice demonstrate extended B cell survival and follicular lymphoproliferation. *Cell* **57**: 79–88.

McDonnell JM, Fushman D, Milliman CL, Korsmeyer SJ and Cowburn D (1999) Solution structure of the proapoptotic molecule BID: a structural basis for apoptotic agonists and antagonists. *Cell* **96**: 625–634.

Motoyama N, Wang FP, Roth KA *et al.* (1995) Massive cell death of immature hematopoietic cells and neurons in bcl-x-deficient mice. *Science* **267**: 1506–1510.

Rinkenberger JL, Horning S, Klocke B, Roth K and Korsmeyer SJ (2000) Mcl-1 deficiency results in peri-implantation embryonic lethality. *Genes & Development* **14**: 23–27.

Sattler M, Liang H, Nettesheim D *et al.* (1997) Structure of bcl-x(l)-bak peptide complex – recognition between regulators of apoptosis. *Science* **275**: 983–986.

Suzuki M, Youle RJ and Tjandra N (2000) Structure of Bax: coregulation of dimer formation and intracellular localization. *Cell* **103**: 645–654.

Tse C, Shoemaker AR, Adickes J *et al.* (2008) ABT-263: a potent and orally bioavailable Bcl-2 family inhibitor. *Cancer Research* **68**: 3421–3428.

Tsujimoto Y, Finger LR, Yunis J, Nowell PC and Croce CM (1984) Cloning of the chromosome breakpoint of neoplastic B cells with the t(14;18) chromosome translocation. *Science* **226**: 1097–1099.

Vaux DL, Cory S and Adams JM (1988) Bcl-2 gene promotes haemopoietic cell survival and cooperates with c-myc to immortalize pre-B cells. *Nature* **335**: 440–442.

Vaux DL, Weissman IL and Kim SK (1992) Prevention of programmed cell death in *Caenorhabditis elegans* by human bcl-2. *Science* **258**: 1955–1957.

Veis DJ, Sorenson CM, Shutter JR and Korsmeyer SJ (1993) Bcl-2-deficient mice demonstrate fulminant lymphoid apoptosis, polycystic kidneys, and hypopigmented hair. *Cell* **75**: 229–240.

White E (2006) Mechanisms of apoptosis regulation by viral oncogenes in infection and tumorigenesis. *Cell Death and Differentiation* **13**: 1371–1377.

Yagi OK, Akiyama Y, Nomizu T *et al.* (1998) Proapoptotic gene bax is frequently mutated in hereditary nonpolyposis colorectal cancers but not in adenomas. *Gastroenterology* **114**: 268–274.

Zhong Q, Gao W, Du F and Wang X (2005) Mule/ARF-BP1, a BH3-only E3 ubiquitin ligase, catalyzes the polyubiquitination of Mcl-1 and regulates apoptosis. *Cell* **121**: 1085–1095.

Zong WX, Edelstein LC, Chen C, Bash J and Gelinas C (1999) The prosurvival Bcl-2 homolog Bfl-1/A1 is a direct transcriptional target of NF-kappaB that blocks TNFalpha-induced apoptosis. *Genes & Development* **13**: 382–387.

Further Reading

Adams JM and Cory S (2007) The Bcl-2 apoptotic switch in cancer development and therapy. *Oncogene* **26**: 1324–1337.

Del Gaizo Moore V and Letai A (2008) Rational design of therapeutics targeting the BCL-2 family: are some cancer cells primed for death but waiting for a final push? *Advances in Experimental Medicine and Biology* **615**: 159–175.

Fletcher JI and Huang DC (2008) Controlling the cell death mediators Bax and Bak: puzzles and conundrums. *Cell Cycle* **7**: 39–44.

Lessene G, Czabotar PE and Colman PM (2008) BCL-2 family antagonists for cancer therapy. *Nature Reviews. Drug Discovery* **7**: 989–1000.

Reed JC (2008) Bcl-2-family proteins and hematologic malignancies: history and future prospects. *Blood* **111**: 3322–3330.

BH3-only Proteins

Lina Happo, *The Walter and Eliza Hall Institute of Medical Research, Melbourne, Victoria, Australia*

Andreas Strasser, *The Walter and Eliza Hall Institute of Medical Research, Melbourne, Victoria, Australia*

Clare L Scott, *The Walter and Eliza Hall Institute of Medical Research, Melbourne, Victoria, Australia*

Advanced article

Article Contents

- Introduction
- Noxa and Puma
- Bid
- Bim/Bod/Bcl2L11
- Bik/Nbk/Blk
- Bad
- Bmf
- Hrk/DP5
- BH3-Mimetic Drugs
- Conclusions

The deregulation of programmed cell death, apoptosis, is a major contributor to the development of cancer and can impair the response of tumour cells to anticancer therapy. The Bcl-2 family of proteins are critical regulators of apoptosis. BH3 (Bcl-2 homologous 3)-only proteins are pro-apoptotic members that

share with each other and the wider Bcl-2 family only the BH3 domain that is critical for their killing capacity. These BH3-only proteins promote cell death by directly or indirectly activating Bax and Bak, in a cell-type and death stimulus-specific manner. Studies of gene-targeted mice that lack two or more BH3-only proteins are beginning to unravel the overlapping functions of these apoptosis initiators. The activation or mimicking of these proteins is expected to improve the treatment of patients suffering from cancer or autoimmune diseases.

Introduction

The deregulation of programmed cell death, apoptosis, is a major contributor to the development of cancer and certain other diseases and can impair the response of tumour cells to anticancer therapy (Adams and Cory, 2007). Apoptotic pathways are tightly regulated to prevent unnecessary death and play critical roles in a multitude of physiological and disease contexts. Two distinct mammalian apoptotic pathways exist – 'death receptor' ('extrinsic') and 'Bcl-2-regulated' ('mitochondrial' or 'intrinsic') apoptosis signalling, each activating different initiator caspases but converging at the level of effector caspases (**Figure 1**). The 'Bcl-2 regulated' apoptotic pathway is regulated by three subgroups of proteins from the Bcl-2 family (**Figure 2**). Pro-survival members, such as BCL-2, are essential for cell survival; the pro-apoptotic BAX/BAK proteins are critical for activation of the downstream events in cell demolition; whereas the pro-apoptotic BH3-only proteins are required for initiation of apoptosis signalling. This article describes the functions of the BH3-only proteins in development, normal cellular homeostasis and disease. **See also**: Apoptosis: Regulatory Genes and Disease; Drug Discovery in Apoptosis; Structure and Function of IAP and Bcl-2

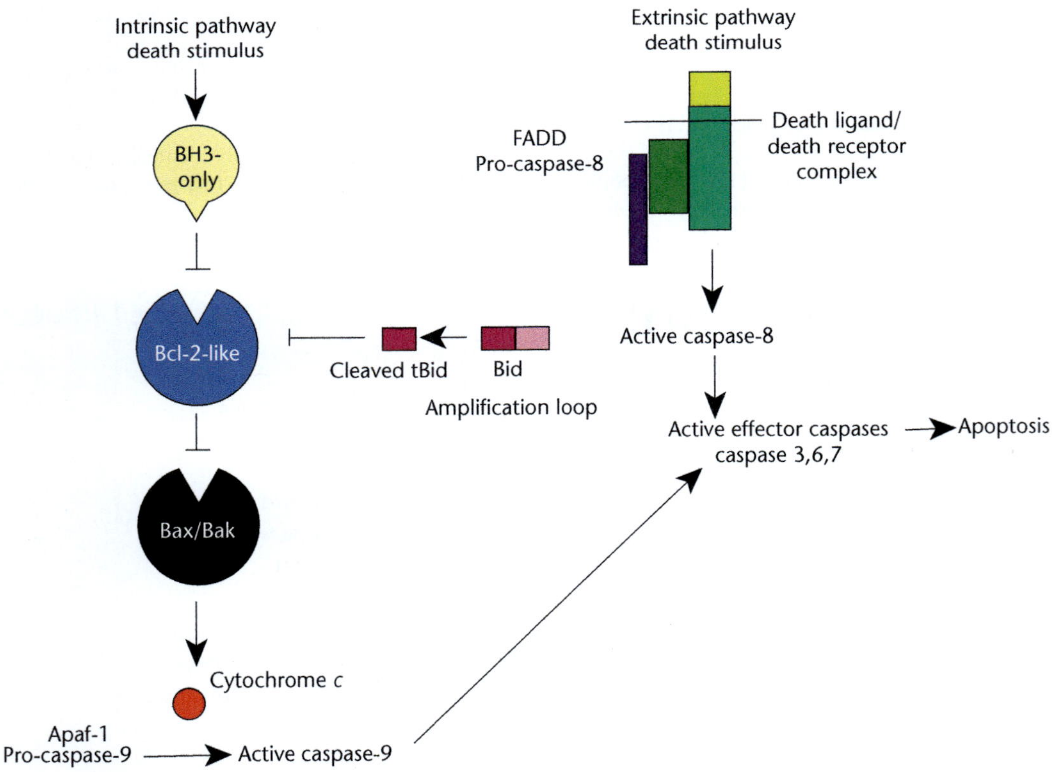

Figure 1 Two distinct mammalian apoptotic pathways exist – 'death receptor' ('extrinsic') and 'Bcl-2-regulated' ('mitochondrial' or 'intrinsic') apoptosis signalling. Each activate different initiator caspases but converge at the level of effector caspases. The 'Bcl-2-regulated' apoptotic pathway is triggered by developmental cues or a broad range of cell stressors (e.g. growth factor deprivation, γ-irradiation) and is regulated by the interplay of the pro- and antiapoptotic members of the Bcl-2 family of proteins.

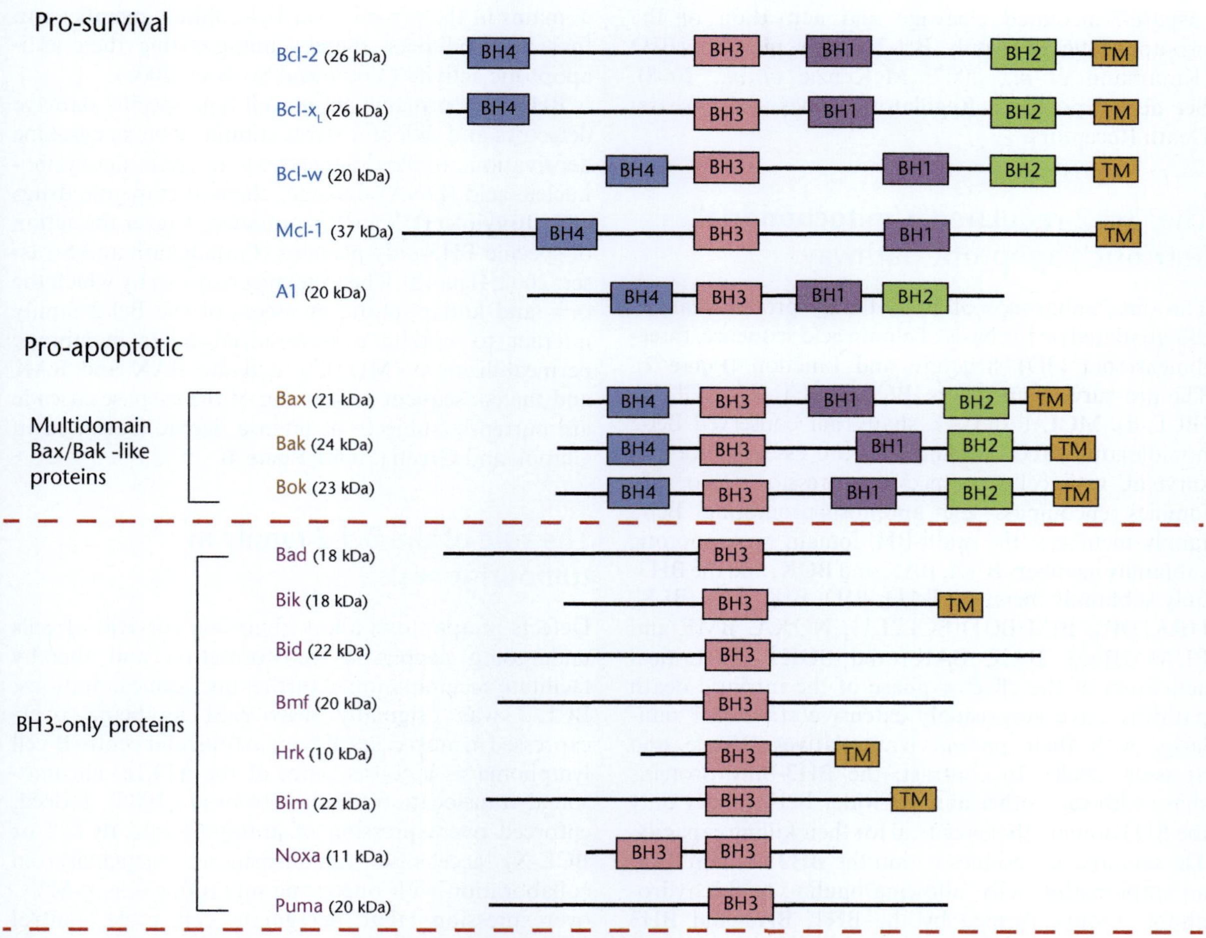

Figure 2 The Bcl-2 protein family consists of three subgroups that can be differentiated on the basis of amino acid sequence, 3D structure and function. The pro-survival members, Bcl-2, Bcl-x$_L$, Bcl-w (Bcl-B), Mcl-1 and A1, share four conserved Bcl-2 homologous (BH) domains and are essential for cell survival, with cell-type-specific expression. Two subfamilies encompass the apoptosis-promoting Bcl-2 family members: the multi-BH domain pro-apoptotic subfamily members Bax, Bak and Bok, which are critical for the execution phase of apoptosis, and the BH3-only subfamily members Bad, Bid, Bik/Nbk/Blk, Hrk/DP5, Bim/Bod/Bcl2L11, Noxa, Bmf and Puma/Bbc3, which are essential for initiation of apoptosis.

Proteins; The Bcl-2 Family Proteins – Key Regulators and Effectors of Apoptosis.

Overview of the apoptotic machinery

The 'Bcl-2-regulated' apoptotic pathway is triggered by developmental cues or cell stressors and is regulated by the interplay of the pro- and antiapoptotic members of the Bcl-2 family of proteins (Youle and Strasser, 2008; **Figure 1**). Such interplay results in activation of the pro-apoptotic BAX/BAK proteins, followed by release of cytochrome *c* and other apoptogenic factors from the mitochondria resulting in the formation of the apoptosome, which promotes activation of the 'initiator caspase', caspase-9. The 'death receptor-induced' pathway is induced by the interaction of cell surface death receptors, such as Fas or tumour necrosis factor (TNF)-R1, with ligands belonging to the TNF family (e.g. FasL or TNFα), leading to the activation of caspase-8 (Adams and Cory, 2007). The initiator caspases, caspase-8 and -9, activate, via proteolytic cleavage, the 'effector caspases'-3, -6 and -7, causing cell destruction. The two pathways are largely independent of one another (Youle and Strasser, 2008). In hepatocytes and pancreatic β cells, however, the extrinsic and intrinsic apoptotic pathways can intersect through caspase-8-mediated cleavage and activation of the pro-apoptotic BH3-only Bcl-2 family member, BID (Kaufmann *et al.*, 2007; McKenzie *et al.*, 2008). **See also**: Apoptosis: Regulatory Genes and Disease; Death Receptors

The 'Bcl-2-regulated' ('mitochondrial', 'intrinsic') apoptotic pathway

The three subgroups of Bcl-2 family proteins can be differentiated on the basis of amino acid sequence, three-dimensional (3D) structure and function (**Figure 2**). The pro-survival members, BCL-2, BCL-X_L, BCL-W, (BCL-B), MCL-1 and A1, share four conserved Bcl-2 homologous (BH) domains and are essential for cell survival, with cell-type-specific expression. Two subfamilies encompass the apoptosis-promoting Bcl-2 family members: the multi-BH domain pro-apoptotic subfamily members BAX, BAK and BOK, and the BH3-only subfamily members BAD, BID, BIK/NBK/BLK, HRK/DP5, BIM/BOD/BCL2L11, NOXA, BMF and PUMA/Bbc3. BAX, BAK (and BOK), the critical activators of the effector phase of the intrinsic death pathway have surprisingly extensive structural similarity with their pro-survival relatives (Youle and Strasser, 2008). In contrast, the BH3-only proteins share with each other and the wider Bcl-2 family only the BH3 domain that is critical for their killing capacity. The amino acid residues within the BH3 domain form an amphipathic helix, allowing binding to the hydrophobic groove formed by the BH1, BH2 and BH3 domains in the pro-survival Bcl-2 family members, to form heterodimers, thereby antagonizing their anti-apoptotic activity (Youle and Strasser, 2008).

BH3-only proteins act as cell-type-specific damage detectors and different stress stimuli, such as cytokine deprivation, anoikis, oncogene activation, deoxyribonucleic acid (DNA) damage, chemotherapeutic drugs and ultraviolet (UV) or γ-irradiation, trigger the action of specific BH3-only proteins (Puthalakath and Strasser, 2002; **Figure 3**). The exact mechanisms by which the pro- and antiapoptotic members of the Bcl-2 family interact to initiate mitochondrial outer membrane permeabilization (MOMP), activate BAX and BAK and the consequent unleashing of the caspase cascade are currently subjects of intense debate (reviewed in Chipuk and Green (2008); **Figure 4**).

The role of the Bcl-2 family in tumourigenesis

Defects in apoptosis allow abnormal survival of cells undergoing neoplastic transformation and thereby facilitate accumulation of further oncogenic mutations. BCL-2 was originally discovered as being overexpressed in most cases of human follicular centre B-cell lymphoma as a consequence of the *t*(14,18) chromosomal translocation (Tsujimoto *et al.*, 1985). Indeed, enforced overexpression of anti-apoptotic BCL-2 or BCL-X_L accelerates tumourigenesis, particularly in collaboration with oncogenic mutations (e.g. c-MYC overexpression) that deregulate cell cycle control

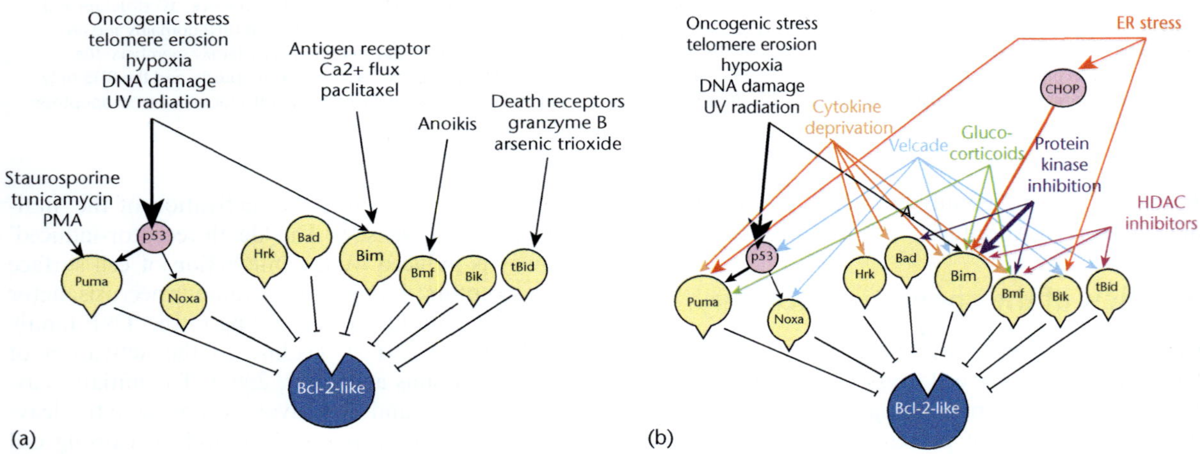

Figure 3 Different apoptotic stimuli activate different BH3-only proteins via distinct signalling pathways. (a) Some death stimuli activate predominantly one BH3-only protein. (b) Other death stimuli appear to elicit cell-type-specific induction of an array of BH3-only proteins. This may reflect the wiring of the particular cell type, the availability of particular BH3-only proteins or may reflect the requirement for combinatorial signalling through multiple BH3-only proteins to allow blockade of all pro-survival Bcl-2 family members present in a particular cell type.

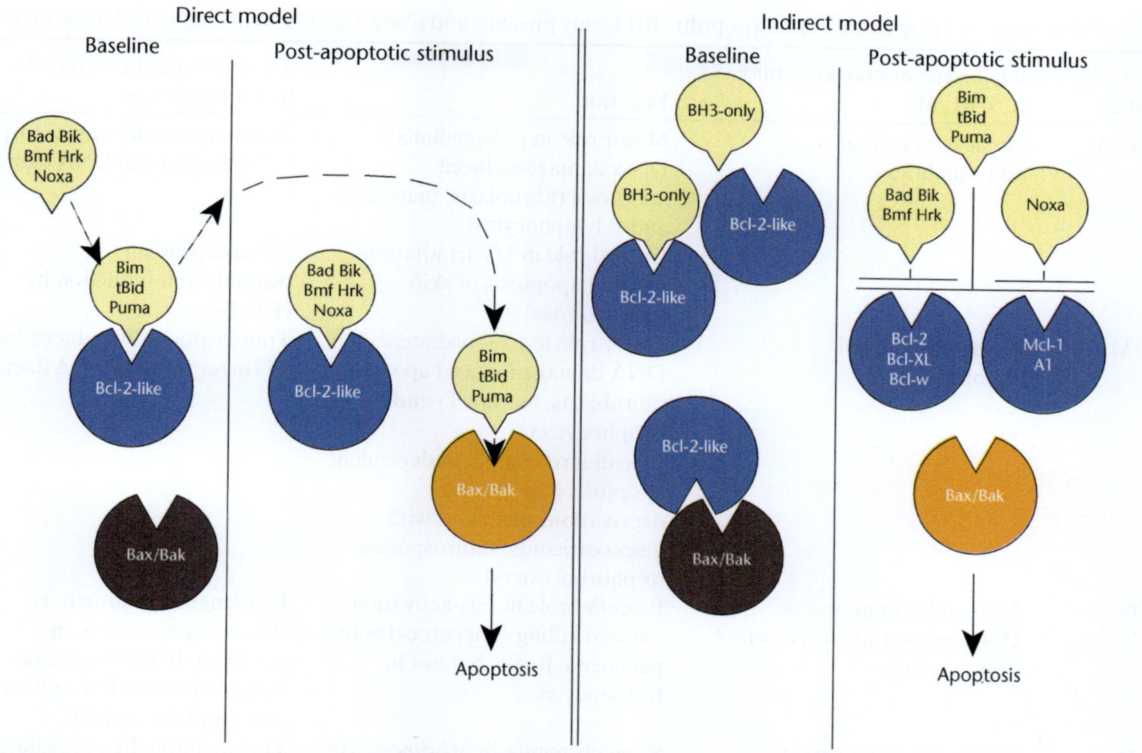

Figure 4 Two competing models have been proposed to explain how BH3-only proteins activate Bax/Bak to unleash the downstream effector phases of apoptosis. According to the 'direct model', in healthy cells certain so-called direct activator BH3-only proteins (particularly Bid, Bim and possibly Puma) are kept in check by binding to the pro-survival Bcl-2 family members. Apoptotic stimuli induce so-called indirect activator BH3-only proteins (e.g. Bad, Bik and Bmf), which by binding to the pro-survival Bcl-2 family members unleash Bid, Bim and Puma, thereby allowing them to bind and directly activate (orange) Bax and Bak from an inactive state (brown). According to the 'indirect model', in healthy cells Bax and Bak are kept in an inactive state (brown) by binding to the pro-survival Bcl-2 family members. Apoptotic stimuli activate BH3-only proteins (in a cell death stimulus- and cell-type-specific manner). By binding to the pro-survival Bcl-2 family members, the BH3-only proteins cause release and activation of Bax and Bak (orange) indirectly. According to this model, for a cell to be committed to undergo apoptosis all pro-survival Bcl-2 family members present must be neutralized by BH3-only proteins. It is important to note that BH3-only proteins differ in their potency to trigger apoptosis and this is due to their differences in binding affinities for the different pro-survival Bcl-2 family members. Bim, Puma and Bid bind all (or most) these proteins with high affinity. In contrast, Bad only binds to Bcl-2, Bcl-x_L and Bcl-w, whereas Noxa only binds to Mcl-1 and A1.

(Strasser *et al.*, 1990) and promote resistance to γ-irradiation and a broad range of cytotoxic drugs (Strasser *et al.*, 1991; Strasser *et al.*, 1994). This oncogenic property of anti-apoptotic Bcl-2 family members suggests that pro-apoptotic counterparts, such as the BH3-only proteins, may exert tumour suppressor functions (**Table 1**).

Noxa and Puma

The tumour suppressor, p53, is a (homotetrameric) nuclear phosphoprotein that regulates the transcription of a large number of target genes to control cell cycling, apoptosis, DNA repair, cellular senescence and certain other processes (Vogelstein *et al.*, 2000). Somatic mutations in p53 are found in more than half of all human cancers and its loss is a major driver of tumourigenesis and resistance to chemotherapy (Levine, 1997). The BH3-only genes *noxa* and *puma* (p53 upregulated modulator of apoptosis) have p53 binding sites within their promoters and *puma* was in fact discovered in expression profiling screens for p53 targets. Loss of NOXA, PUMA or even both does not cause developmental abnormalities in mice (Michalak *et al.*, 2008) but investigations of these gene-targeted mice have provided valuable insights into the functions of these BH3-only proteins in apoptosis and tumourigenesis (**Table 1**).

PUMA and to a lesser extent NOXA are key mediators of p53-mediated DNA damage-induced

Table 1 Function and regulation of pro-apoptotic BH3-only proteins and phenotypes of BH3-only gene knockout mice

BH3-only protein	Phenotype of knockout mice (C57BL/6)	Function	Transcriptional/posttranslational regulation
NOXA	No developmental abnormalities	Minor role in p53-mediated DNA damage-induced apoptosis (fibroblasts, mature T- and B lymphocytes)	Transcriptionally induced by p53 in response to DNA damage
		Critical role in UV irradiation-induced apoptosis of skin keratinocytes	p53-independent transcription-induction by HIF-1α
PUMA	No developmental abnormalities	Major role in p53-mediated DNA damage-induced apoptosis (fibroblasts, mature T- and B lymphocytes) Essential role in p53-independent apoptosis (e.g. cytokine deprivation, treatment with glucocorticoids, staurosporine or phorbol esters)	Transcriptionally induced by p53 in response to DNA damage
BID	Normal haematopoiesis Develop spontaneous myeloid hyperplasia	Essential role in Fas-activation-induced killing of hepatocytes and pancreatic β cells but not in lymphocytes	Full-length Bid protein is cleaved by caspase-8 and granzyme B, into its active truncated form, tBid, enhancing pro-apoptotic activity
BIM	Lymphoid and myeloid hyperplasia	Principle regulator of homeostasis in the lymphoid and myeloid compartment	Transcriptionally upregulated by: FOXO3A in response to cytokine withdrawal and CHOP during ER stress-induced apoptosis
	Fatal SLE-like autoimmune disease (C57BL/6x129SV)	Crucial for the negative selection of autoreactive immature T- and B-lymphoid cells Terminator of the T- and B-cell immune response Involved in apoptosis induced by many death stimuli (e.g. cytokine withdrawal, ionomycin, taxol, glucocorticoids and ER stress)	Bim_L and Bim_{EL} negatively regulated by sequestration to the microtubule-associated dynein motor complex. Phosphorylation by ERK targets Bim for proteasomal degradation, whereas JNK-mediated phosphorylation enahances pro-apoptotic activity
		Some contribution to DNA damage-induced apoptosis	Dephosphorylation by PP2A enhances pro-apoptotic activity
BIK	No developmental abnormalities	Some role in ER stress-induced apoptosis	Posttranslational phosphorylation of Bik enhances pro-apoptotic activity
BAD	No developmental abnormalities Spontaneous diffuse B-cell lymphoma (approximately 40% incidence)	Role in apoptosis induced by cytokine deprivation in lymphocytes	Negatively regulated by phosphorylation by Akt and subsequent sequestration to 14-3-3 scaffold proteins

(*Continued*)

Table 1 Continued

BH3-only protein	Phenotype of knockout mice (C57BL/6)	Function	Transcriptional/posttranslational regulation
BMF	No developmental abnormalities Progressive B-lymphoid hyperplasia	Role in glucocorticoid and HDACi-induced apoptosis in lymphocytes	Negatively regulated by sequestration to the cytoskeleton by binding to the myosin V motor complex
HRK	No developmental abnormalities	Role in apoptosis induced by NGF deprivation in neuronal cells	Transcriptionally induced by NGF deprivation in neuronal cells

Notes: ER, endoplasmic reticulum; ERK, extracellular signal-regulated kinase; HDACi, histone deacetylase inhibitor; HIF, hypoxia inducible factor; JNK, c-Jun *N*-terminal kinase; NGF, nerve growth factor; PP2A, protein phosphatase 2A and SLE, systemic lupus erythematosus.

apoptosis in thymocytes, mature T, as well B lymphocytes and fibroblasts (Youle and Strasser, 2008; **Figure 2**). Interestingly, loss of either PUMA or NOXA alone offered less protection from DNA damage-induced apoptosis than p53 deficiency, suggesting a degree of functional redundancy within the two BH3-only proteins (Villunger *et al.*, 2003). Indeed, PUMA/NOXA double deficient mouse embryonic fibroblasts (MEFs) displayed increased resistance to etoposide-induced apoptosis compared to MEFs lacking either protein alone and *puma*$^{-/-}$*noxa*$^{-/-}$ thymocytes were as resistant to γ-irradiation as p53-deficient thymocytes (Michalak *et al.*, 2008).

Perhaps surprisingly, although γ- as well as UV-irradiation both trigger apoptosis via p53 activation, in MEFs and keratinocytes, PUMA is critical for the former but NOXA for the latter (Naik *et al.*, 2007). This may indicate that different forms of DNA damage (double-strand breaks versus pyrimidine dimers) activate *puma* and *noxa* transcription differentially. This may be due to different post-translational modifications in p53 that affect its preference for binding to different target genes or by activation of different additional pathways that operate in parallel to modulate the relative efficiency of p53 target gene activation.

In addition to their role in DNA damage-induced apoptosis mediated by p53, NOXA and PUMA are critical for apoptosis induced by certain p53-independent pathways. *Noxa* transcription has been shown to be directly induced by hypoxia-inducible factor (HIF)-1α, independent of p53 (Kim *et al.*, 2004) and contributes (in a p53-independent manner) to cytokine deprivation-induced apoptosis of natural killer (NK) cells (Huntington *et al.*, 2007). PUMA is essential for apoptosis induced by a range of p53-independent death stimuli, including cytokine deprivation or treatment with glucocorticoids, staurosporine (a broad spectrum kinase inhibitor) or phorbol esters (Youle and Strasser, 2008). Studies with mice lacking both PUMA and BIM showed that these two BH3-only proteins have overlapping functions in glucocorticoid-induced apoptosis of lymphoid cells (Erlacher *et al.*, 2005, 2006).

PUMA upregulation, via p53, is observed in chronic lymphocytic leukaemia (CLL) on treatment with DNA-damaging anticancer drugs (Kuroda and Taniwaki, 2008). Moreover, proteosome inhibitors kill tumour cells retaining wt p53 predominantly by induction of PUMA, whereas in myelomas, melanomas and squamous cell carcinomas (SCC) this class of drug caused apoptosis by activating NOXA via a p53-independent process (Adams and Cory, 2007). Loss of PUMA does not result in spontaneous tumour formation (Michalak *et al.*, 2008); however, Puma deficiency or even Puma knock-down can accelerate *myc*-induced B-cell lymphoma development (Garrison *et al.*, 2008; Hemann *et al.*, 2004; Michalak *et al.*, 2009). Similarly NOXA-deficient and even NOXA/PUMA double deficient mice do not develop tumours (Michalak *et al.*, 2008), but the combined loss of *noxa* and loss of one or both alleles of *puma* in concert with enforced *myc* expression is able to accelerate lymphomagenesis, including in the pre-B-cell compartment (Michalak *et al.*, 2009), although remarkably, not as potently as loss of p53. These results demonstrate that both Puma and Noxa can function as tumour suppressors, particularly in context of an oncogenic lesion that subverts cell cycle control. Importantly, these results also prove that induction of apoptosis constitutes only one of the critical mechanisms by which p53 suppresses tumour development.

The role of NOXA loss in human tumour development is not clear. Although mutations in *noxa* have been reported in some human lymphomas (Mestre-Escorihuela *et al.*, 2007), examination of a large panel of human tumours types, found only one case with a mutation in *noxa* (Labi *et al.*, 2006). In another study, specifically focusing on mutations in *puma* in head and neck squamous cell carcinoma (HNSCC) and NSCLC, no mutations were found by sequence analysis of 30 primary tumours and 10 tumour-derived cell lines. However, loss of heterozygosity at the chromosomal locus of *puma* (19q) was observed in 56% of HNSCCs and 27% of primary lung cancers (Labi *et al.*, 2006). Moreover, in human gliomas and neuroblastomas, this chromosomal locus is frequently lost (Labi *et al.*, 2006).

Downregulation of PUMA protein expression has been associated with malignancy and metastasis in primary human melanomas (Labi *et al.*, 2006); however, it is unclear whether this reduction in expression is causative or merely a consequence of malignancy. Silencing of *puma* has been demonstrated in murine lymphoma (Garrison *et al.*, 2008), suggesting that PUMA loss can be selected during tumourigenesis in multiple ways. Abnormally high expression of micro ribonucleic acids (miRNAs) is observed in many malignancies such as miR-34a in colon cancers and *Ebstein–Barr virus*-encoded miR-BART5 in nasopharyngeal and gastric carcinomas. Abundant expression of these two specific miRNAs has recently been associated with a reduction of *puma* mitochondrial RNA (mRNA) and protein levels, promoting cell survival and thus tumourigenesis (Choy *et al.*, 2008; Yamakuchi *et al.*, 2008). This reinforces the role of PUMA in tumour suppression and its role in anticancer drug resistance remains of great interest.

Bid

BID is a BH3-only protein that was found in a λ-phage expression library (from a murine T-cell hybridoma line) screen for BCL-2- and BAX-binding proteins. Although classified as a BH3-only protein because it only displays homology within the BH3 domain to other members of the Bcl-2 protein family, uniquely, BID shares structural and thus perhaps also functional similarity to BAX and BAK (Youle and Strasser, 2008). The full-length Bid protein is expressed in most tissues but has relatively poor pro-apoptotic activity. BID's killing capacity is increased upon cleavage by caspase-8, and possibly also by granzyme B (a product of cytotoxic T lymphocytes that contributes to target cell killing) into the active form, truncated BID (tBID). tBid can translocate to the mitochondrial outer membrane to interact with Bcl-2 family members to induce cytochrome *c* release (Strasser, 2005).

Mice lacking BID have normal haematopoiesis (Kaufmann *et al.*, 2007). Although $bid^{-/-}$ mice were reported to develop spontaneous myeloid hyperplasia which can progress to fatal malignancy (Zinkel *et al.*, 2003) these results could not be reproduced in another study (Kaufmann *et al.*, 2007). BID-deficient mice are resistant to Fas ligand or anti-Fas antibody-induced hepatocyte killing and fatal hepatitis. BID was also found to be critical for Fas ligand-induced killing of pancreatic β cells (McKenzie *et al.*, 2008). In contrast, lymphocytes lacking Bid are normally sensitive to Fas ligand-induced apoptosis (Kaufmann *et al.*, 2007). Therefore, BID is considered to be the link between the 'death receptor' and the 'Bcl-2-regulated' apoptotic pathways that appears to function in a cell-type-specific manner as an amplifier of the caspase cascade that leads to cell destruction.

In tumour cells, BID has been shown to contribute to the response to histone deacetylase inhibition (HDACi), proteosome inhibitors, arsenic trioxide (ATO) and cisplatin (Kuroda and Taniwaki, 2008). The role of BID in tumourigenesis is unclear, as both increases and decreases in BID protein expression have been implicated in the progression of numerous types of cancers. Although some studies show no correlation between high Bid expression and response to chemotherapy, others have reported poor metastasis-free survival and radiotherapy outcome (Labi *et al.*, 2006). As a result of these contradictory findings, the significance of BID in human cancers is still unclear.

Bim/Bod/Bcl2L11

BIM/BOD was discovered independently by screening of a λ-bacteriophage expression library from a T-cell lymphoma and a yeast two-hybrid screen of ovarian tissue for BCL-2 interacting proteins. BIM is expressed in lymphoid cells, myeloid cells, epithelial cells, both male and female germ cells and specific neuronal populations (O'Connor *et al.*, 1998; O'Reilly *et al.*, 2000). Three major (and some additional minor) BIM isoforms are produced by alternative splicing and they differ markedly in their pro-apoptotic potency: BIM_S is the most potent killer (found only at very low levels in tissues) followed by BIM_L and BIM_{EL} (the most abundant form found in all tissues) (O'Connor *et al.*, 1998; O'Reilly *et al.*, 2000). *bim* has been shown to be transcriptionally upregulated by the forkhead transcription factor, forkhead box O3A (FOXO3a), in the response of haematopoietic cells to cytokine withdrawal (Dijkers *et al.*, 2000) and by C/EBP homologous protein (CHOP) during endoplasmic reticulum (ER) stress-induced apoptosis (Puthalakath *et al.*, 2007). Although it is unclear how MYC induces BIM, abnormally high Bim expression levels have been reported in B-lymphoid cells of pre-malignant Eμ-*myc* transgenic mice (Egle *et al.*, 2004). Similar to *puma*, *bim* expression is modulated by miRNAs. The genomic region encoding the miR-17-92 cluster is often overexpressed or amplified in lymphoma and other cancers. Increased proliferation and reduced apoptosis was observed in lymphocytes with high expression of miR-17-92 and BIM protein levels were found to be suppressed in these cells (Xiao *et al.*, 2008). In addition, absence of miR-17-92 was shown to increase BIM expression and thus inhibit B-cell development (Ventura *et al.*, 2008).

BIM is also regulated by various posttranslational control mechanisms that can either enhance or diminish its pro-apoptotic activity (Puthalakath and Strasser, 2002). First, in healthy cells BIM_L and BIM_{EL} are sequestered to the microtubule-associated dynein motor complex. BIM_S, however, cannot be restrained in this way and this is thought to contribute to its enhanced killing potency (Puthalakath and Strasser, 2002). Second, phosphorylation by ERK (extracellular signal-regulated kinase) targets BIM (particularly BIM_{EL} and BIM_L) for ubiquitination, thereby priming it for proteosomal degradation (Adams and Cory, 2007). In addition, ERK-mediated phosphorylation has also been shown to reduce the pro-apoptotic potency of BIM_{EL} by lowering its affinity for MCL-1 and BCL-X_L (Ewings *et al.*, 2007). Dephosphorylation of BIM_{EL} and BIM_L by protein phosphatase 2A (PP2A) antagonises these processes and thus enhances their pro-apoptotic activity (Puthalakath *et al.*, 2007). Paradoxically, the pro-apoptotic activity of BIM was reported to be increased in neurons by c-Jun *N*-terminal kinase (JNK)-mediated phosphorylation (Putcha *et al.*, 2003) at some of the same amino acid residues that are phosphorylated by ERK, which inhibits BIM's pro-apoptotic activity. These observations may be reconciled if the two kinases have overlapping yet sufficiently different patterns of BIM phosphorylation.

Such tight regulatory controls imposed on BIM may perhaps reflect the diversity of apoptotic roles it plays in a variety of cell types. Indeed BIM is a principal regulator of homeostasis in the lymphoid and myeloid compartment (Bouillet *et al.*, 1999) and is crucial for the negative selection of autoreactive immature T- and B-lymphoid cells (Bouillet *et al.*, 2002; Enders *et al.*, 2003). It has also been shown to be an important terminator of the T- as well as B-cell immune responses (Strasser, 2005) and some contribution has been reported in mammary epithelial cell differentiation. Because of its major role in haematopoietic cell death, BIM deficiency in mice causes lymphoid and myeloid hyperplasia and on a mixed C57BL/6 × 129SV background also fatal systemic lupus erythematosus (SLE)-like autoimmune disease (Bouillet *et al.*, 1999). Cells lacking BIM, such as in lymphocytes, osteoclasts, mast cells, epithelial/endothelial cells and neurons all display resistance to numerous apoptotic stimuli, including cytokine withdrawal, ionomycin (deregulation of calcium flux), taxol and glucocorticoids but not phorbol esters (Strasser, 2005).

Interestingly, although *bim* is not a direct transcriptional target of p53, it still contributes to DNA damage-induced apoptosis (at least in certain cell types), as loss of BIM provided thymocytes and mature B- as well as T-lymphoid cells with significant, albeit relatively minor, resistance to two (normally) p53-dependent apoptotic stimuli, γ-irradiation and etoposide, both *in vivo* (Erlacher *et al.*, 2005) and *in vitro* (Bouillet *et al.*, 1999). Thus, p53 may indirectly activate *bim*, or DNA damage may activate *bim* through a p53-independent pathway that enhances p53-mediated apoptosis. Interestingly, BIM and PUMA co-operate in many cell types to mediate apoptosis triggered by a broad range of cytotoxic stimuli, including cytokine deprivation, γ-irradiation, dexamethasone, staurosporine and ER stressors. In many instances, their combined deficiency provided protection that equalled that seen with BCL-2 overexpression or combined loss of BAX and BAK (Ekoff *et al.*, 2007; Erlacher *et al.*, 2006).

The ability of BIM to function as a tumour suppressor has been demonstrated in a mouse model of renal cancer (Tan *et al.*, 2005) and in Eμ-*myc* transgenic mice in which tumourigenesis was even accelerated by loss of only a single allele of BIM (Egle *et al.*, 2004). The striking *bim* gene dosage effect is also highlighted by the complete prevention of polycystic kidney disease that is observed in BCL-2 deficient mice by loss of a single allele of *bim* (Bouillet *et al.*, 2001). Remarkably, homozygous deletions of *BIM* were identified in a substantial number of human mantle cell lymphomas (Tagawa *et al.*, 2005) and silencing of the *BIM* gene by promoter methylation and mutation were observed in some human B-cell lymphomas (Mestre-Escorihuela *et al.*, 2007). In some cases of Burkitt lymphoma hallmarked by a *t*(8;14) translocation that causes deregulated c-MYC expression; point mutations in c-*MYC* that prevent its capacity to activate BIM have been reported (Hemann *et al.*, 2005). This selection for loss of

BIM in many human cancers clearly substantiates BIM's role as a tumour suppressor. Importantly, BIM is proving to have a critical role in the efficacy of targeted anticancer drugs, in particular those of the oncogenic kinase inhibitor class (Cragg *et al.*, 2008, 2009). **See also**: Oncogenic Kinases in Cancer

Bik/Nbk/Blk

BIK/BLK/NBK was identified in a protein-interaction screen using different Bcl-2-like pro-survival family members as bait. BIK is expressed in haematopoietic and endothelial cells (Coultas *et al.*, 2004). It is predominantly located at the ER (Mathai *et al.*, 2002) and has been reported to contribute to the release of Ca^{2+} from the ER to elicit apoptosis in response to ER stress (Germain *et al.*, 2002; Mathai *et al.*, 2005) and protein synthesis inhibition (Shimazu *et al.*, 2007). Despite being widely expressed, BIK appears to play a largely redundant role in apoptotic responses induced in non-transformed primary cells by many stimuli, including cytokine withdrawal, glucocorticoids, phorbol ester, ionomycin, etoposide and B-cell receptor cross-linking (Coultas *et al.*, 2004) and may therefore have overlapping functions with other more 'potent' BH3-only proteins. Consistent with this, male mice deficient in both BIM and BIK fail to produce mature sperm cells due to the abnormal accumulation of immature progenitor cells that inhibit the differentiation process required to complete spermatogenesis (Coultas *et al.*, 2005). In contrast, BIK overexpression has been shown to increase sensitivity to drug- and death receptor-induced apoptosis of T-lymphoma cells (Daniel *et al.*, 1999) and human prostate cancer cells (Nikrad *et al.*, 2005). Moreover, BIK was shown to mediate B-cell receptor cross-linking induced apoptosis of human B-cell lymphoma (Jiang and Clark, 2001), perhaps indicating that Bik may be a more important mediator of apoptosis in transformed cells.

To date, the cellular triggers and the molecular pathways by which BIK is induced are unclear. The E1A oncogene, a potent apoptosis inducer that depends on wt p53 activity, induced *BIK* mRNA/protein in a human KB epithelial cell line (Mathai *et al.*, 2002). Of note, no direct *BIK* promoter stimulation by p53 has been detected but it remains possible that p53 can indirectly influence *BIK* gene expression or stability of *BIK* mRNA. The BIK protein is post-translationally phosphorylated to enhance its pro-apoptotic activity (Verma *et al.*, 2001) and this is counteracted by proteasomal degradation (Marshansky *et al.*, 2001).

Bik mutations have been found in some B-cell lymphomas and allelic deletion or silencing by methylation of *BIK* has been identified in renal cell carcinoma, which may contribute to drug resistance in patients (Labi *et al.*, 2006). A study of melanoma cell lines revealed elevated BIK expression compared to normal human melanocytes (Oppermann *et al.*, 2005). Thus, BIK may have cell-type-specific roles and may exert particular functions in tumour cells, but these remain to be elucidated in greater detail.

Bad

BAD was discovered in λ-phage and a yeast-two-hybrid screen as a $BCL\text{-}X_L$-binding protein. In healthy cells, BAD can be negatively regulated by phosphorylation by the protein kinase AKT, which causes its sequestration to 14-3-3 scaffold proteins (Puthalakath and Strasser, 2002). Although BAD deficient mice develop normally with normal numbers of haematopoietic cells, lymphocytes display mild resistance to cytokine deprivation but normal sensitivity to other apoptotic stimuli (Strasser, 2005). Loss of BAD was reported to cause spontaneous diffuse B-cell lymphoma (approximately 40% incidence) and to accelerate γ-irradiation-induced thymic lymphoma development (Kuroda and Taniwaki, 2008). Moreover, abnormal sequestration of BAD has been hypothesised to be (in part) responsible for the chemo-resistance observed in chronic myeloid leukaemia (CML) cells, in which the hallmark *bcr-abl* fusion protein tyrosine kinase results in constitutive P13K/AKT and MAPK (mitogen activated kinase-like protein) signalling to maintain phosphorylation and thus inactivation of Bad. Consistent with this notion, Bad cooperates with Bim in the killing of Bcr–Abl transformed murine myeloid progenitors treated with the Bcr–Abl kinase inhibitor Gleevec (Kuroda *et al.*, 2006).

Bmf

BMF was identified in a yeast-two-hybrid screen for MCL-1 interacting proteins and is expressed in many tissues. BMF is dispensable for normal embryonic development, but appears to have a role in homeostasis of lymphoid cells, as BMF-deficient mice display progressive B-lymphoid hyperplasia and accelerated thymic lymphoma development following γ-irradiation (Labi *et al.*, 2008). Lymphocytes lacking BMF were abnormally resistant to glucocorticoid and

Table 2 Current and ongoing ABT-263 clinical trials

ClinicalTrial. Gov Identifier http://clinicaltrials.gov	Design	Tumor types, patient numbers	Outcome
NCT00408811	Open label ABT-263	Relapsed/refractory leukaemia, lymphoma	Safety, MTD, PFS, ORR, OS, TTP
Phase I/II	Dose escalation		
	Day1-14, q21 days, up to continuous for 1 year or until PD	110 patients	
NCT 00445198	Open label ABT-263	Nonhaematological malignancy	Safety, DLT, MTD, pharmacokinetics
Phase I/II	Nonrandomized	Small cell lung cancer (40 patients)	Extended safety analysis
	Dose escalation		Preliminary Efficacy Study
	Day1-14, q21	76 patients (total)	
NCT 00481091	Open label ABT-263	Relapsed or refractory chronic lymphocytic leukaemia	DLT, MTD, pharmakokinetics
Phase I/II	Nonrandomized		Safety
	Dose escalation		Preliminary efficacy study
	Day1-14, q21 versus continuous dosing	72 patients	
NCT 00406809	Open label ABT-263	Lymphoma	DLT, MTD, pharmakokinetics
Phase I/II	Nonrandomized	Follicular lymphoma	Safety
	Dose escalation		Preliminary efficacy study
	Day1-14, q21 versus continuous dosing until PD	110 patients	
NCT 00788684	Open label	Lymphoproliferative disorders	DLT, MTD
Phase I	Nonrandomized	CD20 positive	PFS, RR, response duration, OS
	Dose escalation		
	ABT-263 continuous dosing until PD	24 patients	
	+ Rituximab IV weekly × 4	Not recruiting at time of publication	
NCT 00743028	Open label	Chronic lymphocytic leukaemia, leukaemias, lymphomas	Bioavailability, pharmacokinetics, safety
Phase I	Randomized		
	ABT-263 current liquid formulation versus capsule formulation	36 patients	

Notes: DLT, dose limiting toxicity; MTD, maximum tolerated dose; ORR, objective response rate; OS, overall survival; PD, progressive disease; PFS, progression-free survival and TTP, time to progression.

HDACi-induced apoptosis (Labi *et al.*, 2008). Co-operation between BIM and BMF was recently found to be crucial in glucocorticoid-induced apoptosis of ALL cells (Ploner *et al.*, 2008).

Like BIM, BMF can be negatively regulated by sequestration to the cytoskeleton; in the case of BMF, this occurs via binding to the myosin V motor complex (Puthalakath and Strasser, 2002). Three splice variants exist; *bmf-I*, *bmf-II* and *bmf-III*, of which *bmf-I* is the only isoform that contains a BH3 domain and is capable of inducing apoptosis. *Bmf-I* transcription was increased following treatment of CML cell lines with Gleevec, and thus BMF may contribute to this response in concert with BIM and BAD. The remaining isoforms, BMF-II and BMF-III did not elicit apoptosis, but instead increased colony formation when overexpressed in HeLa cells, perhaps by acting as a dominant-negative protein (Kuroda and Taniwaki, 2008). Loss of the chromosomal region that contains *bmf* (15q14/15) is associated with advanced human breast, lung and colon cancers (Labi *et al.*, 2006), but no conclusive evidence exists that loss of BMF plays a role in the development or progression of these cancers.

Hrk/DP5

HRK/DP5 was discovered in protein-interaction screens using Bcl-2 pro-survival family members as bait. HRK is widely expressed during embryogenesis, but in adults its expression is restricted to certain cell populations within the central and peripheral nervous systems (Coultas *et al.*, 2007; Imaizumi *et al.*, 1997). Although HRK has been reported to be expressed in humans in certain non-neuronal tissues, such as the pancreas and haematopoietic cells (Coultas *et al.*, 2007; Inohara *et al.*, 1997), loss of Hrk had no impact on the haematopoietic system (Coultas *et al.*, 2007). In neuronal cells, *hrk* is transcriptionally induced by nerve growth factor (NGF) deprivation and motor neurons from mice lacking HRK are protected from death induced by axotomy *in vivo* (Imaizumi *et al.*, 2004) and growth factor deprivation in culture (Coultas *et al.*, 2007; Imaizumi *et al.*, 2004). *Hrk* loss of heterozygosity has been detected in astrocytomas and glioblastomas, where a decrease in *hrk* expression correlated with an increase in tumour grade. Moreover, transcriptional repression of the *hrk* gene promoter by hypermethylation has been observed in some cases of primary central nervous system lymphomas (Labi *et al.*, 2006). Similarly, methylation and thus inactivation of the *hrk* gene transcription start site has been reported in 24% of colorectal and 17% of gastric cancers (Labi *et al.*, 2006).

BH3-Mimetic Drugs

BH3-only proteins have a crucial function in protecting cells from malignant transformation and in the response of tumour cells to anticancer therapy. Therefore, mimicking BH3-only proteins represents a promising strategy for enhancing the effects of conventional anticancer therapy (Adams and Cory, 2007; Letai, 2008).The 'BH3-mimetic' class of anticancer agents mimics BH3-only protein function to trigger apoptosis. Since different BH3-only proteins have been found to preferentially bind to particular pro-survival Bcl-2-like proteins (Chen *et al.*, 2005), there is potential for designing drugs that can target specific pro-survival Bcl-2 proteins which may minimize collateral damage to healthy cells. **See also**: Drug Discovery in Apoptosis

The small molecules, ABT-737 (Oltersdorf *et al.*, 2005), and its close relative, ABT-263 (Tse *et al.*, 2008), are the only BH3-mimetics which are proven to fit the published criteria for true BH3-mimetics (Lessene *et al.*, 2008). Structural studies have been published for binding of ABT-737 bound to BCL-XL (Lessene *et al.*, 2008). ABT-737 and ABT-263 are effective as single agents in treating certain tumour types (e.g. CLL and small cell lung cancer, SCLC), which have dependency on Bcl-2 (van Delft *et al.*, 2006). Synergistic responses have been observed between ABT-737 and conventional as well as novel cancer agents, including a number of targeted therapeutics, in particular inhibitors of the oncogenic kinases (e.g. Bcr-Abl and EGF-R) (Cragg *et al.*, 2008; Kuroda *et al.*, 2007; Cragg *et al.*, 2009), with exciting implications for the clinic. ABT-263, which is an orally available drug, is currently in early Phase I/II clinical trials in haematologic malignancies and SCLC (**Table 2**; Kuroda and Taniwaki, 2008; Cragg *et al.*, 2009).

Conclusions

BH3-only proteins are essential for initiation of appropriate cell-type- and stimulus-specific apoptosis. This pathway can be perturbed in a myriad of ways, and this has particular impact on tumourigenesis and the response of tumour cells to anti-cancer therapy. A detailed understanding of the functions of these major players in health and disease is required to develop strategies for their therapeutic manipulation. In many instances, combinatorial effects of multiple BH3-only proteins may only be unravelled by using gene-targeted mice that lack two or perhaps even more of these proteins. Although *noxa* and *puma*, as direct transcriptional targets of the tumour suppressor p53, have important roles in DNA damage-induced apoptosis, BIM appears to be central to many physiologic and pathologic processes. BID is critical for 'death receptor'-induced killing of certain, albeit not all, cell types, whereas the other BH3-only proteins appear to exert more restricted contributory functions. The activation or mimicking of these proteins is expected to improve the treatment of patients suffering from cancer or autoimmune diseases, whereas the blockade of these proteins may be beneficial in the management of certain degenerative diseases that are characterized by unwanted killing of essential cells.

References

Adams JM and Cory S (2007) The Bcl-2 apoptotic switch in cancer development and therapy. *Oncogene* **26**: 1324–1337.

Bouillet P, Cory S, Zhang L-C, Strasser A and Adams JM (2001) Degenerative disorders caused by Bcl-2 deficiency are prevented by loss of its BH3-only antagonist Bim. *Developmental Cell* **1**: 645–653.

Bouillet P, Metcalf D, Huang DCS *et al.* (1999) Proapoptotic Bcl-2 relative Bim required for certain apoptotic responses, leukocyte homeostasis, and to preclude autoimmunity. *Science* **286**: 1735–1738.

Bouillet P, Purton JF, Godfrey DI *et al.* (2002) BH3-only Bcl-2 family member Bim is required for apoptosis of autoreactive thymocytes. *Nature* **415**: 922–926.

Chen L, Willis SN, Wei A *et al.* (2005) Differential targeting of pro-survival Bcl-2 proteins by their BH3-only ligands allows complementary apoptotic function. *Molecular Cell* **17**: 393–403.

Chipuk JE and Green DR (2008) How do BCL-2 proteins induce mitochondrial outer membrane permeabilization? *Trends in Cell Biology* **18**: 157–164.

Choy EY, Siu KL, Kok KH *et al.* (2008) An *Epstein–Barr virus*-encoded microRNA targets PUMA to promote host cell survival. *Journal of Experimental Medicine* **205**: 2551–2560.

Coultas L, Bouillet P, Loveland KL *et al.* (2005) Concomitant loss of proapoptotic BH3-only Bcl-2 antagonists Bik and Bim arrests spermatogenesis. *EMBO Journal* **24**: 3963–3973.

Coultas L, Bouillet P, Stanley EG *et al.* (2004) Proapoptotic BH3-only Bcl-2 family member Bik/Blk/Nbk is expressed in hemopoietic and endothelial cells but is redundant for their programmed death. *Molecular and Cellular Biology* **24**: 1570–1581.

Coultas L, Terzano S, Thomas T *et al.* (2007) Hrk/DP5 contributes to the apoptosis of select neuronal populations but is dispensable for haematopoietic cell apoptosis. *Journal of Cell Science* **120**: 2044–2052.

Cragg MS, Harris C, Strasser A and Scott CL (2009) Synergy between ABT-737 and targeted therapies. *Nature Reviews Cancer* **9**: 321–326.

Cragg MS, Jansen ES, Cook M *et al.* (2008) Treatment of B-RAF mutant human tumor cells with a MEK inhibitor requires Bim and is enhanced by a BH3 mimetic. *Journal of Clinical Investigation* **118**: 3651–3659.

Daniel PT, Pun KT, Ritschel S *et al.* (1999) Expression of the death gene Bik/Nbk promotes sensitivity to drug-induced apoptosis in corticosteroid-resistant T-cell lymphoma and prevents tumor growth in severe combined immunodeficient mice. *Blood* **94**: 1100–1107.

van Delft MF, Wei AH, Mason KD *et al.* (2006) The BH3 mimetic ABT-737 targets selective Bcl-2 proteins and efficiently induces apoptosis via Bak/Bax if Mcl-1 is neutralized. *Cancer Cell* **10**: 389–399.

Dijkers PF, Medema RH, Lammers JJ, Koenderman L and Coffer PJ (2000) Expression of the pro-apoptotic Bcl-2 family member Bim is regulated by the forkhead transcription factor FKHR-L1. *Current Biology* **10**: 1201–1204.

Egle A, Harris AW, Bouillet P and Cory S (2004) Bim is a suppressor of Myc-induced mouse B cell leukemia. *Proceedings of the National Academy of Sciences of the USA* **101**: 6164–6169.

Ekoff M, Kaufmann T, Engstrom M *et al.* (2007) The BH3-only protein Puma plays an essential role in cytokine deprivation-induced apoptosis of mast cells. *Blood* **110**: 3209–3217.

Enders A, Bouillet P, Puthalakath H *et al.* (2003) Loss of the pro-apoptotic BH3-only Bcl-2 family member Bim inhibits BCR stimulation-induced apoptosis and deletion of autoreative B cells. *Journal of Experimental Medicine* **198**: 1119–1126.

Erlacher M, Laabi V, Manzl C *et al.* (2006) Puma cooperates with Bim, the rate-limiting BH3-only protein in cell death during lymphocyte development, in apoptosis induction. *Journal of Experimental Medicine* **203**: 2939–2951.

Erlacher M, Michalak EM, Kelly PN *et al.* (2005) BH3-only proteins Puma and Bim are rate-limiting for {gamma}-radiation and glucocorticoid-induced apoptosis of lymphoid cells *in vivo*. *Blood* **106**: 4131–4138.

Ewings KE, Hadfield-Moorhouse K, Wiggins CM *et al.* (2007) ERK1/2-dependent phosphorylation of BimEL promotes its rapid dissociation from Mcl-1 and Bcl-xL. *EMBO Journal* **26**: 2856–2867.

Garrison SP, Jeffers JR, Yang C *et al.* (2008) Selection against PUMA gene expression in Myc-driven B-cell lymphomagenesis. *Molecular and Cellular Biology* **28**: 5391–5402.

Germain M, Mathai JP and Shore GC (2002) BH-3-only BIK functions at the endoplasmic reticulum to stimulate cytochrome c release from mitochondria. *Journal of Biological Chemistry* **277**: 18053–18060.

Hemann MT, Bric A, Teruya-Feldstein J *et al.* (2005) Evasion of the p53 tumour surveillance network by tumour-derived MYC mutants. *Nature* **436**: 807–811.

Hemann MT, Zilfou JT, Zhao Z *et al.* (2004) Suppression of tumorigenesis by the p53 target PUMA. *Proceedings of the National Academy of Sciences of the USA* **101**: 9333–9338.

Huntington ND, Puthalakath H, Gunn P *et al.* (2007) Interleukin 15-mediated survival of natural killer cells is determined by interactions among Bim, Noxa and Mcl-1. *Nature Immunology* **8**: 856–863.

Imaizumi K, Benito A, Kiryu-Seo S *et al.* (2004) Critical role for DP5/Harakiri, a Bcl-2 homology domain 3-only Bcl-2 family member, in axotomy-induced neuronal cell death. *Journal of Neuroscience* **24**: 3721–3725.

Imaizumi K, Tsuda M, Imai Y *et al.* (1997) Molecular cloning of a novel polypeptide, DP5, induced during programmed neuronal death. *Journal of Biological Chemistry* **272**: 18842–18848.

Inohara N, Ding L, Chen S and Núñez G (1997) *harakiri*, a novel regulator of cell death, encodes a protein that activates apoptosis and interacts selectively with survival-promoting proteins Bcl-2 and Bcl-X_L. *EMBO Journal* **16**: 1686–1694.

Jiang A and Clark EA (2001) Involvement of Bik, a proapoptotic member of the Bcl-2 family, in surface IgM-mediated B cell apoptosis. *Journal of Immunology* **166**: 6025–6033.

Kaufmann T, Tai L, Ekert PG *et al.* (2007) The BH3-only protein Bid is dispensable for DNA damage- and replicative stress-induced apoptosis or cell-cycle arrest. *Cell* **129**: 423–433.

Kim JY, Ahn HJ, Ryu JH, Suk K and Park JH (2004) BH3-only protein Noxa is a mediator of hypoxic cell death induced by hypoxia-inducible factor 1alpha. *Journal of Experimental Medicine* **199**: 113–124.

Kuroda J, Kimura S, Strasser A *et al.* (2007) Apoptosis-based dual molecular targeting by INNO-406, a second generation Bcr-Abl inhibitor, and ABT-737, an inhibitor of antiapoptotic Bcl-2 proteins against Bcr-Abl-positive leukemia. *Cell Death and Differentiation* **14**: 1667–1677.

Kuroda J, Puthalakath H, Cragg MS *et al.* (2006) Bim and Bad mediate imatinib-induced killing of Bcr/Abl+leukemic cells, and resistance due to their loss is overcome by a BH3 mimetic. *Proceedings of the National Academy of Sciences* **103**: 14907–14912.

Kuroda J and Taniwaki M (2008) Involvement of BH3-only proteins in hematologic malignancies. *Critical Reviews in Oncology/Hematology*, doi: 10.1016/j.critrevonc.2008.10.004.

Labi V, Erlacher M, Kiessling S *et al.* (2008) Loss of the BH3-only protein Bmf impairs B cell homeostasis and accelerates gamma irradiation-induced thymic lymphoma development. *Journal of Experimental Medicine* **205**: 641–655.

Labi V, Erlacher M, Kiessling S and Villunger A (2006) BH3-only proteins in cell death initiation, malignant disease and anticancer therapy. *Cell Death and Differentiation* **13**: 1325–1328.

Lessene G, Czabotar PE and Colman PM (2008) BCL-2 family antagonists for cancer therapy. *Nature Reviews Drug Discovery* **7**: 989–1000.

Letai AG (2008) Diagnosing and exploiting cancer's addiction to blocks in apoptosis. *Nature Reviews of Cancer* **8**: 121–132.

Levine AJ (1997) p53, the cellular gatekeeper for growth and division. *Cell* **88**: 323–331.

Marshansky V, Wang X, Bertrand R *et al.* (2001) Proteasomes modulate balance among proapoptotic and antiapoptotic Bcl-2 family members and compromise functioning of the electron transport chain in leukemic cells. *Journal of Immunology* **166**: 3130–3142.

Mathai JP, Germain M, Marcellus RC and Shore GC (2002) Induction and endoplasmic reticulum location of BIK/NBK in response to apoptotic signaling by E1A and p53. *Oncogene* **21**: 2534–2544.

Mathai JP, Germain M and Shore GC (2005) BH3-only BIK regulates BAX,BAK-dependent release of Ca2+ from endoplasmic reticulum stores and mitochondrial apoptosis during stress-induced cell death. *Journal of Biological Chemistry* **280**: 23829–23836.

McKenzie MD, Carrington EM, Kaufmann T *et al.* (2008) Proapoptotic BH3-only protein Bid is essential for death receptor-induced apoptosis of pancreatic beta-cells. *Diabetes* **57**: 1284–1292.

Mestre-Escorihuela C, Rubio-Moscardo F, Richter JA *et al.* (2007) Homozygous deletions localize novel tumor suppressor genes in B-cell lymphomas. *Blood* **109**: 271–280.

Michalak EM, Jansen ES, Happo L *et al.* (2009) Puma and to a lesser extent Noxa are suppressors of Myc-induced lymphomagenesis. *Cell Death and Differentiation* **16**: 684–696.

Michalak EM, Villunger A, Adams JM and Strasser A. (2008) In several cell types the tumour suppressor p53 induces apoptosis largely via Puma but Noxa can contribute. *Cell Death and Differentiation* **15**: 1019–1029.

Naik E, Michalak EM, Villunger A, Adams JM and Strasser A (2007) UV-radiation triggers apoptosis of fibroblasts and skin keratinocytes mainly via the BH3-only protein Noxa. *Journal of Cell Biology* **176**: 415–424.

Nikrad M, Johnson T, Puthalalath H *et al.* (2005) The proteasome inhibitor bortezomib sensitizes cells to killing by death receptor ligand TRAIL via BH3-only proteins Bik and Bim. *Molecular and Cancer Therapy* **4**: 443–449.

O'Connor L, Strasser A, O'Reilly LA *et al.* (1998) Bim: a novel member of the Bcl-2 family that promotes apoptosis. *EMBO Journal* **17**: 384–395.

O'Reilly LA, Cullen L, Visvader J *et al.* (2000) The pro-apoptotic BH3-only protein Bim is expressed in hemopoietic, epithelial, neuronal and germ cells. *American Journal of Pathology* **157**: 449–461.

Oltersdorf T, Elmore SW, Shoemaker AR *et al.* (2005) An inhibitor of Bcl-2 family proteins induces regression of solid tumours. *Nature* **435**: 677–681.

Oppermann M, Geilen CC, Fecker LF *et al.* (2005) Caspase-independent induction of apoptosis in human melanoma cells by the proapoptotic Bcl-2-related protein Nbk/Bik. *Oncogene* **24**: 7369–7380.

Ploner C, Rainer J, Niederegger H *et al.* (2008) The BCL2 rheostat in glucocorticoid-induced apoptosis of acute lymphoblastic leukemia. *Leukemia* **22**: 370–377.

Putcha GV, Le S, Frank S *et al.* (2003) JNK-mediated BIM phosphorylation potentiates BAX-dependent apoptosis. *Neuron* **38**: 899–914.

Puthalakath H, O'Reilly LA, Gunn P *et al.* (2007) ER stress triggers apoptosis by activating BH3-only protein Bim. *Cell* **129**: 1337–1349.

Puthalakath H and Strasser A (2002) Keeping killers on a tight leash: transcriptional and post-translational control of the pro-apoptotic activity of BH3-only proteins. *Cell Death and Differentiation* **9**: 505–512.

Shimazu T, Degenhardt K, Nur EKA *et al.* (2007) NBK/BIK antagonizes MCL-1 and BCL-XL and activates BAK-mediated apoptosis in response to protein synthesis inhibition. *Genes & Development* **21**: 929–941.

Strasser A (2005) The role of BH3-only proteins in the immune system. *Nature Reviews of Immunology* **5**: 189–200.

Strasser A, Harris AW, Bath ML and Cory S (1990) Novel primitive lymphoid tumours induced in transgenic mice by cooperation between *myc* and *bcl*-2. *Nature* **348**: 331–333.

Strasser A, Harris AW and Cory S (1991) *Bcl-2* transgene inhibits T cell death and perturbs thymic self-censorship. *Cell* **67**: 889–899.

Strasser A, Harris AW, Jacks T and Cory S (1994) DNA damage can induce apoptosis in proliferating lymphoid cells via p53-independent mechanisms inhibitable by Bcl-2. *Cell* **79**: 329–339.

Tagawa H, Karnan S, Suzuki R *et al.* (2005) Genome-wide array-based CGH for mantle cell lymphoma: identification of homozygous deletions of the proapoptotic gene BIM. *Oncogene* **24**: 1348–1358.

Tan TT, Degenhardt K, Nelson DA *et al.* (2005) Key roles of BIM-driven apoptosis in epithelial tumors and rational chemotherapy. *Cancer Cell* **7**: 227–238.

Tse C, Shoemaker AR, Adickes J *et al.* (2008) ABT-263: a potent and orally bioavailable Bcl-2 family inhibitor. *Cancer Research* **68**: 3421–3428.

Tsujimoto Y, Cossman J, Jaffe E and Croce CM (1985) Involvement of the *bcl*-2 gene in human follicular lymphoma. *Science* **228**: 1440–1443.

Ventura A, Young AG, Winslow MM *et al.* (2008) Targeted deletion reveals essential and overlapping functions of the mIR-17 ~ 92 family of miRNA clusters. *Cell* **132**: 875–886.

Verma S, Zhao L and Chinnadurai G (2001) Phosphorylation of the pro-apoptotic protein bik. Mapping of phosphorylation sites and effect on apoptosis. *Journal of Biological Chemistry* **276**: 4671–4676.

Villunger A, Michalak EM, Coultas L *et al.* (2003) p53- and drug-induced apoptotic responses mediated by BH3-only proteins Puma and Noxa. *Science* **302**: 1036–1038.

Vogelstein B, Lane D and Levine AJ (2000) Surfing the p53 network. *Nature* **408**: 307–310.

Xiao C, Srinivasan L, Calado DP *et al.* (2008) Lymphoproliferative disease and autoimmunity in mice with increased miR-17–92 expression in lymphocytes. *Nature Immunology* **9**: 404–414.

Yamakuchi M, Ferlito M and Lowenstein CJ (2008) miR-34a repression of SIRT1 regulates apoptosis. *Proceedings of the National Academy of Sciences of the USA* **105**: 13421–13426.

Youle RJ and Strasser A (2008) The BCL-2 protein family: opposing activities that mediate cell death. *Nature Review. Molecular Cell Biology* **9**: 47–59.

Zinkel SS, Ong CC, Ferguson DO *et al.* (2003) Proapoptotic BID is required for myeloid homeostasis and tumor suppression. *Genes & Development* **17**: 229–239.

Further Reading

Boyd JM, Gallo GJ, Elangovan B *et al.* (1995) Bik, a novel death-inducing protein shares a distinct sequence motif with Bcl-2 family proteins and interacts with viral and cellular survival-promoting proteins. *Oncogene* **11**: 1921–1928.

Han J, Sabbatini P and White E (1996) Induction of apoptosis by human Nbk/Bik, a BH3-containing protein that interacts with E1B 19K. *Molecular and Cellular Biology* **16**: 5857–5864.

Hsu SY, Lin P and Hsueh AJW (1998) BOD (Bcl-2-related ovarian death gene) is an ovarian BH3 domain-containing proapoptotic Bcl-2 protein capable of dimerization with diverse antiapoptotic Bcl-2 members. *Molecular Endocrinology* **12**: 1432–1440.

Nakano K and Vousden KH (2001) *PUMA*, a novel proapoptotic gene, is induced by p53. *Molecular Cell* **7**: 683–694.

Oda E, Ohki R, Murasawa H *et al.* (2000) Noxa, a BH3-only member of the bcl-2 family and candidate mediator of p53-induced apoptosis. *Science* **288**: 1053–1058.

Puthalakath H, Villunger A, O'Reilly LA *et al.* (2001) Bmf: a pro-apoptotic BH3-only protein regulated by interaction with the myosin V actin motor complex, activated by anoikis. *Science* **293**: 1829–1832.

Wang K, Yin X- M, Chao DT, Milliman CL and Korsmeyer SJ (1996) BID: a novel BH3 domain-only death agonist. *Genes & Development* **10**: 2859–2869.

Yang E, Zha J, Jockel J *et al.* (1995) Bad, a heterodimeric partner for Bcl-x_L and Bcl-2, displaces Bax and promotes cell death. *Cell* **80**: 285–291.

Yu J, Zhang L, Hwang PM, Kinzler KW and Vogelstein B (2001) PUMA induces the rapid apoptosis of colorectal cancer cells. *Molecular Cell* **7**: 673–682.

Mitochondrial Outer Membrane Permeabilization

Melissa J Parsons, *St. Jude Children's Research Hospital, Memphis, Tennessee, USA*
Douglas R Green, *St. Jude Children's Research Hospital, Memphis, Tennessee, USA*

Advanced article

Article Contents

- Introduction
- Intrinsic Apoptotic Signalling
- The Bcl-2 Family Controls MOMP
- There are Two Models to Explain the Induction of MOMP
- The Role of Mitochondrial Fission and Fusion in MOMP
- The Role of Mitochondrial Bioenergetics in MOMP
- MOMP Controlled from the Inside
- Summary

Defects in apoptosis have dire consequences and can contribute to the development of autoimmune disorders, neurological diseases and cancer. At the centre of the intrinsic apoptotic signalling pathway lies the mitochondrion, which is not only the bioenergetic centre of the cell, but also the cell's reservoir of pro-death factors such as cytochrome *c*. These proteins reside in the mitochondrial intermembrane space (IMS), and their release from the IMS induces a signalling cascade that leads to the demise of the cell. The critical event governing the release of pro-apoptotic molecules from the IMS is mitochondrial outer membrane permiabilization (MOMP). Although two models exist to explain the execution of MOMP, both models incorporate the widely accepted idea that MOMP is achieved through the coordinated actions of pro- and antiapoptotic members of the Bcl-2 (B-cell lymphoma) family of proteins. In addition, non-Bcl-2 family proteins, mitochondrial dynamics and mitochondrial bioenergetics are also involved in MOMP.

Introduction

The term 'cell death' can have multiple meanings, and there is more than one way for a cell to die. The manner in which the cell dies, however, can take many forms, which have distinct associated morphological characteristics, physiological roles and cellular mechanisms (Kroemer *et al.*, 2009). Apoptosis is one such form of cell death, and plays a very important role in development and tissue homeostasis (i.e. the balance between cell proliferation and cell death). A disruption in tissue homeostasis contributes to many human diseases. For example, aberrant apoptosis can lead to neurodegenerative diseases such as Alzheimer and Huntington disease, whereas blocks in apoptosis can contribute to carcinogenesis by allowing tissues and organs to grow uncontrollably and ignore normal antigrowth signals (Hanahan and Weinberg, 2000). **See also**: Apoptosis: Regulatory Genes and Disease

The physiologic stimuli and morphologic features of apoptosis are very different than those of other forms of cell death. For example, one form of cell death termed 'necrosis' is typically caused by acute injury or infection (Hanahan and Weinberg, 2000; Kroemer *et al.*, 2009). Necrosis is characterized by swelling and bursting of the cell, resulting in the release of cellular contents onto neighbouring cells and a subsequent inflammatory response. Apoptosis, however, is a process resulting in cell death due to the activation of a genetic programme that causes cells to lose viability before they lose membrane integrity (Kroemer *et al.*, 2009). The morphologic features of apoptosis include chromatin condensation and deoxyribonucleic acid (DNA) fragmentation, collapse of the cytoskeleton, cell shrinkage and membrane blebbing (Kroemer *et al.*, 2009). As opposed to necrosis, which is messy and chaotic, apoptosis is a neat and orderly signalling cascade.

Caspases are the central machinery of apoptosis

Molecularly, apoptosis is carried out by caspases, which are cysteine-dependent aspartate-directed proteases (Riedl and Shi, 2004). Caspases are produced in the cell as zymogens, which are inactive in the cytoplasm until they are cleaved at specific aspartate residues (Riedl and Shi, 2004). Of the 14 mammalian caspase proteins that have been identified, only 7 are thought to function primarily in the execution of apoptosis, whereas the rest have roles in immune responses or have unknown functions (Riedl and Shi, 2004). Apoptotic caspases have been classified into one of the two main groups, initiator caspases and effector caspases. Initiator caspases are upstream of effector caspases and respond to apoptotic stimuli following particular molecular pro-death cues. Effector caspases are activated on by upstream initiator caspases and are responsible for proteolytic cleavage of 'death substrates', which gives rise to the morphological features of apoptosis. Once effector caspases are activated, they execute apoptosis through proteolytic cleavage of a number of structural and regulatory proteins within the cell. These proteins include structural proteins, such as actin and nuclear lamin, regulatory proteins, such as p21 and Rb and proteins involved in DNA metabolism and repair, such as poly ADP ribose polymerase (PARP), among many others (Earnshaw *et al.*, 1999). Cleavage of these proteins results in the physical and morphological manifestations of apoptosis. **See also**: Caspases and Cell Death; Caspases, Substrates and Sequential Activation; Dismantling the Apoptotic Cell; The Apoptosome: The Executioner of Mitochondria-mediated Apoptosis

Apoptosis can be broadly defined by two main pathways

At the molecular level, apoptosis occurs through two main pathways, the extrinsic, or death receptor-mediated pathway, and the intrinsic, or mitochondrial-mediated pathway. Intrinsic apoptotic signalling occurs in response to stimuli such as DNA damage, growth-factor withdrawal and exposure to certain chemotherapeutic agents, all of which result in the release of cytochrome *c* and other pro-death factors from the intermembrane space (IMS) of the mitochondria and subsequent downstream signalling through the initiator caspase-9. The extrinsic pathway, as its name implies, occurs in response to external signals. This pathway is also referred to as the 'death receptor pathway', as it is mediated by ligation of cell-surface death receptors to their cognate ligands. Ligation of death receptors induces subsequent downstream signalling through the initiator caspase-8. Activation of either intrinsic or extrinsic apoptotic signalling results in execution of apoptosis, although the particular details of each pathway are quite different. Although both pathways are physiologically important, the focus here will be on intrinsic apoptotic signalling, the defining feature of which is mitochondrial outer membrane permeabilization, or 'MOMP'. **See also**: Mitochondria Fusion and Fission

Intrinsic Apoptotic Signalling

Intrinsic apoptotic signalling occurs in response to internal insults. Following such insults, permeabilization of the outer mitochondrial membrane (OMM) occurs, which allows for release of pro-death factors such as cytochrome *c* into the cytosol. It should be noted that MOMP, which results from a signalling programme that leads to the controlled formation of large pores in the OMM and subsequent release of pro-death molecules from the mitochondrial IMS, is distinct from mitochondrial membrane permeability transition, which refers to an increase in the permeability of the mitochondrial membrane due to the opening of permeability transition pores (see later discussion). There are approximately 16 proteins released from the IMS during MOMP; some of these proteins function to promote cell death, whereas others are considered 'innocent bystanders' or play an unknown role in apoptosis (Saelens *et al.*, 2004). In general, the pro-death factors that are released into the cytosol promote cell death by promoting caspase activation. For example, in the presence of adenosine triphosphate (ATP), cytochrome *c* forms a complex with apoptosis-activating factor 1 (Apaf-1), which allows for recruitment of the initiator caspase procaspase-9 in a 1:1 molar ratio (Zou *et al.*, 1999). This complex of Apaf-1, procaspase-9 and cytochrome *c* is known as the 'apoptosome', a multimeric complex that allows for activation of caspase-9 through induced proximity (Zou *et al.*, 1999). Following its activation in the apoptosome, active caspase-9 is released, where it goes on to cleave downstream effector caspases such as caspase-3 and caspase-7 (Zou *et al.*, 1999). Another pro-apoptotic protein released on MOMP is Smac/DIABLO (second mitochondrial activator of caspases/direct IAP biding protein with low pI) which functions to promote apoptosis by neutralizing inhibitor of apoptosis proteins (IAPs) following its release into the cytosol, thereby liberating executioner caspases, which allows for their activation by caspase-9 (Du *et al.*, 2000). Since executioner caspase activation occurs following MOMP and the release of pro-death factors from the IMS, and this leads to dismantling of structural and regulatory proteins within the cell (as discussed in the section on Introduction), MOMP is often considered the 'point of no return' for a dying cell. However, even in situations where caspase activation is blocked, the cell will eventually undergo a caspase-independent form of cell death following MOMP, as its bioenergetic network (i.e. mitochondria) is functionally disrupted during the process of permeabilizing the OMM (discussed later). Thus, the factors that control MOMP quite literally control the life and death of the cell. **See also**: The Apoptosome: The Executioner of Mitochondria-mediated Apoptosis

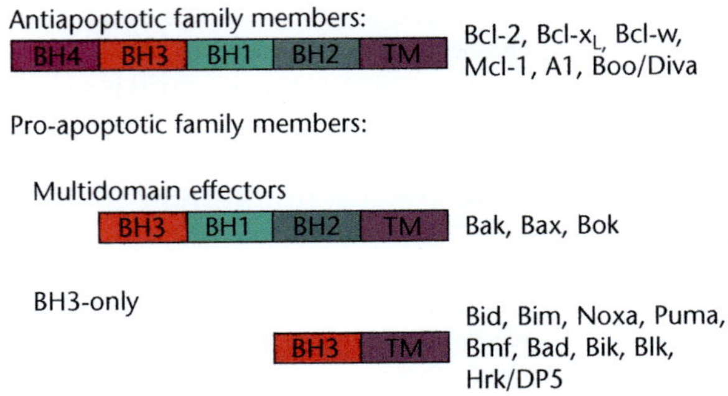

Figure 1 The Bcl-2 family. The Bcl-2 protein family is broken into antiapoptotic family members, which contain four BH homology domains, and pro-apoptotic family members, which contain either three BH homology domains (multidomain effectors) or only the third BH domain (BH3-only). Illustrated are cartoon schematics of the different groups, and a list of the members of each group. Most members have a C-terminal hydrophobic domain (TM) that aids association with the mitochondrial membrane. Exceptions are A1 and many of the BH3-only proteins (Bid, Bad, Bmf, Noxa and Puma).

The Bcl-2 Family Controls MOMP

The Bcl-2 (B-cell lymphoma) family of proteins is central to the regulation of MOMP. The founding member of this family is a potent antiapoptotic molecule that blocks MOMP. A large number of proteins structurally related to Bcl-2 have been identified that either negatively or positively influence mitochondrial release of cytochrome *c*. These proteins thus belong to the 'Bcl-2 family' (**Figure 1**). The structural similarity these proteins share is in the form of one or more Bcl-2 homology (BH) domains, and the proteins themselves can be separated into three functional groups based on the number of BH domains they contain. The antiapoptotic members of the Bcl-2 family (Bcl-2, Bcl-x_L, Bcl-w, A1 and Mcl-1) contain four BH domains and serve to inhibit MOMP and cytochrome *c* release by binding to and sequestering pro-apoptotic family members (Cory and Adams, 2002). The two other functional groups are pro-apoptotic and are classified as multidomain pro-apoptotic proteins (Bax, Bak and Bok) because they contain BH domains 1–3, or BH3-only proteins (Bim, Bid, Bik, Bad, Bmf, Hrk, Noxa and Puma) because they contain only the third BH domain (Cory and Adams, 2002; Reed, 2006). **See also**: The Bcl-2 Family Proteins - Key Regulators and Effectors of Apoptosis

It is well established that multidomain Bcl-2 proteins are required for MOMP. Although the role of Bok in MOMP and apoptosis remains a mystery (Reed, 2006), the other two multidomain proteins Bak and Bax are absolutely necessary for MOMP and mitochondrial-mediated apoptosis, and are therefore referred to as pro-apoptotic effectors (Knudson *et al.*, 1995; Lindsten *et al.*, 2000; Reed, 2006; Wei *et al.*, 2001). Though $bak^{-/-}$ or $bax^{-/-}$ mice appear developmentally and phenotypically normal (although $bax^{-/-}$ mice do have minor defects relating to spermatogenesis and slightly elevated sympathetic neurons, motor neurons and lymphocytes), $bak^{-/-}$ $bax^{-/-}$ mice die embryonically due to an inability to complete the apoptotic processes that are required to remove excess cells during development (Knudson *et al.*, 1995; Lindsten *et al.*, 2000; Wei *et al.*, 2001). Furthermore, cells from these mice are resistant to a variety of apoptotic stimuli known to rely on the intrinsic pathway. These data illustrate that Bak and Bax are functionally redundant, and necessary for MOMP and intrinsic apoptosis.

In the cell, Bak is thought to be constitutively associated with the OMM, whereas Bax is generally cytosolic and translocates to the OMM during apoptosis (Reed, 2006). Following an apoptotic stimulus, Bak and Bax become activated through conformational changes that lead to the formation of homo-oligomers, which are thought to insert into the OMM, thereby inducing its permeabilization. These homo-oligomers presumably form pores in the OMM which allow for the release of apoptogenic molecules from the mitochondrial IMS, thereby engaging cytoplasmic cell death machinery and apoptosis. Thus, a critical step in the control of apoptosis is the activation of Bak and Bax.

The BH3-only proteins are also very important for MOMP. These proteins do not directly permeabilize the OMM, but rather are thought to neutralize the antiapoptotic proteins and/or activate Bak and Bax (discussed later). These proteins are regulated in a variety of ways. Following an apoptotic stimulus, BH3-only proteins can have unique subcellular localizations (i.e. Bim and Bmf), can be posttranslationally modified (i.e. Bad) or can be transcriptionally upregulated (i.e. Puma and Noxa) (Nakano and Vousden, 2001; Oda *et al.*, 2000; Puthalakath *et al.*, 2001; Zha *et al.*, 1996). The complexity of Bcl-2 family protein regulation reflects the importance of inducing MOMP at precisely the correct time to prevent having too much or too little MOMP. **See also**: BH3-Only Proteins

The regulation of Bcl-2 family proteins is further complicated by the specific binding patterns seen among family members. A series of elegant peptide studies (Chipuk *et al.*, 2008) using the BH3 domains of the different pro- and antiapoptotic family members illustrated that certain members preferentially interact with other members (**Figure 2**). For example, Noxa is only capable of interacting with myeloid cell leukemia sequence 1 (Mcl-1) and A1. Bad, however, interacts only with Bcl-2, Bcl-x_L and Bcl-w, whereas Bid and Bim can interact with all of the antiapoptotic Bcl-2 proteins. The preferential interactions between family members become very important when we consider the two competing models for Bak and Bax activation (discussed later).

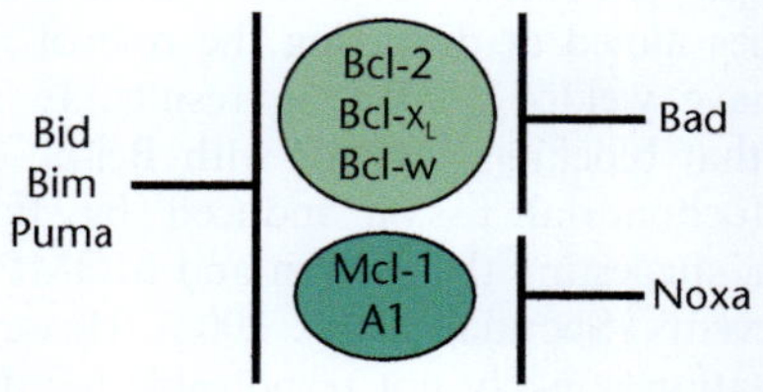

Figure 2 Bcl-2 family protein interactions. The BH3-only proteins Bid, Bim and Puma can neutralize all antiapoptotic Bcl-2 proteins. BH3-only proteins such as Bad can neutralize only Bcl-2, Bcl-x_L and Bcl-w, whereas Noxa can only neutralize Mcl-1 and A1.

There are Two Models to Explain the Induction of MOMP

How Bak and Bax are activated during apoptotic signalling is controversial, and currently, there are two models to explain the regulation of Bak and Bax, which may not be mutually exclusive (**Figure 3**; Chipuk *et al.*, 2008). The direct activator model suggests that Bak and Bax are inactive until they are directly engaged by certain BH3-only proteins. This direct binding induces a conformational change that allows for oligomerization and insertion into the OMM, thereby leading to MOMP. Only a subset of the BH3-only proteins, mainly Bid and Bim, are thought to contain direct

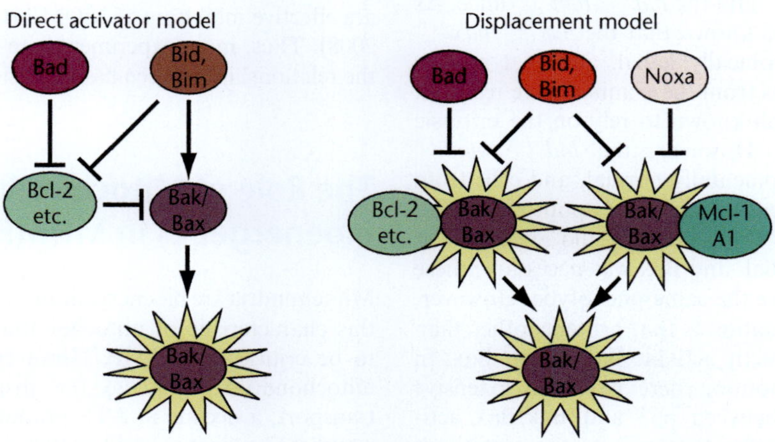

Figure 3 Two competing models for Bak/Bax activation. *Left*, the direct activator model states that Bid and/or Bim are responsible for directly activating Bak and/or Bax (indicated by the yellow star) to induce MOMP, whereas the other BH3-only proteins (such as Bad) function to promote Bak/Bax activation by neutralizing the antiapoptotic family members. *Right*, the displacement model states that Bak and Bax are constitutively in an active conformation, and that sequestration by antiapoptotic proteins is the main mechanism by which Bak/Bax oligomerization and MOMP are prevented. Neutralization of the antiapoptotic proteins by BH3-only proteins releases active Bak and Bax to induce MOMP.

activator activity. The main evidence for this comes from studies with peptides and purified proteins, which illustrate that Bid and Bim, but not the other BH3-only proteins, can induce conformational changes in Bak and Bax and form pores in liposomes (which recapitulate the lipid environment of an OMM) in the absence of other mitochondrial or Bcl-2 family proteins (Chipuk *et al.*, 2008). Similar studies have also demonstrated that the other BH3-only proteins (i.e. Bad, Bmf and Noxa) are not capable of directly activating Bak or Bax, but are capable of neutralizing antiapoptotic Bcl-2 proteins that have sequestered pro-apoptotic proteins (Chipuk *et al.*, 2008). Therefore, in the direct activator model, Bid and Bim are responsible for directly activating Bak and/or Bax, whereas the other BH3-only proteins function to promote Bak/Bax activation by neutralizing the antiapoptotic family members. (Chipuk *et al.*, 2008).

The displacement model, however, suggests that Bak and Bax do not need to be directly engaged by BH3-only proteins but rather are constitutively in an active conformation. In this model, sequestration by antiapoptotic proteins is the main mechanism by which Bak/Bax oligomerization and MOMP are prevented. Neutralization of the antiapoptotic proteins by BH3-only proteins releases active Bak and Bax, which is how MOMP occurs. In this model, Bid and Bim are the most potent inducers of MOMP, because they have the capacity to interact with all of the antiapoptotic Bcl-2 proteins. One of the main arguments for the displacement model and against the direct activator model surrounds the phenotypes of the $bak^{-/-}$ $bax^{-/-}$ and the $bid^{-/-}$ $bim^{-/-}$ mice. As discussed earlier, it is known that the $bak^{-/-}$ $bax^{-/-}$ phenotype is embryonically lethal, and that murine embryonic fibroblasts from these animals are resistant to a variety of stimuli known to rely on the intrinsic apoptotic pathway. However, the $bid^{-/-}$ $bim^{-/-}$ phenotype is developmentally normal, and cells from these mice remain sensitive to some apoptotic stimuli (Willis *et al.*, 2007). This finding would argue that if direct activation of Bak and Bax was necessary, these two mice would share the same phenotype. However, an alternate interpretation is that proteins other than Bid and Bim can directly activate Bak and/or Bax. In keeping with this notion, there has been extensive work on the role between p53 and Bak/Bax activation, which concludes that p53 in the cytoplasm contains direct activator activity (Chipuk *et al.*, 2004). It is likely that, in addition to p53, other unidentified non-Bcl-2 proteins or conditions exist to modulate the Bcl-2 family and MOMP.

The Role of Mitochondrial Fission and Fusion in MOMP

The regulation of MOMP is further impacted by mitochondrial dynamics. Although we typically refer to mitochondria as a single organelle, in cells, a single mitochondrion is actually part of a large tubular network of many mitochondria. This network routinely undergoes fission and fusion to either shorten or elongate the network depending on the energetic needs of the cell. Dynamin-related GTPases (guanosine triphosphatases) are responsible for regulating fission and fusion of mitochondria, with dynamin-related protein 1 (Drp1) and human fission 1 (hFis1) playing a major role in fission, and Mitofusins 1 and 2 regulating mitochondrial fusion (Chen *et al.*, 2003; Smirnova *et al.*, 2001; Yoon *et al.*, 2003). Coincident with MOMP, mitochondria undergo extensive fission and this has been thought to play an important role in the process of MOMP. However, studies aimed at dissecting the role of fission in MOMP have yielded conflicting results. It has been observed that blocking MOMP with Bcl-x_L does not block mitochondrial fission induced by Bax overexpression, suggesting that fission and MOMP are two separate events (Sheridan *et al.*, 2008). However, Bax oligomerization is likely not responsible for fission, as *depleting* Bak and Bax from cells can induce fission. In this particular study, and contrary to the findings of Sheridan *et al.*, Bak and Bax seem capable of promoting mitochondrial *fusion* through activating mitofusin 2 (Mfn2) (Karbowski *et al.*, 2006). In addition to these studies, it has also been shown that novel inhibitors of Drp1 are effective inhibitors of MOMP (Cassidy-Stone *et al.*, 2008). Thus, more experiments are needed to delineate the relationship between fission, fusion and MOMP.

The Role of Mitochondrial Bioenergetics in MOMP

Mitochondria are bioenergetically active organelles, and this characteristic of mitochondria is widely assumed to be critical in MOMP. However, overt changes in mitochondrial physiology (i.e. disruption of electron transport, a decline in ATP production, loss of transmembrane potential and reactive oxidative species production) occur only following MOMP, and are at least partially dependent on caspases (Ricci *et al.*, 2004). The major cause of changes in mitochondrial physiology involves cleavage of NADH dehydrogenase

 Cell Death © 2010, John Wiley & Sons, Ltd.

(ubiquinone) FE-S protein 1 (NDUFS1), the 75-kDa subunit of respiratory complex I, by active caspases following MOMP (Ricci *et al.*, 2004). The resultant reactive oxygen species (ROS) production by permeabilized mitochondria impacts on immunological effects by the dying cell (Kazama *et al.*, 2008). Thus, bioenergetics more affects the cellular response to MOMP, as opposed to contributing to the initiation or execution of MOMP.

MOMP Controlled from the Inside

It is becoming clear that not only fission/fusion of the mitochondria can impact MOMP, but also events within the mitochondria themselves can have an effect as well. The inner mitochondrial membrane involutes into the matrix; the involution points are referred to as mitochondrial cristae junctions. Remodelling of cristae junctions appears to have some impact on MOMP. A recent study (Yamaguchi *et al.*, 2008) examined the disassembly of protein complexes of optic atrophy 1 (Opa1), a protein that is located at cristae junctions, and found that Bid and Bim, which can gain access to the inner membrane during MOMP, appear to perturb Opa1. This causes changes in the organization of the cristae, which can facilitate mobilization of cytochrome *c* from the IMS, through the pores in the OMM, and into the cytosol. In another study (Chao *et al.*, 2008), genetic ablation of HCLS1 associated protein X 1 (Hax1), HtrA2/Omi (Htra serine peptidase 2) and presenilin associated, rhomboid-like (Parl) which are IMS proteins, were compared. HtrA2/Omi and Parl are proteases, whereas Hax1 contains two BH domains and may function as an antiapoptotic protein (Chao *et al.*, 2008). All three knockout animals displayed identical neurodegenerative phenotypes. Further biochemical investigation showed that Hax1 interacts with Parl to process HtrA2/Omi to function in a manner required for the survival of T lymphocytes and some neurons, potentially explaining the similarity in the phenotypes of the knockout mice. Thus it seems that factors within the mitochondria may be playing an important role in the response to signals originating from outside the mitochondria.

Mitochondrial permeability transition (MPT), which describes an increase in the permeability of the mitochondrial membrane to small molecules (i.e. less than 1.5 kDa), is also considered by some to influence MOMP (Tsujimoto and Shimizu, 2007). MPT is caused by the opening of permeability transition pores, which are protein pores that form where the inner and OMM meet (Tsujimoto and Shimizu, 2007). The nature of these pores is unknown, although they likely consist of the proteins adenine nucleotide translocase (ANT), the mitochondrial inner membrane protein transporter (Tim), the protein transporter at the outer membrane (Tom), the outer membrane voltage-dependent anion channel (VDAC) and cyclophilin D (Tsujimoto and Shimizu, 2007). Of these, only cyclophilin D is generally agreed to be important for MPT. A number of factors, including high calcium and ROS, can cause MPT, which can cause mitochondrial matrix swelling and MOMP. However, the notion that MPT is a causal event in MOMP is controversial. For example, defective MPT in cyclophilin D knockout mice has no effect on MOMP in development or in response to BH3 proteins; however, these animals show defects in ischaemia-/reperfusion-induced necrosis (Tsujimoto and Shimizu, 2007). Therefore, while it is likely that MPT plays some role in mitochondrial-mediated cell death events, the precise role and physiologic conditions remain obscure.

Summary

Mitochondria are highly dynamic organelles that are integral to the life and death of the cell. Specifically, permeabilization of the OMM is induced in response to, and is necessary for, apoptotic cell death. The Bcl-2 family of proteins, as well as non-Bcl-2 proteins on the inside and the outside of the mitochondria, all play a role in promoting or inhibiting MOMP. It is lucky for us, then, that this complex process is tightly regulated by the multitude of signalling pathways described earlier.

References

Cassidy-Stone A, Chipuk JE, Ingerman E *et al.* (2008) Chemical inhibition of the mitochondrial division dynamin reveals its role in Bax/Bak-dependent mitochondrial outer membrane permeabilization. *Developmental Cell* **14**(2): 193–204.

Chao JR, Parganas E, Boyd K *et al.* (2008) Hax1-mediated processing of HtrA2 by Parl allows survival of lymphocytes and neurons. *Nature* **452**(7183): 98–102.

Chen H, Detmer SA, Ewald AJ *et al.* (2003) Mitofusins Mfn1 and Mfn2 coordinately regulate mitochondrial fusion and are essential for embryonic development. *Journal of Cell Biology* **160**(2): 189–200.

Chipuk JE, Fisher JC, Dillon CP *et al.* (2008) Mechanism of apoptosis induction by inhibition of the anti-apoptotic

BCL-2 proteins. *Proceedings of the National Academy of Sciences of the USA* **105**(51): 20327–20332.

Chipuk JE, Kuwana T, Bouchier-Hayes L *et al.* (2004) Direct activation of Bax by p53 mediates mitochondrial membrane permeabilization and apoptosis. *Science* **303**(5660): 1010–1014.

Cory S and Adams JM (2002) The Bcl2 family: regulators of the cellular life-or-death switch. *Nature Reviews. Cancer* **2**(9): 647–656.

Du C, Fang M, Li Y, Li L and Wang X (2000) Smac, a mitochondrial protein that promotes cytochrome *c*-dependent caspase activation by eliminating IAP inhibition. *Cell* **102**(1): 33–42.

Earnshaw WC, Martins LM and Kaufmann SH (1999) Mammalian caspases: structure, activation, substrates, and functions during apoptosis. *Annual Review of Biochemistry* **68**: 383–424.

Hanahan D and Weinberg RA (2000) The hallmarks of cancer. *Cell* **100**(1): 57–70.

Karbowski M, Norris KL, Cleland MM, Jeong SY and Youle RJ (2006) Role of Bax and Bak in mitochondrial morphogenesis. *Nature* **443**(7112): 658–662.

Kazama H, Ricci JE, Herndon JM *et al.* (2008) Induction of immunological tolerance by apoptotic cells requires caspase-dependent oxidation of high-mobility group box-1 protein. *Immunity* **29**(1): 21–32.

Knudson CM, Tung KS, Tourtellotte WG, Brown GA and Korsmeyer SJ (1995) Bax-deficient mice with lymphoid hyperplasia and male germ cell death. *Science* **270**(5233): 96–99.

Kroemer G, Galluzzi L, Vandenabeele P *et al.* (2009) Classification of cell death: recommendations of the Nomenclature Committee on Cell Death 2009. *Cell Death and Differentiation* **16**(1): 3–11.

Lindsten T, Ross AJ, King A *et al.* (2000) The combined functions of proapoptotic Bcl-2 family members bak and bax are essential for normal development of multiple tissues. *Molecular Cell* **6**(6): 1389–1399.

Nakano K and Vousden KH (2001) PUMA, a novel proapoptotic gene, is induced by p53. *Molecular Cell* **7**(3): 683–694.

Oda E, Ohki R, Murasawa H *et al.* (2000) Noxa, a BH3-only member of the Bcl-2 family and candidate mediator of p53-induced apoptosis. *Science* **288**(5468): 1053–1058.

Puthalakath H, Villunger A, O'Reilly LA *et al.* (2001) Bmf: a proapoptotic BH3-only protein regulated by interaction with the myosin V actin motor complex, activated by anoikis. *Science* **293**(5536): 1829–1832.

Reed JC (2006) Proapoptotic multidomain Bcl-2/Bax-family proteins: mechanisms, physiological roles, and therapeutic opportunities. *Cell Death and Differentiation* **13**(8): 1378–1386.

Ricci JE, Munoz-Pinedo C, Fitzgerald P *et al.* (2004) Disruption of mitochondrial function during apoptosis is mediated by caspase cleavage of the p75 subunit of complex I of the electron transport chain. *Cell* **117**(6): 773–786.

Riedl SJ and Shi Y (2004) Molecular mechanisms of caspase regulation during apoptosis. *Nature Reviews. Molecular Cell Biology* **5**(11): 897–907.

Saelens X, Festjens N, Vande Walle L *et al.* (2004) Toxic proteins released from mitochondria in cell death. *Oncogene* **23**(16): 2861–2874.

Sheridan C, Delivani P, Cullen SP and Martin SJ (2008) Bax- or Bak-induced mitochondrial fission can be uncoupled from cytochrome *c* release. *Molecular Cell* **31**(4): 570–585.

Smirnova E, Griparic L, Shurland DL and van der Bliek AM (2001) Dynamin-related protein Drp1 is required for mitochondrial division in mammalian cells. *Molecular Biology of the Cell* **12**(8): 2245–2256.

Tsujimoto Y and Shimizu S (2007) Role of the mitochondrial membrane permeability transition in cell death. *Apoptosis* **12**(5): 835–840.

Wei MC, Zong WX, Cheng EH *et al.* (2001) Proapoptotic BAX and BAK: a requisite gateway to mitochondrial dysfunction and death. *Science* **292**(5517): 727–730.

Willis SN, Fletcher JI, Kaufmann T *et al.* (2007) Apoptosis initiated when BH3 ligands engage multiple Bcl-2 homologs, not Bax or Bak. *Science* **315**(5813): 856–859.

Yamaguchi R, Lartigue L, Perkins G *et al.* (2008) Opa1-mediated cristae opening is Bax/Bak and BH3 dependent, required for apoptosis, and independent of Bak oligomerization. *Molecular Cell* **31**(4): 557–569.

Yoon Y, Krueger EW, Oswald BJ and McNiven MA (2003) The mitochondrial protein hFis1 regulates mitochondrial fission in mammalian cells through an interaction with the dynamin-like protein DLP1. *Molecular and Cellular Biology* **23**(15): 5409–5420.

Zha J, Harada H, Yang E, Jockel J and Korsmeyer SJ (1996) Serine phosphorylation of death agonist BAD in response to survival factor results in binding to 14-3-3 not BCL-X(L). *Cell* **87**(4): 619–628.

Zou H, Li Y, Liu X and Wang X (1999) An APAF-1.cytochrome *c* multimeric complex is a functional apoptosome that activates procaspase-9. *Journal of Biological Chemistry* **274**(17): 11549–11556.

Further Reading

Cartron PF, Gallenne T, Bougras G *et al.* (2004) The first alpha helix of Bax plays a necessary role in its ligand-induced activation by the BH3-only proteins Bid and PUMA. *Molecular Cell* **16**(5): 807–818.

Chen L, Willis SN, Wei A *et al.* (2005) Differential targeting of prosurvival Bcl-2 proteins by their BH3-only ligands allows complementary apoptotic function. *Molecular Cell* **17**(3): 393–403.

Chipuk JE and Green DR (2008) How do BCL-2 proteins induce mitochondrial outer membrane permeabilization? *Trends in Cell Biology* **18**(4): 157–164.

Cohen GM (1997) Caspases: the executioners of apoptosis. *Biochemical Journal* **326**(part 1): 1–16.

Fesik SW (2005) Promoting apoptosis as a strategy for cancer drug discovery. *Nature Reviews. Cancer* **5**(11): 876–885.

Kerr JF, Wyllie AH and Currie AR (1972) Apoptosis: a basic biological phenomenon with wide-ranging implications in tissue kinetics. *British Journal of Cancer* **26**(4): 239–257.

Kim H, Rafiuddin-Shah M, Tu HC *et al.* (2006) Hierarchical regulation of mitochondrion-dependent apoptosis by BCL-2 subfamilies. *Nature Cell Biology* **8**(12): 1348–1358.

Kuwana T, Bouchier-Hayes L, Chipuk JE *et al.* (2005) BH3 domains of BH3-only proteins differentially regulate Bax-mediated mitochondrial membrane permeabilization both directly and indirectly. *Molecular Cell* **17**(4): 525–535.

Letai AG (2008) Diagnosing and exploiting cancer's addiction to blocks in apoptosis. *Nature Reviews. Cancer* **8**(2): 121–132.

Letai A, Bassik MC, Walensky LD *et al.* (2002) Distinct BH3 domains either sensitize or activate mitochondrial apoptosis, serving as prototype cancer therapeutics. *Cancer Cell* **2**(3): 183–192.

Mitochondria Fusion and Fission

Giovanni Benard, *University of Maryland Biotechnology Institute, Baltimore, Maryland, USA*

Guihong Peng, *University of Maryland Biotechnology Institute, Baltimore, Maryland, USA*

Mariusz Karbowski, *University of Maryland Biotechnology Institute, Baltimore, Maryland, USA*

Mitochondrial structural dynamics is regulated by the fusion or fission of these organelles. Recently published evidence indicates the vital role of mitochondrial fusion and fission in cellular physiology, including progression of apoptosis. These reports indicate that in addition to intimate link between mitochondrial morphogenesis machineries and regulation of mitochondrial steps in apoptosis, certain proteins vital for the regulation of mitochondrial steps in apoptosis can also regulate mitochondrial fusion and fission in healthy cells. In this article, we focus on the regulation of mitochondrial network dynamics. The emerging evidence indicating that proteins implicated in mitochondrial network dynamics are vital for the mitochondrial steps in apoptosis is presented here, as well. Furthermore, the data demonstrating an unexpected role for the B-cell lymphoma (Bcl)-2 family members in the regulation of mitochondrial morphogenesis are also discussed.

Advanced article

Article Contents

- Introduction
- Regulation of Mitochondrial Fusion and Fission
- Cellular role of Mitochondrial Network Organization: An Overview
- Role of Mitochondrial Dynamics in Cell Death
- Acknowledgements

Introduction

Ultrastructure of mitochondria

Mitochondria are the key organelles for energy production within the cell. They house and integrate a number of important metabolic pathways including β-oxidation of fatty acids, tricarboxylic acid cycle (TCA cycle), iron–sulfur cluster biogenesis and oxygen metabolism. These pathways synergize into the production of adenosine triphosphate (ATP) from adenosine diphosphate (ADP) and phosphate. Besides this canonical role, mitochondria are fundamental for, or participate in, various other processes including Ca^{2+} buffering, reactive oxygen species (ROS) generation and the regulation of apoptosis. **See also**: From Reactive Oxygen and Nitrogen Species to Therapy; Mitochondrial Outer Membrane Permeabilization

This multifaceted function of mitochondria is possible due to the highly regulated compartmentalization of these organelles. A mitochondrion exhibits a complex structure based on two bilayer membranes, the outer mitochondrial membrane (OMM) and the inner mitochondrial membrane (IMM). The IMM surrounds a mitochondrial lumen called the matrix, which contains mitochondrial deoxyribonucleic acid (mtDNA), numerous proteins (e.g. those participating in TCA cycle and mitochondrial DNA maintenance) and a variety of metabolites. The IMM serves as the highly impermeable barrier separating the mitochondrial matrix from the intermembrane space. The active transport of metabolites and proteins through the IMM depends on several membrane-associated protein systems, including ATP synthase, the transporter of the inner membrane (TIM) complex, adenylic nucleotides transporter (ANT) and four respiratory complexes. The highly regulated IMM permeability allows the maintenance of the membrane potential ($\Delta\Psi_m$) that is essential for various mitochondrial functions, including oxidative phosphorylation. In addition, recent advances in imaging revealed that invaginations of the IMM represent another specialized mitochondrial subcompartment named cristae that are separated from inner boundary IMM by the so-called cristae junctions (Frey and Mannella, 2000). Consistent with their major role in oxidative phosphorylation, cristae are enriched in oxidative phosphorylation complexes and cytochrome *c*. Like the IMM, the OMM is relatively impermeable except for small molecules and ions. Through various channels, including the voltage-dependent anionic channel (VDAC) or transporter of the outer membrane (TOM) complex, this membrane also actively participates in the exchange of ions and proteins. Significantly, the OMM serves as a mitochondrial platform important for synchronizing mitochondrial function with extra-mitochondrial signals. The association of various B-cell lymphoma (Bcl)-2 family proteins with the OMM exemplifies this feature. Discrete stress signals are integrated on the OMM by pro- and antiapoptotic Bcl-2 family proteins resulting in the activation of the subsequent caspase cascade and progression of apoptosis. **See also**: Apoptosis: Regulatory Genes and Disease; Mitochondrial Outer Membrane Permeabilization; Structure and Function of IAP and Bcl-2 Proteins; The Bcl-2 Family Proteins – Key Regulators and Effectors of Apoptosis

Dynamic network-like organization of mitochondria in the cell

In addition to the intricate internal structure, mitochondria exhibit a complex organization within the cell. The mitochondrial architecture oscillates between long tubules and small round vesicles, resulting in a dynamic organizational equilibrium (**Figure 1** and **Figure 2**). This equilibrium is achieved through the continuously occurring but opposing processes of mitochondrial fusion and fission (Nunnari *et al.*, 1997). The relative contribution of each process determines the overall degree of continuity, branching and the average size of mitochondria within the cell (**Figure 2**). Mitochondrial fusion and fission are tightly regulated, and under physiological conditions mitochondrial fusion is evenly counterbalanced by fission. Although considerable knowledge on the machineries that regulates mitochondrial fusion and fission have been obtained by the analyses of mitochondrial morphology in mutants of the budding yeast *Saccaromices cerevisiae*, the focus of this article is on mitochondrial morphogenesis in mammalian cells. The regulation of these processes in yeast has been reviewed elsewhere. In the following section, we will briefly discuss mitochondrial morphogenesis proteins (**Figure 3**; **Table 1**) and mechanisms of mitochondrial fusion and fission in mammalian cells. This will be followed by the review of recent work connecting mitochondrial morphogenesis proteins with the progression of the mitochondrial steps in apoptosis, as well as exciting findings showing that certain proteins associated with apoptosis regulation

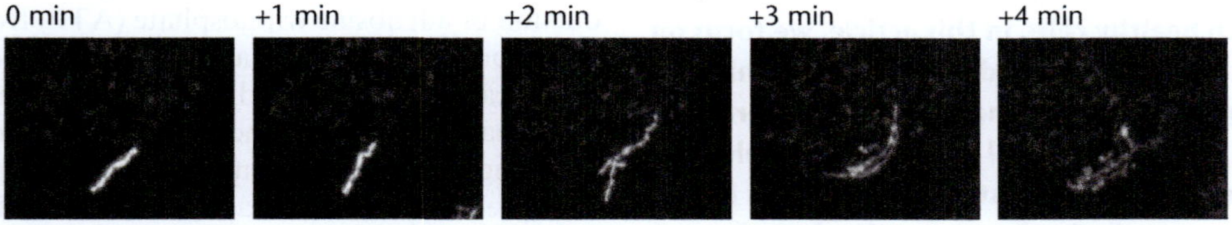

Figure 1 Visualization of mitochondrial remodelling and fusion using mito-PAGFP. The region of interest (ROI) in a mito-PAGFP (Karbowski *et al.*, 2004) transfected mouse embryonic fibroblast was photoactivated with 413-nm light followed by time-lapse acquisition of images using 488-nm laser excitation. Note that between 1 and 2 min the redistribution of mito-PAGFP from activated to nonactivated mitochondria occurs, indicating mitochondrial fusion.

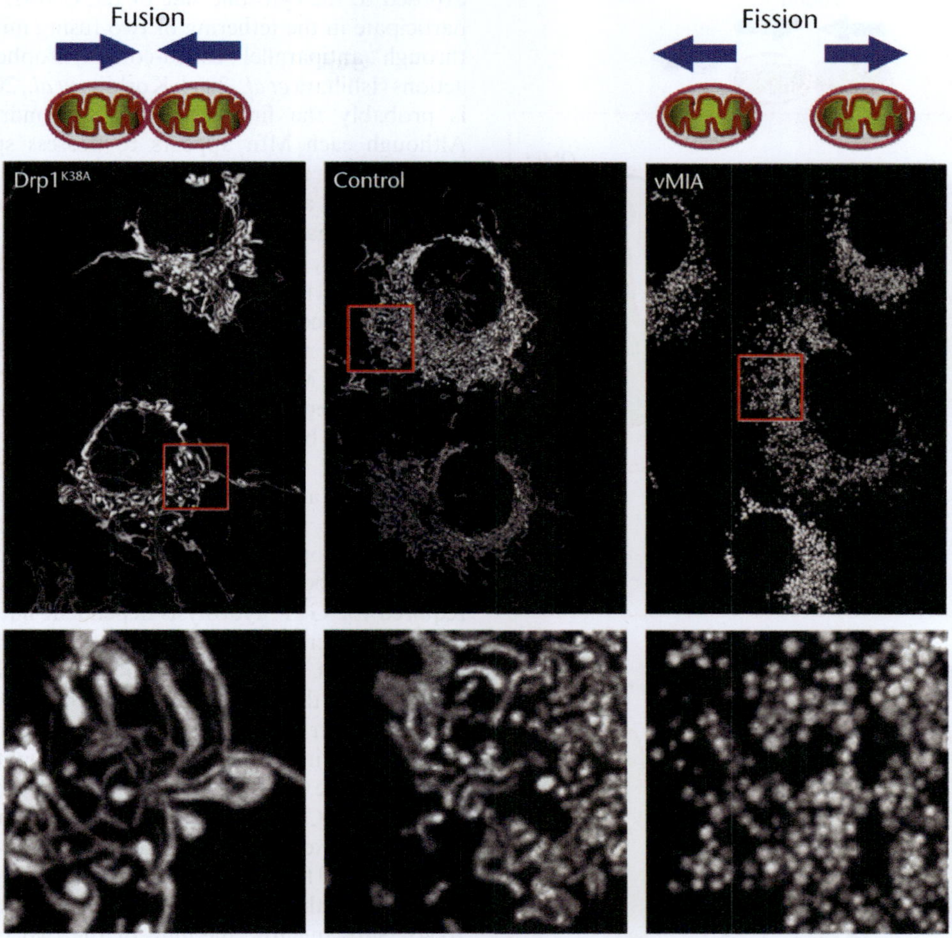

Figure 2 Mitochondrial fusion and fission regulate cellular organization of these organelles. The mitochondrial architecture oscillates between long tubules and small round vesicles (control), resulting in a dynamic organizational equilibrium. This equilibrium is achieved through the continuously occurring but opposing processes of mitochondrial fusion and fission. The relative contribution of each process determines the overall degree of continuity, branching and the average size of mitochondria within the cell. The effects of Drp1^{K38A} and vMIA on mitochondrial morphology, which when overexpressed inhibits mitochondrial fission or fusion, respectively, exemplify this feature. Note highly elongated and interconnected mitochondria in Drp1^{K38A} expressing cells and short, vesicular mitochondria induced by expression of vMIA. Cos-7 cells transfected with mito-YFP alone, or together with Drp1^{K38A} or vMIA are shown.

can also participate in the regulation of mitochondrial fusion and fission in healthy cells.

Regulation of Mitochondrial Fusion and Fission

Mitochondrial Fusion

The unique feature of mitochondrial fusion is the necessity of merging double bilayer systems from the two fusing mitochondria (**Figure 3**). Although under certain conditions the OMM can fuse without subsequent fusion of the IMM (Malka *et al.*, 2005), the fusion events of the OMM and the IMM are usually synchronized. In mammalian cells, dissipation of the mitochondrial membrane potential ($\Delta\Psi_m$) appears to selectively inhibit IMM fusion (Malka *et al.*, 2005). This, in addition to the identification of yeast mitochondrial fusion intermediates *in vitro* (Meeusen *et al.*, 2004), suggests that the OMM and IMM may fuse in successive and independent reactions. Mitochondrial fusion is mediated by two large

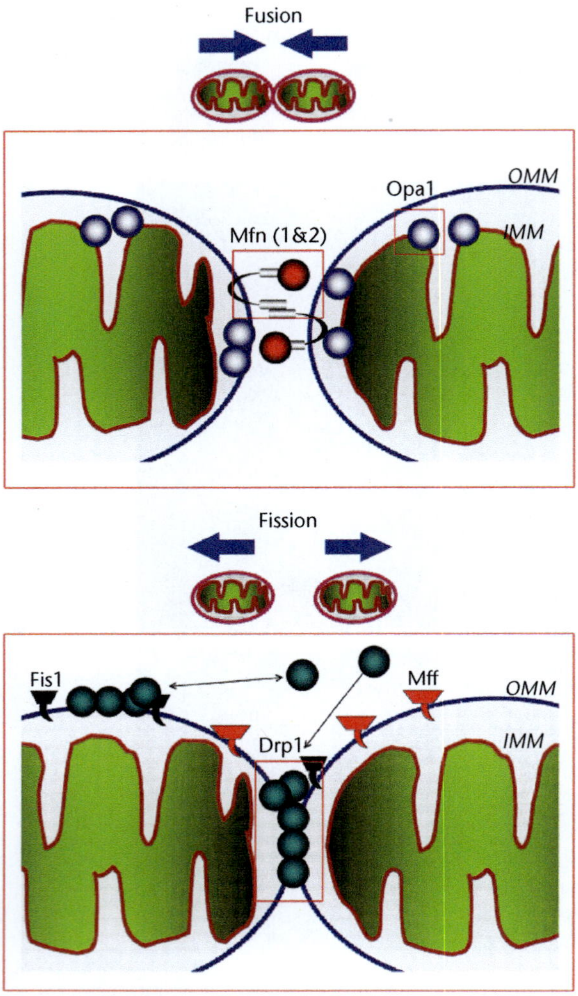

Figure 3 Regulation of mitochondrial morphology by fusion and fission. An overview of proteins vital for mitochondrial fusion and fission. The multimeric dynamin-like GTPase Drp1 is a critical protein mediating mitochondrial fission. Under normal growth conditions a major cellular fraction of Drp1 localizes to the cytosol, and the translocation of Drp1 to the OMM is needed to initiate mitochondrial fission. The OMM-associated pool of Drp1 form punctate foci that often colocalize with the progressive mitochondrial fission sites. Mitochondrial fusion is mediated by two large GTPases in the OMM, mitofusins (Mfn) 1 and 2, and by optic atrophy 1 (Opa1), a dynamin related protein in the IMM.

guanosine triphosphatases (GTPases) in the OMM, mitofusins (Mfn) 1 and 2 (Chen *et al.*, 2003) and by optic atrophy 1 (Opa1) (Delettre *et al.*, 2000), a dynamin-related protein in the IMM (**Figure 3**). Mfns are inserted in the OMM through two transmembrane domains, and both the *N*- (GTPase domain) and the *C* (heptad repeat regions, HR2)-terminal parts of these proteins are exposed to the cytosolic side of the OMM. The HR2s participate in the tethering of two fusing mitochondria through antiparallel coiled-coil hydrophobic interactions (Ishihara *et al.*, 2004; Koshiba *et al.*, 2004), which is probably the first step in mitochondrial fusion. Although each Mfn appears to possess specific and nonredundant roles (Ishihara *et al.*, 2004), the presence of both Mfn1 and Mfn2 is imperative for the maintenance of normal rates of mitochondrial fusion. Single knockouts of either Mfn1 or Mfn2 exhibit decreases in mitochondrial fusion rates (Chen *et al.*, 2003; Karbowski *et al.*, 2004) and increase mitochondrial fragmentation. However, the resulting mitochondrial shapes and sizes are Mfn1 or Mfn2 knockdown specific (Chen *et al.*, 2003). The depletion of Mfn1 induces small vesicular mitochondria broadly dispersed in the cell, whereas, the lack of Mfn2 results in larger mitochondrial vesicles concentrated around the nucleus (Chen *et al.*, 2003). Consequently, the role of each Mfn protein in mitochondrial fusion regulation might differ. Supporting this notion, it has been shown that Mfn1 might specifically be required for GTP hydrolysis-dependent mitochondrial tethering, whereas Mfn2 is less efficient in this process, and might act as a signalling GTPase (Ishihara *et al.*, 2004). Other than mitochondrial tethering Mfn-mediated molecular events are currently not clear. An intermediate step in mitochondrial fusion might involve a pool of soluble Opa1 present in the intermembrane space (Detmer and Chan, 2007), which stabilize the transient Opa1 complexes. This mechanism, like Mfn-dependent mitochondrial fusion steps, is energy dependent, which is provided by the hydrolysis of GTP (Meeusen *et al.*, 2004). Subsequently, merging of the two inner membranes is mediated by the physical interaction between IMM-localized Opa1. Mfn- and Opa1-dependnt steps in mitochondrial fusion are likely synchronized by direct interactions between these proteins. As in case of Mfns, knockdown of Opa1 leads to inhibition of mitochondrial fusion, and mitochondrial fragmentation (Olichon *et al.*, 2003), indicating the essential role of this protein in mitochondrial fusion.

In addition to Mfn1, Mfn2 and Opa1, various other proteins are implicated in the regulation of mitochondrial fusion in mammalian cells, including Mfn-binding protein (MiB) (Eura *et al.*, 2006), mitochondria-associated phospholipase D (mito-PLD) (Choi *et al.*, 2006) and Stomatin-like protein 2 (Stoml2 also known as SLP2) (Hajek *et al.*, 2007). Furthermore, some exciting advances in the understanding of fusion regulation including the role of IMM-associated proteases in Opa1 processing and regulation have been made.

Table 1 Major factors implicated in the regulation of mitochondrial fusion and fission in mammalian cells

Gene product	Regulated process	Localization	Loss-of-function phenotype	Effect on apoptosis
OPA1	Fusion	IMM	– Fragmented mitochondria – Decrease in mitochondrial ATP production, $\Delta\Psi_m$ and oxygen consumption – Instability of mtDNA	– Downregulation sensitizes cells to apoptosis, overexpression delays apoptosis
Mitofusins (Mfn1 and Mfn2)	Fusion	OMM	– Fragmented mitochondria – $\Delta\Psi_m$ changes – Decline in mitochondrial function – Developmental defects	– Downregulation sensitizes cells to apoptosis, overexpression delays apoptosis
MiB	Fusion	OMM	– Fragmented mitochondria	N/A
Mito-PLD	Fusion	OMM	– Defects in mitochondrial tethering	N/A
Drp1	Fission	OMM, peroxisomes and cytosol	– Elongated mitochondria – Peroxisome fission defects	– Downregulation or inhibition delays cytochrome *c* release and apoptosis
Fis1	Fission	OMM and peroxisomes	– Elongated mitochondria	– Downregulation delays apoptosis
Mff	Fission	OMM	– Elongated mitochondria	– Downregulation delays apoptosis
GDAP1	Fission	OMM	– Elongated mitochondria – Neuronal demyelization	N/A
MARCH5	Fission (fusion)	OMM and peroxisomes	– Elongated mitochondria; mitochondrial fragmentation also reported	N/A
MTP18	Fission	OMM	– Elongated mitochondria	– Downregulation sensitizes cells to apoptosis
Bcl-x_L, Bcl-w	Fission and Fusion	OMM (ER)	– Mitochondrial elongation – Developmental defects	– Antiapoptotic
Bax, Bak	Fusion (fission?)	OMM and cytosol (Bax)	– Reduced mitochondrial fusion – Developmental defects	– Pro-apoptotic

Mitochondrial fission

The multimeric dynamin-like GTPase Drp1 is a critical protein-mediating mitochondrial division. It harbours multiple motifs including GTP-binding, middle, and GTPase effector domains (GED) that are important for both intramolecular and intermolecular interactions. Drp1 is broadly expressed in the organism and appears to be essential for mitochondrial division in all tested phyla. Drp1 knockdown or expression of dominant negative mutants of this protein (e.g. GTPAse domain mutant Drp1^{K38A} (**Figure 2**), or phosphomimetic mutant Drp1^{S637D}) lead to inhibition of mitochondrial division and formation of abnormally elongated mitochondria, as well as highly interconnected mitochondrial networks (Chang and Blackstone, 2007; Smirnova *et al.*, 2001). Under normal growth conditions a major cellular fraction of Drp1 localizes to the cytosol, and the translocation of Drp1 to the OMM is needed to initiate mitochondrial division. The OMM associated pool of Drp1 form punctate foci that often colocalize with the progressive mitochondrial fission sites (Smirnova *et al.*, 2001) (**Figure 3**). These structures are stabilized by SUMO (small ubiquitin-like modifier)ylation (Harder *et al.*, 2004) and destabilized by ubiquitination (Karbowski *et al.*, 2007), suggesting that these two posttranslational modifications of Drp1 are critical for initiation and/or progression of the mitochondrial fission reaction. Furthermore, phosphorylation of Drp1 is also critical for the mitochondrial activity of this protein. The precise mechanism by which Drp1 promotes the mitochondrial division in mammal cells is not clear. However, *in vitro* studies in yeast have demonstrated the possibility that Dnm1p (a yeast homologue of Drp1) and Drp1 form helicoidal structures that wrap around the mitochondrial tubule and mediate fission of the OMM (Ingerman *et al.*, 2005). Also, based on the mitochondrial division studies in *S. cerevisiae*, it has been proposed that in mammalian cells Fis1, an OMM anchored protein, acts as a mitochondrial receptor for Drp1 (Yoon *et al.*, 2003; **Figure 3**). Yet, although the role of Fis1 in division of mammalian mitochondria might indeed be conserved (James *et al.*, 2003; Yoon *et al.*, 2003), the mechanism of Fis1 action in mammalian cells is not clear. Even though Fis1 and Drp1 were coimmunoprecipitated from crosslinked cell lysates (Yoon *et al.*, 2003), changes in Fis1 expression do not influence mitochondrial association of Drp1.

Recently, a number of additional factors required for, or implicated in, mitochondrial division has been reported. These include Mff (for mitochondrial fission factor) (Gandre-Babbe and van der Bliek, 2008) and membrane-associated RING-CH 5 (MARCH5), a really interesting new gene (RING) finger E3 ubiquitin ligase of the OMM (Karbowski *et al.*, 2007; Yonashiro *et al.*, 2006). Furthermore, various signalling pathways, including cyclic adenosine monophosphate (cAMP)-dependent protein kinase, Ca^{2+}/calmodulin-dependent protein kinase I alpha (CaMKIalpha) and cyclin-dependent kinase (Cdk1/cyclin B), dependent phopshorylation of Drp1, and aforementioned Drp1 SUMOylation (Harder *et al.*, 2004) and ubiquitination (Yonashiro *et al.*, 2006) have also been found to participate in mitochondrial fission regulation.

Cellular role of Mitochondrial Network Organization: An Overview

Published studies on the role of mitochondria and mitochondrial network remodelling, as we describe in this section, draw attention to the importance of mitochondrial morphogenesis and proteins regulating this process in cell and organism function, and embryonic development. Yet they also reveal how little is known about the mechanisms by which mitochondrial network dynamics and specific mitochondrial morphogenesis proteins participate in the regulation of cellular homeostasis. The relation between cell physiology and mitochondrial morphogenesis is extremely intricate and mutual. For example, decreases in bioenergetic activity of mitochondria induce disturbances in mitochondrial network dynamics, and vice versa alterations in mitochondrial network, for example, inhibition of mitochondrial fusion, result in defects in mitochondrial energy production (Chen *et al.*, 2005). Changes in mitochondrial membrane potential, concentration of ATP and ROS generation also govern mitochondrial shape (Benard *et al.*, 2007). It has been proposed that mitochondrial fusion facilitate intramitochondrial redistribution and complementation of mitochondrial DNA, as well as matrix contents and membrane proteins. Furthermore, the cellular trafficking of the whole mitochondria appears to relay on mitochondrial dynamics (Varadi *et al.*, 2004). As a consequence, an impairment of mitochondrial dynamics can induce abnormal accumulation of mitochondria in certain subcellular compartments, for example, axons or perinuclear areas. This might be particularly relevant for pathologies associated with mutations of mitochondrial morphogenesis proteins (Opa1, Mfns and Drp1), where such amassing of mitochondria in axons has been observed. Furthermore, elimination of dysfunctional mitochondria also depends on local inhibition of mitochondrial fusion that enables the isolation of aberrant mitochondria from the bulk mitochondrial network. The isolated dysfunctional mitochondria are then targeted for autophagic degradation, as showed by Twig *et al.* (2008). In addition to lowered fusion of mitochondria, activation of mitochondrial fission might also be needed for mitochondrial elimination as demonstrated in the *C. elegans* model, where activity of Drp1 appears to facilitate mitochondrial elimination in dying cells (Breckenridge *et al.*, 2008).

The published data also suggest that the regulation of mitochondrial fusion and fission might have a significant impact on early development. The embryonic lethality induced by knockouts or mutations of Mfn1, Mfn2 (Chen *et al.*, 2003) and Opa1 (Davies *et al.*, 2007), indicate that the maintenance of a dynamic mitochondrial network is also vital for embryogenesis in mammals. Further supporting this notion, the *heterozygous* dominant negative mutation of Drp1 in one human subject has been linked to various birth defects, including microcephaly, abnormal brain development, optic atrophy and hypoplasia, lactic acidaemia, and long-chain fatty acids accumulation, leading to death at 37 days *postpartum* (Waterham *et al.*, 2007). The data show that specific fusion proteins can have a different impact on development, suggesting that in addition to shared functions these proteins may also function independently (Chen *et al.*, 2003). For example, although Mfn1 and Mfn2 are ubiquitously expressed, the impact of embryonic defects in $Mfn1^{-/-}$ and $Mfn2^{-/-}$ differ (Chen *et al.*, 2003). In $Mfn2^{-/-}$ mice a disruption in placental development, most obviously in the paucity of trophoblast giant cells, has been reported (Chen *et al.*, 2003). Since similar defects were not obvious in $Mfn1^{-/-}$ cells, one may conclude that Mfn1 and Mfn2 are required for particular but different developmental transitions. There is no available data on mitochondrial network dynamics in stem cells, yet the images from published studies show that mitochondria in stem cells form rod-like structures highly enriched in perinuclear areas (Lonergan *et al.*, 2006). It is likely that under conditions when major alterations of mitochondrial structure and function need to occur (e.g. differentiation of stem cell into cardiomyocytes or neurons), the proper balancing of mitochondrial fusion and fission might be more critical than under continuous division. Indeed, mitochondria appear to play a critical role in the maintenance of stem cell physiology and their ability to differentiate. Mouse embryonic stem cells sorted for low and high $\Delta\Psi_m$ were indistinguishable morphologically and by the expression of pluripotency markers. Yet, cells with higher mitochondrial metabolism (high $\Delta\Psi_m$) retained high division rates and tended to form tumours, whereas those with lower metabolic capacities (low $\Delta\Psi_m$) tend to differentiate into other cell types.

Role of Mitochondrial Dynamics in Cell Death

Apoptosis, a genetically driven form of cell death, results in a highly organized dismantling of dying cells. Decomposition of apoptotic cells is mediated by a family of proteases, called caspases that in some modes of apoptosis are regulated by anti- and pro-apoptotic proteins from the Bcl-2 family. The main target of the Bcl-2 family in mammalian cells is the OMM. Bcl-2 proteins can either induce OMM permeabilization and thus apoptosis (e.g. Bax, Bak and Bok) or inhibit it and promote cell survival (e.g. Bcl-2, Bcl-x_L and Mcl-1). The ratio of pro-apoptotic versus antiapoptotic Bcl-2 family members is a critical factor in regulating susceptibility to cell death, and regulation of Bcl-2 family proteins, a way to initiate or modulate the apoptotic signal cascade in response to various stimuli, is under stringent control. Various regulatory mechanisms, including regulation of Bcl-2 family protein expression, as well as diverse posttranslational modifications regulate the balance of the activities of pro- and antiapoptotic Bcl-2 family proteins in healthy cells, and on induction of apoptosis. Recent reports suggest, that in addition to Bcl-2 family proteins, factors implicated in the regulation of mitochondrial dynamics are also important for the control of the mitochondrial steps in apoptosis, and vice versa certain Bcl-2 family members participate in the regulation of mitochondrial fusion and/or fission.

See also: Mitochondrial Outer Membrane Permeabilization; Structure and Function of IAP and Bcl-2 Proteins; The Bcl-2 Family Proteins – Key Regulators and Effectors of Apoptosis

Role of Bcl-2 Family in the Regulation of Mitochondrial Network Dynamics

A number of published reports indicate a mechanistic link between the core mitochondrial fusion and fission regulating proteins (e.g. Drp1, Mfn2 and Opa1) (**Figure 3**) and proteins from Bcl-2 family. The antiapoptotic Bcl-2 family protein Bcl-x_L, as well as its *C. elegans* counterpart, localizes to the mitochondria, and as was recently shown their expression levels are critical for mitochondrial morphogenesis (Berman *et al.*, 2009; Tan *et al.*, 2008). Consistent with a modulatory, but not essential role of CED (CEll Death abnormality)-9 cells lacking this protein have no alteration in mitochondrial size or ultrastructure, yet they appear more sensitive to mitochondrial fragmentation. By contrast, increased CED-9 expression in these cells produces highly interconnected mitochondria (Tan *et al.*, 2008). This mitochondrial phenotype was partially suppressed by overexpression of DRP (dynamin-related protein)-1 and depended on the BH3-binding pocket of CED-9, indicating that in *C. elegans* CED-9 might directly regulate DRP-1 (Tan *et al.*, 2008). In addition, as shown by Delivani *et al.* (2006) overexpression of CED-9 induced mitochondrial elongation in mammalian cells. Furthermore, as suggested by immunoprecipitation data showing that overexpressed CED-9 interacts with Mfn2, this might depend on the fusion-stimulating activity of CED-9 (Delivani *et al.*, 2006). Although, the physiological significance of these data is not obvious, it suggests that direct interaction of CED-9 with mitochondrial fusion protein might also contribute to the mitochondrial elongation observed in CED-9 overexpressing worms. This possibility has not been addressed yet.

However, as shown by Berman *et al.* (2009), Bcl-x_L a mammalian homologue of CED-9 is also implicated in the regulation of both fusion and fission of mitochondrial network in mammalian neurons. In Bcl-$x_L^{-/-}$ cortical neurons, mitochondrial size was significantly reduced, as compared to wild-type cells. However, overexpression of Bcl-x_L induced mitochondrial elongation that depended on the increased fusion rates of these organelles (Berman *et al.*, 2009). Yet these elegant time-lapse microscopy studies, utilizing mitochondrial matrix targeted photoactivable green fluorescent protein (mito-PAGFP), also revealed that mitochondrial fission in Bcl-x_L overexpressing cells is stimulated, suggesting a role for Bcl-x_L in the regulation of both fusion and fission of mitochondria. In addition to this feature, it has also been found that Bcl-x_L controls the mitochondrial biomass and energy-producing capacity of mitochondria under normal growth conditions, before the cell is faced with a life-or-death decision. Thus, it has been concluded that Bcl-x_L-dependent mitochondrial biogenesis might be critical for balancing mitochondrial network dynamics (Berman *et al.*, 2009). It has also been reported that activity of Bcl-w, another pro-survival Bcl-2 family protein, is required for the regulation of mitochondrial length in Purkinje cell processes (Liu and Shio, 2008). Yet, unlike the case of Bcl-x_L, quantitative electron microscopy revealed a significant increase in mitochondrial length in *Bcl-w*$^{-/-}$ dendrites, as compared to wild-type mouse dendrites. Furthermore, mitochondria in *Bcl-w*$^{-/-}$ mice often contained points where they became constricted, suggesting that the elongation of mitochondria might be due to the inhibition or slowdown of the mitochondrial fission process. Based on this it has been proposed that mitochondrial fission occurring during neuronal growth might be critically important for dendrite development and synapse formation, and that it can be regulated by Bcl-2 family protein(s) (Liu and Shio, 2008).

In addition to Bcl-x_L and Bcl-w, Bax and Bak, two pro-apoptotic proteins essential for the execution of an apoptotic signal relayed by other Bcl-2 family members, participate in the regulation of mitochondrial network dynamics (Brooks *et al.*, 2007; Karbowski *et al.*, 2006). As in the case of Bcl-$x_L$$^{-/-}$ neurons, the reduction of mitochondrial size, as well as lowered fusion rates, have been detected in Bax/Bak double knockout mouse embryonic fibroblasts (Bax/Bak$^{-/-}$ MEFs) (Karbowski *et al.*, 2006), indicating a role for these proteins in the regulation of mitochondrial fusion. It is not known whether fission of mitochondria in Bax/Bak$^{-/-}$MEFs proceed at the normal rate. Yet, the fact that mitochondrial defects in these cells can be rescued more efficiently by Mfn overexpression-induced stimulation of mitochondrial fusion than by expression of a dominant negative mutant of Drp1 (Drp1^{K38A}) and thus by inhibition of mitochondrial division (Karbowski *et al.*, 2006), suggests that, like Bcl-x_L, Bax and Bak might also participate in the coordination of both mitochondria fusion and fission. Notably, Bax and Drp1 colocalize on the mitochondria in apoptotic cells (Karbowski *et al.*, 2002; Wasiak *et al.*, 2007). Inhibition or downregulation of Drp1 did not affect mitochondrial association of Bax, but led to the accumulation of this protein on the constricted mitochondrial sites (Karbowski *et al.*, 2002), perhaps the sites of fission initiation, accumulating due to lack of Drp1-dependent membrane scission. These data suggest that under apoptotic conditions Bax might also regulate the mitochondrial fission process.

How the Bcl-2 family proteins intersect mitochondrial fusion and fission pathways is currently not clear. It has been shown that Bax and Bak can influence the assembly of the Mfn2-containing protein complexes and change their submitochondrial distribution and membrane mobility-properties that correlate with different GTP-bound states of Mfn2 (Karbowski *et al.*, 2006). Immunoprecipitation studies also revealed that in healthy cells Mfn1 and Mfn2 interact with Bak (Brooks *et al.*, 2007). Consistent with the high correlation of mitochondrial fusion inhibition with apoptotic activation of Bax and Bak (Karbowski *et al.*, 2004) and the fusion-stimulating role of these proteins in healthy cells, the interaction of Bak with Mfn2 was no longer detectable on induction of apoptosis (Brooks *et al.*, 2007). Since other mitochondrial morphogenesis proteins, including Fis1 and Drp1, can interact with and possibly be regulated by Bcl-2 family proteins (James *et al.*, 2003; Li *et al.*, 2008; Liu and Shio, 2008), one could envision that direct molecular interactions between core mitochondrial morphogenesis proteins (e.g. Drp1 and Mfn2) and Bcl-2 family members participate in mitochondrial morphogenesis regulation.

In addition to the Mfn activity-related roles of Bax and Bak, the expression and likely conformation of these proteins (e.g. pre-apoptotic and apoptotic) can influence trafficking of fission protein Drp1 between the cytosol and the OMM (Wasiak *et al.*, 2007). During apoptosis, in synchrony with mitochondrial activation of Bax and before $\Delta\Psi_m$ collapse, Drp1 is stabilized on the mitochondrial fission sites. The increased stability of Drp1 at mitochondrial fission sites is promoted by SUMOylation of Drp1 that occurs in a Bax- and Bak-dependent manner (Wasiak *et al.*, 2007). Since this process does not depend on Fis1-, Bax- and Bak-dependent regulation of Drp1 in apoptotic cells might differ from the fission mechanism in healthy cells.

Interestingly, the work by Martin and colleagues (Sheridan *et al.*, 2008) indicates that overexpression of Bcl-x_L as well as Mcl-1, another antiapoptotic Bcl-2 family protein, can induce mitochondrial fragmentation, and that this depends on expression levels of other Bcl-2 family proteins. Mcl-1 overexpression-induced mitochondrial fragmentation is dramatically facilitated in Bax/Bak RNAi (ribonucleic acid interference) cells, whereas Bcl-x_L overexpression can also fragment mitochondria in cells expressing normal levels of Bax and Bak (Sheridan *et al.*, 2008). Thus, these data suggest that the specific activity (e.g. stimulation of fusion or fission) of certain Bcl-2 family members in healthy cells is strongly influenced by expression patterns of other Bcl-2 family proteins. Therefore, as in case of apoptosis regulation, the overall activity of different members of the Bcl-2 family might be critical for their function in mitochondrial morphogenesis. Since pro-apoptotic Bax and Bak and antiapoptotic Bcl-x_L increase mitochondrial length in healthy cells, these proteins might act cooperatively to regulate the steady state of the mitochondrial network before apoptosis induction. The opposite activities of these proteins in the apoptotic pathway, for example, pro-survival (Bcl-x_L, Bcl-w) and pro-death (Bax and Bak), could have evolved from the common role of these proteins in mitochondrial morphogenesis regulation.

In conclusion, the works described in this part support the role for the Bcl-2 family in the regulation of mitochondrial network dynamics and mitochondrial biogenesis in healthy cells. Obviously, further studies are needed to clarify the molecular mechanisms by which Bcl-2 family proteins influence mitochondrial network dynamics. Yet, as discussed in the subsequent section the data also suggest that the effects of Bax, Bak and Bcl-x_L on mitochondria in healthy cells might also contribute to their roles in the regulation of the mitochondrial steps in apoptosis. Furthermore, the intimate link between mitochondrial morphogenesis machineries and the regulation of the mitochondrial steps in apoptosis is also strengthened by a number of studies describing apoptosis modulating roles for mitochondrial fusion and fission proteins.

Role of core mitochondrial fusion and fission proteins in progression of apoptosis

Although mitochondrial fragmentation induced by certain triggers, including protonophores and elevated Ca^{2+}, does not necessarily lead to activation of the mitochondrial steps in apoptosis, mitochondrial fragmentation is a common feature of stress-induced apoptosis in a broad number of cell types and regardless of the apoptotic trigger used. The discovery that apoptosis might be functionally linked to alterations in mitochondrial network dynamics, through the activity of Drp1 (Frank *et al.*, 2001), stimulated an extensive research effort, capitalizing on the availability of molecular and imaging tools to precisely modulate and monitor mitochondrial network alterations. It has been shown that inhibition of mitochondrial fragmentation by Drp1^{K38A}, a dominant negative mutant of Drp1, resulted in delaying release of cytochrome *c* from the mitochondria to the cytosol and consequently inhibited apoptosis (Frank *et al.*, 2001). Notably, the role for Drp1 in apoptosis execution has also been confirmed by the studies using the *C. elegans* (Jagasia *et al.*, 2005) and *D. melanogaster* (Goyal *et al.*, 2007) models of developmental cell death, indicating that Drp1 role in cell death is evolutionally conserved. Albeit, the study by Breckenridge *et al.* (2008) indicated that not apoptosis regulation but elimination of dysfunctional mitochondria might be the main cell death-related function of Drp1 in *C. elegans*.

Importantly, several studies have also shown the ability of other mitochondrial morphogenesis proteins than Drp1 to accelerate or delay the apoptotic response of the cell (Gandre-Babbe and van der Bliek, 2008; James *et al.*, 2003; Olichon *et al.*, 2003), indicating that alterations in mitochondrial network dynamics might be linked to the apoptotic signalling cascade. Based on these studies, one might conclude that mitochondrial fragmentation facilitates cell death, and inhibition of this process counteracts apoptosis activation. Consistently, mitochondrial fragmentation correlates with apoptotic activation of Bax and Bak and with OMM permeabilization induced by these proteins (Karbowski *et al.*, 2002, 2004). As in cells with reduced Drp1 activity, inhibition of mitochondrial division, through silencing of the other known fission proteins Fis1 and

Mff or overexpression of fusion proteins Mfn1 and Mfn2, reduces cell sensitivity to apoptosis and inhibits cytochrome *c* release. Notably, Sugioka *et al.* (2004) have demonstrated that Bax and Bak oligomerization, a step in Bax and Bak activation linked to the OMM permeabilization, can be inhibited by Mfn1 and Mfn2 overexpression. These data, as well as molecular interaction of Mfn2 and Bak (Brooks *et al.*, 2007), as discussed in the section on Role of Bcl-2 family in the regulation of mitochondrial network dynamics, suggest that apoptotic activation of Bax (and Bak) might be linked with mitochondrial fusion modulation. This notion is further supported by the data showing that viral mitochondrial-associated inhibitor of apoptosis (vMIA) can interact with Bax, Bak and Mfn2, and can sequester these proteins in apoptotically nonactive protein complexes (Karbowski *et al.*, 2006). Significantly, mitochondrial fusion is also inhibited in vMIA expressing cells (**Figure 2**). However, it has been shown that Opa1 downregulation-induced inhibition of mitochondria fusion and subsequent mitochondrial fragmentation leads to unprompted apoptosis and sensitizes cells to various cell death triggers (Olichon *et al.*, 2003). This is associated with spontaneous release of cytochrome *c*, $\Delta\Psi_m$ collapse and apoptosis (Olichon *et al.*, 2003). Since silencing of Opa1 also induces Bax activation on the OMM it is likely that Opa1 deficiency-dependent apoptosis relies on changes in the Bcl-2 family proteins.

Despite increasing knowledge about the regulation of mitochondrial fragmentation during cell death, the debate is still open as to how and to what degree mitochondrial network dynamics alterations participate in the decision phase of mitochondrial apoptosis. For example the mechanism by which stimulated fission of the mitochondria triggers release of cytochrome *c* is not known. In addition, the effects of mitochondrial fission inhibition on the apoptosis progression vary significantly between different published reports. It is also not clear why inhibition of mitochondrial division has a much stronger effect on cytochrome *c* release than the release of other apoptotic proteins, for example, SMAC/Diablo (second mitochondria-derived activator of caspase/direct IAP binding protein with LOw pl), from the mitochondria to the cytosol. These discrepancies are likely caused by vastly different approaches used to modulate mitochondrial dynamics used in these studies, as well as different times of treatments with apoptosis-inducing compounds. Furthermore, despite very extensive effort, the general mechanism of OMM permeabilization that releases cytochrome *c* and what kind, if any, of mitochondrial priming that has to occur before cytochrome *c* is released is also not well understood. Some of the proposed mechanisms of how mitochondrial morphogenesis proteins influence the apoptotic activities of Bcl-2 family proteins, including mitochondrial cristae remodeling and direct effect of mitochondrial morphogenesis proteins on the OMM permeabilization, are discussed later.

Cristae remodelling is a process where mitochondrial cristae fuse and the narrow cristae junctions, structures connecting cristae with the IMM, open. This process is thought to enable more efficient release of cytochrome *c*, likely through the mobilization of this protein in the proximity of the OMM. As reported by Scorrano and colleagues, proteolysis of Opa1 appears to be a prerequisite for apoptotic cristae changes (Cipolat *et al.*, 2006). Active Opa1 prevents cristae opening, perhaps by stabilizing cristae junctions. During apoptosis this form of Opa1 is diminished in the mitochondria by the concurrent actions of proteolytic enzymes (Cipolat *et al.*, 2006) and Opa1 release from the mitochondria to the cytosol (Arnoult *et al.*, 2003; Griparic *et al.*, 2007). Several proteases have been identified to be involved in the processing of Opa1, including the presinilin-associated rhomboid-like protein (PARL), Yme1, paraplegin and the m-AAA protease complex. However, implication of these enzymes in the regulation of apoptosis-specific cristae remodeling and apoptosis itself is not clear. Knockdowns of these proteases lead in some cases to apoptosis inhibition or promotion of cell death or have no effect. Yet, the antiapoptotic form of Opa1 appears to be absent in PARL$^{-/-}$ cells, suggesting a critical role for this protease in Opa1-dependent apoptotic cristae remodelling (Cipolat *et al.*, 2006). Furthermore, mammalian cells lacking PARL do not show altered mitochondrial morphology, but enhanced ability to release cytochrome *c* from the IMM to the cytosol (Cipolat *et al.*, 2006). This suggests that PARL might specifically be implicated in the apoptotic Opa1-dependent regulation of cristae morphology, perhaps through processing of a specific, cristae junction-associated subset of Opa1. Consistently, PARL knockout did not affect activation of Bax- and Bak-dependent mitochondrial permeabilization, but increased cell death, further suggesting that cristae structure is a major determinant of abnormal cell death in PARL$^{-/-}$ cells (Cipolat *et al.*, 2006). Certain Bcl-2 family members also regulate Opa1-dependent changes in mitochondrial cristae. The recent works by Newmeyer and colleagues (Yamaguchi *et al.*, 2008) have demonstrated that mitochondrial cristae remodeling require the activities of Bax and Bak, but can be experimentally separated from OMM permeabilization, another process that depends on these proteins. Of note, these data further support roles for Bax and Bak in the regulation of mitochondrial morphogenesis, as discussed in the section on Role of Bcl-2 family in the regulation of mitochondrial network dynamics.

However, the recent work by Sun *et al.* (2007) has challenged the importance of cristae remodelling in the apoptosis progression. This correlative microscopy study revealed that under certain conditions caspase inhibition could block cristae changes without affecting cytochrome *c* release. Furthermore, Arnoult *et al.* (2005) observed that, compared to control cells, cytochrome *c*-GFP release was approximately 60 s faster in Opa1 RNAi cells. Opa1 processing and cristae remodelling in normal cells is already relatively fast. Thus, although OPA1-dependent cristae remodelling appears to modulate cytochrome *c* release, the significance of this step for the general apoptotic response of the cell needs to be addressed in more detail. It is unclear how the strong pro-apoptotic effect of PARL deficiency could result from the apparently small changes in the cytochrome *c* release kinetics.

The direct role of mitochondrial morphogenesis proteins in the regulation of the OMM permeabilization is also supported by published evidence. As mentioned earlier, during apoptosis Bax and Bak colocalize with Drp1 at discrete submitochondrial foci (Karbowski *et al.*, 2002; Wasiak *et al.*, 2007). The chemical inhibition of Drp1 activity blocked Bid-activated Bax and Bak-dependent cytochrome *c* release from mitochondria (Cassidy-Stone *et al.*, 2008). Thus, it has been concluded that Drp1 directly regulates mitochondrial permeabilization independent of Drp1-dependent division of mitochondria (Cassidy-Stone *et al.*, 2008). Since chemical inhibition of Drp1 affects release of not only cytochrome *c*, but also SMAC/Diablo it is likely that Drp1 directly regulates OMM permeabilization. Furthermore, our preliminary data indicate that although Drp1^{K38A}-dependent inhibition of Drp1 does not affect mitochondrial accumulation of Bax and submitochondrial clustering of this protein, the oligomerization of Bax is significantly reduced. Thus, it is likely that Drp1 activity might directly influence certain steps of apoptotic activation of Bax. Fis1, another fission protein, does not affect stabilization of Drp1 on the mitochondria in apoptotic cells, although it is thought that Fis1 serves as a Drp1 receptor to mediate mitochondrial fission under normal growth condition. These data suggest that apoptotic activity of Drp1, including regulation and molecular interactions, might vary between normal cell growth and apoptotic conditions.

Acknowledgements

The authors also gratefully acknowledge financial support from National Institute of General Medical Science RO1 GM083131 (M.K).

References

Arnoult D, Gaume B, Karbowski M *et al.* (2003) Mitochondrial release of AIF and EndoG requires caspase activation downstream of Bax/Bak-mediated permeabilization. *EMBO Journal* **22**: 4385–4399.

Arnoult D, Grodet A, Lee YJ, Estaquier J and Blackstone C (2005) Release of OPA1 during apoptosis participates in the rapid and complete release of cytochrome *c* and subsequent mitochondrial fragmentation. *Journal of Biological Chemistry* **280**: 35742–35750.

Benard G, Bellance N, James D *et al.* (2007) Mitochondrial bioenergetics and structural network organization. *Journal of Cell Science* **120**: 838–848.

Berman SB, Chen Y-B, Qi B *et al.* (2009) Bcl-xL increases mitochondrial fission, fusion, and biomass in neurons. *Journal of Cell Biology* **184**: 707–719.

Breckenridge DG, Kang BH, Kokel D *et al.* (2008) *Caenorhabditis elegans* drp-1 and fis-2 regulate distinct cell-death execution pathways downstream of ced-3 and independent of ced-9. *Molecular Cell* **31**: 586–597.

Brooks C, Wei Q, Feng L *et al.* (2007) Bak regulates mitochondrial morphology and pathology during apoptosis by interacting with mitofusins. *Proceedings of the National Academy of Sciences of the USA* **104**: 11649–11654.

Cassidy-Stone A, Chipuk JE, Ingerman E *et al.* (2008) Chemical inhibition of the mitochondrial division dynamin reveals its role in Bax/Bak-dependent mitochondrial outer membrane permeabilization. *Developmental Cell* **14**: 193–204.

Chang CR and Blackstone C (2007) Drp1 phosphorylation and mitochondrial regulation. *EMBO Report* **8**: 1088–1089, author reply 1089-90.

Chen H, Chomyn A and Chan DC (2005) Disruption of fusion results in mitochondrial heterogeneity and dysfunction. *Journal of Biological Chemistry* **280**: 26185–26192.

Chen H, Detmer SA, Ewald AJ *et al.* (2003) Mitofusins Mfn1 and Mfn2 coordinately regulate mitochondrial fusion and are essential for embryonic development. *Journal of Cell Biology* **160**: 189–200.

Choi SY, Huang P, Jenkins GM *et al.* (2006) A common lipid links Mfn-mediated mitochondrial fusion and SNARE-regulated exocytosis. *Nature Cell Biology* **8**: 1255–1262.

Cipolat S, Rudka T, Hartmann D *et al.* (2006) Mitochondrial rhomboid PARL regulates cytochrome *c* release during apoptosis via OPA1-dependent cristae remodeling. *Cell* **126**: 163–175.

Davies VJ, Hollins AJ, Piechota MJ *et al.* (2007) Opa1 deficiency in a mouse model of autosomal dominant optic atrophy impairs mitochondrial morphology, optic nerve structure and visual function. *Human Molecular Genetics* **16**: 1307–1318.

Delettre C, Lenaers G, Griffoin JM *et al.* (2000) Nuclear gene OPA1, encoding a mitochondrial dynamin-related protein, is mutated in dominant optic atrophy. *Nature Genetics* **26**: 207–210.

Delivani P, Adrain C, Taylor RC, Duriez PJ and Martin SJ (2006) Role for CED-9 and Egl-1 as regulators of mitochondrial fission and fusion. *Molecular Cell* **21**: 761–773.

Detmer SA and Chan DC (2007) Functions and dysfunctions of mitochondrial dynamics. *Nature Reviews. Molecular Cell Biology* **8**: 870–879.

Eura Y, Ishihara N, Oka T and Mihara K (2006) Identification of a novel protein that regulates mitochondrial fusion by modulating mitofusin (Mfn) protein function. *Journal of Cell Science* **119**: 4913–4925.

Frank S, Gaume B, Bergmann-Leitner ES *et al.* (2001) The role of dynamin-related protein 1, a mediator of mitochondrial fission, in apoptosis. *Developmental Cell* **1**: 515–525.

Frey TG and Mannella CA (2000) The internal structure of mitochondria. *Trends in Biochemical Sciences* **25**: 319–324.

Gandre-Babbe S and van der Bliek AM (2008) The novel tail-anchored membrane protein Mff controls mitochondrial and peroxisomal fission in mammalian cells. *Molecular Biology of the Cell* **19**: 2402–2412.

Goyal G, Fell B, Sarin A, Youle RJ and Sriram V (2007) Role of mitochondrial remodeling in programmed cell death in *Drosophila melanogaster*. *Developmental Cell* **12**: 807–816.

Griparic L, Kanazawa T and van der Bliek AM (2007) Regulation of the mitochondrial dynamin-like protein Opa1 by proteolytic cleavage. *Journal of Cell Biology* **178**: 757–764.

Hajek P, Chomyn A and Attardi G (2007) Identification of a novel mitochondrial complex containing mitofusin 2 and stomatin-like protein 2. *Journal of Biological Chemistry* **282**: 5670–5681.

Harder Z, Zunino R and McBride H (2004) Sumo1 conjugates mitochondrial substrates and participates in mitochondrial fission. *Current Biology* **14**: 340–345.

Ingerman E, Perkins EM, Marino M *et al.* (2005) Dnm1 forms spirals that are structurally tailored to fit mitochondria. *Journal of Cell Biology* **170**: 1021–1027.

Ishihara N, Eura Y and Mihara K (2004) Mitofusin 1 and 2 play distinct roles in mitochondrial fusion reactions via GTPase activity. *Journal of Cell Science* **117**: 6535–6546.

Jagasia R, Grote P, Westermann B and Conradt B (2005) DRP-1-mediated mitochondrial fragmentation during EGL-1-induced cell death in *C. elegans*. *Nature* **433**: 754–760.

James DI, Parone PA, Mattenberger Y and Martinou JC (2003) hFis1, a novel component of the mammalian mitochondrial fission machinery. *Journal of Biological Chemistry* **278**: 36373–36379.

Karbowski M, Arnoult D, Chen H *et al.* (2004) Quantitation of mitochondrial dynamics by photolabeling of individual organelles shows that mitochondrial fusion is blocked during the Bax activation phase of apoptosis. *Journal of Cell Biology* **164**: 493–499.

Karbowski M, Lee YJ, Gaume B *et al.* (2002) Spatial and temporal association of Bax with mitochondrial fission sites, Drp1, and Mfn2 during apoptosis. *Journal of Cell Biology* **159**: 931–938.

Karbowski M, Neutzner A and Youle RJ (2007) The mitochondrial E3 ubiquitin ligase MARCH5 is required for Drp1 dependent mitochondrial division. *Journal of Cell Biology* **178**: 71–84.

Karbowski M, Norris KL, Cleland MM, Jeong SY and Youle RJ (2006) Role of Bax and Bak in mitochondrial morphogenesis. *Nature* **443**: 658–662.

Koshiba T, Detmer SA, Kaiser JT *et al.* (2004) Structural basis of mitochondrial tethering by mitofusin complexes. *Science* **305**: 858–862.

Li H, Chen Y, Jones AF *et al.* (2008) Bcl-xL induces Drp1-dependent synapse formation in cultured hippocampal neurons. *Proceedings of the National Academy of Sciences of the USA* **105**: 2169–2174.

Liu QA and Shio H (2008) Mitochondrial morphogenesis, dendrite development, and synapse formation in cerebellum require both Bcl-w and the glutamate receptor delta2. *PLoS Genetics* **4**: e1000097.

Lonergan T, Brenner C and Bavister B (2006) Differentiation-related changes in mitochondrial properties as indicators of stem cell competence. *Journal of Cell Physiology* **208**: 149–153.

Malka F, Guillery O, Cifuentes-Diaz C *et al.* (2005) Separate fusion of outer and inner mitochondrial membranes. *EMBO Report* **6**: 853–859.

Meeusen S, McCaffery JM and Nunnari J (2004) Mitochondrial fusion intermediates revealed *in vitro*. *Science* **305**: 1747–1752.

Nunnari J, Marshall WF, Straight A *et al.* (1997) Mitochondrial transmission during mating in *Saccharomyces cerevisiae* is determined by mitochondrial fusion and fission and the intramitochondrial segregation of mitochondrial DNA. *Molecular Biology of the Cell* **8**: 1233–1242.

Olichon A, Baricault L, Gas N *et al.* (2003) Loss of OPA1 perturbates the mitochondrial inner membrane structure and integrity, leading to cytochrome *c* release and apoptosis. *Journal of Biological Chemistry* **278**: 7743–7746.

Sheridan C, Delivani P, Cullen SP and Martin SJ (2008) Bax- or Bak-induced mitochondrial fission can be uncoupled from cytochrome *c* release. *Molecular Cell* **31**: 570–585.

Smirnova E, Griparic L, Shurland DL and van der Bliek AM (2001) Dynamin-related protein Drp1 is required for mitochondrial division in mammalian cells. *Molecular Biology of the Cell* **12**: 2245–2256.

Sugioka R, Shimizu S and Tsujimoto Y (2004) Fzo1, a protein involved in mitochondrial fusion, inhibits apoptosis. *Journal of Biological Chemistry* **279**: 52726–52734.

Sun MG, Williams J, Munoz-Pinedo C *et al.* (2007) Correlated three-dimensional light and electron microscopy reveals transformation of mitochondria during apoptosis. *Nature Cell Biology* **9**: 1057–1065.

Tan FJ, Husain M, Manlandro CM *et al.* (2008) CED-9 and mitochondrial homeostasis in *C. elegans* muscle. *Journal of Cell Science* **121**: 3373–3382.

Twig G, Elorza A, Molina AJ *et al.* (2008) Fission and selective fusion govern mitochondrial segregation and elimination by autophagy. *EMBO Journal* **27**: 433–446.

Varadi A, Johnson-Cadwell LI, Cirulli V *et al.* (2004) Cytoplasmic dynein regulates the subcellular distribution of mitochondria by controlling the recruitment of the fission factor dynamin-related protein-1. *Journal of Cell Science* **117**: 4389–4400.

Wasiak S, Zunino R and McBride HM (2007) Bax/Bak promote sumoylation of DRP1 and its stable association with mitochondria during apoptotic cell death. *Journal of Cell Biology* **177**: 439–450.

Waterham HR, Koster J, van Roermund CW *et al.* (2007) A lethal defect of mitochondrial and peroxisomal fission. *New England Journal of Medicine* **356**: 1736–1741.

Yamaguchi R, Lartigue L, Perkins G *et al.* (2008) Opa1-mediated cristae opening is Bax/Bak and BH3 dependent, required for apoptosis, and independent of Bak oligomerization. *Molecular Cell* **31**: 557–569.

Yonashiro R, Ishido S, Kyo S *et al.* (2006) A novel mitochondrial ubiquitin ligase plays a critical role in mitochondrial dynamics. *EMBO Journal* **25**: 3618–3626.

Yoon Y, Krueger EW, Oswald BJ and McNiven MA (2003) The mitochondrial protein hFis1 regulates mitochondrial fission in mammalian cells through an interaction with the dynamin-like protein DLP1. *Molecular and Cellular Biology* **23**: 5409–5420.

Further Reading

Benard G and Karbowski M (2009) Mitochondrial fusion and division: regulation and role in cell viability. *Seminars in Cell & Developmental Biology* **20**(3): 365–374.

Chan DC (2006) Mitochondria: dynamic organelles in disease, aging, and development. *Cell* **125**: 1241–1252.

Neuspiel M, Zunino R, Gangaraju S, Rippstein P and McBride H (2005) Activated mitofusin 2 signals mitochondrial fusion, interferes with Bax activation, and reduces susceptibility to radical induced depolarization. *Journal of Biological Chemistry* **280**: 25060–25070.

Rojo M, Legros F, Chateau D and Lombes A (2002) Membrane topology and mitochondrial targeting of mitofusins, ubiquitous mammalian homologs of the transmembrane GTPase Fzo. *Journal of Cell Science* **115**: 1663–1674.

Schieke SM, Ma M, Cao L *et al.* (2008) Mitochondrial metabolism modulates differentiation and teratoma formation capacity in mouse embryonic stem cells. *Journal of Biological Chemistry* **283**: 28506–28512.

Shaw JM and Nunnari J (2002) Mitochondrial dynamics and division in budding yeast. *Trends in Cell Biology* **12**: 178–184.

Vogel F, Bornhovd C, Neupert W and Reichert AS (2006) Dynamic subcompartmentalization of the mitochondrial inner membrane. *Journal of Cell Biology* **175**: 237–247.

Westermann B (2008) Molecular machinery of mitochondrial fusion and fission. *Journal of Biological Chemistry* **283**: 13501–13505.

Youle RJ and Strasser A (2008) The BCL-2 protein family: opposing activities that mediate cell death. *Nature Reviews. Molecular Cell Biology* **9**: 47–59.

Death Receptors

Peter H Krammer, *German Cancer Research Center, Heidelberg, Germany*

Inna N Lavrik, *German Cancer Research Center, Heidelberg, Germany*

Advanced article

Article Contents

- Introduction
- Structure
- Death Ligands and Initiation of Death Receptor Signalling
- Caspases in Death Receptor Signalling
- CD95/Fas/APO-1 Signalling
- TRAIL- R1/TRAIL-R2 Signalling
- TNF Signalling
- DR3 and DR6 signalling
- Modulation of DR-induced Apoptosis

Apoptosis or programmed cell death is a common property of all multicellular organisms. It can be triggered by a number of factors including ultraviolet (UV) or γ-irradiation, chemotherapeutic drugs or growth factor withdrawal. A death signal can either be induced by death receptors (extrinsic pathway) or via the mitochondria (intrinsic pathway). The death receptor (DR) family is the subfamily of the tumour necrosis factor receptor (TNFR) superfamily. Stimulation of the death receptors results in the transduction of either apoptotic or survival signals. Here we present a general overview of death receptor signalling and describe the main mechanisms leading to activation of death receptor signalling pathways.

Introduction

Six members of the tumour necrosis factor receptor (TNFR) superfamily are distinguished by a region of approximately 80 amino acids length at their intracellular part that was termed death domain (DD). These receptors are referred to as death receptors (DR). Six members of the DR family have been characterized so far: TNF-R1 (DR1, CD120a, p55, p60), CD95 (DR2, apoptosis antigen-1 (APO-1)/FAS, factor-activating Exo S), DR3 (APO-3, LARD, lymphocyte-associated receptor death; TNFR-associated membrane protein, TRAMP; WSL1), TNF-related apoptosis-inducing ligand (TRAIL)-R1 (APO-2, DR4), TRAIL-R2 (DR5, KILLER, TRICK2, TRAIL receptor inducer of cell killing) and DR6 (**Figure 1**) (Ashkenazi and Dixit, 1998; Krammer, 2000). Stimulation of DR results in the transduction of either apoptotic or survival signals. **See also**: Apoptosis: Regulatory Genes and Disease; Death Receptors at the Molecular Level: Therapeutic Implications

Structure

All members of the DR family are Type I transmembrane proteins comprising three main parts: the extracellular, the transmembrane and the intracellular part (Bodmer *et al.*, 2002; Ashkenazi and Dixit, 1998). Extracellular domains of DR are characterized by the presence of the so-called cysteine rich repeats (**Figure 1**). In addition, extracellular domains are highly glycosylated. An important feature of the intracellular domains is the presence of the DD, which plays the central role in the signal transduction. DD are formed by six α helixes, located in antiparallel orientation following the so-called Greek key topology.

Death Ligands and Initiation of Death Receptor Signalling

The DR signalling cascade is typically triggered by a corresponding death ligand (DL). The DL family includes tumour necrosis factor (TNF), CD95 ligand (CD95L; also known as FASL), TL1A and TRAIL. Following DR stimulation a number of molecules are recruited to the DD of a receptor, which are responsible for starting a signalling cascade. DL also interact with decoy receptors (DcR) that do not possess DD and, correspondingly, do not form a signalling complex (Ashkenazi and Dixit, 1999). Four DcR have been described to date: TRAIL-R3 (DcR1), TRAIL-R4 (DcR2), DcR3 and osteoprotegerin (OPG) (**Figure 1**). TRAIL-R3 (DcR1), TRAIL-R3 (DcR2) and OPG bind TRAIL, whereas DcR3 interacts with CD95L.

There are two types of signalling complexes formed at the DR. The first type of complex is named death-inducing signalling complex (DISC), which is formed at CD95 and TRAIL receptors (Lavrik *et al.*, 2005a).

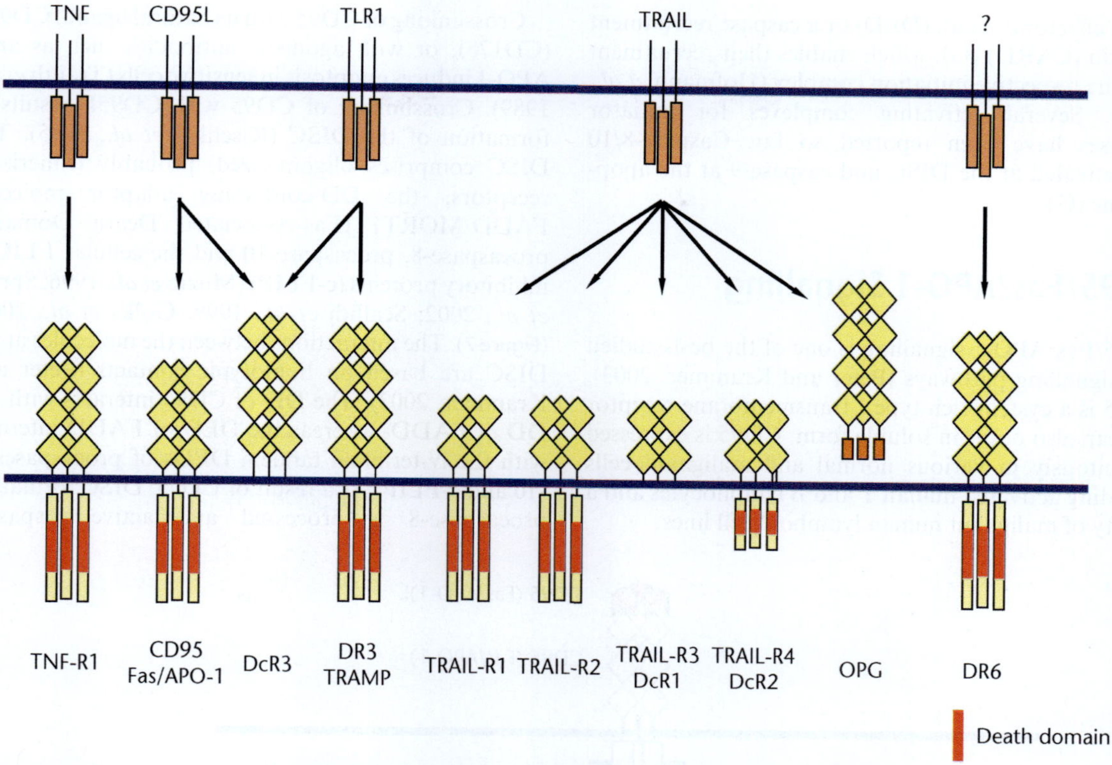

Figure 1 Death receptors. Death receptors and corresponding death ligands are shown. Death domains (DD) are shown in red. DR-death receptors. DcR-decoy receptors.

Formation of the DISC leads to initiation of the apoptotic signal. The second type of complexes is formed at the TNF-R1, DR3 and DR6 and leads to induction of nonapoptotic pathways, such as the nuclear factor-κB (NFκB) pathway.

Caspases in Death Receptor Signalling

Caspases play the central role in the transduction of DR apoptotic signals (Thornberry and Lazebnik, 1998; Salvesen, 2002). Caspases, a family of cysteinyl aspartate-specific proteases, are synthesized as zymogens with a prodomain of variable length followed by the large subunit (p20) and the small subunit (p10) (Fuentes-Prior and Salvesen, 2004). The caspases are activated through proteolysis at specific aspartate (D) residues that are located between the prodomain, the p20 and p10 subunits. This results in the generation of mature active caspases that consist of heterotetramers $p20_2$-$p10_2$. Subsequently, active caspases specifically process various substrates that are implicated in apoptosis and inflammation. Depending on the structure of the prodomain and their function, caspases are typically divided into initiator caspases and effector caspases (Fuentes-Prior and Salvesen, 2004). Caspases with large prodomains are referred to as initiator caspases, whereas caspases with a short prodomain of 20–30 amino acids are named effector caspases. **See also**: Caspases and Cell Death; Caspases, Substrates and Sequential Activation; Dismantling the Apoptotic Cell; The Apoptosome: The Executioner of Mitochondria-mediated Apoptosis

All caspases are produced in cells as catalytically inactive zymogens and must undergo proteolytic processing and activation during apoptosis. The effector caspases are activated by initiator or apical caspases. It is generally accepted that apical caspase activation takes place in large protein complexes bringing together several caspase zymogens (Boatright *et al*., 2003). All initiator caspases are characterized by the presence of a member of the 'death domain superfamily' containing a death effector domain (DED) or a caspase recruitment domain (CARD) (G), which enables their recruitment into the respective initiation complex (Hofmann *et al.*, 1997). Several activating complexes for initiator caspases have been reported so far. Caspase-8/10 are activated at the DISC and caspase-9 at the apoptosome (G).

CD95/Fas/APO-1 Signalling

CD95/Fas/APO-1 signalling is one of the best-studied DR signalling pathways (Peter and Krammer, 2003). CD95 is a cystein-rich type I transmembrane receptor and can also occur in soluble form. CD95 is expressed ubiquitously in various normal and malignant cells including activated human T and B lymphocytes and a variety of malignant human lymphoid cell lines.

Crosslinking of CD95 with its natural ligand, CD95L (CD178), or with agonistic antibodies such as anti-APO-1 induces apoptosis in sensitive cells (Trauth *et al.*, 1989). Crosslinking of CD95 with CD95L results in formation of the DISC (Kischkel *et al.*, 1995). The DISC comprises oligomerized, probably trimerized, receptors, the DD-containing adaptor molecule FADD/MORT1 (Fas-Associated Death Domain), procaspase-8, procaspase-10 and the cellular FLICE-inhibitory proteins (c-FLIP) (Muzio *et al.*, 1996; Sprick *et al.*, 2002; Scaffidi *et al.*, 1999; Golks *et al.*, 2005) (**Figure 2**). The interactions between the molecules at the DISC are based on homotypic contacts (Peter and Krammer, 2003). The DD of CD95 interacts with the DD of FADD, whereas the DED of FADD interacts with the *N*-terminal tandem DEDs of procaspases-8, -10 and c-FLIP. As a result of CD95, DISC formation procaspase-8 is processed and active caspase-8

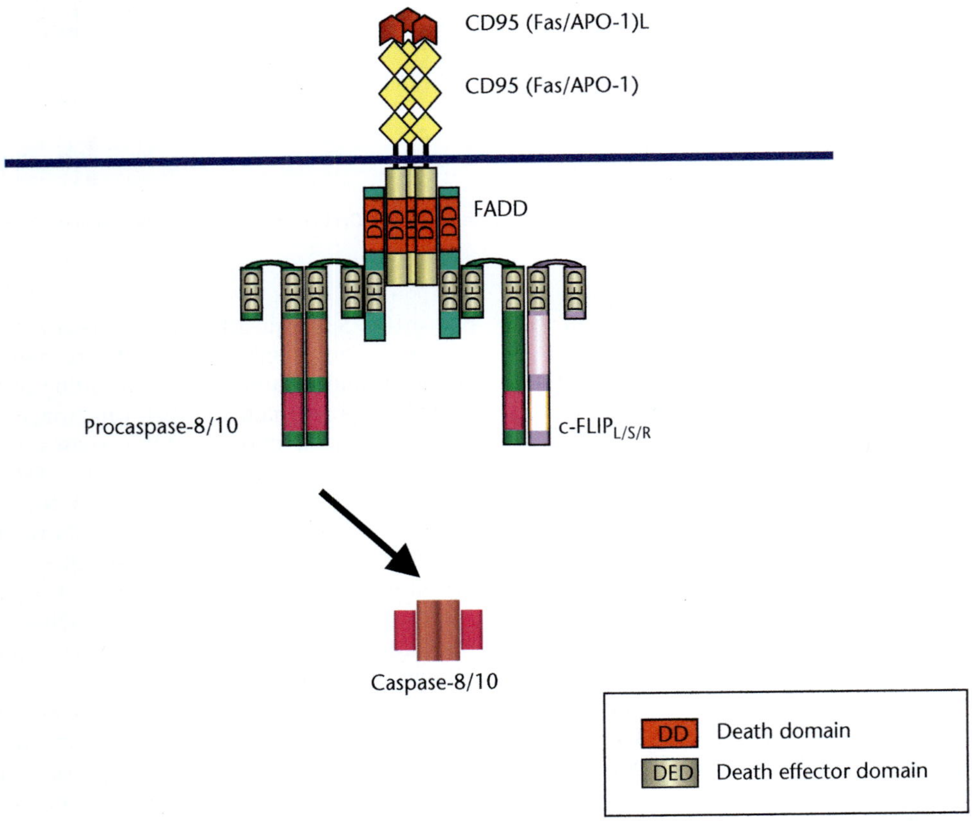

Figure 2 CD95 death-inducing signaling complex (DISC). The DISC consists of CD95, (depicted in yellow), FAS-associated death domain, FADD, (depicted in light blue), procaspase-8/procaspase-10, (depicted in green) and cellular FLICE-inhibitory proteins, c-FLIP, (depicted in violet). The interactions between the molecules at the DISC are based on homotypic contacts. The death domain (DD) of CD95 interacts with the DD of FADD, whereas the death effector domain (DED) of FADD interacts with the N-terminal tandem DEDs of procaspase-8, procaspase-10 and c-FLIP. DD are shown in red; DED are shown in light yellow.

heterotetramer $p10_2$-$p18_2$ is formed. Subsequently, active caspase-8 initiates the caspase cascade. Procaspase-10 is also activated at the DISC, forming an active heterotetramer. However, whether caspase-10 can trigger cell death in the absence of caspase-8 in response to CD95 stimulation is controversial.

Two types of apoptosis CD95 signalling pathways have been established (Scaffidi *et al.*, 1998) (**Figure 3**). Type I cells are characterized by high levels of DISC formation and high amounts of active caspase-8. Activated caspase-8 directly leads to activation of downstream, effector caspases. In Type II cells, there are lower levels of CD95 DISC formation and, thus, lower levels of active caspase-8. In this case, signalling requires an amplification loop. This amplification loop involves the cleavage by caspase-8 of the Bcl-2-family protein Bid to generate truncated (t) Bid and subsequent tBid-mediated release of cytochrome C (cyt C) from mitochondria. The release of cyt C from mitochondria results in apoptosome formation followed by the activation of procaspase-9, which in turn cleaves downstream, effector caspases-3, -6 and -7. Type II CD95 signalling might be blocked by Bcl-2 family members such as Bcl-2 and Bcl-x_L. **See also**: BH3-Only Proteins; The Bcl-2 Family Proteins – Key Regulators and Effectors of Apoptosis

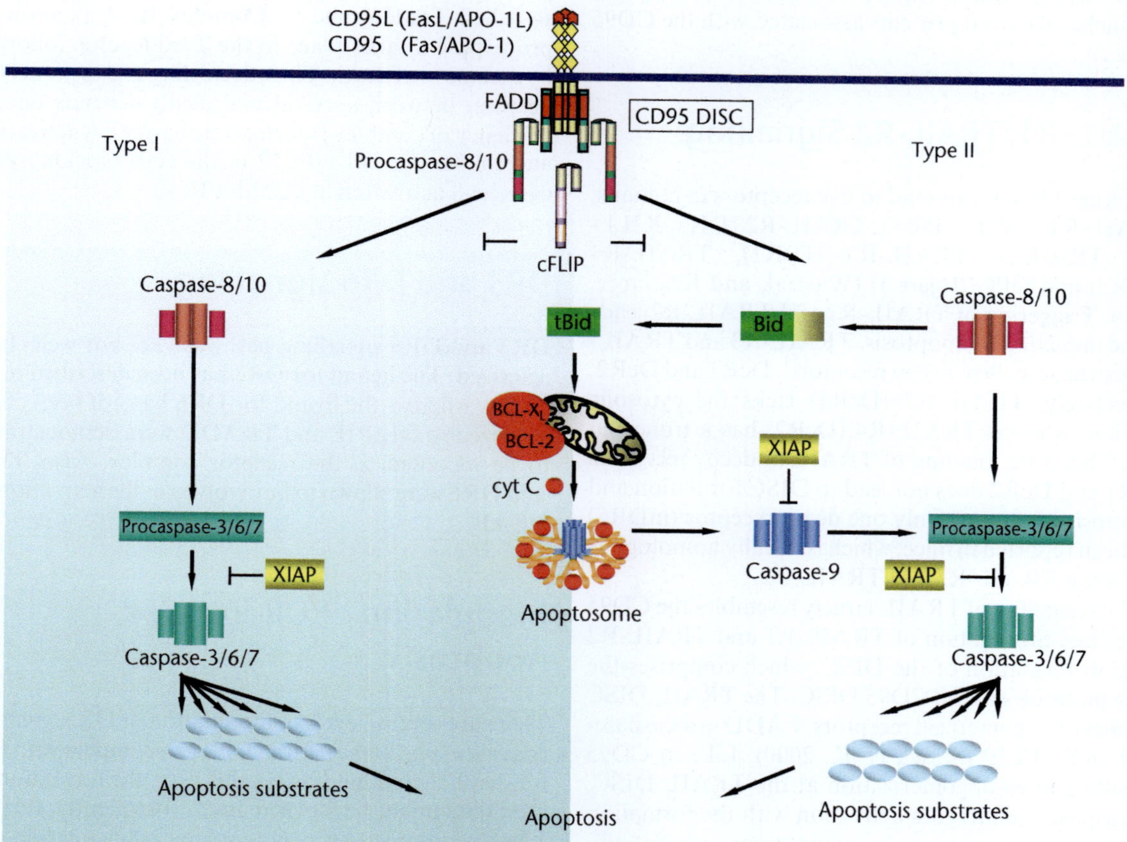

Figure 3 CD95 signaling pathways. Stimulation with ligand (CD95L/FasL/APO-1L) leads to formation of the CD95 death-inducing signaling complex (DISC), where activation of caspase-8 and caspase-10 takes place. Caspase-8-mediated apoptosis can occur in two ways according to the Type I/Type II model. Type I cells are characterized by high levels of DISC formation and increased amounts of active caspase-8. Activated caspase-8 directly leads to the activation of downstream effector caspase-3, caspase-6 and caspase-7. In Type II cells, there are lower levels of CD95 DISC formation and, thus, lower levels of active caspase-8. In this case, signaling requires an additional amplification loop that involves the cleavage by caspase-8 of the Bcl-2-family protein Bid to generate truncated (t) Bid and subsequent tBid-mediated release of cytochrome C (cyt C) from mitochondria. The release of cyt C from mitochondria results in apoptosome formation followed by the activation of initiator procaspase-9, which in turn cleaves downstream effector caspases. Type II CD95 signaling might be blocked by Bcl-2 family members such as Bcl-2 and Bcl-x_L.

In addition, to the CD95 DISC formed at the membrane a secondary DED-containing protein complex is formed, which plays an amplification role in caspase activation (Lavrik *et al.*, 2008). This complex was named complex II. Complex II comprises procaspase-8, FADD and c-FLIP proteins and does not contain CD95. Also internalization of CD95 has been reported to play a role in amplification of the apoptotic signal (Lee *et al.*, 2006; Feig *et al.*, 2007).

CD95 is not only a potent apoptosis inducer but is also capable of activating multiple nonapoptotic pathways involving NF-κB, Erk1/2, p38 and JNK (Peter *et al.*, 2007). However, the exact molecular mechanisms of nonapoptotic signalling via CD95 remains unclear and possibly involves signal transduction through as yet uncharacterized proteins associated with the CD95 DISC.

TRAIL- R1/TRAIL-R2 Signalling

Cytokine TRAIL can bind to five receptors in humans: TRAIL-R1 (APO-2, DR4), TRAIL-R2 (DR5, KILLER, TRICK2), TRAIL-R3 (DcR1), TRAIL-R4 (DcR2) and OPG (**Figure 1**) (Walczak and Krammer, 2000). Triggering of TRAIL-R1 and TRAIL-R2 leads to the induction of apoptosis. TRAIL-R3 and TRAIL-R4 are the so-called 'decoy receptors', DcR1 and DcR2, respectively. TRAIL-R3 (DcR1) lacks the cytosolic domain, whereas TRAIL-R4 (DcR2) has a truncated DD. Therefore, binding of TRAIL to decoy receptors DcR1 and DcR2 does not lead to DISC formation and apoptosis initiation. Only one death receptor (mDR5) has been reported in mice, which is equally homologous to human TRAIL-R1 and TRAIL-R2.

The signalling of TRAIL largely resembles the CD95 signalling. Stimulation of TRAIL-R1 and TRAIL-R2 leads to formation of the DISC, which comprises the same molecules as the CD95 DISC. The TRAIL DISC consists of oligomerized receptors, FADD, procaspase-8/10 and c-FLIP (Sprick *et al.*, 2000). Like in CD95 signalling after oligomerization at the TRAIL DISC, procaspase-8 undergoes activation with the formation of an active caspase-8 heterotetramer, which subsequently triggers the apoptotic cascade. Procaspase-10 is also activated at the TRAIL DISC.

Addition of TRAIL to the cells has been also reported to induce NF-κB, JNK, p38 and other nonapoptotic pathways. Receptor-interacting protein (RIP) and TRAF2 were reported to be essential for the initiation of the nonapoptotic pathways via TRAIL signalling pathway.

TNF Signalling

TNF signalling differs from CD95- or TRAIL signalling. Recently several groups have demonstrated that TNF stimulation leads to formation of two independent complexes. Complex I, formed at the membrane, comprises TNF-R1, RIP1, TRADD, TRAF2 and probably other not yet identified molecules (Micheau and Tschopp, 2003). Complex I triggers the NF-κB signalling pathway via recruitment of the IκB kinase (IKK) complex and activates Jun *N*-terminal kinase (JNK) in a TRAF-2-dependent way. Complex I is devoid of FADD and procaspase-8. It was shown that complex I translocates to the cytosol where FADD, procaspases-8/10 and FLIP are recruited to form the so-called Traddosome or Complex II. Activation of procaspase-8 takes place in the Traddosome followed by execution of death. Thus, according to this model the decision between survival and death depends on the efficiency of complex II formation, caspase-8 activation and the amount of c-FLIP in the cells blocking procaspase-8 activation at complex II.

DR3 and DR6 signalling

DR3 and DR6 signalling pathways are not well characterized. The ligand for DR3 has been described to be TL1A, whereas the ligand for DR6 has not been characterized so far. RIP and TRADD were demonstrated to be recruited to the receptor complex. Also, DR3 and DR6 were shown to be involved in the activation of NF-κB.

Modulation of DR-induced Apoptosis

There are several levels of modulation of DR signalling (Lavrik *et al.*, 2005b). First, DcR can compete with DR for death ligand binding. In this way, the formation of the DISC might be blocked and, subsequently, further steps in apoptosis such as procaspase-8 activation will not be possible.

The next level of blocking of DR-induced apoptosis is by c-FLIP proteins (Krueger *et al.*, 2001b). C-FLIP, also known as FLAME-1/I-FLICE/CASPER/CASH/MRIT/CLARP/Usurpin, is a well-described inhibitor of death receptor-mediated apoptosis. On the messenger ribonucleic acid (mRNA) level, c-FLIP exists in multiple splice variants whereas on the protein level only three isoforms, c-FLIP$_L$ (L = Long), c-FLIP$_S$ (S = Short) and c-FLIP$_R$, (R = Raji) have been described (**Figure 4**) (Krammer *et al.*, 2007; Krueger *et al.*, 2001a). All three c-FLIP isoforms contain two death effector domains (DED) that are structurally similar to the *N*-terminal part of procaspase-8. C-FLIP$_L$ also contains catalytically inactive caspase-like domains (p20 and p12) (**Figure 4**).

C-FLIP proteins are recruited to the DISC and complex II by DED interactions. Short c-FLIP isoforms, c-FLIP$_S$ and c-FLIP$_R$ block caspase-8 activation at the DISC and complex II, and thereby DR-induced apoptosis (Krammer *et al.*, 2007). The role of c-FLIP$_L$ in the DR-induced apoptosis is more complex. It has been shown that depending on the c-FLIP$_L$ concentration at the DISC, c-FLIP$_L$ can act either as an antiapoptotic molecule, functioning in a way analogous to c-FLIP$_S$, or as a proapoptotic molecule, facilitating the activation of procaspase-8 at the DISC (Micheau *et al.*, 2002; Chang *et al.*, 2002). This proapoptotic role goes along with the phenotype of c-FLIP-deficient mice, which are characterized by heart failure and death at embryonic day 10.5 (Yeh *et al.*, 2000). The same phenotype is described for caspase-8- and FADD-deficient mice (Yeh *et al.*, 1998; Varfolomeev *et al.*, 1998).

Further downstream, XIAPs (X-linked Inhibitor of Apoptosis Protein) inhibit effector caspase activation by direct binding to caspase-3, -7 and -9. In addition, there are inhibitors of XIAP molecules such as Smac which can, in turn, inhibit the action of XIAPs and thereby enable the apoptosis.

At the mitochondrial level, the ratio between pro- and antiapoptotic Bcl-2-family members defines life/death decisions. Overexpression of antiapoptotic Bcl-2 family members such as Bcl-2 and Bcl-x$_L$ block cyt C release from mitochondria and, subsequently, apoptosome formation and caspase-9 activation required for DR-induced apoptosis in Type II cells. **See also**: Inhibitor of Apoptosis (IAP) and BIR-containing Proteins

Our current understanding of DR signalling has opened possibilities for the design of new strategies of targeting DR pathways. This would allow the treatment of a number of diseases associated with defects in DR signalling. Understanding of this complex regulation in DR-induced apoptosis and the role of the particular inhibitors is also a challenge for the newly developing field of systems biology (Bentele *et al.*, 2004).

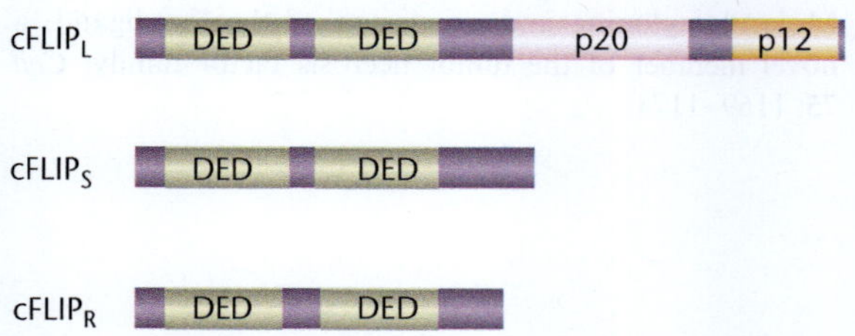

Figure 4 Scheme of c-FLIP isoforms. The scheme represents three isoforms of c-FLIP: c-FLIP$_{Long}$ (c-FLIP$_L$), c-FLIP$_{Short}$ (c-FLIP$_S$) and c-FLIP$_{Raji}$ (c-FLIP$_R$). Death effector domain (DED) is presented in light yellow. Catalytically inactive caspase-like domains of c-FLIP$_L$: p20 and p12 are indicated.

References

Ashkenazi A and Dixit VM (1998) Death receptors: signaling and modulation. *Science* **281**: 1305–1308.

Ashkenazi A and Dixit VM (1999) Apoptosis control by death and decoy receptors. *Current Opinion in Cell Biology* **11**: 255–260.

Bentele M, Lavrik I, Ulrich M *et al.* (2004) Mathematical modeling reveals threshold mechanism in CD95-induced apoptosis. *Journal of Cell Biology* **166**: 839–851.

Boatright KM, Renatus M, Scott FL *et al.* (2003) A unified model for apical caspase activation. *Molecular Cell* **11**: 529–541.

Bodmer JL, Schneider P and Tschopp J (2002) The molecular architecture of the TNF superfamily. *Trends in Biochemical Sciences* **27**: 19–26.

Chang DW, Xing Z, Pan Y *et al.* (2002) c-FLIP(L) is a dual function regulator for caspase-8 activation and CD95-mediated apoptosis. *EMBO Journal* **21**: 3704–3714.

Feig C, Tchikov V, Schutze S and Peter ME (2007) Palmitoylation of CD95 facilitates formation of SDS-stable receptor aggregates that initiate apoptosis signaling. *EMBO Journal* **26**: 221–231.

Fuentes-Prior P and Salvesen GS (2004) The protein structures that shape caspase activity, specificity, activation and inhibition. *Biochemical Journal* **384**: 201–232.

Golks A, Brenner D, Fritsch C, Krammer PH and Lavrik IN (2005) c-FLIPR, a new regulator of death receptor-induced apoptosis. *Journal of Biological Chemistry* **280**: 14507–14513.

Hofmann K, Bucher P and Tschopp J (1997) The CARD domain: a new apoptotic signalling motif. *Trends in Biochemical Sciences* **22**: 155–156.

Kischkel FC, Hellbardt S, Behrmann I *et al.* (1995) Cytotoxicity-dependent APO-1 (Fas/CD95)-associated proteins form a death-inducing signaling complex (DISC) with the receptor. *EMBO Journal* **14**: 5579–5588.

Krammer PH (2000) CD95's deadly mission in the immune system. *Nature* **407**: 789–795.

Krammer PH, Arnold R and Lavrik IN (2007) Life and death in peripheral T cells. *Nature Reviews Immunology* **7**: 532–542.

Krueger A, Baumann S, Krammer PH and Kirchhoff S (2001a) FLICE-inhibitory proteins: regulators of death receptor-mediated apoptosis. *Molecular and Cellular Biology* **21**: 8247–8254.

Krueger A, Schmitz I, Baumann S, Krammer PH and Kirchhoff S (2001b) Cellular flice-inhibitory protein splice variants inhibit different steps of caspase-8 activation at the cd95 death-inducing signaling complex. *Journal of Biological Chemistry* **276**: 20633–20640.

Lavrik I, Golks A and Krammer PH (2005a) Death receptor signaling. *Journal of Cell Science* **118**: 265–267.

Lavrik IN, Golks A and Krammer PH (2005b) Caspases: pharmacological manipulation of cell death. *Journal of Clinical Investigation* **115**: 2665–2672.

Lavrik IN, Mock T, Golks A *et al.* (2008) CD95 stimulation results in the formation of a novel death effector domain protein-containing complex. *Journal of Biological Chemistry* **283**: 26401–26408.

Lee KH, Feig C, Tchikov V *et al.* (2006) The role of receptor internalization in CD95 signaling. *EMBO Journal* **25**: 1009–1023.

Micheau O, Thome M, Schneider P *et al.* (2002) The long form of FLIP is an activator of caspase-8 at the Fas death-inducing signaling complex. *Journal of Biological Chemistry* **277**: 45162–45171.

Micheau O and Tschopp J (2003) Induction of TNF receptor I-mediated apoptosis via two sequential signaling complexes. *Cell* **114**: 181–190.

Muzio M, Chinnaiyan AM, Kischkel FC *et al.* (1996) FLICE, a novel FADD-homologous ICE/CED-3-like protease, is recruited to the CD95 (Fas/APO-1) death-inducing signaling complex. *Cell* **85**: 817–827.

Peter ME, Budd RC, Desbarats J *et al.* (2007) The CD95 receptor: apoptosis revisited. *Cell* **129**: 447–450.

Peter ME and Krammer PH (2003) The CD95(APO-1/Fas) DISC and beyond. *Cell Death and Differentiation* **10**: 26–35.

Salvesen GS (2002) Caspases: opening the boxes and interpreting the arrows. *Cell Death and Differentiation* **9**: 3–5.

Scaffidi C, Fulda S, Srinivasan A *et al.* (1998) Two CD95 (APO-1/Fas) signaling pathways. *EMBO Journal* **17**: 1675–1687.

Scaffidi C, Schmitz I, Krammer PH and Peter ME (1999) The role of c-FLIP in modulation of CD95-induced apoptosis. *Journal of Biological Chemistry* **274**: 1541–1548.

Sprick MR, Rieser E, Stahl H *et al.* (2002) Caspase-10 is recruited to and activated at the native TRAIL and CD95 death-inducing signalling complexes in a FADD-dependent manner but can not functionally substitute caspase-8. *EMBO Journal* **21**: 4520–4530.

Sprick MR, Weigand MA, Rieser E *et al.* (2000) FADD/MORT1 and caspase-8 are recruited to TRAIL receptors 1 and 2 and are essential for apoptosis mediated by TRAIL receptor 2. *Immunity* **12**: 599–609.

Thornberry NA and Lazebnik Y (1998) Caspases: enemies within. *Science* **281**: 1312–1316.

Trauth BC, Klas C, Peters AM *et al.* (1989) Monoclonal antibody-mediated tumor regression by induction of apoptosis. *Science* **245**: 301–305.

Varfolomeev EE, Schuchmann M, Luria V *et al.* (1998) Targeted disruption of the mouse Caspase 8 gene ablates cell death induction by the TNF receptors, Fas/Apo1, and DR3 and is lethal prenatally. *Immunity* **9**: 267–276.

Walczak H and Krammer PH (2000) The CD95 (APO-1/Fas) and the TRAIL (APO-2L) apoptosis systems. *Experimental Cell Research* **256**: 58–66.

Yeh WC, Itie A, Elia AJ *et al.* (2000) Requirement for Casper (c-FLIP) in regulation of death receptor-induced apoptosis and embryonic development. *Immunity* **12**: 633–642.

Yeh WC, Pompa JL, McCurrach ME *et al.* (1998) FADD: essential for embryo development and signaling from some, but not all, inducers of apoptosis. *Science* **279**: 1954–1958.

Further Reading

Kischkel FC, Lawrence DA, Tinel A *et al.* (2001) Death receptor recruitment of endogenous caspase-10 and apoptosis initiation in the absence of caspase-8. *Journal of Biological Chemistry* **2**: 2.

Suda T, Takahashi T, Golstein P and Nagata S (1993) Molecular cloning and expression of the Fas ligand, a novel member of the tumor necrosis factor family. *Cell* **75**: 1169–1178.

Death Receptors at the Molecular Level: Therapeutic Implications

Advanced article

Article Contents

- Introduction
- Death Receptors as Potential Therapeutic Targets
- TRAIL/TRAIL-Receptors – Lessons from Studies at the Molecular Level
- Targeting TRAIL Death Receptors for Cancer Therapy

Marion MacFarlane, *MRC Toxicology Unit, University of Leicester, Leicester, UK*

Since the discovery that activation of a subset of cell surface receptors within the tumour necrosis factor (TNF) receptor superfamily could trigger apoptosis, several members of the TNF superfamily including TNF, CD95L and TNF-related apoptosis-inducing ligand (TRAIL) have been identified as potentially important targets for cancer therapy. Although systemic administration of TNF or CD95L causes severe toxic side effects, thus hampering their potential application in the clinic, the discovery of TRAIL and its cognate death receptors, TRAIL-R1 and TRAIL-R2, have provided an exciting new opportunity for selective targeting of tumour cells. TRAIL receptor activation has emerged as the most promising approach for death receptor-targeted therapy while inducing minimal toxicity in the majority of normal cells. Intensive research, including detailed analysis of death ligand–death receptor pairs at the structural level have enabled the development of several different approaches aimed at selective targeting of TRAIL-R1/TRAIL-R2 in tumour cells.

Introduction

Extrinsic apoptosis signals are initiated by binding of death ligands to specific pro-apoptotic 'death receptors' on the cell surface. Death receptors belong to the tumour necrosis factor receptor (TNF-R) superfamily and to date six members have been characterized, the best studied of these being TNF-R1, CD95 (APO-1 or Fas), TRAIL-R1 and TRAIL-R2. Since the discovery of this class of cell surface receptors and their corresponding death ligands, several members of the TNF superfamily including CD95L, TNF and TNF-related apoptosis-inducing ligand (TRAIL) have been identified as important targets for cancer therapy (Wiley *et al.*, 1995; reviewed in Ashkenazi, 2008). Although activation of CD95 or TNF-R1 can induce apoptosis in tumour cells, systemic administration of CD95L or TNF causes severe toxic side-effects, therefore hampering their potential application in the clinic. With the discovery of TRAIL and its cognate death receptors, TRAIL-R1 and TRAIL-R2, a new opportunity for death receptor targeting arose (reviewed in Ashkenazi and Dixit, 1998). For reasons that are still not well understood, tumour cells are more susceptible than normal cells to the cytotoxic effects of TRAIL. Consequently, TRAIL-R activation has emerged as the most promising approach for death receptor-targeted cancer therapy due to its remarkable feature of selectively inducing apoptosis in tumours *in vivo* without causing toxicity to the majority of normal cells (Walczak *et al.*, 1999). Further studies demonstrated that TRAIL, in combination with a wide range of conventional chemotherpeutics or irradiation, acts synergistically in the killing of tumour cells and importantly this can be achieved in the absence of any overt additional side effects (Ashkenazi *et al.*, 1999). These findings have motivated extensive research efforts aimed at increasing our understanding of death receptor activation/signalling, including detailed analysis of several death ligand–death receptor pairs at the structural level. These studies have enabled the development of different approaches aimed at selective death receptor targeting culminating in the current examination of several TRAIL-R agonists in Phase I and II clinical trials for the treatment of a variety of human tumours (as illustrated in **Table 1**). **See also**: Death Receptors

Death Receptors as Potential Therapeutic Targets

The rationale for targeting apoptosis in the treatment of cancer is primarily based on the observation that

Table 1 Development status of TRAIL death receptor agonists

Agent	Treatment schedule	Disease	Study phase
rhApo2L/TRAIL	Apo2L/TRAIL	Solid tumours/NHL	Phase Ia
	Apo2L/TRAIL + Rituximab	NHL	Phases Ib/II
	Apo2L/TRAIL + Chemotherapy	NSCLC	Phases Ib/II
HGS-ETR1	HGS-ETR1	NHL; CRC; NSCLC	Phase II
(Mapatumumab)	HGS-ETR1 + Chemotherapy/BTZ	Advanced solid tumours; MM	Phases Ib/II
HGS-ETR2	HGS-ETR2	Advanced solid tumours	Phase I
(Lexatumumab)	HGS-ETR2 + Chemotherapy	Solid and haematological tumours	Phase Ib
Apomab	Apopmab	Advanced solid tumours	Phase I
	Apomab + Avastin; Apomab + CD20	NSCLC; NHL	Phase II
AMG655	AMG655	NSCLC; CRC	Phase I
LBY135	LBY135 LBY135 + Chemotherapy	Advanced solid tumours	Phases I/II
TRA-8 (CS-1008)	TRA-8	Solid tumours and lymphoma	Phase I

Notes: BTZ, Bortezomib; CRC, colorectal carcinoma; MM, multiple myeloma; NHL, non-Hodgkin's lymphoma; NSCLC, non-small-cell lung cancer (reviewed in Ashkenazi and Herbst, 2008; Johnstone *et al.*, 2008)

apoptosis is deregulated in cancer cells but not in normal cells (Hanahan and Weinberg, 2000; Lowe *et al.*, 2004). Consequently, if pro-apoptotic pathways could be reactivated aberrant tumour cells may be more susceptible than normal cells. Indeed, the promotion of apoptosis is an important component of traditional anticancer therapies, including chemotherapeutic drugs and radiotherapy. However, p53 inactivation often renders tumour cells resistant to conventional therapies, and the lack of specificity of these therapies often results in systemic toxicity. These limitations of traditional therapy have driven the search for novel approaches that can circumvent mechanisms of drug resistance while specifically targeting tumour cells.

In this respect, the extrinsic pathway is uniquely attractive as a drug development target for several reasons. First, death receptors are widely expressed in tumours; thus death receptor agonists may exhibit a broad spectrum of anticancer activity. Second, because the extrinsic pathway triggers apoptosis independently of p53, targeting death receptors provides a strategy to kill tumours regardless of their p53 status. Furthermore, death receptor agonists could be useful not only for monotherapy but also in combination with traditional therapies or other therapies. For example, Bcl-2 family antagonists can cooperate with death receptor agonists by enhancing extrinsic–intrinsic pathway crosstalk, whereas inhibitor of apoptosis (IAP) antagonists might synergise by promoting caspase activation. **See also**: Inhibitor of Apoptosis (IAP) and BIR-containing Proteins; The Bcl-2 Family Proteins – Key Regulators and Effectors of Apoptosis

In principle, all of the above could apply generally to death receptors; however, death receptor subfamilies exhibit some important key differences. One of these is the degree of selectivity for tumour versus normal cells associated with the activation of different death receptor subfamilies. For example, the main function of TNFα is to stimulate proinflammatory gene expression through TNF-R1-mediated activation of nuclear factor-κB (NFκB), although TNFα can trigger apoptosis under certain circumstances, for example, when NFκB activation is blocked. As a result, the pro-inflammatory effects of TNFα on normal tissues have severely hampered the clinical development of TNFα-based approaches for systemic therapy. Nevertheless, TNFα has been used successfully to treat unresectable soft tissue sarcoma by isolated limb perfusion which facilitates local TNFα administration (Grunhagen *et al.*, 2006). **See also**: Death Receptors

Following the discovery of CD95 (Trauth *et al.*, 1989; Yonehara *et al.*, 1989), hopes were high that agonists of CD95 would provide powerful novel agents for the treatment of cancer. However, it was subsequently demonstrated that systemic administration of agonistic CD95 antibodies or CD95L led to massive hepatocyte apoptosis and lethal liver damage in animal models (Ogasawara *et al.*, 1993; Nagata, 1997). Thus, although CD95 represented the most potent physiologically occurring extracellular apoptosis inducer known, agonists of CD95 were deemed unsuitable for clinical investigation due to their severe systemic toxicity. To date, several approaches have been used to try and reinforce CD95 action at the tumour site while circumventing harmful side effects, including the development of cell surface antigen-restricted activation of CD95L-based fusion proteins. This approach is based on the fact that soluble CD95L alone has very little bioactivity but becomes bioactive when bound to an extracellular matrix. Thus, several fusion proteins have been generated where CD95L is connected to an antibody that specifically recognizes tumour cells or tumour stroma, such as the tumour stroma marker, FAP (fibroblast activation protein), or the T-cell leukaemia-associated antigen, CD7 (Samel *et al.*, 2003; Bremer *et al.*, 2006). In a very recent study, soluble CD95L was combined with Rituximab, a CD20-specific chimeric monoclonal antibody that specifically activates antibody-dependent cytotoxicity but also induces apoptosis by cross-linking of its target antigen, CD20. Using this approach, Bremer *et al.* (2008) demonstrated that the scFvRit:sCD95L fusion protein potently induced CD20-restricted apoptosis in various B cell lines while exhibiting no systemic toxicity in nude mouse models. Although several strategies aimed at harnessing CD95 action at the tumour site look promising none of these have to date been tested in clinical trials.

In contrast to TNF and CD95L, TRAIL selectively kills a variety of tumour cell lines while sparing the majority of normal cells. Currently there are few agents that are truly cancer cell-specific in terms of efficacy and induction of cell death; consequently this unique feature among TNF family death-inducing ligands makes TRAIL a promising tool for anticancer therapy (Walczak *et al.*, 1999; Ashkenazi *et al.*, 1999). Despite this unique phenomenon being realized more than a decade ago, the development of TRAIL as an anticancer agent was delayed due to reported hepatocyte toxicity. Although some recombinant forms of TRAIL were shown to be toxic to hepatocytes and other normal cells, these effects are thought to be related to the particular recombinant forms of the protein used rather than TRAIL itself (Lawrence *et al.*, 2001; Ganten *et al.*, 2006). However, a recent study suggests that caution should still be taken in terms of future TRAIL therapy in patients with inflammatory liver disease (Volkmann *et al.*, 2007). Despite these initial concerns about potential hepatotoxicity, optimized death receptor agonists targeting TRAIL-R1 or TRAIL-R2 have been well tolerated in preclinical safety models and in Phase I clinical trials (see later).

The molecular basis for the apparent tumour-selective activity of TRAIL remains to be fully elucidated and may relate to multiple factors. These include the sensitization of cancer cells to apoptosis induction by common oncogenes such as MYC and RAS, overexpression of specific *O*-glycosyl transferase enzymes that hyperglycosylate TRAIL-R1/R2 in tumours thereby promoting ligand-induced receptor clustering, and in some instances differential expression of the decoy receptors, TRAIL-R3 (DcR1) and TRAIL-R4 (DcR2) in tumours versus normal tissues. Furthermore, although expression of TRAIL-R1 and TRAIL-R2 is detectable in several tissues, some studies have suggested that it is generally higher in tumour cells compared with normal tissues (reviewed in Ashkenazi and Herbst, 2008).

In summary, in terms of death receptors as potential therapeutic targets, the pathway involving TRAIL and TRAIL-R1/R2 is clearly the most promising. As a result, extensive research efforts have been aimed at optimally targeting the pro-apoptotic death receptors, TRAIL-R1/TRAIL-R2 for potential cancer therapy. This has led to the development of TRAIL receptor agonists that target both TRAIL-R1 and TRAIL-R2 or selectively trigger apoptosis via one or other receptor (as illustrated in **Figure 1**).

TRAIL/TRAIL-Receptors – Lessons from Studies at the Molecular Level

TRAIL/Apo2L was discovered independently by two laboratories in the mid-1990s (Wiley *et al.*, 1995; Pitti *et al.*, 1996). TRAIL is expressed on the surface of natural killer cells and cytotoxic T cells, and loss-of-function studies in mice suggest that endogenous TRAIL plays a role in the killing of virus-infected or malignant cells by these immune effector cells. Recent findings suggest that TRAIL also plays a role in modulating memory T cells (reviewed in Johnstone *et al.*, 2008).

Immune cells express endogenous TRAIL as a Type II transmembrane protein of 281 amino acids; however, shedding of the extracellular *C*-terminal domain of the protein can also occur, thus releasing soluble TRAIL. Importantly, the very first TRAIL death receptor agonists were based on various recombinant versions of this endogenous soluble ligand. Indeed, for preclinical studies several recombinant TRAIL variants were generated, including versions with various exogenous polypeptide tags (thus aiding with purification of the recombinant protein), as well as a clinical grade untagged version of soluble TRAIL called recombinant

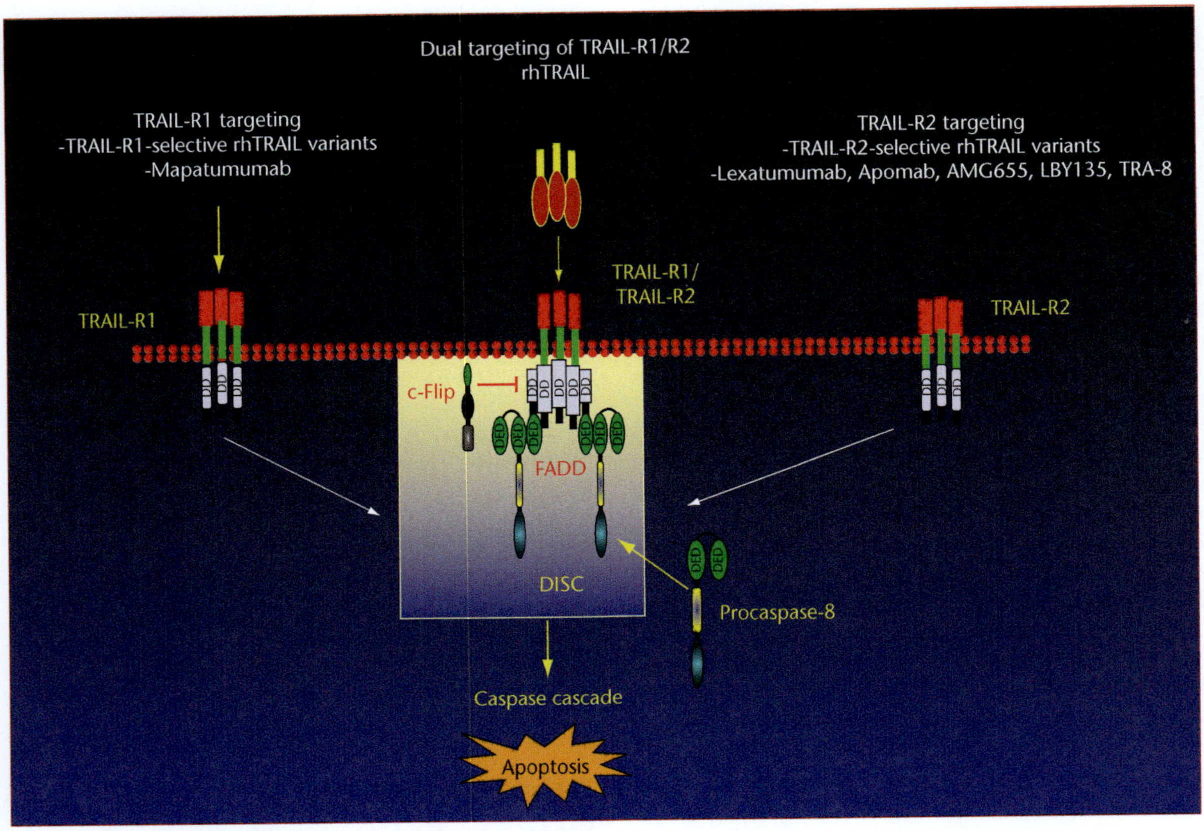

Figure 1 Potential Therapeutic approaches to target TRAIL death receptor activation. The main types of TRAIL receptor agonists discussed in this article include ligand- and antibody-based protein agents which induce apoptosis via TRAIL receptor-mediated activation of the caspase cascade. Recombinant human Apo2L/TRAIL interacts with the death receptors, TRAIL-R1 and TRAIL-R2 (as well as the decoy receptors, TRAIL-R3/R4), whereas receptor-selective rhTRAIL variants and the monoclonal agonist antibodies are monospecific for either TRAIL-R1 or TRAIL-R2 (see text and **Table 1** for further detail).

human Apo2L/TRAIL (rhApo2L/TRAIL), now being jointly developed by Genentech and Amgen (Ashkenazi *et al.*, 2008).

rhApo2L/TRAIL comprises amino acids 114–281 of the endogenous TRAIL molecule and is produced in *Escherichia coli* without any exogenous tag. The optimization of rhApo2L/TRAIL for clinical development was further enhanced by X-ray crystallographic studies of rhApo2L/TRAIL in complex with TRAIL-R2. Structural analysis of death ligand–death receptor pairs has the distinct advantage in that it reveals not only important conformational information but also identifies potentially key binding sites for ligand-induced receptor activation and thus induction of apoptosis. In this respect, the development of rhApo2L/TRAIL as well as several other potential TRAIL receptor agonists has been made possible by X-ray crystallography and structural modelling of TRAIL in complex with either TRAIL-R1 or TRAIL-R2, respectively. The crystal structure of the complex between rhApo2L or TRAIL and the extracellular domain of TRAIL-R2 revealed that soluble TRAIL is a homotrimeric molecule (as illustrated in **Figure 2**; Hymowitz *et al.*, 1999; Mongkolsapaya *et al.*, 1999). However, a key discovery and a unique feature of the cytokine TRAIL is the presence of a central zinc atom that coordinates the sulfhydryl groups of three unpaired cysteines, located at position 230 of each subunit (as illustrated in **Figure 2**). Indeed, it was reported that addition of zinc to the bacterial cell culture media and to purification buffers during recombinant protein production enabled nearly stoichiometric zinc coordination, thereby stabilizing the trimeric protein structure and maintaining solubility. It was also around this time that concerns were raised with respect to the apparent sensitivity of hepatocytes, and some other normal cell types, to certain tagged and

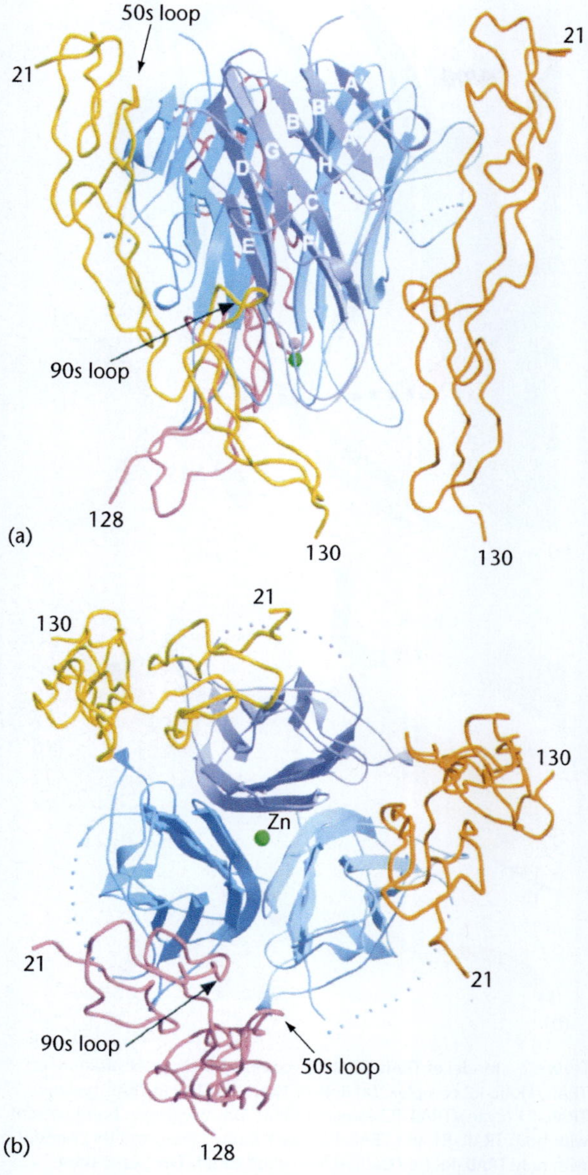

Figure 2 Crystal structure of the complex between rhApo2L/TRAIL and the extracellular domain of TRAIL-R2. The rhApo2L/TRAIL trimer is shown as a ribbon rendering in gradations of blue, and the three receptors are rendered as tubes in yellow and orange colours. The zinc atom that coordinates the sulfhydryl groups of three unpaired cysteines, located at position 230 of each subunit, is shown in green. β strands and relevant loops are labelled (see text for further detail). (a) Side view. In this orientation, the membrane of the receptor-containing cell is at the bottom of the figure. (b) Axial view. View down the 3-fold axis of the complex, perpendicular to (a) (Hymowitz *et al.*, 1999). Reproduced by permission of Elsevier (Cell Press).

non-zinc optimized variants of recombinant TRAIL. The view was that normal cells might require a higher degree of receptor crosslinking to trigger apoptosis, and it appears that such effects may be more evident with certain tagged or non-zinc optimized TRAIL variants, which reportedly tend to aggregate (Ganten *et al.*, 2006). As a result of these concerns, the version of rhApo2L/TRAIL currently in clinical development comprises amino acids 114–281, without any potentially oligomerizing exogenous tag, and has been purified at a neutral pH, in the presence of zinc to help stabilize the trimeric ligand structure (Ashkenazi *et al.*, 2008).

X-ray crystallographic analysis of the complex between Apo2L/TRAIL and the extracellular domain of TRAIL-R2 revealed that the ligand binds three receptor molecules, with each ligand subunit contacting two receptors, via cysteine-rich domain (CRD) 2 and CRD3 of TRAIL-R2. Essentially, two 'receptor loops' mediate most of the interactions, dividing the interface into two distinct patches: the '50s loop' (residues 51–65) and the '90s loop' (residues 91–104) (as illustrated in **Figure 2**). In patch A, the 90s loop interacts with a cluster of Apo2L residues around Gln-205 near the bottom of the trimer, whereas patch B is formed by the 50s loop of TRAIL-R2 and Apo2L/TRAIL residues around Tyr-216 near the top of the trimer (as illustrated in **Figure 2**). Structural analysis of the Apo2L/TRAIL–TRAIL-R2 complex further revealed that patch B involves general hydrophobic features of TNF-like ligands and appears to be important for binding of ligand–receptor complexes throughout the TNF superfamily, whereas patch A appears to control the specificity and cross-reactivity among different TNF superfamily members.

Although many cancer cell lines express both TRAIL-R1 and TRAIL-R2 (and in some cases decoy receptors) on their cell surface, until recently the relative contribution of TRAIL-R1/R2 to recombinant TRAIL-induced apoptosis in tumour cells was largely unknown. Furthermore, while rhApo2L/TRAIL would be predicted to bind to and induce apoptosis through either TRAIL-R1 or TRAIL-R2 or both (as illustrated in **Figure 1**), in some contexts the pro-apoptotic activity of this ligand could potentially be reduced by its inherent ability to also bind to the 'decoy' receptors TRAIL-R3/R4. To explore this question, we and others have generated TRAIL death receptor-selective TRAIL variants that specifically mediate apoptosis via TRAIL-R1/R2 (as illustrated in **Figure 1** and **Table 1**). Importantly, recombinant TRAIL variants capable of selectively targeting TRAIL-R1 or TRAIL-R2, but not the decoy receptors, could provide better tumour-specific therapies by avoiding potential decoy receptor-mediated antagonism, and thus could represent alternatives to existing agonistic TRAIL receptor antibodies (see later; detailed in **Table 1**).

Using a novel approach that enabled phage display of mutated trimeric proteins, Kelley *et al.* (2005) successfully generated a TRAIL-R2-selective variant, termed Apo2L.DR5-8, by mutation of six key residues in recombinant Apo2L/TRAIL and demonstrated good bioactivity of Apo2L.DR5-8 in several TRAIL-sensitive cell lines. Using the same approach, another TRAIL variant was designed which instead specifically bound to TRAIL-R1 (Apo2L.DR4-8). However, when this TRAIL-R1-specific mutant was tested in a panel of TRAIL-sensitive cell lines only weak bioactivity was observed. This led the authors to conclude that in the majority of tumour cell lines TRAIL signals to apoptosis via triggering of TRAIL-R2 rather than TRAIL-R1 (Kelley *et al.*, 2005). However, this suggestion was at odds with some earlier observations made in primary tumour cells from patients with chronic lymphocytic leukaemia (CLL); these TRAIL-resistant primary tumour cells could be sensitized to TRAIL-R-specific agonistic antibodies that specifically targeted TRAIL-R1 but not TRAIL-R2; thus, it would appear that CLL cells signal to apoptosis via TRAIL-R1 (MacFarlane *et al.*, 2005a). Furthermore, by omitting one (Y189A) of the six amino acid substitutions that had previously been reported to confer specificity for TRAIL-R1 (Kelley *et al.*, 2005), we designed a receptor-selective recombinant TRAIL variant (TRAIL.R1-5) that exhibited selectivity for TRAIL-R1 but was still biologically active in a number of target cells, including primary CLL cells (MacFarlane *et al.*, 2005b).

To gain further insight into the effect of TRAIL mutations on the binding of TRAIL to TRAIL-R1 and thus aid the design of a biologically active TRAIL-R1-selective mutant, we generated a structural model of the TRAIL/TRAIL-R1 complex and compared it with the crystal structure of the TRAIL/TRAIL-R2 complex (as illustrated in **Figure 3**). The comparison was facilitated by the sequence similarity between the two proteins; TRAIL-R1 and TRAIL-R2 share 64% amino acid sequence identity in their extracellular domains. When compared to wild-type TRAIL, the substitutions Y213W; S215D and two further substitutions, N199V; K201R, resulted in a TRAIL mutant that showed some selectivity for signalling via TRAIL-R1 compared with TRAIL-R2 (MacFarlane *et al.*, 2005b). Analysis of the TRAIL/TRAIL-R1 (our model) and the TRAIL/TRAIL-R2 (Hymowitz *et al.*, 1999) interface suggests that this small increase in TRAIL-R1-selectivity may have been due to the substitution N199V (but not K201R). In TRAIL/TRAIL-R2, this substitution is predicted to cause the loss of two hydrogen bonds (to the side chain of Arg-104 and to the main chain

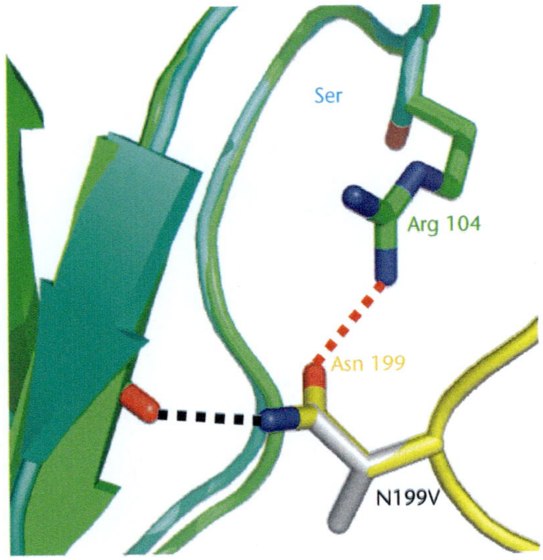

(a)

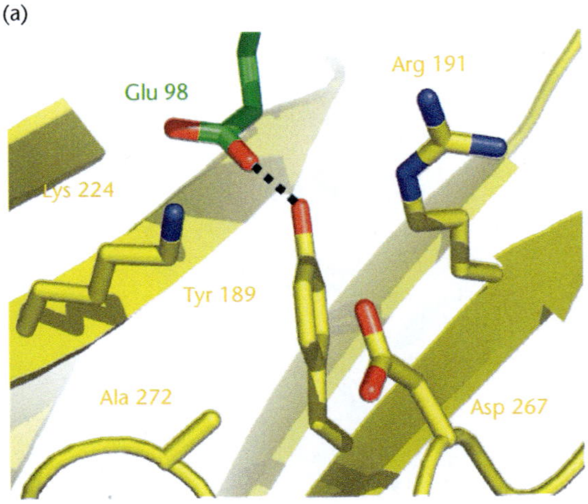

(b)

Figure 3 Model of TRAIL/TRAIL-R1 complex and crystal structure of TRAIL/TRAIL-R2 complex. (a) Role of TRAIL Asn-199 in TRAIL (yellow)/TRAIL-R1 (cyan)/TRAIL-R2 (green) interactions. Hydrogen bond present with both TRAIL-R1 and TRAIL-R2 (black dashed line), and that present with only TRAIL-R2 (red dashed line) is indicated. The loss of these hydrogen bonds with the TRAIL substitution N199V is also illustrated. (b) Role of TRAIL Tyr-189 in TRAIL (yellow)/TRAIL-R1/R2 (green) interactions. Hydrogen bond from this tyrosine to the conserved glutamate in TRAIL-R1/R2 (dashed line) is indicated. Residues in TRAIL involved in hydrophobic interactions with Tyr-189, that is, interactions lost in Y189A-substituted TRAIL, are also shown (see text for further detail) (MacFarlane *et al.*, 2005b). Reproduced by permission of American Association for Cancer Research.

carbonyl of Cys-125). In contrast, in TRAIL/TRAIL-R1, the N199V substitution would result in the loss of only one interprotein hydrogen bond, because in TRAIL-R1 the equivalent to Arg-104 is a much shorter Ser residue, which cannot bond hydrogen to Asn-199 (as illustrated in **Figure 3a**).

Our structural model of TRAIL/TRAIL-R1 also provided some insight into the potential importance of the Y189A substitution, which was present in the original Apo2L.DR4-8 variant that exhibited low bioactivity (Kelley *et al.*, 2005), but which we omitted to obtain the biologically active TRAIL-R1-specific variant, TRAIL.R1-5. The importance of Tyr-189 in the binding of TRAIL-R1 and TRAIL-R2 could be rationalized through direct or indirect effects. Tyr-189 forms a hydrogen bond to a conserved Glu in both TRAIL-R1 and TRAIL-R2 (corresponding to Glu-98 in TRAIL-R2); thus, the substitution Y189A removes this hydrogen bond in both the TRAIL/TRAIL-R1 and TRAIL/TRAIL-R2 complexes (as illustrated in **Figure 3b**). Substitution of Y189A also results in the removal of hydrophobic interactions to Arg-191, Asp-267, Ala-272 and Lys-224 close to the surface of TRAIL (as illustrated in **Figure 3b**). Thus, the substitution Y189A may indirectly affect ligand binding per se by distorting the surface of TRAIL (MacFarlane *et al.*, 2005b).

Using the alternative approach of computational protein design methods, van der Sloot *et al.* (2006) found that mutation of only two residues, Asp-269 to Histidine and Glu-195 to Arginine (D269H/E195R) generated a TRAIL variant with >100-fold higher preference for TRAIL-R2 over TRAIL-R1 when compared to wild-type TRAIL. Importantly, the D269H/E195R variant demonstrated positive anti-tumour activity in TRAIL-R2 responsive tumour cell lines without any adverse toxicity in nontransformed cells (van der Sloot *et al.*, 2006). Using exactly the same approach, a TRAIL-R1-selective variant was subsequently generated by a single amino acid substitution at Asp-218 as this residue was predicted to be important for selectivity towards TRAIL-R1. The best-performing TRAIL-R1-selective variant, D218H, was highly selective for TRAIL-R1, exhibited a reduced binding affinity for TRAIL-R2, and pro-apoptotic activity in TRAIL-R1 responsive tumour cell lines comparable to that seen with recombinant wild-type TRAIL (Tur *et al.*, 2008). The recombinant TRAIL variants derived from these computational protein design methods clearly demonstrate that, in the case of TRAIL-R1- or TRAIL-R2-specific variants, only one or two amino acid substitutions are required to obtain TRAIL-R1 or TRAIL-R2 selectivity, respectively. In view of a potential use of these TRAIL-R1/R2-selective TRAIL variants as anticancer therapeutics, one would predict that having fewer mutations relative to the wild-type sequence would be more favourable as it would reduce the risk of an immunogenic response.

In addition to rhApo2L/TRAIL, which targets both TRAIL-R1 and TRAIL-R2, the other class of TRAIL receptor agonists which are currently being developed are the monoclonal antibodies, which like the receptor-selective TRAIL variants described earlier, display agonistic activity towards TRAIL-R1 or TRAIL-R2 (as illustrated in **Figure 1**). Importantly, with the exception of the TRAIL-R1-targeting mAb (monoclonal antibody), Mapatumumab, all other anti-TRAIL receptor antibodies currently in clinical development target TRAIL-R2 rather than TRAIL-R1 (as illustrated in **Table 1**). The reason for this is not entirely clear, but may be based on initial studies indicating that TRAIL-R2 is more highly expressed on some tumour cells or, as already discussed earlier, the suggestion by Kelley and colleagues that signalling to apoptosis via TRAIL-R2 may be more potent than via TRAIL-R1 (Kelley *et al.*, 2005).

Although several agonistic antibodies are currently in Phase II clinical trials (as illustrated in **Figure 1** and **Table 1**), very little detailed structural information is available regarding the structure of these antibodies in complex with TRAIL-R1/R2. The exception to this is the TRAIL-R2-targeting antibody, Apomab, being developed by Genentech, for which detailed information was recently published (Adams *et al.*, 2008). Apomab is a fully human agonistic monoclonal antibody, originally isolated by the phage display approach, which exhibits selective pro-apoptotic activity in TRAIL-R2 expressing tumour cell lines. To determine how Apomab binds to TRAIL-R2 at the atomic level, X-ray crystallographic analysis was performed of the complex between the Fab fragment of Apomab and the extracellular domain of TRAIL-R2 (Adams *et al.*, 2008). The solved structure revealed an interaction epitope in Apomab that exhibits significant overlap with that of Apo2L/TRAIL, contacting CRD2 and CRD3 on TRAIL-R2 (as illustrated in **Figure 4a**). However, while the ligand's contact site is divided into two patches (as illustrated in **Figure 2**), the region that Apomab contacts is more continuous. In the case of rhApo2L/TRAIL, the trimeric subunit structure is the same as the endogenous ligand and is therefore thought to mimic the natural mode of receptor engagement. In this context, it is therefore interesting to speculate on precisely how homodimeric agonistic antibodies might activate TRAIL-R2. Comparison of crystallographic structures of TRAIL-R2 in complex

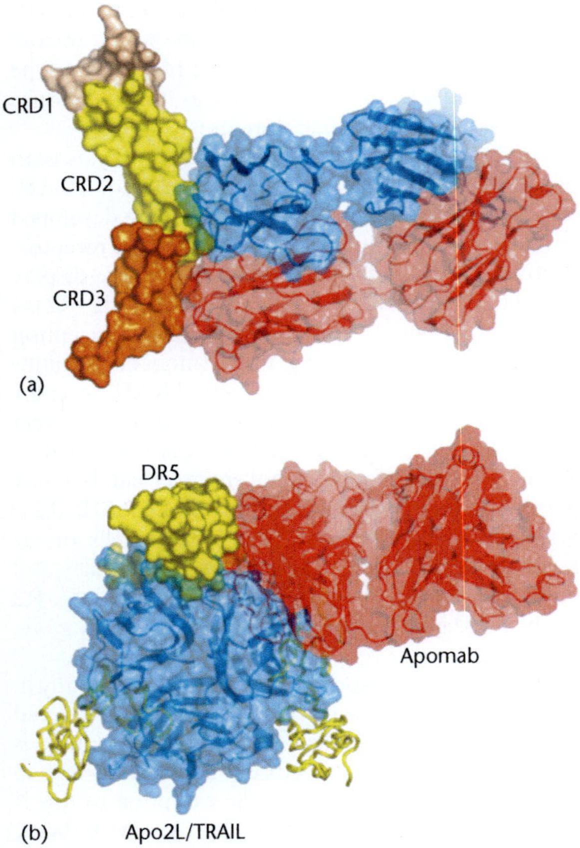

Figure 4 Crystal structure of the Apomab Fab fragment in complex with the TRAIL-R2 extracellular domain. (a) Apomab binds at the junction between CRD2 and CRD3. TRAIL-R2 (brown, yellow and orange) and the Apomab Fab (light chain blue and heavy chain red) are shown as molecular interfaces. (b) Apo2L/TRAIL (blue) and Apomab (red) bind to overlapping yet distinct sites on TRAIL-R2 (yellow). The structure of the Apomab Fab/TRAIL-R2 complex is overlaid on the previously solved Apo2L/TRAIL-R2 complex (see **Figure 2**). For clarity, only one copy of TRAIL-R2 is shown. Additional copies of TRAIL-R2 that would bind at the Apo2L/TRAIL monomer–monomer interfaces are shown as yellow backbone ribbons (Adams *et al.*, 2008). Reproduced by permission of Nature Publishing Group.

with rhApo2L/TRAIL or Fab fragments of Apomab (as illustrated in **Figure 4b**), as well as other TRAIL-R2 antibodies with little or no agonistic activity, indicate that the membrane proximal CRD3 region is dynamic *in vivo* and may be stabilized by Apomab in a specific arrangement that promotes TRAIL-R2 activation. Clearly, data obtained from structural analysis of an agonistic antibody in complex with a pro-apoptotic TRAIL receptor has provided important new insights into the potential mechanisms that lead to apoptosis, and have even helped to shape the development of Apomab for cancer therapy. In this respect, we await with interest the publication of more detailed information on other TRAIL-R agonistic antibodies currently under development.

Targeting TRAIL Death Receptors for Cancer Therapy

As described earlier, since the discovery of TRAIL and its receptors, two major classes of death receptor agonists have been developed: recombinant human Apo2L/TRAIL, which activates both receptors, and receptor-selective TRAIL mutants or monoclonal agonistic antibodies which activate either TRAIL-R1 or TRAIL-R2 (as illustrated in **Figure 1** and **Table 1**). Several TRAIL receptor agonists are currently being tested in Phase I and II clinical trials, including one recombinant ligand (rhApo2L/TRAIL), one anti-TRAIL-R1 agonistic antibody (Mapatumumab) and five anti-TRAIL-R2 agonistic antibodies (Lexatumumab, Apomab, AMG655, TRA-8/CS-1008 and LBY135). A summary of clinical studies with pro-apoptotic receptor agonists, including details of treatment schedules and their developmental status is provided in **Table 1**. **See also**: Apoptosis: Inherited Disorders

In contrast, the receptor-selective TRAIL variants described earlier have only been tested in preclinical tumour models including tumour cell lines, mouse xenograft models and primary tumour cells from CLL patients cultured *ex vivo*. However, these studies, as well as the extensive preclinical data obtained with rhApo2L/TRAIL and TRAIL-R1/R2 agonistic antibodies, have produced a number of important findings that need careful consideration in the future optimization of TRAIL-R-targeted therapies. In this regard, one of the most significant findings is that, while death receptor agonists may be useful as single agents against tumours that are particularly sensitive to their pro-apoptotic effects, there is now overwhelming evidence that TRAIL-R agonists will need to be employed in combination with conventional or other novel therapies to increase their efficacy and utility in primary tumours (Johnstone *et al.*, 2008; Dyer *et al.*, 2007; Ashkenazi and Herbst, 2008). Furthermore, to trigger apoptosis, tumour cells of different tissue origin may benefit from selective targeting of TRAIL-R1 or TRAIL-R2, thus highlighting the potential application of the various receptor-specific TRAIL-based therapies described earlier in combination with current established therapies.

In preclinical studies, rhApo2L/TRAIL induced apoptosis in various cancer cell lines, including those with p53 mutations, without affecting normal cells. In addition, co-administration of rhApo2L/TRAIL with various chemotherapeutics was shown to result in either additive or synergistic activity. rhApo2L/TRAIL also displayed antitumour activity *in vivo* in mouse xenograft models of human cancer derived from colon, lung and pancreatic carcinomas, and multiple myeloma when given either as a single agent or in combination with conventional therapeutics (Ashkenazi *et al.*, 2008). Despite promising data showing the efficacy of TRAIL in many tumour cell lines, when sensitivity to TRAIL was examined in preclinical models using primary patient material, tumour cells obtained from patients with CLL and MCL were found to be resistant to TRAIL. In every CLL patient tested, B cells freshly isolated from patients and cultured *ex vivo* were completely resistant to TRAIL, when used as a single agent (MacFarlane *et al.*, 2002). However, it was subsequently discovered that pre-treatment of CLL cells with subtoxic concentrations of histone deacetylase inhibitors (HDACi) could consistently sensitize these primary tumour cells to TRAIL-induced apoptosis, thus highlighting that combination regimens will almost certainly be required for future TRAIL therapy of lymphoid malignancies (Dyer *et al.*, 2007).

Using primary tumour cells obtained from patients with CLL, we also examined the efficacy of the receptor-specific TRAIL variants generated within our laboratory that specifically target TRAIL-R1 or TRAIL-R2 (see earlier discussion). To our surprise, we found that only TRAIL variants that bind to TRAIL-R1 were active at inducing apoptosis in HDACi-sensitized CLL cells (MacFarlane *et al.*, 2005b). Intriguingly, this finding was consistent with our earlier findings, using humanized therapeutic monoclonal antibodies against TRAIL-R1/R2 (Human Genome Sciences; HGS), again in primary tumour cells from CLL patients. In this case, only antibodies that targeted TRAIL-R1 (mapatumumab; HGS-ETR1) and not TRAIL-R2 (lexatumumab; HGS-ETR2) were effective at inducing apoptosis in HDACi-sensitized CLL cells (MacFarlane *et al.*, 2005a). However, when antibodies to TRAIL-R2 were Fc crosslinked, before tumour cell exposure, signalling to apoptosis via TRAIL-R2 was partially restored. This finding presumably reflects the higher degree of receptor crosslinking/aggregation required to trigger apoptosis via TRAIL-R2, and further highlights the need for careful evaluation of individual death receptor agonists in different tumour cell settings (Dyer *et al.*, 2007; Natoni *et al.*, 2007).

Importantly, agonistic antibodies against TRAIL-R1/R2, like rhApo2L/TRAIL, exhibit selectivity for tumour cells over normal cells and have also been shown to slow the growth of tumours in xenograft tumour models with no apparent systemic toxicity (reviewed in Johnstone *et al.*, 2008). Compared to rhApo2L/TRAIL, antibodies have the advantage of having a relatively long half-life and can use additional mechanisms for cell killing through antibody-dependent cellular toxicity and complement-dependent cytotoxicity mechanisms, mediated by the Fc portion of the antibodies. Although there are several key differences between the two classes of TRAIL death receptor agonists currently being developed, it is difficult to predict what impact such differences might have on the tolerability and clinical efficacy of these agents. Because of differences in their molecular size and composition there are likely to be significant variations in pharmacokinetics and pharmacodynamics between TRAIL or its variants and agonistic antibodies to TRAIL-R1/R2. This could influence the duration and frequency of tumour exposure, tumour penetration and effects on normal tissue. Furthermore, while recombinant TRAIL activates both death receptors, receptor-specific TRAIL variants and the agonistic antibodies under development are monospecific. As highlighted earlier, we have shown that in certain tumours such as haematological malignancies TRAIL-R1 is more responsive to triggering for apoptosis than TRAIL-R2, and so in some cases stimulation of either one, or both, receptors might be more beneficial.

Determination of the three-dimensional structures of the natural ligand or antibody-based TRAIL-R agonists complexed to their protein targets has aided the future development of receptor-specific TRAIL-based therapies. Clearly, TRAIL death receptor agonists provide an exciting opportunity to attack tumour cells on the basis of their inherent apoptotic vulnerability; however, as clinical trials progress, it will also be important to investigate what determines tumour sensitivity to these agents, with the ultimate aim of optimizing treatment for different cancers and for individual patients while minimizing unwanted toxicities.

References

Adams C, Totpal K, Lawrence D *et al.* (2008) Structural and functional analysis of the interaction between the agonistic monoclonal antibody Apomab and the proapoptotic receptor DR5. *Cell Death and Differentiation* **15**: 751–761.

Ashkenazi A (2008) Targeting the extrinsic apoptosis pathway in cancer. *Cytokine Growth Factor Review* **19**: 325–331.

Ashkenazi A and Dixit VM (1998) Death receptors: signaling and modulation. *Science* **281**: 1305–1308.

Ashkenazi A and Herbst RS (2008) To kill a tumor cell: the potential of proapoptotic receptor agonists. *Journal of Clinical Investigation* **118**: 1979–1990.

Ashkenazi A, Holland P and Eckhardt SG (2008) Ligand-based targeting of apoptosis in cancer: the potential of recombinant human apoptosis ligand 2/tumor necrosis factor-related apoptosis-inducing ligand (rhApo2L/TRAIL). *Journal of Clinical Oncology* **26**: 3621–3630.

Ashkenazi A, Pai RC, Fong S *et al.* (1999) Safety and antitumor activity of recombinant soluble Apo2 ligand. *Journal of Clinical Investigation* **104**: 155–162.

Bremer E, ten Cate B, Samplonius DF, de Leij LF and Helfrich W (2006) CD7-restricted activation of Fas-mediated apoptosis: a novel therapeutic approach for acute T-cell leukemia. *Blood* **107**: 2863–2870.

Bremer E, ten Cate B, Samplonius DF *et al.* (2008) Superior activity of fusion protein scFvRit:sFasL over cotreatment with rituximab and Fas agonists. *Cancer Research* **68**: 597–604.

Dyer MJ, MacFarlane M and Cohen GM (2007) Barriers to effective TRAIL-targeted therapy of malignancy. *Journal of Clinical Oncology* **25**: 4505–4506.

Ganten TM, Koschny R, Sykora J *et al.* (2006) Preclinical differentiation between apparently safe and potentially hepatotoxic applications of TRAIL either alone or in combination with chemotherapeutic drugs. *Clinical Cancer Research* **12**: 2640–2646.

Grunhagen DJ, de Wilt JH, Graveland WJ, van Geel AN and Eggermont AM (2006) The palliative value of tumor necrosis factor alpha-based isolated limb perfusion in patients with metastatic sarcoma and melanoma. *Cancer* **106**: 156–162.

Hanahan D and Weinberg RA (2000) The hallmarks of cancer. *Cell* **100**: 57–70.

Hymowitz SG, Christinger HW, Fuh G *et al.* (1999) Triggering cell death: the crystal structure of Apo2L/TRAIL in a complex with death receptor 5. *Molecular Cell* **4**: 563–571.

Johnstone RW, Frew AJ and Smyth MJ (2008) The TRAIL apoptotic pathway in cancer onset, progression and therapy. *Nature Reviews. Cancer* **8**: 782–798.

Kelley RF, Totpal K, Lindstrom SH *et al.* (2005) Receptor-selective mutants of apoptosis-inducing ligand 2/tumor necrosis factor-related apoptosis-inducing ligand reveal a greater contribution of death receptor (DR) 5 than DR4 to apoptosis signaling. *Journal of Biological Chemistry* **280**: 2205–2212.

Lawrence D, Shahrokh Z, Marsters S *et al.* (2001) Differential hepatocyte toxicity of recombinant Apo2L/TRAIL versions. *Nature Medicine* **7**: 383–385.

Lowe SW, Cepero E and Evan G (2004) Intrinsic tumour suppression. *Nature* **432**: 307–315.

MacFarlane M, Harper N, Snowden RT *et al.* (2002) Mechanisms of resistance to TRAIL-induced apoptosis in primary B cell chronic lymphocytic leukaemia. *Oncogene* **21**: 6809–6818.

MacFarlane M, Inoue S, Kohlhaas SL *et al.* (2005a) Chronic lymphocytic leukemic cells exhibit apoptotic signaling via TRAIL-R1. *Cell Death and Differentiation* **12**: 773–782.

MacFarlane M, Kohlhaas SL, Sutcliffe MJ, Dyer MJ and Cohen GM (2005b) TRAIL receptor-selective mutants signal to apoptosis via TRAIL-R1 in primary lymphoid malignancies. *Cancer Research* **65**: 11265–11270.

Mongkolsapaya J, Grimes JM, Chen N *et al.* (1999) Structure of the TRAIL-DR5 complex reveals mechanisms conferring specificity in apoptotic initiation. *Nature Structural Biology* **6**: 1048–1053.

Nagata S (1997) Apoptosis by death factor. *Cell* **88**: 355–365.

Natoni A, MacFarlane M, Inoue S *et al.* (2007) TRAIL signals to apoptosis in chronic lymphocytic leukaemia cells primarily through TRAIL-R1 whereas cross-linked agonistic TRAIL-R2 antibodies facilitate signalling via TRAIL-R2. *British Journal of Haematology* **139**: 568–577.

Ogasawara J, Watanabe-Fukunaga R, Adachi M *et al.* (1993) Lethal effect of the anti-Fas antibody in mice. *Nature* **364**: 806–809.

Pitti RM, Marsters SA, Ruppert S *et al.* (1996) Induction of apoptosis by Apo-2 ligand, a new member of the tumor necrosis factor cytokine family. *Journal of Biological Chemistry* **271**: 12687–12690.

Samel D, Muller D, Gerspach J *et al.* (2003) Generation of a FasL-based proapoptotic fusion protein devoid of systemic toxicity due to cell-surface antigen-restricted activation. *Journal of Biological Chemistry* **278**: 32077–32082.

van der Sloot AM, Tur V, Szegezdi E *et al.* (2006) Designed tumor necrosis factor-related apoptosis-inducing ligand variants initiating apoptosis exclusively via the DR5 receptor. *Proceedings of the National Academy of Sciences of the USA* **103**: 8634–8639.

Trauth BC, Klas C, Peters AM *et al.* (1989) Monoclonal antibody-mediated tumor regression by induction of apoptosis. *Science* **245**: 301–305.

Tur V, van der Sloot AM, Reis CR *et al.* (2008) DR4-selective tumor necrosis factor-related apoptosis-inducing ligand (TRAIL) variants obtained by structure-based design. *Journal of Biological Chemistry* **283**: 20560–20568.

Volkmann X, Fischer U, Bahr MJ *et al.* (2007) Increased hepatotoxicity of tumor necrosis factor-related apoptosis-inducing ligand in diseased human liver. *Hepatology* **46**: 1498–1508.

Walczak H, Miller RE, Ariail K *et al.* (1999) Tumoricidal activity of tumor necrosis factor-related apoptosis- inducing ligand *in vivo* [see comments]. *Nature Medicine* **5**: 157–163.

Wiley SR, Schooley K, Smolak PJ *et al.* (1995) Identification and characterization of a new member of the TNF family that induces apoptosis. *Immunity* **3**: 673–682.

Yonehara S, Ishii A and Yonehara M (1989) A cell-killing monoclonal antibody (anti-Fas) to a cell surface antigen co-downregulated with the receptor of tumor necrosis factor. *Journal of Experimental Medicine* **169**: 1747–1756.

Further Reading

Ashkenazi A (2002) Targeting death and decoy receptors of the tumour-necrosis factor superfamily. *Nature Reviews. Cancer* **2**: 420–430.

Ashkenazi A (2008) Directing cancer cells to self-destruct with pro-apoptotic receptor agonists. *Nature Review of Drug Discovery* **7**: 1001–1012.

Buchsbaum DJ, Forero-Torres A and LoBuglio AF (2007) TRAIL-receptor antibodies as a potential cancer treatment. *Future Oncology* **3**: 405–409.

Fesik SW (2005) Promoting apoptosis as a strategy for cancer drug discovery. *Nature Reviews. Cancer* **5**: 876–885.

Fulda S and Debatin KM (2004) Exploiting death receptor signaling pathways for tumor therapy. *Biochimica et Biophysica Acta* **1705**: 27–41.

Gajewski TF (2007) On the TRAIL toward death receptor-based cancer therapeutics. *Journal of Clinical Oncology* **25**: 1305–1307.

Hymowitz SG, O'Connell MP, Ultsch MH *et al.* (2000) A unique zinc-binding site revealed by a high-resolution X-ray structure of homotrimeric Apo2L/TRAIL. *Biochemistry* **39**: 633–640.

MacFarlane M (2003) TRAIL-induced signalling and apoptosis. *Toxicological Letters* **139**: 89–97.

Papenfuss K, Cordier SM and Walczak H (2008) Death receptors as targets for anti-cancer therapy. *Journal of Cellular and Molecular Medicine* **12**: 2566–2585.

Smyth MJ, Takeda K, Hayakawa Y *et al.* (2003) Nature's TRAIL – on a path to cancer immunotherapy. *Immunity* **18**: 1–6.

Death Receptor-induced Necroptosis

Wim Declercq, *Molecular Signaling and Cell Death Unit, Department for Molecular Biomedical Research, VIB-Ghent University, Ghent, Belgium*

Franky Van Herreweghe, *Molecular Signaling and Cell Death Unit, Department for Molecular Biomedical Research, VIB-Ghent University, Ghent, Belgium*

Tom Vanden Berghe, *Molecular Signaling and Cell Death Unit, Department for Molecular Biomedical Research, VIB-Ghent University, Ghent, Belgium*

Peter Vandenabeele, *Molecular Signaling and Cell Death Unit, Department for Molecular Biomedical Research, VIB-Ghent University, Ghent, Belgium*

Advanced article

Article Contents

- Introduction
- Death Receptor-induced Necroptosis Signalling Complexes
- Death Receptor-induced Necroptosis, Dying from too much Respiration
- The Way to Necroptosis is Paved with Many Pathways
- Physiological Role of Necroptosis
- Conclusions and Perspectives
- Acknowledgements

Recent studies on cellular *in vitro* models have made it clear that death receptor (DR)-induced necrosis, also called necroptosis, is a programmed form of necrotic cell death. In contrast, accidental necrosis is mainly induced by physicochemical stress, such as heath or chemicals. Many DRs signal to apoptosis or necroptosis depending on whether caspases are active or not. The kinase activities of receptor-interacting protein 1 (RIP1) and RIP3 have emerged as crucial regulators of DR-induced necroptosis. The precise execution mechanisms during necroptosis are still poorly understood but involve different cellular compartments such as the mitochondria, the lysosomes and the cell membrane. *In vivo*, necroptosis occurs mainly in pathophysiological processes such as ischaemia-reperfusion injury in heart, brain and kidneys, viral infection and pancreatitis, and is capable of killing tumour cells that have developed strategies to evade apoptosis. Thus, detailed knowledge of necroptotic signalling may be exploited in therapeutic strategies.

Introduction

Cells can die in many ways (Kroemer *et al.*, 2009). Three major morphologies of cell death have been described: apoptosis, necrosis and cell death associated with autophagy. The present paradigm is that caspase-dependent apoptosis is the predominant cell death pathway. Apoptosis involves a sequence of specific morphological changes in the dying cell (**Figure 1a**): reduction of the cellular volume (pyknosis), condensation of the chromatin and nuclear fragmentation (karyorhexis), followed by breakage of cells into membrane-contained apoptotic bodies containing a variety of cytoplasmic organelles and nuclear fragments (Kerr *et al.*, 1972), which are engulfed by neighbouring cells and phagocytes. In mammalian cells, the apoptotic response is mediated either by an intrinsic or by an extrinsic pathway, in which caspase activation plays a crucial role. Necrosis is characterized by swelling of the endoplasmic reticulum, mitochondria and cytoplasm, with subsequent rupture of the plasma membrane and lysis of the cells (**Figure 1c**; Schweichel and Merker, 1973). Necrosis refers to an accidental form of cell death due to physicochemical damage, whereas necroptosis is a regulated form of necrosis that prevails in the absence of caspase activation (Degterev *et al.*, 2008; Hitomi *et al.*, 2008; Vercammen *et al.*, 1998). Autophagy is foremost a survival mechanism that is activated in cells subjected to nutrient or obligate growth factor deprivation. When cellular stress continues, cell death may occur from autophagy alone or in association with features of apoptotic or necrotic cell death, depending on the stimulus and cell type. The way how cells die matters for subsequent intercellular communication, immunological consequences, tissue repair and regeneration. Although initially the idea has been propagated to associate apoptosis with tolerogenic cell death and necrosis with immunogenic cell death, recent findings have challenged this too simple bifurcation and provide a more dynamic immunologic and tissue repair outcome. **See also**: Apoptosis: Regulatory Genes and Disease; Autophagy in Non-Mammalian Systems; The Apoptosome: The Executioner of Mitochondria-mediated Apoptosis

The precise execution mechanisms during necroptosis are still poorly understood and involve multiple pathways (Festjens *et al.*, 2006b): (1) at the mitochondrial level increased bioenergetics, mitochondrial complex I-mediated reactive oxygen species (ROS) production and calcium overload are implicated; (2) lysosomal membrane permeabilization and release of cathepsins contribute to membrane permeabilization http://merops.sanger.ac.uk/; (3) activation of enzymatic systems such as phospholipase A_2 and NADPH (nicotinamide adenine dinucleotide phosphate reduced form) oxidase at the cell membrane contribute to necroptosis (Festjens *et al.*, 2006b). Recent reports have put the protein kinase receptor-interacting protein 1 (RIP1) and RIP3 at the centre-stage of tumour necrosis factor (TNF)-induced necroptosis. These kinases constitute a necrosome signalling complex that in the absence of caspase activation mediates necroptotic signalling. In addition, the development of RIP1 kinase inhibitors, the necrostatins, allows to control necroptosis in many experimental pathologies.

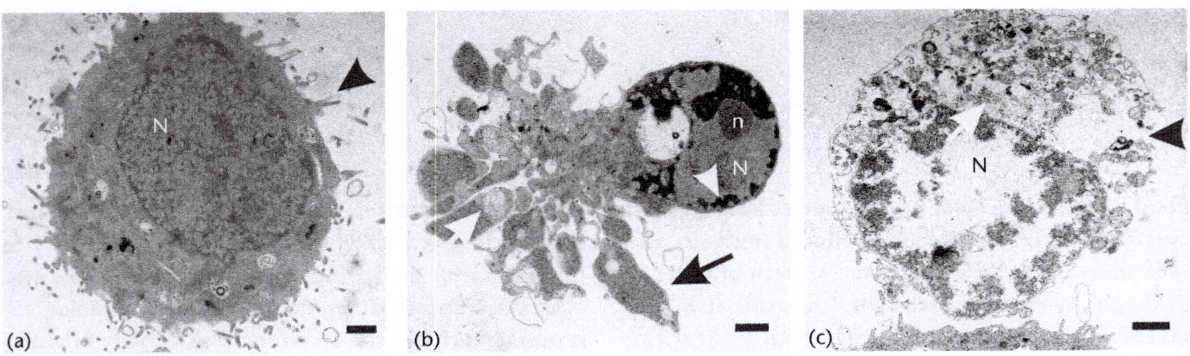

Figure 1 Cell morphology of apoptotic and necrotic cells by transmission electron microscopy. (a) Unstimulated L929sAhFas fibrosarcoma cell. The cell shows microvilli protruding from the entire surface (arrowhead), a smoothly outlined nucleus with chromatin in the form of heterochromatin and well-preserved cytoplasmic organelles. (b) Apoptotic L929sAhFas cell (treated with agonistic anti-Fas for 1 h) with condensed and marginated (arrowhead) chromatin. Note the nucleolus (n) and damaged mitochondria (arrow). (c) Necrotic L929sAhFas cell (treated with TNF for 7 h) with clumps of chromatin with ill-defined edges, swollen and completely disrupted mitochondria (arrow) and loss of plasma membrane integrity (arrowhead). Scale bars: 1 μm, N, nucleus. Pictures are courtesy of Katharina D'Herde and Dmitri Krysko (Ghent University).

Death Receptor-induced Necroptosis Signalling Complexes

TNF induces either apoptosis, necrosis or gene activation in most cell types, whereas Fas and TNF-related apoptosis-inducing ligand (TRAIL) mainly induce apoptosis. Many cell lines are protected against TNF-induced apoptosis by zVAD (a pan-caspase inhibitor) treatment, whereas others are sensitized to undergo TNF-induced necroptosis when caspases are inhibited by the drug (Festjens *et al.*, 2007). Several cellular models of necrotic cell death induced by DRs and Toll-like receptors (TLRs), in the absence of caspase activation, have pinpointed the crucial role of receptor-interacting protein 1 kinase (RIP1) in necrotic cell death (**Table 1**). RIP serine/threonine kinase family members are essential sensors of cellular stress (Meylan and Tschopp, 2005). They share a homologous *N*-terminal kinase domain but have different recruitment domains (**Figure 2a**). The death domain of the RIP1 kinase binds to the death domain of death receptors, such as TNFR1, Fas, TNF-related apoptosis-inducing ligand receptors 1 and 2 (TRAILR1 and 2), and to death domain-containing adaptor proteins, such as TNF receptor-associated death domain (TRADD) and Fas-associated death domain (FADD), which are required for caspase-8 activation and apoptosis. The intermediate domain of RIP1 contains an RIP homotypic interaction motif (RHIM) that enables homotypic interaction with RIP3. **See also:** Death Receptors; Death Receptors at the Molecular Level: Therapeutic Implications

TNF-induced cell death and nuclear factor-κB (NF-κB) activation are initiated from distinct, dynamically formed signalling complexes (Schutze *et al.*, 2008; Wilson *et al.*, 2009). TNFR1 exists as preformed aggregated receptors due to the presence of a pre-ligand assembly domain (PLAD) in the extracellular cysteinrich domain 1 (Chan, 2000). Binding of TNF to TNFR1 causes a conformational change of the receptor leading to the recruitment of TRADD, which in turn mediates the recruitment of RIP1, inhibitor of apoptoses 1 and 2 (cIAP1 and 2), and TNF receptor-associated factor 2 (TRAF2), forming the membrane proximal complex I. To date, RIP3 has not been found in this complex (Cho *et al.*, 2009; He *et al.*, 2009). The ubiquitination state of RIP1 determines whether it functions as a pro-survival scaffold molecule governing NF-κB and mitogen-activated protein kinase (MAPK) activation, or a kinase that promotes cell death. TNFR1 triggers cIAP1- and cIAP2-mediated polyubiquitination linked through lysine 63 (K63) of ubiquitin on lysine K377 of the RIP1 intermediate domain (Bertrand *et al.*, 2008; Mahoney *et al.*, 2008; Varfolomeev *et al.*, 2008). K63-polyubiquitinated RIP1 then allows downstream activation of MAPKs and NF-κB (**Figure 2b**, complex I). Consistent with a role for RIP1 as a survival signalling molecule, mice lacking RIP1 display extensive apoptosis and die at 1–3 days of age (Kelliher *et al.*, 1998). Moreover, cultured cells lacking RIP1 are highly sensitive to TNF-induced cell death in the presence of the protein synthesis inhibitor cycloheximide. Remarkably, the RIP1 kinase activity is not required for survival signalling, but crucial for FasL-, TNF- and TRAIL-induced necrotic cell death in conditions preventing caspase activation (Holler *et al.*, 2000). In addition, necrostatin-1 (Nec-1), an allosteric RIP1 kinase inhibitor, inhibited DR-induced necroptosis in different cellular models, indicating that necroptotic signalling is mediated by a common mechanism (Degterev *et al.*, 2008). Recently, it became clear that the expression of RIP3 makes cells permissive to undergo necroptosis on TNF treatment, requiring its kinase activity and RHIM domain (Cho *et al.*, 2009; He *et al.*, 2009; Zhang *et al.*, 2009). RIP3 overexpression has been shown to be able to induce both apoptotic and necrotic cell death (Feng *et al.*, 2007). However, mice lacking RIP3 do not exhibit defects in development, NF-κB activation, or apoptosis (Newton *et al.*, 2004).

During a second phase after TNFR1 triggering, several adaptor proteins are reshuffled to form secondary cytosolic complexes (Schutze *et al.*, 2008; Wilson *et al.*, 2009). RIP1 K63-ubiquitination is negatively regulated by the deubiquitinases cylindromatosis (CYLD) and A20 (Wilson *et al.*, 2009). In most conditions, TNFR1 triggering induces a TRADD-dependent complex II that attracts FADD and caspase-8 to initiate apoptosis (**Figure 2b**, complex II). Because both TNF-induced apoptosis and necroptosis (in the presence of the pan-caspase inhibitor zVAD) are blocked in TRADD-deficient cells (Ermolaeva *et al.*, 2008; Pobezinskaya *et al.*, 2008), this complex or a downstream-derived complex can trigger both apoptotic and necrotic signalling pathways. Under conditions of caspase blockage, RIP1 and RIP3 are recruited to complex II (Cho *et al.*, 2009; He *et al.*, 2009). RIP3 is probably activated and autophosphorylated at Ser199 within the complex (He *et al.*, 2009). The sequence of events is not yet clear. *In vivo*, RIP1 and RIP3 phosphorylation is clearly mutually dependent (Cho *et al.*, 2009; He *et al.*, 2009), and is implicated in stable complex II formation (Cho *et al.*, 2009). Reducing cellular levels of cIAP1 and cIAP2 by

Table 1 List of *in vitro* cellular necroptotic assays

In vitro models	Stimulus	Conditions	References
L929sA (mouse fibrosarcoma cells)	TNF	By default	Fiers *et al.* (1999)
	Poly I:C	In presence of IFN-γ	Hitomi *et al.* (2008)
JURKAT (T cells)	TNF, TRAIL	In presence of cycloheximide (inhibition of translation)	Holler *et al.* (2000)
	TNF	When FADD-deficient	
	FasL	In casp8$^{-/-}$ or in presence of zVAD.fmk or CrmA	
Activated human T cells	FasL Phytohaemagglutinin	In presence of zVAD.fmk	
HT29 (human colon cancer cells)	TRAIL	With acidic extracellular pH	Meurette *et al.* (2007)
macrophages	LPS	When caspase-8 and NFκB activity is inhibited	Ma *et al.* (2005)
		When overexpressing disease-associated mutant NALP3	Willingham *et al.* (2007)
	Shigella flexneri	By default	Willingham *et al.* (2007)
IBMK (immortalized baby mouse kidney epithelial cells)	Metabolic stress (ischaemia/glucose deprivation)	When BAX/BAK and autophagy deficient	Degenhardt *et al.* (2006)
LLC-PK(1) (kidney proximal tubular epithelial cells)	High glucose	By default	Harwood *et al.* (2007)
Hepatocytes	Calcium (ionophore) and ROS overload		Nakagawa *et al.* (2005)
Fibroblasts	DNA-damage		Xu *et al.* (2006)
Vero (kidney epithelial cells from the African green monkey)	West Nile virus	Upon high virus titer infection	Chu and Ng (2003)

Notes: Chu JJ and Ng ML (2003) The mechanism of cell death during West Nile virus infection is dependent on initial infectious dose. *Journal of Genetic Virology* **84**: 3305–3314; Degenhardt K, Mathew R, Beaudoin B *et al.* (2006) Autophagy promotes tumor cell survival and restricts necrosis, inflammation, and tumorigenesis. *Cancer Cell* **10**: 51–64; Fiers W, Beyaert R, Declercq W and Vandenabeele P (1999) More than one way to die: apoptosis, necrosis and reactive oxygen damage. *Oncogene* **18**: 7719–7730; Harwood SM, Allen DA, Raftery MJ and Yaqoob MM (2007) High glucose initiates calpain-induced necrosis before apoptosis in LLC-PK1 cells. *Kidney International* **71**: 655–663; Hitomi J, Christofferson DE, Ng A *et al.* (2008) Identification of a molecular signaling network that regulates a cellular necrotic cell death pathway. *Cell* **135**: 1311–1323; Holler N, Zaru R, Micheau O *et al.* (2000) Fas triggers an alternative, caspase-8-independent cell death pathway using the kinase RIP as effector molecule. *Nature Immunology* **1**: 489–495; Ma Y, Temkin V, Liu H and Pope RM (2005) NF-kappaB protects macrophages from lipopolysaccharide-induced cell death: the role of caspase 8 and receptor-interacting protein. *Journal of Biological Chemistry* **280**: 41827–41834; Meurette O, Rebillard A, Huc L *et al.* (2007) TRAIL induces receptor-interacting protein 1-dependent and caspase-dependent necrosis-like cell death under acidic extracellular conditions. *Cancer Research* **67**: 218–226; Nakagawa T, Shimizu S, Watanabe T *et al.* (2005) Cyclophilin D-dependent mitochondrial permeability transition regulates some necrotic but not apoptotic cell death. *Nature* **434**: 652–658; Willingham SB, Bergstralh DT, O'Connor W *et al.* (2007) Microbial pathogen-induced necrotic cell death mediated by the inflammasome components CIAS1/cryopyrin/NLRP3 and ASC. *Cell Host & Microbe* **2**: 147–159; Xu Y, Huang S, Liu ZG and Han J (2006) Poly(ADP-ribose) polymerase-1 signaling to mitochondria in necrotic cell death requires RIP1/TRAF2-mediated JNK1 activation. *Journal of Biological Chemistry* **281**: 8788–8795.

genetic ablation or by depleting them by using Smac mimetics (small molecules that mimic the action of the Smac/DIABLO protein; Sun *et al.*, 2008) results in a reduction in RIP1 K63-ubiquitination and TNF-induced NF-κB activation (Bertrand *et al.*, 2008; Mahoney *et al.*, 2008; Varfolomeev *et al.*, 2008), and causes RIP1 to switch from functioning as a pro-survival scaffold molecule to a caspase-dependent pro-apoptotic or an RIP3-dependent pro-necroptotic (in the presence of zVAD) kinase (Bertrand *et al.*, 2008; He *et al.*, 2009; Wang *et al.*, 2008). However, in all other conditions of DR-induced apoptosis, RIP1 is not required for apoptosis induction. **See also**: Inhibitor of Apoptosis (IAP) and BIR-containing Proteins

In the absence of caspase inhibitors, RIP1 and RIP3 are cleaved by caspase-8 in complex II (Cho *et al.*, 2009; Feng *et al.*, 2007). Cleavage of RIP1 under apoptotic conditions could serve to ensure 'silent' apoptotic death by preventing necroptosis and RIP1-mediated gene activation. In T cells, which require limited caspase-8 activation to proliferate, caspase-8 ablation results in antigen- or mitogen-activated necroptosis that depends on RIP1 (Ch'en *et al.*, 2008), and probably also on RIP3 (Cho *et al.*, 2009). Furthermore, cleavage of RIP3 was shown to abolish its ability to activate caspase-independent cell death (Feng *et al.*, 2007), suggesting that zVAD-mediated sensitization would be mainly due to its ability to prevent caspase-dependent RIP1 or 3 cleavage. However, in view of its off-target effects on the mitochondrial adenosine nucleotide translocator (ANT), zVAD could have additional necrosis-sensitizing effects (Vandenabeele *et al.*, 2006).

Death Receptor-induced Necroptosis, Dying from too much Respiration

From the early 1990s TNF-induced necroptosis has been linked with the energy metabolism and ROS production. Mitochondria are the major intracellular source of ROS, which are mainly generated at complex I and III of the respiratory chain. Complex I-mediated ROS production were shown to be crucial for TNF-induced necroptosis (Schulze-Osthoff *et al.*, 1992). This suggests that TNF signalling affects the mitochondrial ROS production and ROS scavenging mechanisms (Festjens *et al.*, 2006b), although also RIP1-dependent activation of membrane-associated Nox1, the major NADPH oxidase responsible for TNF-induced generation of superoxide anions has been implicated in TNF-induced necroptosis (**Figure 2**) (Kim *et al.*, 2007). Upon TNF stimulation Nox1 gets activated through an RIP1-dependent signalling complex containing TRADD, NOXO1 (NADPH oxidase organizer 1) and the small GTPase (guanosine triphosphatase) Rac1 (Kim *et al.*, 2007). Knockdown of Nox1 delays necrosis at least, in mouse fibroblasts (Kim *et al.*, 2007).

The accumulation of mitochondrial ROS is also RIP1-dependent (Lin *et al.*, 2004) and can be blocked by the lipophilic ROS scavenger butylated hydroxyanisole (BHA) and by complex I inhibitors (Festjens *et al.*, 2006b; Lin *et al.*, 2004). Therefore, it is conceivable that RIP1 directly or indirectly targets the mitochondria. Indeed, it has been demonstrated that TNF stimulation leads to localization of RIP1 to the mitochondria and reduces the interaction between ANT and CypD, resulting in adenosine triphosphate (ATP) depletion and induction of necrosis. Moreover, overexpression of CypD blocked TNF-induced necrotic cell death (Temkin *et al.*, 2006) but enhanced necrosis triggered by Ca^{2+} overload or oxidative stress (Li *et al.*, 2004), suggesting the involvement of different signalling mechanisms in necrosis, depending on the stimulus. Indeed, cells deficient in CypD are as sensitive as wildtype cells to TNF-induced cell death but are less sensitive to necrotic cell death induced by Ca^{2+} overload or oxidative stress (Baines *et al.*, 2005; Nakagawa *et al.*, 2005). In addition, $CypD^{-/-}$ mice were protected against heart and brain failure on IR and Ca^{2+} influx.

Mitochondria-derived ROS are absolutely required for TNF- and double-stranded ribonucleic acid (dsRNA)-induced necroptosis in L929 cells (Festjens *et al.*, 2007). Furthermore, caspase inhibition enhanced TNF-induced ROS production and necroptosis. In contrast, ROS quenching did not prevent necroptosis in HT-29 cells treated with a combination of TNF, Smac mimetic and zVAD (He *et al.*, 2009), indicating that ROS production is only required for necroptosis in some cell types or conditions. Recent work demonstrated the crucial role of RIP3 kinase activity in linking TNFR1-associated events, bioenergetics and increased ROS production in several cell types. In the immunoprecipitated RIP3 complex under necrotic conditions (TNF/zVAD-fmk), seven metabolic enzymes were found (Zhang *et al.*, 2009). The cytosolic glycogen phosphorylase (PYGL), cytosolic glutamate-ammonia ligase (GLUL) and mitochondrial matrix glutamate dehydrogenase 1 (GLUD1) directly interacted with RIP3. Glutamate-ammonia ligase catalyses the condensation of glutamate and ammonia to form glutamine. Glutamine is imported in the mitochondria, deaminated to glutamate by glutaminase and converted to α-ketoglutarate by glutamate dehydrogenase 1, which then feeds and enhances the citric acid cycle. Glycogen phosphorylase catalyses the degradation of glycogen to glucose-1-phosphate, which is subsequently converted into the glucose-6-phosphate intermediate that fuels glycolysis directly. Wild-type RIP3, but not kinase-dead RIP3, was shown to enhance PYGL, GLUL and GLUD1 activity *in vitro* and in cells (Zhang *et al.*, 2009; **Figure 3**), suggesting that these metabolic enzymes could be direct substrates of RIP3 kinase. Knockdown of PYGL,

GLUL or GLUD1 partially reduced the level of TNF/zVAD-fmk-mediated ROS production and necroptosis. These results further substantiate the previously reported role of glycolysis and glutaminolysis in TNF-induced necroptosis using chemical inhibitors and defined media (Goossens *et al.*, 1996). In view of the role of RIP3 in metabolic regulation, one may envisage that perhaps the primary goal of RIP3 signalling is to increase the cell's energy levels to promote cell survival or repair following toxic insult, injury or infection. This could imply that necroptosis is a byproduct of overstimulation or dysregulation of this pathway, similar to the role played by excess autophagy in promoting cell death instead of survival.

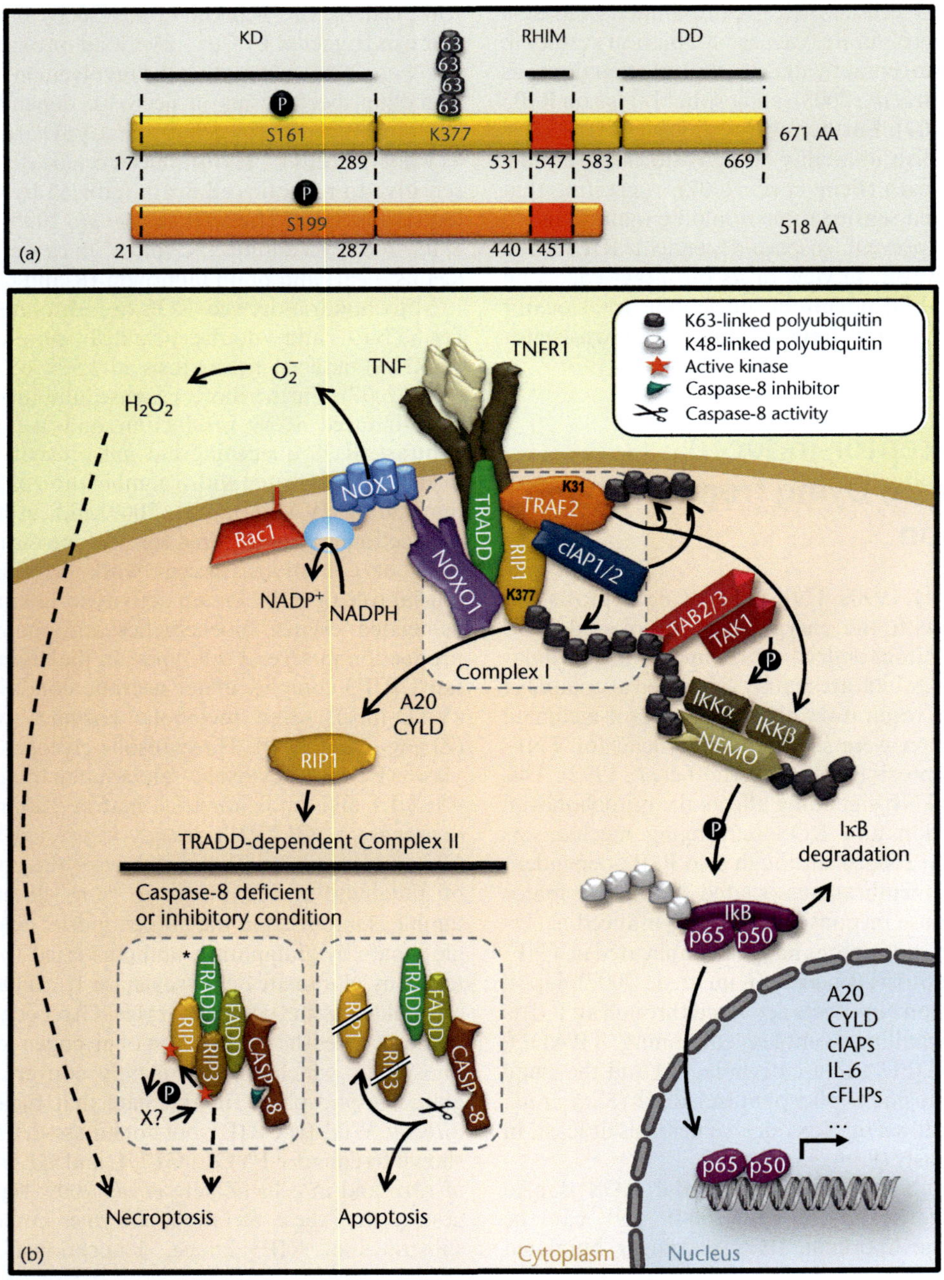

The Way to Necroptosis is Paved with Many Pathways

In the past few years, several mediators of necroptosis were identified, including calcium, lipid metabolism, lysosomes, the noncaspase proteases cathepsins and the mitochondrial permeability transition pore (mPTP) (Festjens *et al.*, 2006b). In most necrotic cell death models, these mediators are highly interconnected (**Figure 3**).

Calcium contributes to DR-induced necrotic cell death in several ways. Calcium enhances mitochondrial ROS production through calcium-induced activation of key enzymes of the Krebs cycle (Griffiths and Rutter, 2009). Calcium is also needed for translocation and subsequent activation of phospholipase A_2 (PLA_2), an esterase that is responsible for the liberation of arachidonic acid (AA) from arachidonate-containing phospholipids. Once released, the AA can be further converted by dioxygenases like lipoxygenases (LOXs). LOXs augment oxidative stress in the cell mainly through production of conjugated lipid hydroperoxides (LOOH) from free or esterified AA. Formation of LOOHs in the cell or organelle membranes induces their destabilization, a key feature of necrosis. The involvement of the $cPLA_2$/LOX pathway was demonstrated in TNF-induced necroptosis (Festjens *et al.*, 2006a).

In addition, lipid peroxidation of the lysosomal membrane leads to lysosomal membrane permeabilization (LMP) (Boya and Kroemer, 2008). LMP contributes to necrotic cell death in several ways. First, it induces ROS directly through release of iron so that hydroxyl radical-generating Fenton-type reactions are accelerated. Alternatively, LMP is inducing ROS indirectly through activation of PLA_2. Second, LMP induces the release of lysosome-contained proteases into the cytosol, such as cathepsins, which mediate DR-induced necrosis. Involvement of cathepsins in other models of necrotic cell death has also been demonstrated (Willingham *et al.*, 2007). A connection between calcium homeostasis and LMP was found in the model of TNF-induced necrosis in L929 cells. TNF induces a moderate increase of intracellular Ca^{2+} that provokes an increase in both lysosome number and size. These oversized lysosomes undergo LMP and induce plasma membrane collapse and cell death.

In vivo, necrotic cell death is typically observed after I/R of heart and brain, a model that is typically partially RIP1 dependent (Degterev *et al.*, 2005). The production of lactic acid through anaerobic glycolysis, with a consequent drop in intracellular pH, increases Ca^{2+} influx by activating acid-sensing ion channels in the cell membrane (Galluzzi *et al.*, 2009). After replenishment of cells with oxygen during reperfusion, Ca^{2+} enters the re-energized mitochondria, stimulates the Krebs cycle, and induces ROS production. This scenario is optimal for the opening of the mPTP (Odagiri *et al.*, 2009). This results in massive influx of water and efflux of glutathione and matrix pyridine nucleotides (NAD(P)H) from the mitochondria, causing inhibition of oxidative phosphorylation, and depolarization of the inner membrane, which, in turn, induces hydrolysis of all ATP by the mitochondrial ATPase, creating a permissive condition for necroptosis (Leist *et al.*, 1997). The exact molecular structure of the mPTP remains elusive (Baines, 2009). A model portrays it as a pore that forms at sites of contact between the inner and the outer

Figure 2 The necrosome, a new player in death receptor-induced necroptosis. (a) Schematic representation of the different domains in human RIP1 and RIP3. We indicated the number of amino acids corresponding to each domain, the modifications important for NFκB activation (K63-linked polyubiquitin chains on K377 in RIP1) or necroptosis (phosphorylation on RIP1 S161 and RIP3 S199). The interaction between RIP1 and RIP3 is mediated by the RHIM domain. Abbreviations: KD, kinase domain; RHIM, RIP homotypic interaction motif; DD, death domain.
(b) TNF-induced necroptosis signalling complexes. Upon TNFR1 stimulation, TRADD provides, by binding RIP1, TRAF2 and cIAP1 and cIAP2, a scaffold for the assembly of complex I at the plasma membrane. This complex is crucial for activating NFκB and MAPK pathways. K63-linked polyubiquitination of RIP1 by cIAPs results in interaction of RIP1 with the TAK1 (TGF (transforming growth factor)-β-activated kinase 1)/TAB2 (TAK-1 binding protein 2)/3 complex. K63-linked TRAF2 polyubiquitination can also recruit the TAK1/TAB2/3 complex. TAK1 activates the IKK complex, containing IKK-α, IKK-β and NEMO/IKKγ, which in turn phosphorylates IκB and results in its K48 polyubiquitination and proteasomal degradation. Once freed from its inhibitor, NFκB translocates to the nucleus and induces transcription. In a negative feedback loop, NFκB-mediated upregulation of A20 and CYLD targets RIP1 for K63 deubiquitination, thereby abolishing its ability to activate NFκB. After receptor internalization TRADD-dependent secondary cytosolic complexes are formed (complex II). The establishment of complex II involves FADD-mediated recruitment and activation of caspase-8, leading to RIP1 and RIP3 cleavage. In conditions where apoptosis is blocked by experimental (e.g. zVAD) or physiological conditions (e.g. viral inhibitors), RIP1 and RIP3 assemble a complex involving FADD and caspase-8. The asterisk indicates that TRADD may also be part of this complex; however, this has not been formally proven. Mutual direct or indirect phosphorylation of RIP1 and RIP3 in the necrosome activates necroptotic signalling. In addition, TNF induces the formation of a TNFR1 signalling complex containing TRADD, RIP1, Nox1/NOXA1 and the small GTPase Rac1. Recruitment of Nox1 complex to TNFR1 is dependent on the adaptor proteins TRADD and RIP1 kinase. Upon activation of Nox1, superoxide anions and consecutive hydrogen peroxides are generated eventually contributing to necroptosis. See text for details.

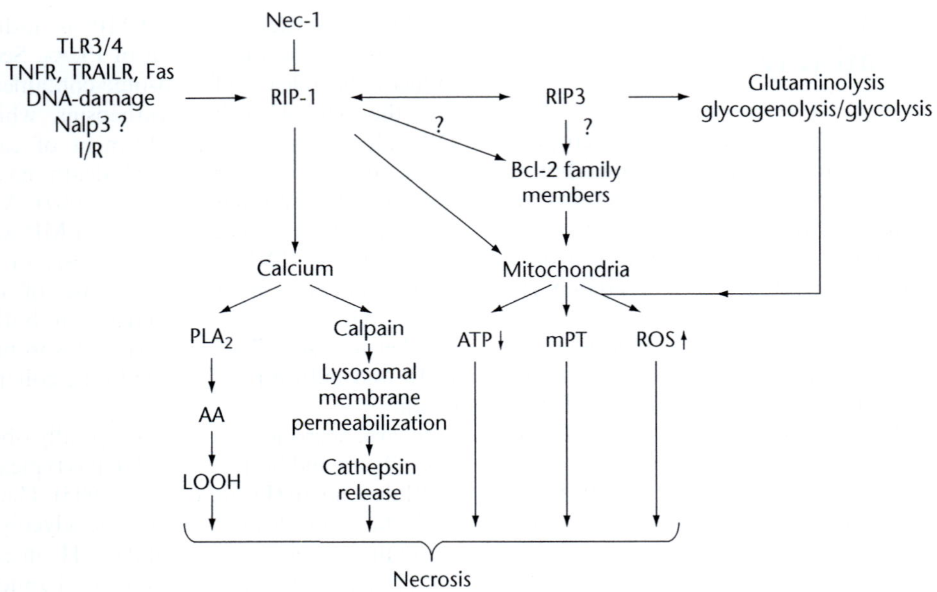

Figure 3 Different intracellular signalling events contribute to necroptosis. The kinase activity of RIP1 is needed to induce necroptosis induced by death receptors and several other stimuli. Necrotic cell death depends on the kinase activity of RIP1, which is inhibitable by the compound Nec-1, and on a phosphorylation loop between RIP1 and RIP3. The main players in the propagation of necroptosis are calcium and mitochondria. Calcium controls the activation of PLA$_2$ and calpains, inducing lipid peroxidation (LOOH) and cathepsin release from lysosomes, respectively http://merops.sanger.ac.uk/. Mitochondria contribute to necroptosis by excessive ROS production, mitochondrial permeability transition (mPT) or ATP depletion. RIP3 mediates enhanced glycolysis through glycogenolysis, and glutaminolysis, stimulating Krebs cycle and ROS production. See text for details. In addition, Bcl-2 family members also contribute to necroptotic signalling, probably by affecting the mitochondrial function.

mitochondrial membrane and spans both membranes. It is believed to consist of voltage dependent anion channel (VDAC) located at the outer membrane, ANT located at the inner membrane, and CyP-D, a peptidylprolyl *cis–trans* isomerase located in the matrix. Knockout studies prove that Cyp-D is essential for mPTP but ANT and VDAC apparently are not (Juhaszova *et al.*, 2008) (cross reference to chapter 17-Green). **See also**: Mitochondrial Outer Membrane Permeabilization

Intriguingly, Bcl-2-modifying factor (Bmf), a proapoptotic Bcl-2 protein family member, is required for death receptors-induced necroptosis (Hitomi *et al.*, 2008). Furthermore, knockdown of the membrane-associated tyrosine/threonine kinase 1, which interacts with the antiapoptotic protein Bcl-2, protects FADD-deficient Jurkat cells against TNF-induced necroptosis (Cho *et al.*, 2009), suggesting that Bcl-2 protein family members function in necroptosis, in addition to their roles in apoptosis. **See also**: BH3-Only Proteins; The Bcl-2 Family Proteins - Key Regulators and Effectors of Apoptosis

Physiological Role of Necroptosis

Several pathologies, such as neurodegeneration, ischaemia-reperfusion and infection, are associated with necrotic cell death. Excitotoxicity is implicated in stroke and many neurodegenerative disorders, for example, Alzheimer (AD), Huntington (HD) and Parkinson diseases (PD) (http://en.wikipedia.org/wiki/Excitotoxicity). Oxidative stress and mitochondrial dysfunction occur early in all major neurodegenerative diseases, and there is strong evidence that it has a causal role in pathogenesis (Galluzzi *et al.*, 2009). During normal ageing, the brain accumulates iron, copper and zinc, which increase oxidative stress, via the Fenton reaction, which can lead to necrotic cell death (Doraiswamy and Finefrock, 2004). Consequently, iron chelation and ROS scavenging seem promising for future treatment of patients with neurodegenerative diseases or acute myocardial infarction (reviewed in Vanlangenakker *et al.*, 2008).

The partially protective effects of necrostatins in experimental models of neurodegenerative diseases, ischaemic brain and heart injuries and head trauma

 Cell Death © 2010, John Wiley & Sons, Ltd.

suggest that endogenously produced TNF in conditions of cellular stress may contribute in an autocrine or a paracrine way to stimulate necroptosis (Degterev *et al.*, 2005; Lim *et al.*, 2007; Smith *et al.*, 2007; You *et al.*, 2008). This form of cell death may occur only when caspase activation is impaired or when caspases are inhibited. *In vivo* TNF provokes a systemic inflammatory response syndrome that leads to shock, organ failure and death. Similar to the *in vitro* sensitization of TNF-induced necroptosis by the addition of zVAD, zVAD exacerbated the *in vivo* TNF toxicity in an experimental mouse model by enhancing oxidative stress and mitochondrial damage, resulting in hyperacute haemodynamic collapse, kidney failure and death (Cauwels *et al.*, 2003). The zVAD sensitization of TNF-induced lethality was abrogated by the addition of ROS scavengers, such as butylated hydroxyanisole or tempol, or by the addition of a $cPLA_2$ inhibitor. Nevertheless, none of these compounds affected mouse TNF toxicity as such. This could indicate that enhanced necroptosis or respiration determines the sensitivity for *in vivo* TNF toxicity.

Notably, viruses often express anti-apoptotic proteins that prevent caspase-8 recruitment or activation at the death inducing complex to prolong cell viability and to facilitate viral replication. $RIP3^{-/-}$ T-cells infected with Vaccinia virus (VV), which encodes the viral caspase inhibitor B13R/Spi2, exhibited significantly reduced sensitivity to TNF-induced necroptosis (Cho *et al.*, 2009). The requirement for RIP3 in induction of cell death in VV-infected cells suggests that *in vivo* the TNF-mediated protection against VV infection acts via RIP3-dependent necroptosis. Indeed, $RIP3^{-/-}$ mice showed increased viral titers and succumbed to the VV infection, whereas wild-type mice were protected (Cho *et al.*, 2009). In addition, cerulein-treated wild-type mice showed multiple areas of pancreas acinar cell loss and necrosis, but these effects were largely prevented in $RIP3^{-/-}$ littermates (He *et al.*, 2009; Zhang *et al.*, 2009). *In vivo*, necroptotic cell death probably serves two functions. First, it acts as a backup mechanism to kill virus-infected cells if the apoptotic pathway is blocked. Second, in an immunological context, the release of ATP and other intracellular DAMPs (danger-associated molecular patterns) from necroptotic dying cells, could serve as an endogenous adjuvant boosting the innate immune system. **See also**: Microbial Inhibitors of Apoptosis.

Conclusions and Perspectives

Recently it became clear that in the absence of sufficient caspase-8 activity, DR-induced necroptosis acts as a physiological form of cell death. In addition, an *in vivo* role for RIP1/RIP3-mediated necroptosis in anti-viral responses has been established. Importantly, in the absence of caspase-8 activity, the RIP1/RIP3 necrosome regulates the molecular bifurcation between necroptosis and survival. Furthermore, apoptosis actively prevents necroptotic signalling by caspase-mediated cleavage of RIP1 and RIP3. Apparently, apoptotic pathways are mainly controlled by the proteolytic caspase network, whereas necroptotic signalling cascades are controlled by a kinase cascade. The latter may become a useful biochemical marker to define necrotic cell death. However, necrotic cell death is not due to a single well-described signalling pathway but results from the interplay between several signalling events. The RIP3 cascade connects necroptosis with energy metabolism and increased ROS production. This could open new avenues for the development of kinase inhibitors to prevent pathological cell death in diseases such as ischaemia reperfusion damage during organ transplantation, cardiac infarction, stroke and traumatic brain injury. In part, this has already been initiated with the development of the necrostatin RIP1 inhibitors. Alternatively, the further development of antioxidant therapy to decrease ROS production and to upregulate the endogenous antioxidant defence network will be of clinical importance.

Acknowledgements

This research has been supported by VIB, UGhent, FP6 ApopTrain, MRTN-CT-035624; FP6 Epistem, LSHB-CT-2005-019067; Apo-Sys FP7-200767, IAP 6/18, FWO-Vlaanderen (3G.0218.06 and G.0226.09) and BOF-GOA – 12.0505.02. T.V.B. holds a grant of the FWO. P.V. holds a Methusalem grant from the Flemish Government.

References

Baines CP (2009) The mitochondrial permeability transition pore and ischemia-reperfusion injury. *Journal of Molecular and Cellular Cardiology* **46**: 969–977.

Baines CP, Kaiser RA, Purcell NH *et al.* (2005) Loss of cyclophilin D reveals a critical role for mitochondrial permeability transition in cell death. *Nature* **434**: 658–662.

Bertrand MJ, Milutinovic S, Dickson KM *et al.* (2008) cIAP1 and cIAP2 facilitate cancer cell survival by functioning as E3 ligases that promote RIP1 ubiquitination. *Molecular Cell* **30**: 689–700.

Boya P and Kroemer G (2008) Lysosomal membrane permeabilization in cell death. *Oncogene* **27**: 6434–6451.

Cauwels A, Janssen B, Waeytens A, Cuvelier C and Brouckaert P (2003) Caspase inhibition causes hyperacute tumor necrosis factor-induced shock via oxidative stress and phospholipase A2. *Nature Immunology* **4**: 387–393.

Chan FK (2000) The pre-ligand binding assembly domain: a potential target of inhibition of tumour necrosis factor receptor function. *Annals of the Rheumatic Diseases* **59**(suppl 1): i50–i53.

Ch'en IL, Beisner DR, Degterev A *et al.* (2008) Antigen-mediated T cell expansion regulated by parallel pathways of death. *Proceedings of the National Academy of Sciences of the United States of America* **105**: 17463–17468.

Cho YS, Challa S, Moquin D *et al.* (2009) Phosphorylation-driven assembly of the RIP1-RIP3 complex regulates programmed necrosis and virus-induced inflammation. *Cell* **137**: 1112–1123.

Degterev A, Hitomi J, Germscheid M *et al.* (2008) Identification of RIP1 kinase as a specific cellular target of necrostatins. *Nature Chemical Biology* **4**: 313–321.

Degterev A, Huang Z, Boyce M *et al.* (2005) Chemical inhibitor of nonapoptotic cell death with therapeutic potential for ischemic brain injury. *Nature Chemical Biology* **1**: 112–119.

Doraiswamy PM and Finefrock AE (2004) Metals in our minds: therapeutic implications for neurodegenerative disorders. *Lancet Neurology* **3**: 431–434.

Ermolaeva MA, Michallet MC, Papadopoulou N *et al.* (2008) Function of TRADD in tumor necrosis factor receptor 1 signaling and in TRIF-dependent inflammatory responses. *Nature Immunology* **9**: 1037–1046.

Feng S, Yang Y, Mei Y *et al.* (2007) Cleavage of RIP3 inactivates its caspase-independent apoptosis pathway by removal of kinase domain. *Cell Signal* **19**: 2056–2067.

Festjens N, Kalai M, Smet J *et al.* (2006a) Butylated hydroxyanisole is more than a reactive oxygen species scavenger. *Cell Death and Differentiation* **13**: 166–169.

Festjens N, Vanden Berghe T and Vandenabeele P (2006b) Necrosis, a well-orchestrated form of cell demise: signalling cascades, important mediators and concomitant immune response. *Biochimica et Biophysica Acta* **1757**: 1371–1387.

Festjens N, Vanden Berghe T, Cornelis S and Vandenabeele P (2007) RIP1, a kinase on the crossroads of a cell's decision to live or die. *Cell Death and Differentiation* **14**: 400–410.

Galluzzi L, Morselli E, Kepp O and Kroemer G (2009) Targeting post-mitochondrial effectors of apoptosis for neuroprotection. *Biochimica et Biophysica Acta* **1787**: 402–413.

Goossens V, Grooten J and Fiers W (1996) The oxidative metabolism of glutamine. A modulator of reactive oxygen intermediate-mediated cytotoxicity of tumor necrosis factor in L929 fibrosarcoma cells. *Journal of Biological Chemistry* **271**: 192–196.

Griffiths EJ and Rutter GA (2009) Mitochondrial calcium as a key regulator of mitochondrial ATP production in mammalian cells. *Biochimica et Biophysica Acta* **1787**(11): 1324–1333.

He S, Wang L, Miao L *et al.* (2009) Receptor interacting protein kinase-3 determines cellular necrotic response to TNF-alpha. *Cell* **137**: 1100–1111.

Hitomi J, Christofferson DE, Ng A *et al.* (2008) Identification of a molecular signaling network that regulates a cellular necrotic cell death pathway. *Cell* **135**: 1311–1323.

Holler N, Zaru R, Micheau O *et al.* (2000) Fas triggers an alternative, caspase-8-independent cell death pathway using the kinase RIP as effector molecule. *Nature Immunology* **1**: 489–495.

Juhaszova M, Wang S, Zorov DB *et al.* (2008) The identity and regulation of the mitochondrial permeability transition pore: where the known meets the unknown. *Annals of the New York Academy of Sciences* **1123**: 197–212.

Kelliher MA, Grimm S, Ishida Y *et al.* (1998) The death domain kinase RIP mediates the TNF-induced NF-kappaB signal. *Immunity* **8**: 297–303.

Kerr JF, Wyllie AH and Currie AR (1972) Apoptosis: a basic biological phenomenon with wide-ranging implications in tissue kinetics. *British Journal of Cancer* **26**: 239–257.

Kim YS, Morgan MJ, Choksi S and Liu ZG (2007) TNF-induced activation of the Nox1 NADPH oxidase and its role in the induction of necrotic cell death. *Molecular Cell* **26**: 675–687.

Kroemer G, Galluzzi L, Vandenabeele P *et al.* (2009) Classification of cell death: recommendations of the Nomenclature Committee on Cell Death 2009. *Cell Death and Differentiation* **16**: 3–11.

Leist M, Single B, Castoldi AF, Kuhnle S and Nicotera P (1997) Intracellular adenosine triphosphate (ATP) concentration: a switch in the decision between apoptosis and necrosis. *Journal of Experimental Medicine* **185**: 1481–1486.

Li Y, Johnson N, Capano M, Edwards M and Crompton M (2004) Cyclophilin-D promotes the mitochondrial permeability transition but has opposite effects on apoptosis and necrosis. *Biochemical Journal* **383**: 101–109.

Lim SY, Davidson SM, Mocanu MM, Yellon DM and Smith CC (2007) The cardioprotective effect of necrostatin requires the cyclophilin-D component of the mitochondrial permeability transition pore. *Cardiovascular Drugs and Therapy* **21**: 467–469.

Lin Y, Choksi S, Shen HM *et al.* (2004) Tumor necrosis factor-induced nonapoptotic cell death requires receptor-interacting protein-mediated cellular reactive oxygen species accumulation. *Journal of Biological Chemistry* **279**: 10822–10828.

Mahoney DJ, Cheung HH, Mrad RL *et al.* (2008) Both cIAP1 and cIAP2 regulate TNFalpha-mediated NF-kappaB activation. *Proceedings of the National Academy of Sciences of the USA* **105**: 11778–11783.

Meylan E and Tschopp J (2005) The RIP kinases: crucial integrators of cellular stress. *Trends in Biochemical Sciences* **30**: 151–159.

Nakagawa T, Shimizu S, Watanabe T *et al.* (2005) Cyclophilin D-dependent mitochondrial permeability transition regulates some necrotic but not apoptotic cell death. *Nature* **434**: 652–658.

Newton K, Sun X and Dixit VM (2004) Kinase RIP3 is dispensable for normal NF-kappa Bs, signaling by the B-cell and T-cell receptors, tumor necrosis factor receptor 1, and Toll-like receptors 2 and 4. *Molecular and Cellular Biology* **24**: 1464–1469.

Odagiri K, Katoh H, Kawashima H *et al.* (2009) Local control of mitochondrial membrane potential, permeability transition pore and reactive oxygen species by calcium and calmodulin in rat ventricular myocytes. *Journal of Molecular and Cellular Cardiology* **46**: 989–997.

Pobezinskaya YL, Kim YS, Choksi S *et al.* (2008) The function of TRADD in signaling through tumor necrosis factor receptor 1 and TRIF-dependent Toll-like receptors. *Nature of Immunology* **9**: 1047–1054.

Schulze-Osthoff K, Bakker AC, Vanhaesebroeck B *et al.* (1992) Cytotoxic activity of tumor necrosis factor is mediated by early damage of mitochondrial functions. Evidence for the involvement of mitochondrial radical generation. *Journal of Biological Chemistry* **267**: 5317–5323.

Schutze S, Tchikov V and Schneider-Brachert W (2008) Regulation of TNFR1 and CD95 signalling by receptor compartmentalization. *Nature Reviews* **9**: 655–662.

Schweichel JU and Merker HJ (1973) The morphology of various types of cell death in prenatal tissues. *Teratology* **7**: 253–266.

Smith CC, Davidson SM, Lim SY *et al.* (2007) Necrostatin: a potentially novel cardioprotective agent? *Cardiovascular Drugs and Therapy* **21**: 227–233.

Sun H, Nikolovska-Coleska Z, Yang CY *et al.* (2008) Design of small-molecule peptidic and nonpeptidic Smac mimetics. *Accounts of Chemical Research* **41**: 1264–1277.

Temkin V, Huang Q, Liu H, Osada H and Pope RM (2006) Inhibition of ADP/ATP exchange in receptor-interacting protein-mediated necrosis. *Molecular and Cellular Biology* **26**: 2215–2225.

Vandenabeele P, Vanden Berghe T and Festjens N (2006) Caspase inhibitors promote alternative cell death pathways. *Science's STKE* **2006**: pe44.

Vanlangenakker N, Berghe TV, Krysko DV, Festjens N and Vandenabeele P (2008) Molecular mechanisms and pathophysiology of necrotic cell death. *Current Molecular Medicine* **8**: 207–220.

Varfolomeev E, Goncharov T, Fedorova AV *et al.* (2008) c-IAP1 and c-IAP2 are critical mediators of tumor necrosis factor alpha (TNFalpha)-induced NF-kappaB activation. *Journal of Biological Chemistry* **283**: 24295–24299.

Vercammen D, Brouckaert G, Denecker G *et al.* (1998) Dual signaling of the Fas receptor: initiation of both apoptotic and necrotic cell death pathways. *Journal of Experimental Medicine* **188**: 919–930.

Wang L, Du F and Wang X (2008) TNF-alpha induces two distinct caspase-8 activation pathways. *Cell* **133**: 693–703.

Willingham SB, Bergstralh DT, O'Connor W *et al.* (2007) Microbial pathogen-induced necrotic cell death mediated by the inflammasome components CIAS1/Cryopyrin/NLRP3 and ASC. *Cell Host & Microbe* **2**: 147–159.

Wilson NS, Dixit V and Ashkenazi A (2009) Death receptor signal transducers: nodes of coordination in immune signaling networks. *Nature Immunology* **10**: 348–355.

You Z, Savitz SI, Yang J *et al.* (2008) Necrostatin-1 reduces histopathology and improves functional outcome after controlled cortical impact in mice. *Journal of Cerebral Blood Flow and Metabolism* **28**: 1564–1573.

Zhang DW, Shao J, Lin J *et al.* (2009) RIP3, an energy metabolism regulator that switches TNF-induced cell death from apoptosis to necrosis. *Science* **325**: 332–336.

Further Reading

Degterev A and Yuan J (2008) Expansion and evolution of cell death programmes. *Nature Reviews. Molecular Cell Biology* **9**: 378–390.

Galluzzi L and Kroemer G (2008) Necroptosis: a specialized pathway of programmed necrosis. *Cell* **135**: 1161–1163.

Green DR, Ferguson T, Zitvogel L and Kroemer G (2009) Immunogenic and tolerogenic cell death. *Nature Reviews* **9**: 353–363.

Zong WX and Thompson CB (2006) Necrotic death as a cell fate. *Genes & Development* **20**: 1–15.

Inhibitor of Apoptosis (IAP) and BIR-containing Proteins

David L Vaux, *La Trobe University, Victoria, Australia*

Advanced article

Article Contents

- Introduction
- Structure of the BIR fold
- RING Domains
- UBA domains
- XIAP
- ML-IAP
- cIAP1 and cIAP2
- IAPs and Cancer
- IAP-antagonist Compounds
- Survivin and BRUCE
- NAIP

The inhibitor of apoptosis (IAP) family of proteins are characterized by presence of one or more baculoviral IAP Repeat (BIR) domains, so they are also termed BIRCs, for 'BIR-containing' proteins. The first IAPs were identified in insect-attacking viruses, which use them to inhibit defensive apoptosis of the host cell. Subsequently, other BIR-containing proteins have been found in organisms from yeasts to mammals. The human IAP XIAP can block apoptosis by directly inhibiting caspases, whereas cIAP1 and cIAP2 inhibit apoptosis by reducing caspase-activating signals from tumour necrosis factor (TNF) receptor superfamily members. As the genes for some IAPs are amplified in cancers, they appear to be able to act as oncogenes. IAP-antagonist compounds modelled on the IAP antagonist Smac/Diablo are being developed for the treatment of cancer, and are currently undergoing clinical trials.

Introduction

The first inhibitor of apoptosis (IAP) was found by Lois Miller and colleagues as a baculoviral gene that could prevent insect cell apoptosis in response to infection with a virus that had mutations in the gene for the antiapoptotic caspase inhibitor p35 (Crook *et al.*, 1993). Because the IAP was able to complement for loss of p35, to keep the infected cell alive during viral replication, it was clear that this baculoviral IAP was, like p35, an inhibitor of apoptosis. However, these results did not necessarily mean that OpIAP and p35 inhibited cell death by the same mechanism.

Miller and colleagues noticed that the baculoviral IAPs bore two domains they termed baculoviral IAP repeats (BIRs) (Birnbaum *et al.*, 1994).

The next BIR-containing protein to be identified was the human protein NAIP (BIRC1) (Roy *et al.*, 1995), whose gene lies next to *SMN*, the gene mutated in spinal muscular atrophy (SMA). Because *NAIP* was a candidate gene for SMA, and was found to bear a BIR, it was proposed to function as a neuronal apoptosis inhibitory protein, hence its name. Subsequently, it was shown that mutations to SMN are responsible for SMA, and good evidence has been obtained in mice to suggest that NAIP functions to promote innate immunity against certain bacterial infections, rather than to inhibit cell death (Diez *et al.*, 2003; Wright *et al.*, 2003).

The mammalian BIRCs cIAP1 and cIAP2 recognized as BIR-bearing proteins by sequence analysis, but also as proteins that associated with TRAF1 and TRAF2 and the cytoplasmic domain of tumour necrosis factor (TNF) receptor 2 (Rothe *et al.*, 1995). Other BIRCs, including BRUCE, Survivin, XIAP, ML-IAP and others were identified in searches for sequences encoding BIRs (Uren *et al.*, 1998) (**Box 1**). **See also**: Apoptosis: Regulatory Genes and Disease; Death Receptors; Death Receptors at the Molecular Level: Therapeutic Implications

Survivin, BRUCE and ML-IAP were classified as IAPs when they were found by sequence analysis to encode BIR domains (**Figure 1**, **Table 1**).

Structure of the BIR fold

All IAPs bear one or more BIR domains. The baculoviral IAPs typically bear two BIRs. These domains were

Box 1 Timeline of research on IAPs

1993	IAPs identified in baculoviruses in Lois Miller's lab (Crook *et al.*, 1993)
1995	NAIP, a candidate gene for spinal muscular atrophy found by Alex McKenzie's group, is found to have BIRs (Roy *et al.*, 1995)
1994–1996	Identification of IAP antagonists in the fly, Reaper, Grim and HID (Chen *et al.*, 1996; Grether *et al.*, 1995; White *et al.*, 1994)
1995–1996	Identification of RING and BIR-bearing cellular IAPs in mammals and *Drosophila* (Duckett *et al.*, 1996; Hay *et al.*, 1995; Liston *et al.*, 1996; Rothe *et al.*, 1995; Uren *et al.*, 1996)
1997	Identification of Survivin (Ambrosini *et al.*, 1997). (It was first cloned three years earlier in antisense orientation as EPR-1 (Altieri, 1994))
1997	Discovery that XIAP could directly inhibit caspases (Deveraux *et al.*, 1997)
1997	Demonstration that Reaper-induced apoptosis could be blocked by IAPs (Vucic *et al.*, 1997)
1998	BRUCE is identified in a search for genes encoding ubiquitin-conjugating enzymes (Hauser *et al.*, 1998)
1998	HID and Grim directly bind to *Drosophila* IAPs (Vucic *et al.*, 1998)
1999–2000	Demonstration that homologues of survivin are required for chromosome segregation in mitosis (Fraser *et al.*, 1999; Speliotes *et al.*, 2000; Uren *et al.*, 1999; Uren *et al.*, 2000)
2000	Cloning of mammalian IAP-binding protein Smac/Diablo (Du *et al.*, 2000; Verhagen *et al.*, 2000)
2000	Binding of amino-termini of insect and mammalian IAP-binding proteins to BIRs of IAPs. (Silke *et al.*, 2000; Wu *et al.*, 2000)
2007	Description of Smac-mimetic IAP-antagonist compounds, and demonstration that their main targets are cIAP1 and cIAP2, and that they cause cell death by a mechanism involving autocrine TNF (Gaither *et al.*, 2007; Petersen *et al.*, 2007; Varfolomeev *et al.*, 2007; Vince *et al.*, 2007)
2008	BRUCE shown to function in cytokinesis (Pohl and Jentsch, 2008)

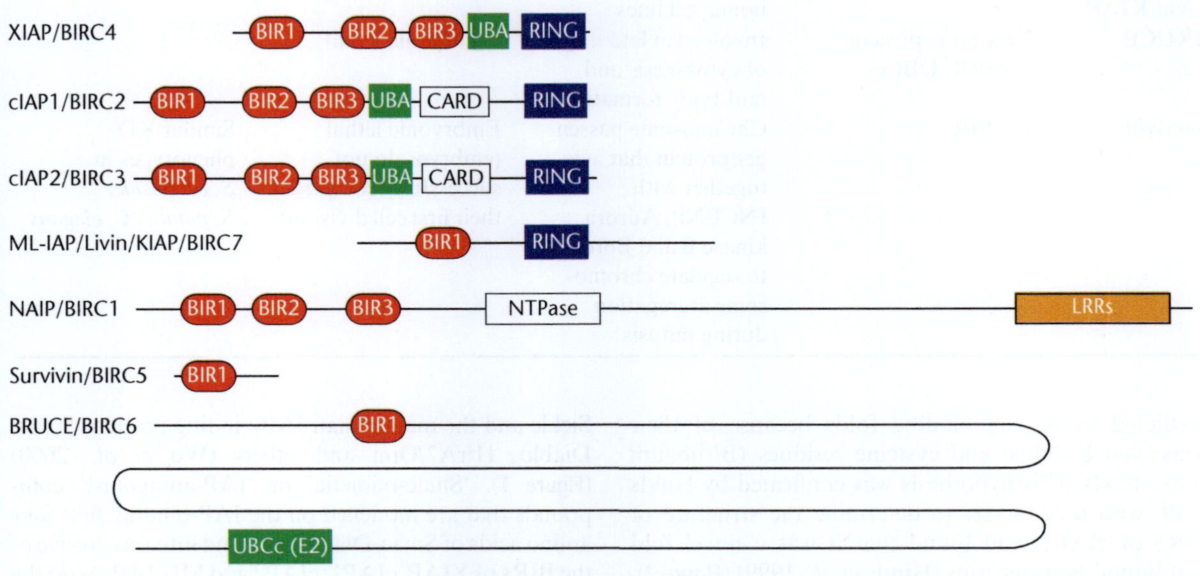

Figure 1 The defining characteristic of all IAPs is the presence of one or more baculoviral IAP repeats (BIRs; red ovals). BIRs allow binding to a variety of other proteins, such as caspases and IAP antagonists (e.g. BIR2 and BIR3 of cIAPs and XIAP). cIAP1&2 bind to TRAF2 via their BIR1 domains. Many IAPs bear C-terminal RING domains (blue rectangles) that allow them to act as ubiquitin E3 ligases. After dimerization, the RINGs associate with E2s. Several of the IAPs bear UBA domains, which act as receptors for K63-linked poly ubiquitinylated proteins. In addition to BIRs and an NTPase domain, NAIP bears leucine-rich repeats, which are often found in proteins involved in innate immunity. The single BIRs in Survivin and BRUCE resemble each other more than the BIRs of other BIRCs, and both these proteins act mainly during mitosis to allow proper chromosome segregation.

Table 1 Mammalian BIR-containing proteins

Protein	Domains	Function	KO phenotypes	Homologues
NAIP	3 × BIR, NTPase, LRR	Innate immune responses, especially against intracellular bacteria	Decreased resistance to *Legionella* infections	Mice have several expressed NAIP genes
XIAP	3 × BIR, UBA (ubiquitin receptor), RING	Immune response, inhibition of cell death, potent direct inhibitor of caspases	Mild to no phenotype in mice	XIAP mutation in humans leads to immunodeficiency syndrome. *Drosophila* has two XIAP-like genes, DIAP1 and DIAP2 that are essential for development.
cIAP1 & cIAP2	3 × BIR, UBA, CARD, RING	Decrease proapoptotic signals from TNF receptor superfamily members. Reduce activation of non-canonical NFκB signalling pathways. cIAP1 & 2 can bind to caspases but do not directly inhibit them	cIAP1 and cIAP2 KO mice have mild phenotypes, presumably due to mutual redundancy	KO of the only cIAP1/2 homologue in Zebrafish is embryonic lethal, with failure of endothelial cells to maintain integrity (Santoro *et al.*, 2007)
ML-IAP/ Livin/KIAP	BIR, RING	Expressed in melanoma cell lines	Not yet described	
BRUCE	Very large protein, 1 × BIR, UBCc	Involved in late stages of cytokinesis and mid-body formation	Embryonic lethal	
Survivin	1 × BIR	Chromosome passenger protein that acts together with INCENP, Aurora kinase B and Borealin to regulate chromosome segregation during mitosis	Embryonic lethal (embryos do not successfully complete their first cell division)	Similar KO phenotypes in *S. cerevisiae*, *S. pombe*, *C. elegans*

predicted to be zinc-binding folds because of their conserved histidine and cysteine residues (Birnbaum *et al.*, 1994). This hypothesis was confirmed by Hinds *et al.* who used NMR to determine the structure of BIR3 of cIAP1, and found that it was a novel fold that bound two zinc ions (Hinds *et al.*, 1999) (**Figure 2**). **See also**: Inhibitor of Apoptosis (IAP) and BIR-containing Proteins

The surface of most BIRs has a groove that can bind the amino-termini of IAP antagonists, such as the proapoptotic *Drosophila* proteins Reaper, Grim, HID and Sickle and the mammalian IAP-binding proteins Smac/Diablo, HtrA2/Omi and others (Wu *et al.*, 2000) (**Figure 1**). 'Smac-mimetic' or 'IAP-antagonist' compounds that are modelled on the IAP-binding first four amino acids of Smac/Diablo also bind into this groove of the BIRs of XIAP, cIAP1, cIAP2 and ML-IAP, as do the amino-termini of processed caspases 3, 7 and 9 (Srinivasula *et al.*, 2001), but caspase 3 can also interact with the region flanking BIR2 of XIAP as well, in a manner that blocks the caspase's catalytic cysteine (Sun *et al.*, 2000).

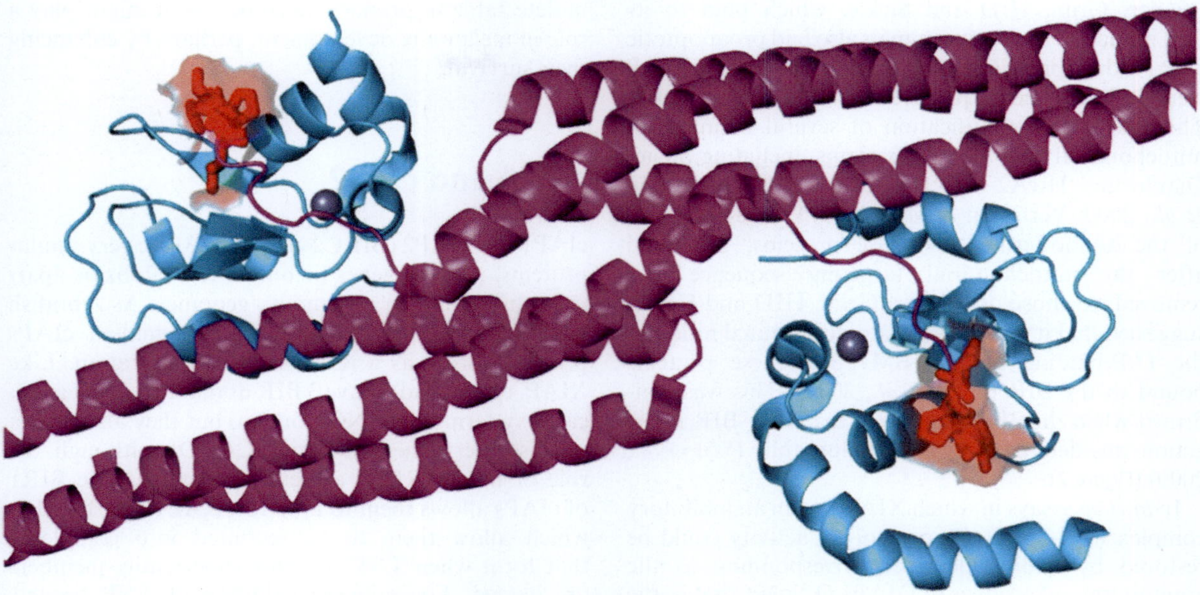

Figure 2 Complex of Smac/DIABLO dimer with BIR3s from XIAP. The BIRs are shown in blue; note the zinc ions. The residues (AVPI ...) at the processed amino-terminus of smac/DIABLO (red) bind into a groove of the surface of the BIRs. Synthetic IAP antagonist compounds have been designed that mimic the *N*-terminus of smac/DIABLO, and are therefore sometimes termed 'smac-mimetics'. The remainder of smac/DIABLO is composed of extended α helices (purple). (This figure is based on the structure produced by (Wu *et al.*, 2000).

RING Domains

Several IAPs, including XIAP, cIAP1, cIAP2 and ML-IAP in humans, DIAP1 and DIAP2 in *Drosophila* and the original baculoviral IAPs, bear a carboxy-terminal RING domain, and are therefore able to act as ubiquitin E3 ligases (Vaux and Silke, 2005). This means these IAPs can function by directly binding to other proteins, as well as by ubiquitylating them. Ubiquitin tags on proteins can cause them to be targeted for destruction by the proteasome, but, depending on the number and linkage of the ubiquitin monomers, can also stabilize proteins, or alter their binding properties.

UBA domains

XIAP, cIAP1, cIAP2 and DIAP2 (but not ML-IAP nor *Drosophila* DIAP1) bear a ubiquitin-associated domain (UBA) after BIR3 (Gyrd-Hansen *et al.*, 2008). UBAs are a domain that consists of three α helices that enable proteins to bind K63-linked ubiquitin chains. During signalling from TNF superfamily receptors, proteins such as RIP1 and NEMO become K63 ubiquitylated, which would allow them to bind to the UBAs of IAPs.

XIAP

XIAP, cIAP1, cIAP2, ML-IAP form a subgroup of BIR-containing proteins as all of them also bear RING domains, like the original baculoviral IAPs. Humans also have another XIAP-like gene, hILP-2 (BIRC8), but as expression of this gene is restricted to the testes, and the protein appears to be unstable (Shin *et al.*, 2005).

XIAP (officially designated *BIRC4*) is so-called because its gene is on the X chromosome of mammals. XIAP is the only IAP that is a potent, direct inhibitor of the proteolytic activity of caspases and is able to inhibit caspase 3 and caspase 7 with nanomolar activity. The key interaction with caspases involves the region amino terminal to BIR2 of XIAP (Sun *et al.*, 2000), but there is also an interaction of the processed amino-terminus of the caspase with the groove of BIR3.

Like XIAP, DIAP1 from *Drosophila* has three BIR domains and a carboxy-terminal RING. DIAP1 is antagonized by the cytoplasmic proapoptotic proteins

Reaper, Grim, HID and Sickle, which bind to its BIR domains. To see if mammals also had proapoptotic IAP-binding proteins, the molecules binding to XIAP were isolated and sequenced by mass spectrometry. This led to the identification of several mammalian mitochondrial IAP-binding proteins, including Smac/Diablo and HtrA2/Omi (Du *et al.*, 2000; Verhagen *et al.*, 2000; Verhagen *et al.*, 2007). The resemblance of the amino-termini of these proteins, generated after the mitochondrial targeting sequence was removed, to those of Reaper, Grim, HID and Sickle, suggested that it was via these amino-terminal residues, the IAP-binding motif (IBM) that these proteins bound to the BIR (Silke *et al.*, 2000). This was confirmed when the 3D structure of the IBM–BIR interaction was determined by crystallography (Wu *et al.*, 2000) (**Figure 2**).

In *in vitro* assays in which XIAP was in an inhibitory complex with caspase 3, proteolytic activity could be restored by adding peptides corresponding to the amino-terminal residues of DIABLO/Smac, that is, the IBM. This prompted the design of peptido-mimetic IAP antagonist compounds. It was hoped that such compounds might free active caspases from inhibition by XIAP, thereby leading to apoptosis of cancer cells.

Several other proteins have been reported to interact with XIAP. Many of these are mitochondrial proteins that have amino-terminal IBMs, but there are also cytosolic proteins including cIAP1, TAB1, Xaf1, ILPIP, but it is too early to tell the significance of these interactions.

To determine the role of XIAP *in vivo*, XIAP gene knockout mice were generated, but to date they have not been very informative, as the mice have no or only a very subtle phenotype (Harlin *et al.*, 2001). However, the discovery of human pedigrees with mutations in XIAP and a lymphoproliferative syndrome with a deficiency of NKT cells suggests that human XIAP is an important regulator of immunity (Rigaud *et al.*, 2006).

ML-IAP

ML-IAP, also known as Livin/KIAP, and in humans officially named BIRC7, was first identified in complementary deoxyribonucleic acid (cDNA) libraries derived from melanoma cells (Kasof and Gomes, 2001; Vucic *et al.*, 2000). ML-IAP contains a single BIR and RING domain. Its single BIR is most similar to BIR2 of XIAP. Because ML-IAP was readily detectable in many melanoma lines, whereas it was undetectable in primary melanocytes, it might play a role in melanoma development, perhaps by enhancing their survival.

cIAP1 and cIAP2

cIAP1 and cIAP2 (BIRC2 and BIRC3) are very similar proteins, and the genes lie only a few kilobases apart in both the human and mouse genomes. As zebrafish have only one cIAP gene, the mammalian cIAPs probably arose as a result of gene duplication. Like XIAP, cIAP1 and 2 have 3 BIR domains, a UBA and a carboxy-terminal RING domain, but they also bear a caspase recruitment domain (CARD), although the role of the CARD is currently unknown. The BIR1 of cIAPs allows them to bind to TRAF1 and TRAF2, which allow them to be recruited into complexes that form when TNF receptor superfamily members are ligated. For example, addition of TNF triggers formation of a complex containing TRAFs and cIAPs that transmit signals leading to activation of nuclear factor-κB (NFκB) and AP1 transcription factors. However, in resting cells, in the absence of TNF superfamily cytokines, TRAF2 and cIAP1 act together to reduce the levels of NIK, and thereby maintain low levels of activation of the noncanonical NFκB pathway (**Figure 3**).

As both *ciap1* gene knockout and *ciap2* gene knockout mice have relatively subtle phenotypes, but knockdown of the single *ciap* gene in zebrafish is lethal, it is likely that in mice presence of cIAP1 can compensate for loss of *cIAP2* and vice versa.

Analysis of *traf2* and *ciap* gene-deleted mice have revealed some of the aspects of their function in mediating signalling by TNF receptor superfamily members. Ligation of TNFR1, whose cytoplasmic tail bears a TRAF2 interacting domain as well as a death domain, triggers binding of signalling intermediate proteins such as TRAF2, RIP1 and TRADD. TRAF2 recruits cIAP1 and 2, via interactions with BIR1. Several other proteins can be incorporated into the complex, and some, such as IKKγ (NEMO) and RIP1 become ubiquitylated. As both TRAF2 and the cIAPs bear RING domains, it is likely that one or other are involved in ubiquitylating other components of the complex.

Other TNF receptor superfamily members such as those that do not bear a death domain, for example, CD40 or TWEAK, seem to signal differently because although they recruit TRAFs, and via them cIAPs, they do not recruit RIP1. Experiments with TRAF2 and

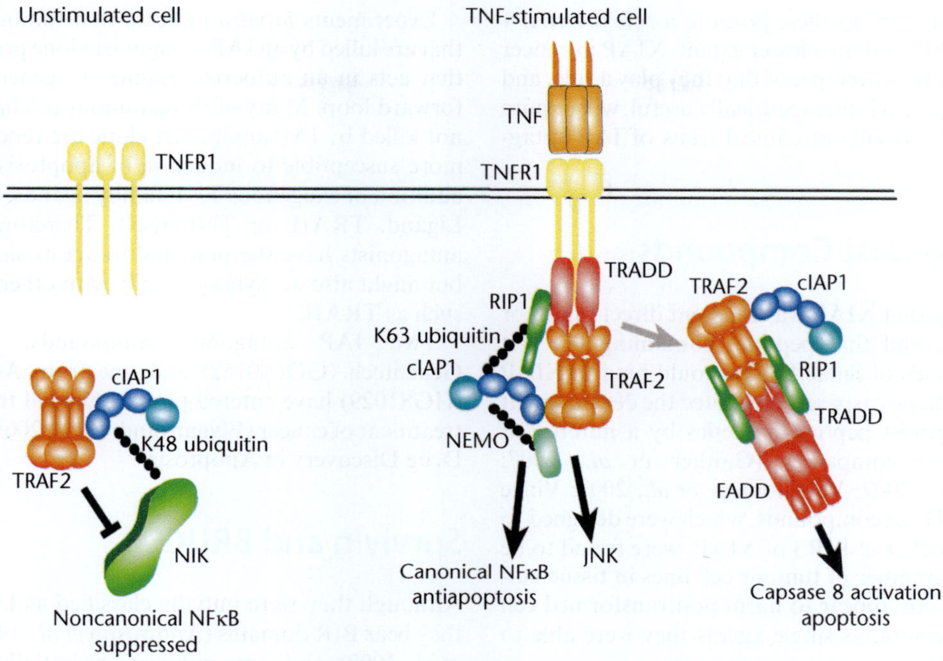

Figure 3 Roles for cIAP1 in TNF signaling. In unstimulated cells (left), cIAP1 (together with TRAF2 & TRAF3) ubiquitylate NIK, preventing it from accumulating, so that activation of noncanonical NFκB p100 is suppressed. When TNF is added to cells (right), a complex forms on the cytoplasmic domains of TNFR1. RIP1, TRAF2 and cIAP1 interact, causing K63-ubiquitylation of NEMO, which activates canonical NFκB, as well as generating signals that activate MAP-kinases. These pathways can lead to induction of antiapoptotic proteins such as A20, FLIP, A1 and cIAP2, that are usually sufficient to prevent death of the cell from apoptosis caused by aggregation of FADD and activation of caspase 8 (far right). However, in cells treated with IAP antagonist compounds, cIAP1 autoubiquitylates and is degraded. This leads to increased levels of NIK, and processing of p100 NFκB2 to p52, but reduced antiapoptotic signalling by the TNF receptors. Cells treated with IAP antagonists, or with NFκB inhibitors, therefore readily activate FADD and caspase 8 and undergo apoptosis when exposed to TNF.

cIAP deleted cells have shown that whereas activation of canonical NFκB *requires* TRAF2 and cIAPs, TRAF2 and cIAPs are also needed to *prevent* spontaneous stabilization of NFκB-inducing kinase (NIK), and activation of noncanonical NFκB. Thus cIAPs and TRAF2 seem to work together to suppress spontaneous noncanonical NFκB activity, whereas mediating inducible canonical NFκB activity.

IAPs and Cancer

The genes for cIAPs, ML-IAP and XIAP have been found to be amplified or altered in a number of types of cancer, suggesting that increased activity of IAPs can be oncogenic. For example, increased expression of XIAP has been found in lung cancer cell lines and acute myeloid leukaemia, and ML-IAP has frequently been detected in melanoma lines.

The genes for cIAP1 and cIAP2 lie within 15kb of each other on chromosome 11. This region is amplified in cancers of the liver, lung and oral squamous epithelium in humans, and the syntenic region in the mouse was amplified in a model of primary liver cancer (Zender *et al.*, 2006).

The *CIAP2* gene is involved in translocations commonly seen in MALT lymphoma, where it is joined to the gene for MALT1/paracaspase (Dierlamm *et al.*, 1999), a protein that transmits signals from lymphocyte antigen receptors to activate p100 noncanonical NFκB. Rather than destabilizing NIK to *reduce* p100 activation, as wild type cIAP1 and 2 do, the cIAP2-MLT fusion protein might act with TRAF2 to cause constitutive *activation* of p100 NFκB, enhancing tumour cell survival.

Curiously, in some lymphoid malignancies, such as multiple myelomas, the genes for cIAP1 and cIAP2 or TRAF2 or TRAF3 are often deleted (Keats *et al.*, 2007). The apparent ability of cIAPs to act as oncogenes in some malignancies but to act as tumour suppressor genes in lymphoid malignancies might be due to the different signals provided by canonical versus non-canonical NFκB activity in differing cell types.

Associations such as these point to a role for cIAP1 and 2, ML-IAP, and, to a lesser extent, XIAP in cancer development. However, proof that they play a role, and that targeting them is therapeutically useful, will require waiting for the results of clinical trials of IAP-antagonist drugs.

IAP-antagonist Compounds

The discovery that XIAP was a potent direct inhibitor of caspase 3, and that peptides containing the first four amino acids of Smac/Diablo could bind to XIAP *in vitro* and release caspase 3 prompted the development of IAP antagonist peptido-mimetics by a number of pharmaceutical companies (Gaither *et al.*, 2007; Petersen *et al.*, 2007; Varfolomeev *et al.*, 2007; Vince *et al.*, 2007). These compounds, which were designed to bind to the pocket of BIR3 of XIAP, were found to be able to kill a number of tumour cell lines in tissue culture, but did not appear to harm nontransformed cell lines. Furthermore, as single agents they were able to cure mice of human tumour xenografts.

Although these IAP-antagonist 'smac-mimetic' compounds were designed to target XIAP, a number of findings suggest that their ability to cause apoptosis of tumour cells is due to their ability to bind to cIAP1/2. When the XIAP gene was knocked out in mice, it resulted in a very subtle phenotype, which might have been due to functional redundancy with cIAP1 and cIAP2, so that if XIAP was not present to inhibit caspases, cIAP1 or cIAP2 could do the job. However, it was subsequently shown that neither cIAP1 nor cIAP2, nor any other mammalian caspase other than XIAP, could directly inhibit the proteolytic activity of caspases (Eckelman and Salvesen, 2006). It is now clear that IAP-antagonist compounds act in large part by binding to cIAP1/2, causing their degradation, thereby increasing proapoptotic signalling from TNFR superfamily members (Gaither *et al.*, 2007; Petersen *et al.*, 2007; Varfolomeev *et al.*, 2007; Vince *et al.*, 2007).

Experiments with IAP antagonist compounds that target both cIAP1 and cIAP2 cause elevation of NIK, the activator of the 'noncanonical' NFκB pathway and make cells very sensitive to killing by TNF (**Figure 3**). This suggests that the normal role of cIAP1 and cIAP2 is to degrade NIK, thereby reducing activation of noncanonical NFκB, and to indirectly reduce activation of caspase 8 when TNF is added, but the exact mechanism by which they do so has not yet been determined.

Experiments *in vitro* indicate that tumour cell lines that are killed by an IAP antagonist alone produce TNF that acts in an autocrine manner to generate a feed-forward loop. Many of those tumour cell lines that are not killed by IAP antagonist alone are rendered much more susceptible to induction of apoptosis caused by addition of exogenous TNF family ligands, such as Fas Ligand, TRAIL or TNF itself. Therefore, the IAP antagonists have the potential to act as single agents, but might also act synergistically with other molecules, such as TRAIL.

Two IAP antagonist compounds, one from Genentech (GDC-0152) and one from Aegera/HGS (HGS1029) have entered phase 1 clinical trials for the treatment of cancer (Flygare and Vucic, 2009). **See also**: Drug Discovery in Apoptosis

Survivin and BRUCE

Although they were initially classified as IAP because they bear BIR domains (Ambrosini *et al.*, 1997; Hauser *et al.*, 1998), and some publications initially supported this assumption, subsequent evidence has shown that these proteins function in the main (if not entirely) in some other process.

Survivin (BIRC5) (see **Box 2**) has a single amino-terminal BIR domain and a carboxy-terminal amphipathic α-helix. It homodimerizes via the turn between the BIR and the α-helix. Gene deletion studies of Survivin and its homologues in the worm *C. elegans* and the yeasts *Schizosaccharomyces pombe* and *Schizosaccharomyces cerevisiae*, have shown that it is required for chromosome segregation during mitosis, functioning in a complex with Aurora B kinase, INCENP and Borealin (Ruchaud *et al.*, 2007; Uren *et al.*, 2000). All of these proteins are known as 'chromosome passenger proteins' because they appear on the centromeres at the start of metaphase, but remain on the metaphase plate during anaphase, and end up in the mid-body at telophase. Survivin is then ubiquitylated and degraded, so that levels in interphase and noncycling cells are very low (Zhao *et al.*, 2000). The restriction of Survivin messenger ribonucleic acid (mRNA) and protein to tissues with actively dividing cells might explain the correlation of Survivin expression with cancers.

Survivin knockout mouse embryos, like BIR-1 depleted *C. elegans*, and Bir mutant *S. pombe*, fail to make any successful cell divisions. Rather than having a role as a cell death inhibitor, Survivin is essential for the survival of dividing cells, and if it is removed or

Box 2 Role of Survivin

The Survivin sequence first appeared in databases in reverse orientation as effector cell protease receptor 1 (Altieri, 1994). When it was found to bear a BIR, it was classed as an IAP (Ambrosini *et al.*, 1997). Survivin is readily detected in cycling cells, such as cancer cells and cytokine-stimulated cells, but is often undetectable in normal adult tissues. The reason for this became clear when it was found that in both fission and budding yeasts, the worm *Caenorhabditis elegans*, *Drosophila* and mice, deletion of the genes for Survivin had the same effect, namely to prevent proper chromosome segregation prior to cytokinesis. Furthermore, in all these organisms, Survivin protein is only present during mitosis. Survivin is now known to be a chromosome passenger protein, that works in concert with other chromosome passenger proteins including aurora kinase B, INCENP and Borealin, to regulate chromosome segregation and cytokinesis (Uren *et al.*, 2000; Yue *et al.*, 2008). If any of these proteins are deleted, cycling cells fail to complete mitosis, activate a checkpoint and undergo apoptosis.

inhibited, cells undergo apoptosis as a response to a mitotic catastrophe.

BRUCE (BIRC6) is an extremely large protein, with a molecular mass of over 500 kD. It has a single BIR near its amino-terminus, and a ubiquitin-conjugating domain (UBC) near its carboxy-terminus (Hauser *et al.*, 1998). During cytokinesis, like Survivin, BRUCE localizes to the mid-body, but BRUCE is not a chromosome passenger protein, because it does not first localize to the centromeres, and it is found on the mid-body ring well before Survivin and the other chromosome passenger proteins that only end up in the mid-body at the very end of telophase (Pohl and Jentsch, 2008). However, like depletion of Survivin, depletion of BRUCE *in vitro* results in failure of proper chromosome segregation, indirectly triggering apoptosis. As overexpressed BRUCE could be coimmunoprecipitated with Survivin, it is possible they act together for some role during mitosis.

NAIP

NAIP (BIRC1) was the first BIR-containing protein to be found other than in viruses. Humans bear one *NAIP* gene, but have several physically linked *NAIP*-like pseudogenes, whereas mice have 5 *naip*-like genes that appear to encode proteins. Genetic studies and gene knockout studies have shown that the *birc1e* (*naip5*) gene in mice confers resistance to *Legionella pneumophila*, the bacterium that causes Legionaires disease. Significantly, a transgene encoding NAIP5 was able to restore resistance against Legionella to a susceptible strain of mice that carried mutant NAIP5 genes (Diez *et al.*, 2003).

NAIP has three BIR domains in its amino-terminal half, and also has an adenosine triphosphatase (ATPase) NACHT domain and a carboxy-terminal region with leucine-rich repeats (LRRs) that are commonly found in proteins of the innate immune system where they act as receptors for bacterial products.

References

Altieri DC (1994) Molecular cloning of effector cell protease receptor-1, a novel cell surface receptor for the protease factor Xa. *Journal of Biological Chemistry* **269**: 3139–3142.

Ambrosini G, Adida C and Altieri DC (1997) A novel anti-apoptosis gene, survivin, expressed in cancer and lymphoma. *Nature Medicine* **3**: 917–921.

Birnbaum MJ, Clem RJ and Miller LK (1994) An apoptosis inhibiting gene from a nuclear polyhedrosis virus encoding a polypeptide with cys/his sequence motif. *Journal of Virology* **68**: 2521–2528.

Chen P, Nordstrom W, Gish B and Abrams JM (1996) Grim, a novel cell death gene in *Drosophila*. *Genes & Development* **10**: 1773–1782.

Crook NE, Clem RJ and Miller LK (1993) An apoptosis inhibiting baculovirus gene with a zinc finger like motif. *Journal of Virology* **67**: 2168–2174.

Deveraux QL, Takahashi R, Salvesen GS and Reed JC (1997) X-linked IAP is a direct inhibitor of cell-death proteases. *Nature* **388**: 300–304.

Dierlamm J, Baens M, Wlodarska I *et al.* (1999) The apoptosis inhibitor gene API2 and a novel 18q gene, MLT, are recurrently rearranged in the t(11;18)(q21;q21) associated with mucosa-associated lymphoid tissue lymphomas. *Blood* **93**: 3601–3609.

Diez E, Lee SH and Gauthier S (2003) Birc1e is the gene within the Lgn1 locus associated with resistance to *Legionella pneumophila*. *Nature Genetics* **33**: 55–60.

Du CY, Fang M, Li YC, Li L and Wang XD (2000) Smac, a mitochondrial protein that promotes cytochrome c-dependent caspase activation by eliminating IAP inhibition. *Cell* **102**: 33–42.

Duckett CS, Nava VE, Gedrich RW *et al.* (1996) A conserved family of cellular genes related to the baculovirus IAP gene and encoding apoptosis inhibitors. *EMBO Journal* **15**: 2685–2694.

Eckelman BP and Salvesen GS (2006) The Human Anti-apoptotic Proteins cIAP1 and cIAP2 bind but do not inhibit caspases. *Journal of Biological Chemistry* **281**: 3254–3260.

Flygare JA and Vucic D (2009) Development of novel drugs targeting inhibitors of apoptosis. *Future Oncology* **5**: 141–144.

Fraser AG, James C, Evan GI and Hengartner MO (1999) Caenorhabditis elegans inhibitor of apoptosis protein (IAP) homologue BIR-1 plays a conserved role in cytokinesis. *Current Biology* **9**: 292–301.

Gaither A, Porter D, Yao Y *et al.* (2007) A Smac mimetic rescue screen reveals roles for inhibitor of apoptosis proteins in tumor necrosis factor-alpha signaling. *Cancer Research* **67**: 11493–11498.

Grether ME, Abrams JM, Agapite J, White K and Steller H (1995) The head involution defective gene of *Drosophila melanogaster* functions in programmed cell death. *Genes & Development* **9**: 1694–1708.

Gyrd-Hansen M, Darding M, Miasari M *et al.* (2008) IAPs contain an evolutionarily conserved ubiquitin-binding domain that regulates NF-kappaB as well as cell survival and oncogenesis. *Nature Cell Biology* **10**: 1309–1317.

Harlin H, Reffey SB, Duckett CS, Lindsten T and Thompson CB (2001) Characterization of XIAP-deficient mice. *Molecular & Cellular Biology* **21**: 3604–3608.

Hauser HP, Bardroff M, Pyrowolakis G and Jentsch S (1998) A giant ubiquitin-conjugating enzyme related to IAP apoptosis inhibitors. *Journal of Cell Biology* **141**: 1415–1422.

Hay BA, Wassarman DA and Rubin GM (1995) *Drosophila* homologs of baculovirus inhibitor of apoptosis proteins function to block cell death. *Cell* **83**: 1253–1262.

Hinds MG, Norton RS, Vaux DL and Day CL (1999) Solution structure of a baculoviral inhibitor of apoptosis (IAP) repeat. *Nature Structural Biology* **6**: 648–651.

Kasof GM and Gomes BC (2001) Livin, a novel inhibitor of apoptosis protein family member. *Journal of Biological Chemistry* **276**: 3238–3246.

Keats JJ, Fonseca R, Chesi M *et al.* (2007) Promiscuous mutations activate the noncanonical NF-kappaB pathway in multiple myeloma. *Cancer Cell* **12**: 131–144.

Liston P, Roy N, Tamai K *et al.* (1996) Suppression of apoptosis in mammalian cells by NAIP and a related family of IAP genes. *Nature* **379**: 349–353.

Petersen SL, Wang L, Yalcin-Chin A *et al.* (2007) Autocrine TNFalpha signaling renders human cancer cells susceptible to Smac-mimetic-induced apoptosis. *Cancer Cell* **12**: 445–456.

Pohl C and Jentsch S (2008) Final stages of cytokinesis and midbody ring formation are controlled by BRUCE. *Cell* **132**: 832–845.

Rigaud S, Fondaneche MC, Lambert N *et al.* (2006) XIAP deficiency in humans causes an X-linked lymphoproliferative syndrome. *Nature* **444**: 110–114.

Rothe M, Pan MG, Henzel WJ, Ayres TM and Goeddel DV (1995) The tnfr2-traf signaling complex contains two novel proteins related to baculoviral-inhibitor of apoptosis proteins. *Cell* **83**: 1243–1252.

Roy N, Mahadevan MS, Mclean M *et al.* (1995) The gene for neuronal apoptosis inhibitory protein is partially deleted in individuals with spinal muscular atrophy. *Cell* **80**: 167–178.

Ruchaud S, Carmena M and Earnshaw WC (2007) Chromosomal passengers: conducting cell division. *Nature Reviews. Molecular Cell Biology* **8**: 798–812.

Santoro MM, Samuel T, Mitchell T, Reed JC and Stainier DY (2007) Birc2 (cIap1) regulates endothelial cell integrity and blood vessel homeostasis. *Nature Genetics* **39**: 1397–1402.

Shin H, Renatus M, Eckelman BP *et al.* (2005) The BIR domain of IAP-like protein 2 is conformationally unstable: implications for caspase inhibition. *Biochemical Journal* **385**: 1–10.

Silke J, Verhagen AM, Ekert PG and Vaux DL (2000) Sequence as well as functional similarity for DIABLO/Smac and Grim, Reaper and Hid? *Cell Death & Differentiation* **7**: 1275.

Speliotes EK, Uren AG, Vaux DL and Horvitz HR (2000) The survivin-like C. elegans BIR-1 protein acts with the Aurora-like kinase AIR-2 to affect chromosomes and the spindle midzone. *Molecular Cell* **6**: 211–223.

Srinivasula SM, Hegde R, Saleh A *et al.* (2001) A conserved XIAP-interaction motif in caspase-9 and Smac/DIABLO regulates caspase activity and apoptosis. *Nature* **410**: 112–116.

Sun CH, Cai ML, Meadows RP *et al.* (2000) NMR structure and mutagenesis of the third Bir domain of the inhibitor of apoptosis protein XIAP. *Journal of Biological Chemistry* **275**: 33777–33781.

Uren AG, Beilharz T, O'Connell MJ *et al.* (1999) Role for yeast inhibitor of apoptosis (IAP)-like proteins in cell division. *Proceedings of the National Academy of Sciences of the USA* **96**: 10170–10175.

Uren AG, Coulson EJ and Vaux DL (1998) Conservation of baculovirus inhibitor of apoptosis repeat proteins (BIRps) in viruses, nematodes, vertebrates and yeasts. *Trends in Biochemical Sciences* **23**: 159–162.

Uren AG, Pakusch M, Hawkins CJ, Puls KL and Vaux DL (1996) Cloning and expression of apoptosis inhibitory protein homologs that function to inhibit apoptosis and/or bind tumor necrosis factor receptor-associated factors. *Proceedings of the National Academy of Sciences of the USA* **93**: 4974–4978.

Uren AG, Wong L, Pakusch M *et al.* (2000) Survivin and the inner centromere protein INCENP show similar cell-cycle localization and gene knockout phenotype. *Current Biology* **10**: 1319–1328.

Varfolomeev E, Blankenship JW, Wayson SM *et al.* (2007) IAP antagonists induce autoubiquitination of c-IAPs, NF-kappaB activation, and TNFalpha-dependent apoptosis. *Cell* **131**: 669–681.

Vaux DL and Silke J (2005) IAPs, RINGs and ubiquitylation. *Nature Reviews. Molecular Cell Biology* **6**: 287–297.

Verhagen AM, Ekert PG, Pakusch M *et al.* (2000) Identification of DIABLO, a mammalian protein that promotes apoptosis by binding to and antagonizing IAP proteins. *Cell* **102**: 43–53.

Verhagen AM, Kratina TK, Hawkins CJ *et al.* (2007) Identification of mammalian mitochondrial proteins that interact with IAPs via N-terminal IAP binding motifs. *Cell Death and Differentiation* **14**: 348–357.

Vince JE, Wong WW, Khan N *et al.* (2007) IAP antagonists target cIAP1 to induce TNFalpha-dependent apoptosis. *Cell* **131**: 682–693.

Vucic D, Kaiser WJ, Harvey AJ and Miller LK (1997) Inhibition of reaper-induced apoptosis by interaction with inhibitor of apoptosis proteins (IAPS). *Proceedings of the National Academy of Sciences of the USA* **94**: 10183–10188.

Vucic D, Kaiser WJ and Miller LK (1998) Inhibitor of apoptosis proteins physically interact with and block apoptosis induced by *Drosophila* proteins hid and grim. *Molecular & Cellular Biology* **18**: 3300–3309.

Vucic D, Stennicke HR, Pisabarro MT, Salvesen GS and Dixit VM (2000) ML-IAP, a novel inhibitor of apoptosis that is preferentially expressed in human melanomas. *Current Biology* **10**: 1359–1366.

White K, Grether ME, Abrams JM *et al.* (1994) Genetic control of programmed cell death in *Drosophila. Science* **264**: 677–683.

Wright EK, Goodart SA, Growney JD *et al.* (2003) Naip5 affects host susceptibility to the intracellular pathogen *Legionella pneumophila. Current Biology* **13**: 27–36.

Wu G, Chai JJ, Suber TL *et al.* (2000) Structural basis of IAP recognition by Smac/DIABLO. *Nature* **408**: 1008–1012.

Yue Z, Carvalho A, Xu Z *et al.* (2008) Deconstructing Survivin: comprehensive genetic analysis of Survivin function by conditional knockout in a vertebrate cell line. *Journal of Cell Biology* **183**: 279–296.

Zender L, Spector MS, Xue W *et al.* (2006) Identification and validation of oncogenes in liver cancer using an integrative oncogenomic approach. *Cell* **125**: 1253–1267.

Zhao J, Tenev T, Martins LM, Downward J and Lemoine NR (2000) The ubiquitin-proteasome pathway regulates survivin degradation in a cell cycle-dependent manner. *Journal of Cell Science* **113**: 4363–4371.

Further Reading

Fulda S (2007) Inhibitor of apoptosis proteins as targets for anticancer therapy. *Expert Review of Anticancer Therapy* **7**: 1255–1264.

LaCasse EC, Mahoney DJ, Cheung HH *et al.* (2008) IAP-targeted therapies for cancer. *Oncogene* **27**: 6252–6275.

O'Riordan MX, Bauler LD, Scott FL and Duckett CS (2008) Inhibitor of apoptosis proteins in eukaryotic evolution and development: a model of thematic conservation. *Developmental Cell* **15**: 497–508.

Varfolomeev E and Vucic D (2008) (Un)expected roles of c-IAPs in apoptotic and NFkappaB signaling pathways. *Cell Cycle* **7**: 1511–1521.

Vucic D (2008) Targeting IAP (inhibitor of apoptosis) proteins for therapeutic intervention in tumors. *Current Cancer Drug Targets* **8**: 110–117.

Structures, Domains and Functions in Cell Death (DD, DED, CARD, PYD)

Hao Wu, *Weill Cornell Medical College, New York, USA*
Yu-Chih Lo, *Weill Cornell Medical College, New York, USA*

Advanced article

Article Contents

- Introduction
- Functions of DDs, DEDs, CARDs and PYDs: Assembly of Caspase-activating Signalling Complexes
- Structures of Isolated Domains and Their Surface Features
- Interactions in the DD Superfamily
- Preferred Modes among the Homotypic Interactions
- Summary
- Acknowledgement

The death domain (DD), death effector domain (DED), caspase-recruitment domain (CARD) and pyrin domain (PYD) are subfamilies of the DD superfamily. By mediating homotypic interactions, these proteins play important roles in the assembly and activation of apoptotic signalling complexes. They are responsible

for caspase recruitment and for formation of oligomeric platforms for signalling. DD superfamily proteins have a common six-helical bundle fold and show different surface features for each of the subfamilies. Most interestingly, the homotypic interactions within each subfamily are mostly mediated by asymmetric contacts, in which different surfaces of the interaction partners are adjacent to each other. The DD superfamily proteins appear to use three common types of asymmetric interactions to assemble into large oligomeric complexes.

Introduction

Apoptosis is an orderly cellular suicide programme, that is, critical for the development and homeostasis of a multicellular organism. Failure to control apoptosis can lead to serious diseases that threaten the existence of the organism (Rathmell and Thompson, 2002). Apoptosis proceeds through characteristic morphological changes that are dependent on caspase activities. Caspases are cysteine proteases that cleave specifically after aspartic acid residues (Riedl and Shi, 2004). Because caspases are the executioners of apoptosis, they are the key players in apoptotic cell death. **See also**: Caspases and Cell Death; Caspases, Substrates and Sequential Activation; Dismantling the Apoptotic Cell

On a molecular level, the death domain (DD) superfamily members are involved in formation of oligomeric signalling complexes for caspase activation. A central paradigm in caspase activation is the assembly of oligomeric signalling complexes in response to internal or external stimuli. In a simplified view, these molecular complexes activate their effectors via 'proximity-induced autoactivation' such as dimerization and proteolytic autoprocessing (Salvesen and Dixit, 1999). The DD superfamily plays a critical role in this assembly by participating in both self-association and other protein:protein interactions.

The DD superfamily is one of the largest and most studied domain superfamilies (McEntyre and Gibson, 2004). It is currently composed of four subfamilies, the DD, the death effector domain (DED), the caspase recruitment domain (CARD) and the pyrin domain (PYD) subfamilies (Reed *et al.*, 2004). DDs, DEDs, CARDs and PYDs are involved in both caspase activation and the related process of nuclear factor-κB (NFκB) activation in host defence. The DD superfamily proteins are evolutionarily conserved in many multicellular organisms such as mammals, *Drosophila* and *Caenorhabditis elegans*. Based on a genome analysis, there are 32 DDs, 7 DEDs, 28 CARDs and 19 PYDs in the human genome (Reed *et al.*, 2004).

Deregulation of caspase activation is related to many human diseases. Most notably, defective receptor-mediated caspase activation and cell death underlies the human genetic disease autoimmune lymphoproliferative syndrome (ALPS) (Poppema *et al.*, 2004; Rieux-Laucat *et al.*, 2003b). When lymphocytes from patients with ALPS are cultured *in vitro*, they are resistant to apoptosis as compared to cells from healthy controls. Most patients with ALPS have mutations in the intracellular DD of the receptor (Rieux-Laucat *et al.*, 2003a).

Functions of DDs, DEDs, CARDs and PYDs: Assembly of Caspase-activating Signalling Complexes

DDs, DEDs, CARDs and PYDs participate in homotypic interactions within the same subfamily. Many proteins involved in apoptosis contain these domains (**Figure 1**). Some examples of these important signalling complexes are presented in the next section.

Death-inducing signalling complex for activation of caspase-8 and caspase-10 and its inhibition by FLIPs

Fas (also known as CD95) is a prototypical member of death receptors that form a subfamily of the tumour necrosis factor (TNF) receptor superfamily to mediate the extrinsic cell death pathway. Members of the TNF superfamily of ligands are mostly trimeric and activate the TNF receptor superfamily by ligand-induced receptor trimerization and higher order oligomerization (Kischkel *et al.*, 1995). The intracellular regions of death receptors contain DDs (Tartaglia *et al.*, 1993). For Fas, on ligand activation, its DD recruits the Fas-associated DD (FADD) adapter protein via a homotypic interaction with the *C*-terminal DD of FADD (Chinnaiyan *et al.*, 1995; Kischkel *et al.*, 1995). FADD also contains an *N*-terminal DED that interacts homotypically with the tandem DED in the prodomain of caspase-8 or -10 (Wajant, 2002). These interactions form the ternary death-inducing signalling complex (DISC)-containing Fas, FADD and caspase-8 or -10

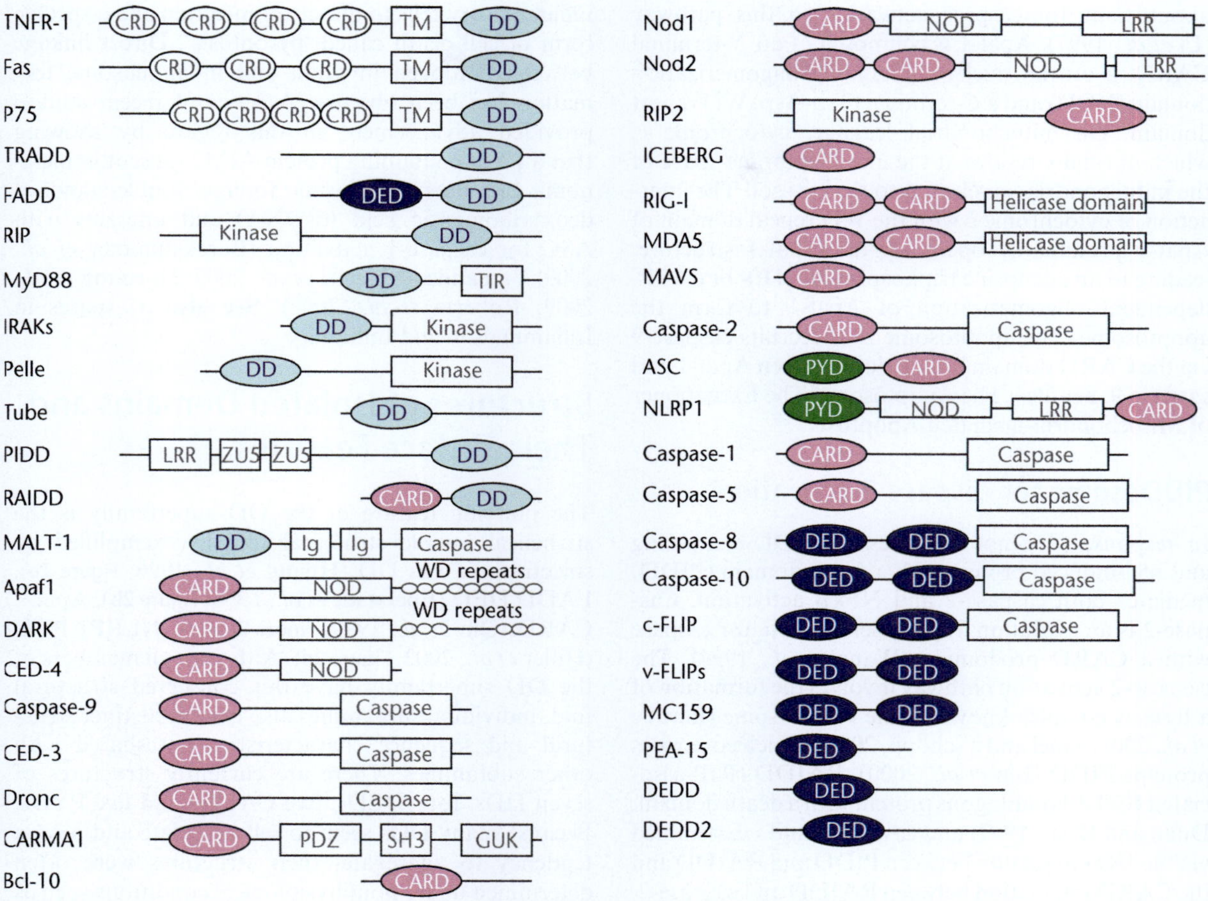

Figure 1 (Figure is also reproduced in colour plate section.) Domain organizations of selective proteins containing the DD superfamily domains. Abbreviations: CARD, caspase recruitment domain; CRD, cysteine-rich domain; DD, death domain; DED, death effector domain; GUK, guanylate kinase-like; LRR, leucine-rich repeat; NOD, nucleotide-binding oligomerization domain; PYD, pyrin domain; TIR, toll/interleukin 1 receptor and TM, transmembrane domain.

(Wajant, 2002). Recruitment of pro-caspases into the DISC initiates caspase proteolytic auto-processing. This liberates active caspase-8 or -10 into the cytoplasm to cleave and activate effector caspases such as caspase-3 and caspase-7, leading to a cascade of events in apoptotic cell death. **See also**: Death Receptors; Death Receptors at the Molecular Level: Therapeutic Implications

Caspase activation by the DISC is inhibited by FLIPs, a family of cellular and viral tandem DED-containing proteins that interact with FADD (Thome and Tschopp, 2001). Cellular FLIPs (c-FLIPs), comprising the long and short isoforms, c-FLIP-L and c-FLIP-S, are tightly regulated in expression in T cells and may be involved in controlling both T-cell activation and death (Thome and Tschopp, 2001). v-FLIPs (viral FLIPs) appear to have evolved to inhibit apoptosis of virally infected host cells and are present in the poxvirus *Molluscum contagiosum virus* (MCV) as proteins MC159 and MC160 (Shisler and Moss, 2001) and in γ-herpesviruses (Thome and Tschopp, 2001). **See also**: Death Receptors

Apoptosome for caspase-9 activation

The intrinsic pathway of apoptotic cell death is induced in a mitochondria-dependent manner in response to intracellular insults. A CARD-containing protein Apaf-1 (apoptosis-activating factor 1) forms the central platform of a molecular complex known as the apoptosome for caspase activation in this pathway (Li *et al.*, 1997). Apaf-1 is composed of an *N*-terminal CARD, a central nucleotide-binding oligomerization domain (NOD) and a *C*-terminal (Trp-Asp) WD repeat domain. On mitochondrial leakage, cytochrome *c*, which normally resides at the intermembrane space of the mitochondria, is released to the cytosol. The interaction of cytochrome *c* with the WD repeat domain of Apaf-1 presumably opens up the Apaf-1 structure, leading to an adenosine triphosphate (ATP)- or dATP-dependent oligomerization of Apaf-1 to form the apoptosome. The apoptosome then recruits caspase-9 via the CARD domain interaction between Apaf-1 and caspase-9. **See also**: The Apoptosome: The Executioner of Mitochondria-mediated Apoptosis

PIDDosome for caspase-2 activation

In response to genotoxic stress, the DD-containing and p53-induced protein with a death domain (PIDD) mediates both caspase-2 and NFκB activation. Caspase-2 is an evolutionarily conserved initiator caspase with a CARD prodomain (Wang *et al.*, 1994). The caspase-2 activation pathway involves the formation of a ternary complex known as the PIDDosome (Berube *et al.*, 2005; Tinel and Tschopp, 2004), which comprises proteins PIDD (Lin *et al.*, 2000), RAIDD (RIP-associated ICH-1 homologous protein with a death domain; Duan and Dixit, 1997) and caspase-2 and is assembled via the DD interaction between PIDD and RAIDD and the CARD interaction between RAIDD and caspase-2. Accumulating data have shown that caspase-2 acts upstream of the mitochondria to initiate apoptosis (Lassus *et al.*, 2002).

Inflammasome for caspase-1 activation

Members of the Nod-like receptor (NLR) family, including NLRP1, NLRP3 and NLRC4, and the adaptor ASC (apoptosis-associated speck-like protein containing a CARD) are critical components of the inflammasome that link microbial and endogenous 'danger' signals to caspase-1 activation (Franchi *et al.*, 2009). NLR family proteins are PYD- and NOD-containing proteins. For example, NLRP1 contains an *N*-terminal PYD, a central NOD for oligomerization, a leucine-rich repeat (LRR) region and a *C*-terminal CARD. ASC contains both a PYD and a CARD. Caspase-1 is recruited to the inflammasome either via ASC or directly to NLR proteins. Once activated, caspase-1 cleaves and activates the cytokine interleukin 1β (IL-1β), leading to recruitment of inflammatory cells to the site of infection and a specific form of cell death called 'pyroptosis'. Direct linkage between pathogen invasion and inflammasome formation has been obscure. A series of recent studies provided advancement in this regard by showing that a PYD-containing protein AIM2 (absent in melanoma 2) senses cytoplasmic foreign double-standard deoxyribonucleic acid (dsDNA) and interacts with ASC for caspase-1 activation (Burckstummer *et al.*, 2009; Fernandes-Alnemri *et al.*, 2009; Hornung *et al.*, 2009; Roberts *et al.*, 2009). **See also**: Caspases in Inflammation and Immunity

Structures of Isolated Domains and Their Surface Features

The unifying feature of the DD superfamily is the six-helical bundle structural fold as exemplified by structures of Fas DD (Huang *et al.*, 1996; **Figure 2a**), FADD DED (Eberstadt *et al.*, 1998; **Figure 2b**), Apaf-1 CARD (Qin *et al.*, 1999; **Figure 2c**) and NLRP1 PYD (Hiller *et al.*, 2003; **Figure 2d**). Although all members of the DD superfamily have this conserved structural fold, individual subfamilies also exhibit distinct structural and sequence characteristics not shared with other subfamilies. There are currently structures of seven DDs, four DEDs, five CARDs and five PYDs. Because many DDs seem to self-associate and have a tendency to aggregate, their structures were often determined under nonphysiological conditions such as extreme pH and/or with 'de-aggregating' mutations (Huang *et al.*, 1996). Of the four DED structures, three are more similar to each other than to other members of the DD superfamily. In contrast, the MC159 DED1, which was solved in the context of a tandem DED, is structurally more divergent from the other known DED structures (Li *et al.*, 2006; Yang *et al.*, 2005). In particular, helix H3 is missing and replaced by a short loop connecting helices H2 and H4. Two additional helices are present, helix H0 at the *N*-terminus and helix H7 that brings the chain to the beginning of DED2. For CARDs, although their topology is identical with the conserved six-helical bundle fold of the DD superfamily, the structures are unique in that helix H1 tends to be either bent or broken into two closely separated H1a and H1b helices (**Figure 2c**). In addition, the orientations and lengths of several helices may be somewhat different among the different CARDs. The PYD structures are divergent among themselves in the H2 and H3 region (Hiller *et al.*, 2003; Liepinsh *et al.*, 2003; Natarajan *et al.*, 2006). Although the PYD of

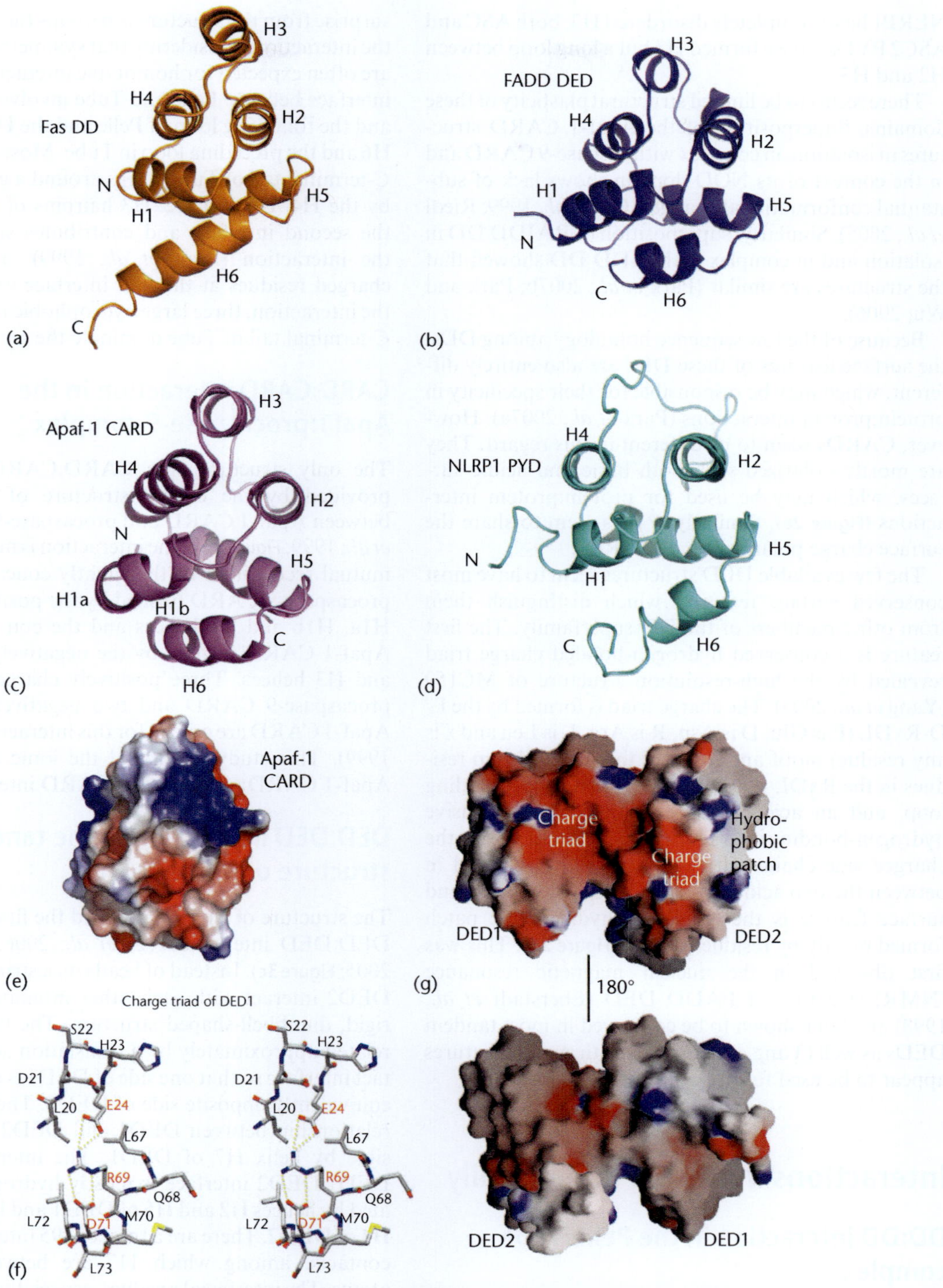

Figure 2 Ribbon diagrams for each subfamily of the DD superfamily: (a) Fas DD, (b) FADD DED, (c) Apaf-1 CARD and (d) NLRP1 PYD. (e) Electrostatic surface representation of Apaf-1 CARD. (f) Stick model for the hydrogen-bonding interactions in the charge triad of MC159 DED1. (g) Electrostatic surface representation of MC159 DED1 and DED2, showing the dumbbell-shaped structure, the rich charges on one face and the location of the hydrophobic patch. Panels (f) and (g) are taken from Yang *et al.* (2005).

NLRP1 has a completely disordered H3, both ASC and ASC2 PYDs have a formed H3 but a long loop between H2 and H3.

There seems to be limited structural plasticity of these domains. Superposition of the Apaf-1 CARD structures in isolation, in complex with caspase-9 CARD and in the context of its NOD domain shows lack of substantial conformational changes (Qin *et al.*, 1999; Riedl *et al.*, 2005). Similarly, superposition of RAIDD DD in isolation and in complex with PIDD DD showed that the structures are similar (Park *et al.*, 2007b; Park and Wu, 2006).

Because of the low sequence homology among DDs, the surface features of these DDs are also entirely different, which may be responsible for their specificity in protein:protein interactions (Park *et al.*, 2007a). However, CARDs seem to be different in this regard. They are mostly polarized with both basic and acidic surfaces, which may be used for protein:protein interactions (**Figure 2e**). Similarly, PYDs seem to share the surface charge polarization of CARDs.

The few available DED structures seem to have most conserved surface features, which distinguish them from other members of the DD superfamily. The first feature is a conserved hydrogen-bonded charge triad revealed by the high-resolution structure of MC159 (Yang *et al.*, 2005). The charge triad is formed by the E/D-RxDL (E is Glu, D is Asp, R is Arg, L is Leu and x is any residue) motif and involves the Arg and Asp residues in the RxDL motif in helix H6 and the preceding loop, and an acidic residue in helix H2. Extensive hydrogen-bonding interactions are observed among the charged side chains with the Arg residue situated in between the two acidic residues (**Figure 2f**). The second surface feature is the conserved hydrophobic patch formed mostly by residues on H2 (**Figure 2g**). This was first observed in the nuclear magnetic resonance (NMR) structure of FADD DED (Eberstadt *et al.*, 1998) and later shown to be conserved in most tandem DEDs as well (Yang *et al.*, 2005). Both surface features appear to be used for protein:protein interactions.

Interactions in the DD Superfamily

DD:DD interaction in the Pelle:Tube complex

The crystal structure of the monomeric Pelle DD:Tube DD complex is the first structure of a complex in the DD superfamily (Xiao *et al.*, 1999; **Figure 3a**). The biggest surprise from the structure is perhaps the asymmetry of the interaction, considering that symmetric interactions are often expected for homotypic interactions. The first interface between Pelle and Tube involves the H4 helix and the following loop of Pelle and the H1–H2 corner, H6 and the preceding loop in Tube. Most strikingly, the *C*-terminal tail of Tube wraps around a groove formed by the H4–H5 and H2–H3 hairpins of Pelle to form the second interface and contributes significantly to the interaction (Xiao *et al.*, 1999). Though many charged residues at the first interface are involved in the interaction, three large hydrophobic residues on the *C*-terminal tail of Tube dominate the second interface.

CARD:CARD interaction in the Apaf1:procaspase-9 complex

The only structure of a CARD:CARD complex is provided by the crystal structure of the complex between Apaf-1 CARD and procaspase-9 CARD (Qin *et al.*, 1999; **Figure 3b**). The interaction is mediated by the mutual recognition of the slightly concave surface of procaspase-9 CARD formed by the positively charged H1a, H1b and H4 helices and the convex surface of Apaf-1 CARD formed by the negatively charged H2 and H3 helices. Three positively charged residues in procaspase-9 CARD and two negatively charged in Apaf-1 CARD are crucial for this interaction (Qin *et al.*, 1999). This study confirmed the ionic nature of the Apaf-1 CARD:procaspase-9 CARD interaction.

DED:DED interaction in the tandem DED structure of MC159 v-FLIP

The structure of MC159 revealed the first glimpse of a DED:DED interaction (Li *et al.*, 2006; Yang *et al.*, 2005; **Figure 3c**). Instead of beads on a string, DED1 and DED2 interact with each other intimately to form a rigid, dumbbell-shaped structure. The two DEDs are related approximately by a translation across the contact interface so that one side of DED1 is contacting the equivalently opposite side of DED2. The translational relationship between DED1 and DED2 is made possible by helix H7 of DED1. The interaction at the DED1:DED2 interface is mostly hydrophobic, mediated by helices H2 and H5 of DED1 and helices H1 and H4 of DED2. There are a total of 195 interfacial atomic contacts, among which 117 are between nonpolar atoms. The interfacial residues, especially those that are completely buried at the DED1:DED2 interface and contribute large surface areas are mostly conserved in tandem DEDs. This suggests that all known tandem

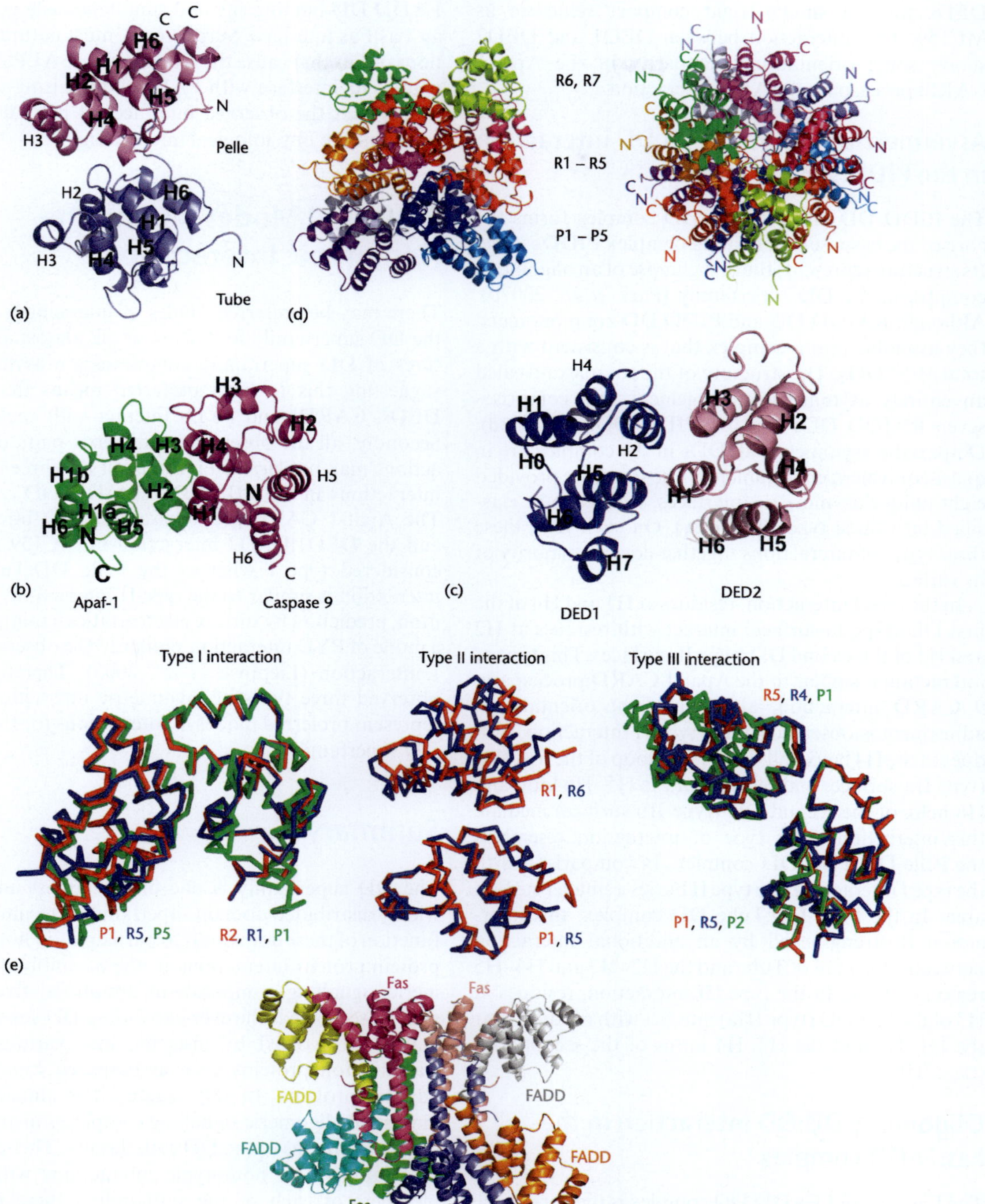

Figure 3 Structural features of interactions in the DD superfamily. (a) Pelle DD:Tube DD complex, as a type II interaction. (b) Apaf-1 CARD:procaspase-9 CARD complex, as a type I interaction. (c) MC159 DED1:DED2 interaction, as a type I interaction. (d) Overview of the PIDD DD:RAIDD DD complex in two orthogonal orientations. (e) Superposition of the eight unique interactions in the PIDD DD:RAIDD DD complex. R:RAIDD DD; P:PIDD DD. There are three type I interactions, two type II interactions and three type III interactions. (f) Tetrameric Fas DD:FADD DD complex. Panels (d) and (e) are taken from Park *et al.* (2007b).

DEDs form a similar rigid compact structure as MC159. This interaction between DED1 and DED2 shows some orientational similarity in the Apaf-1 CARD:procaspase-9 CARD interaction.

Asymmetric oligomeric DD:DD interaction in the PIDD:RAIDD complex

The PIDD DD and RAIDD DD complex forms the core of the caspase-2-activating complex PIDDosome. Its structure represents the first glimpse of an oligomeric complex in the DD superfamily (Park *et al.*, 2007b). Although RAIDD DD and PIDD DD are monomers, they assemble into a complex that is consistent with a total of 12 DDs. The structure of the complex revealed an entirely asymmetric arrangement that comprises seven RAIDD DDs and five PIDD DDs (**Figure 3d**). Despite the asymmetry, all DDs in the complex are in quasi-equivalent environments. The structure provided eight unique asymmetric interfaces, which can be classified into three types (**Figure 3e**). On each DD, these three types of interactions together cover a majority of its surface.

In the type I interaction, residues at H1 and H4 of the first DD (type Ia surface) interact with residues at H2 and H3 of the second DD (type Ib surface). This type of interaction is similar to the Apaf-1 CARD:procaspase-9 CARD interaction, although slight orientational adjustment is observed. In the type II interaction, residues at the H4 helix and the H4–H5 loop of the first DD (type IIa surface) and residues at the H5–H6 loop and H6 helix of the second DD (type IIb surface) mediate this interaction. This type of interaction resembles the Pelle DD:Tube DD complex. In comparison with the type I interaction, the type II buries a smaller surface area. In the Pelle DD:Tube DD complex, this interaction is strengthened by an additional interaction between a long tail of Tube and the H2–H3 and H4–H5 region of Pelle. In the type III interaction, residues at H3 of the first DD (type IIIa) interact with residues near the H1–H2 and the H3–H4 loops of the second DD (type IIIb).

Oligomeric DD:DD interaction in the Fas:FADD complex

The Fas DD and FADD DD complex is the core of the DISC in death receptor signalling. A structure of the complex, which was crystallized at pH 4.0, showed a striking, tetrameric arrangement of four FADD DDs bound to four Fas DDs (Scott *et al.*, 2009; **Figure 3f**). A conformational opening of the Fas DD exposes the FADD DD binding site and simultaneously generates an Fas:Fas interface. Surprisingly, most natural mutations of Fas that cause the human disease ALPS did not map to its interface with FADD DD, raising the possibility that the observed interfaces at pH 4.0 do not represent the physiological interactions.

Preferred Modes among the Homotypic Interactions

There may be preferred modes of interactions among the DD superfamily fold. First of all, almost all structures of DD superfamily complexes are asymmetric, suggesting this to be a preferred means that DDs, DEDs, CARDs and PYDs interact with each other. Secondly, all the observed asymmetric pairs of interactions may be classified into one of the three types of interactions in the PIDD DD:RAIDD DD complex. The Apaf-1 CARD:procaspase-9 CARD interaction and the DED1:DED2 interaction in MC159 may be considered type I whereas the Pelle DD:Tube DD interaction is similar to the type II interaction. In addition, prediction by surface electrostatics has implicated a mode of PYD interaction similar to the observed type I interaction (Liepinsh *et al.*, 2003). Therefore, the observed three types of asymmetric interactions may represent preferred modes of interactions for the entire DD superfamily.

Summary

The DD superfamily is one of the largest and most widely distributed domain superfamily. One important function of these domains is to participate in homotypic protein:protein interactions in the assembly of oligomeric signalling complexes in apoptosis. Evolutionarily, it seems that the ever-expanding DD superfamily may have evolved by inserting into various signal transduction proteins such as caspases, kinases and adapter proteins. In this regard, it is amazing that almost all oligomeric signalling complexes in apoptosis contain domains of the DD superfamily. Through self-associations and homotypic interactions with other members of each of the subfamilies, these proteins often form the platform of these oligomeric assemblies to allow 'proximity' induced caspase and kinase activation.

Many biochemical and structural studies have been performed on these domains. These studies have revealed a conserved six-helical bundle fold of the DD superfamily. Almost all known protein:protein interactions in the superfamily are either self-association or homotypic interactions with other members of the same subfamily. This is somewhat surprising given the structural similarity among the different subfamilies, but may reflect evolutionary circumstances.

Acknowledgement

This work was supported by the National Institute of Health (AI50872 and AI76927). We apologize to all whose work has not been appropriately reviewed or cited due to space limitations.

References

Berube C, Boucher LM, Ma W *et al.* (2005) Apoptosis caused by p53-induced protein with death domain (PIDD) depends on the death adapter protein RAIDD. *Proceedings of the National Academy of Sciences of the USA* **102**: 14314–14320.

Burckstummer T, Baumann C, Bluml S *et al.* (2009) An orthogonal proteomic-genomic screen identifies AIM2 as a cytoplasmic DNA sensor for the inflammasome. *Nature Immunology* **10**: 266–272.

Chinnaiyan AM, O'Rourke K, Tewari M and Dixit VM (1995) FADD, a novel death domain-containing protein, interacts with the death domain of Fas and initiates apoptosis. *Cell* **81**: 505–512.

Duan H and Dixit VM (1997) RAIDD is a new 'death' adaptor molecule. *Nature* **385**: 86–89.

Eberstadt M, Huang B, Chen Z *et al.* (1998) NMR structure and mutagenesis of the FADD (Mort1) death-effector domain. *Nature* **392**: 941–945.

Fernandes-Alnemri T, Yu JW, Datta P, Wu J and Alnemri ES (2009) AIM2 activates the inflammasome and cell death in response to cytoplasmic DNA. *Nature* **458**: 509–513.

Franchi L, Eigenbrod T, Munoz-Planillo R and Nunez G (2009) The inflammasome: a caspase-1-activation platform that regulates immune responses and disease pathogenesis. *Nature Immunology* **10**: 241–247.

Hiller S, Kohl A, Fiorito F *et al.* (2003) NMR structure of the apoptosis- and inflammation-related NALP1 pyrin domain. *Structure* **11**: 1199–1205.

Hornung V, Ablasser A, Charrel-Dennis M *et al.* (2009) AIM2 recognizes cytosolic dsDNA and forms a caspase-1-activating inflammasome with ASC. *Nature* **458**: 514–518.

Huang B, Eberstadt M, Olejniczak ET, Meadows RP and Fesik SW (1996) NMR structure and mutagenesis of the Fas (APO-1/CD95) death domain. *Nature* **384**: 638–641.

Kischkel FC, Hellbardt S, Behrmann I *et al.* (1995) Cytotoxicity-dependent APO-1 (Fas/CD95)-associated proteins form a death-inducing signaling complex (DISC) with the receptor. *EMBO Journal* **14**: 5579–5588.

Lassus P, Opitz-Araya X and Lazebnik Y (2002) Requirement for caspase-2 in stress-induced apoptosis before mitochondrial permeabilization. *Science* **297**: 1352–1354.

Li FY, Jeffrey PD, Yu JW and Shi Y (2006) Crystal structure of a viral FLIP: insights into FLIP-mediated inhibition of death receptor signaling. *Journal of Biological Chemistry* **281**: 2960–2968.

Li P, Nijhawan D, Budihardjo I *et al.* (1997) Cytochrome c and dATP-dependent formation of Apaf-1/caspase-9 complex initiates an apoptotic protease cascade. *Cell* **91**: 479–489.

Liepinsh E, Barbals R, Dahl E *et al.* (2003) The death-domain fold of the ASC PYRIN domain, presenting a basis for PYRIN/PYRIN recognition. *Journal of Molecular Biology* **332**: 1155–1163.

Lin Y, Ma W and Benchimol S (2000) Pidd, a new death-domain-containing protein, is induced by p53 and promotes apoptosis. *Nature Genetics* **26**: 122–127.

McEntyre JR and Gibson TJ (2004) Patterns and clusters within the PSM column in TiBS, 1992–2004. *Trends in Biochemical Science* **29**: 627–633.

Natarajan A, Ghose R and Hill JM (2006) Structure and dynamics of ASC2, a pyrin domain-only protein that regulates inflammatory signaling. *Journal of Biological Chemistry* **281**: 31863–31875.

Park HH and Wu H (2006) Crystal structure of RAIDD death domain implicates potential mechanism of PIDDosome assembly. *Journal of Molecular Biology* **357**: 358–364.

Park HH, Lo YC, Lin SC *et al.* (2007a) The death domain superfamily in intracellular signaling of apoptosis and inflammation. *Annual Review of Immunology* **25**: 561–586.

Park HH, Logette E, Rauser S *et al.* (2007b) Death domain assembly mechanism revealed by crystal structure of the oligomeric PIDDosome core complex. *Cell* **128**: 533–546.

Poppema S, Maggio E and van den Berg A (2004) Development of lymphoma in autoimmune lymphoproliferative syndrome (ALPS) and its relationship to Fas gene mutations. *Leukemia & Lymphoma* **45**: 423–431.

Qin H, Srinivasula SM, Wu G *et al.* (1999) Structural basis of procaspase-9 recruitment by the apoptotic protease-activating factor 1. *Nature* **399**: 549–557.

Rathmell JC and Thompson CB (2002) Pathways of apoptosis in lymphocyte development, homeostasis, and disease. *Cell* **109**(suppl.): S97–S107.

Reed JC, Doctor KS and Godzik A (2004) The domains of apoptosis: a genomics perspective. *Science's STKE* **2004**: re9.

Riedl SJ and Shi Y (2004) Molecular mechanisms of caspase regulation during apoptosis. *Nature Reviews. Molecular Cell Biology* **5**: 897–907.

Riedl SJ, Li W, Chao Y, Schwarzenbacher R and Shi Y (2005) Structure of the apoptotic protease-activating factor 1 bound to ADP. *Nature* **434**: 926–933.

Rieux-Laucat F, Fischer A and Deist FL (2003a) Cell-death signaling and human disease. *Current Opinion in Immunology* **15**: 325–331.

Rieux-Laucat F, Le Deist F and Fischer A (2003b) Autoimmune lymphoproliferative syndromes: genetic defects of apoptosis pathways. *Cell Death and Differentiation* **10**: 124–133.

Roberts TL, Idris A, Dunn JA *et al.* (2009) HIN-200 proteins regulate caspase activation in response to foreign cytoplasmic DNA. *Science* **323**: 1057–1060.

Salvesen GS and Dixit VM (1999) Caspase activation: the induced-proximity model. *Proceedings of the National Academy of Sciences of the USA* **96**: 10964–10967.

Scott FL, Stec B, Pop C *et al.* (2009) The Fas-FADD death domain complex structure unravels signalling by receptor clustering. *Nature* **457**: 1019–1022.

Shisler JL and Moss B (2001) *Molluscum contagiosum virus* inhibitors of apoptosis: the MC159 v-FLIP protein blocks Fas-induced activation of procaspases and degradation of the related MC160 protein. *Virology* **282**: 14–25.

Tartaglia LA, Ayres TM, Wong GH and Goeddel DV (1993) A novel domain within the 55 kd TNF receptor signals cell death. *Cell* **74**: 845–853.

Thome M and Tschopp J (2001) Regulation of lymphocyte proliferation and death by FLIP. *Nature Reviews. Immunology* **1**: 50–58.

Tinel A and Tschopp J (2004) The PIDDosome, a protein complex implicated in activation of caspase-2 in response to genotoxic stress. *Science* **304**: 843–846.

Wajant H (2002) The Fas signaling pathway: more than a paradigm. *Science* **296**: 1635–1636.

Wang L, Miura M, Bergeron L, Zhu H and Yuan J (1994) Ich-1, an Ice/ced-3-related gene, encodes both positive and negative regulators of programmed cell death. *Cell* **78**: 739–750.

Xiao T, Towb P, Wasserman SA and Sprang SR (1999) Three-dimensional structure of a complex between the death domains of Pelle and Tube. *Cell* **99**: 545–555.

Yang JK, Wang L, Zheng L *et al.* (2005) Crystal structure of MC159 reveals molecular mechanism of DISC assembly and FLIP inhibition. *Molecular Cell* **20**: 939–949.

Further Reading

Damiano JS and Reed JC (2004) CARD proteins as therapeutic targets in cancer. *Current Drug Targets* **5**: 367–374.

Hong GS and Jung YK (2002) Caspase recruitment domain (CARD) as a bi-functional switch of caspase regulation and NF-kappaB signals. *Journal of Biochemistry and Molecular Biology* **35**: 19–23.

Valmiki MG and Ramos JW (2009) Death effector domain-containing proteins. *Cell Mol Life Sci* **66**: 814–830.

Yu JW and Shi Y (2008) FLIP and the death effector domain family. *Oncogene* **27**: 6216–6227.

Structure and Function of IAP and Bcl-2 Proteins

Advanced article

Mark G Hinds, *Walter and Eliza Hall Institute of Medical Research, Parkville, Victoria, Australia*

Peter D Mace, *University of Otago, Dunedin, New Zealand*

Catherine L Day, *University of Otago, Dunedin, New Zealand*

Article Contents

- Introduction
- Structural Basis of Bcl-2 Family Action
- Membrane Interactions and the Bcl-2 Family
- Structural Basis of IAP Function
- Caspase Regulation by IAPs
- The RING E3-ligase Activity of IAPs
- IAP and Bcl-2 Antagonists: A Route to New Cancer Therapies?
- Summary

Interactions between pro-apoptotic and pro-survival proteins control the apoptotic programme in cells. Regulation of caspases, the proteolytic enzymes that destroy the cell, is critical and the actions of the inhibitor of apoptosis (IAP) and B-cell lymphoma-2 (Bcl-2) proteins control the life–death switch. IAP proteins can inhibit caspases and signal the destruction of regulatory molecules. In contrast, in mammals, the Bcl-2 family controls mitochondrial integrity and the release of factors that activate caspases or block the action of IAPs. Structures of many of these proteins, and the complexes they form, are now available

and underpin current models of apoptosis. Here we review key features of these structures and highlight how these studies have led to the development of antagonist compounds that allow the pro-survival effects of IAP and Bcl-2 proteins to be negated.

Introduction

Apoptosis, or programmed cell death, is an essential process that is required for homeostasis of multicellular organisms. Genetic and biochemical studies have established the molecular framework by which apoptosis is controlled, and as expected for such an important process it is tightly regulated to prevent inappropriate cell death (Adams, 2003). Two pathways, known as the extrinsic and intrinsic pathways, are responsible for cell death (**Figure 1**). The 'extrinsic' pathway is initiated when specific ligands bind to cell surface death receptors of the tumour necrosis factor receptor (TNFR) family (Ashkenazi and Dixit, 1998). The 'intrinsic' pathway (also known as the mitochondrial or Bcl-2 inhibitable pathway) is instigated in response to cellular stress and triggers disruption of intracellular organelles, and relies on release of pro-apoptotic factors from mitochondria. Initiation of

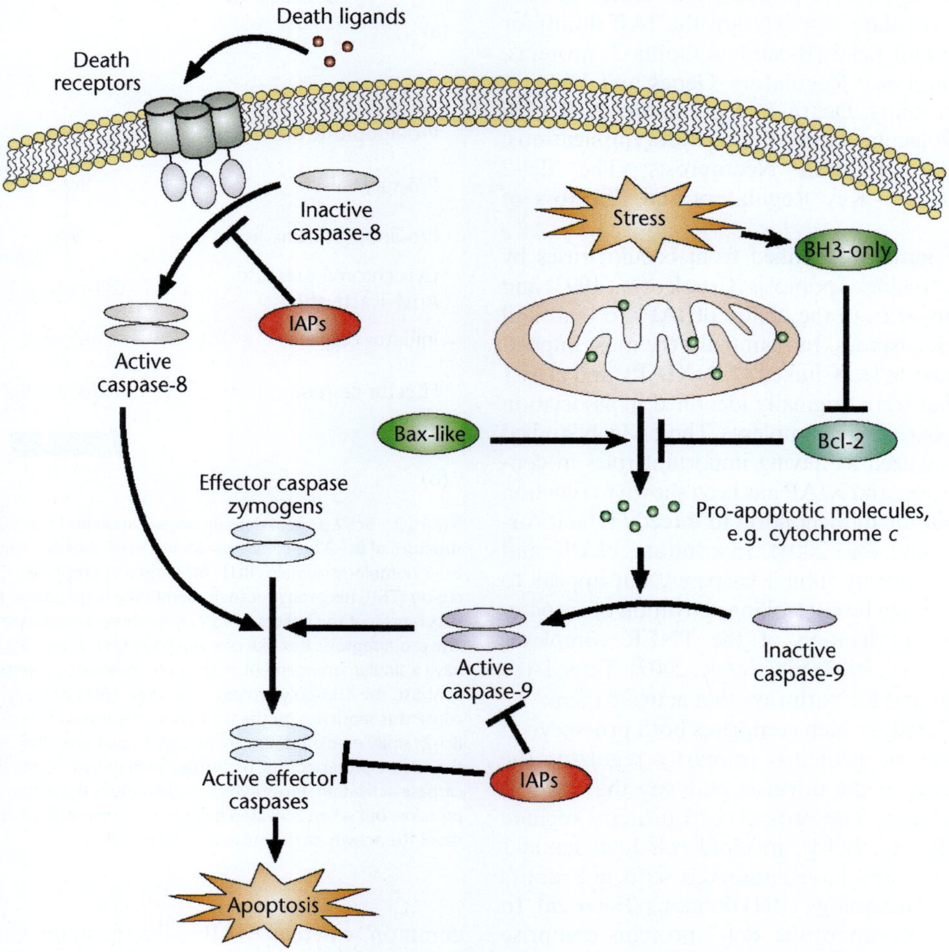

Figure 1 Schematic highlighting of the points at which IAP and Bcl-2 proteins regulate caspase activity. The structural basis for inhibition of caspase-9 and effector caspases by IAPs is well understood. In contrast, inhibition of caspase-8 occurs indirectly and these complexes have not been characterized in detail. Structures are also available for many Bcl-2 proteins, including complexes between different classes, but it remains uncertain how Bcl-2 inhibits Bax, and the structure of the pro-apoptotic Bax (or Bak) complex has eluded analysis.

either the intrinsic or extrinsic pathway ultimately leads to the activation of caspases – a family of cysteine aspartic proteases that cleave vital cellular proteins and destroy the cell (Salvesen and Abrams, 2004). Two initiator caspases, caspase-9 and caspase-8, control the proteolytic cascade, the former is activated by the intrinsic pathway and the latter by the extrinsic pathway following ligand binding to cell surface death receptors. Activation of caspase-8 or caspase-9 leads to proteolytic activation of downstream caspases, or 'effector' caspases, such as caspase-3 and caspase-7, committing cells to death through proteolysis of key substrates (Salvesen and Abrams, 2004). Caspases are poised for action in all cells, even healthy cells, and regulatory molecules prevent their inappropriate activity. Two families of proteins that either directly or indirectly regulate caspases are the IAP (inhibitor of apoptosis) and Bcl-2 (B-cell lymphoma-2) proteins. **See also**: Apoptosis: Regulatory Genes and Disease; BH3-Only Proteins; Death Receptors; Death Receptors at the Molecular Level: Therapeutic Implications; Death Receptor-Induced Necroptosis; The Bcl-2 Family Proteins – Key Regulators and Effectors of Apoptosis

IAPs were initially identified from baculoviruses by their capacity to block apoptosis (Crook *et al.*, 1993) and in part this arises from the ability of IAPs to bind and directly inhibit caspases. In mammals, the most important IAPs appear to be X-linked IAP (XIAP), and cIAP1 and cIAP2 that were originally identified in association with cell surface receptor complexes. These widely studied IAPs are recognized as having important roles in controlling apoptosis, and XIAP has been shown to function downstream of the mitochondria to directly inhibit caspase-9 (Riedl and Shi, 2004). In contrast, cIAP1 and cIAP2 do not directly inhibit caspases, but appear to function principally by controlling activation of caspase-8 in response to ligation of the TNFR complexes (Varfolomeev *et al.*, 2007; Vince *et al.*, 2007). Thus, IAPs block caspases and the pathways that activate them.

The Bcl-2 family, which comprises both pro-survival and pro-apoptotic proteins, primarily regulates the initiating events of the intrinsic pathway that lead to caspase activation. The pro-survival proteins include Bcl-2 itself, Bcl-x_L, Bcl-w, myeloid cell Leukaemia-1 (Mcl-1) and A1, and have conserved sequence motifs known as Bcl-2 homology (BH) domains (**Figure 2a**). In mammals the pro-apoptotic Bcl-2 proteins comprise two distinct groups, the BH3-only proteins (Bim, Bad, Bmf, Bid, Puma, Bik, Hrk and Noxa) and the Bax-like proteins, Bax, Bak and Bok (Youle and Strasser, 2008). The BH3-only proteins have only a BH3 domain in

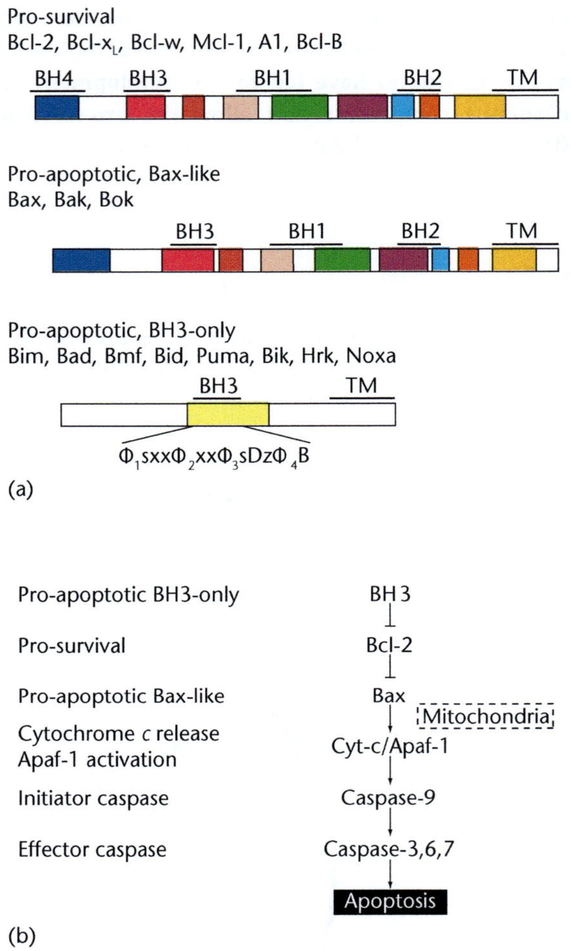

Figure 2 Bcl-2 protein domain organization and function. (a) Domain structure of Bcl-2 proteins. Pro-survival Bcl-2 proteins bear up to four Bcl-2 homology domains BH1–BH4 and a hydrophobic C-terminal region (TM), necessary for their membrane localization. The coloured bars represent the helices $\alpha1$–$\alpha9$ in the three-dimensional structure. The pro-apoptotic Bax-like proteins have BH1-3, and 9 α helices that have a similar arrangement as their pro-survival counterparts. In contrast, the BH3-only proteins bear only a BH3 domain. The consensus sequence for the BH3 domain is shown (see text). (b) The Bcl-2 family mediated caspase-activation pathway. Bax and Bak control the release pro-apoptotic molecules from mitochondria that lead to caspase activation. Pro-survival proteins block the action of the Bax-like proteins, but when activated by apoptotic stimuli BH3-only proteins block the activity of pro-survival Bcl-2 proteins.

common, whereas the Bax-like proteins have multiple BH domains and resemble the pro-survival proteins. Gene knockout studies have shown the absence of *bax* and *bak* makes cells highly resistant to apoptosis indicating that they are essential for commitment to cell death (Wei *et al.*, 2001; Zong *et al.*, 2001). In healthy cells the pro-survival Bcl-2 proteins restrain Bax and Bak, to preserve mitochondrial outer membrane integrity. On receipt of an apoptotic stimulus this restraint is relieved leading to release of cytochrome *c*, caspase activation and the cell's demise (**Figure 2b**).

Structural Basis of Bcl-2 Family Action

The pro-survival proteins are structurally similar, having up to four conserved BH domains (BH1–BH4) and a *C*-terminal hydrophobic region that localizes them to intracellular membranes (**Figure 2a**). Structures of a number of pro-survival Bcl-2 family members have been solved, and all share the Bcl-2 fold (Muchmore *et al.*, 1996). The central pillar of the small helical bundle is the hydrophobic helix α5, on which the surrounding amphipathic α helices pack. A hydrophobic BH3-binding groove is formed from α2, α3 and α4, with α5 and α8 lining the base of the groove (**Figure 3**). Residues on α5 are therefore rendered solvent inaccessible on interaction of BH3 domain ligands, such as Noxa. The pro-survival proteins used for structural studies have mostly had their hydrophobic *C*-terminal residues truncated to improve their solution properties, but when present in Bcl-w they fold back to lie in the binding groove as a helix, α9 (**Figure 3**; Hinds *et al.*, 2003). The multidomain Bax-like pro-apoptotic proteins adopt a similar structure with the *C*-terminal residues having a comparable position, and the BH1–BH3 domains that form the hydrophobic docking site for BH3-only proteins are in proximity in all multidomain Bcl-2 proteins (Sattler *et al.*, 1997; Suzuki *et al.*, 2000).

BH3-only proteins act as sentinels that are released or upregulated following an apoptotic stimulus, and they then bind and inactivate pro-survival Bcl-2 proteins. The BH3 domain of BH3-only proteins forms an amphipathic helix that binds to the surface exposed hydrophobic groove of pro-survival proteins. The core of the BH3 domain, providing the amphipathic interface and key interactions with pro-survival proteins, is a 13-residue sequence motif: $\Phi_1 sxx\Phi_2 xx\Phi_3 sDz\Phi_4 B$, where $\Phi_1 - \Phi_4$ are hydrophobic residues; s, a small residue (Gly, Ala or Ser); x, any residue; D, aspartic acid; z, normally an acidic residue and B, a hydrogen bond acceptor. The Φ_2 residue is a conserved leucine, and the aspartic acid, D, is the only invariant residues in the BH3-domains. The leucine at Φ_2 is buried in the protein–protein interface and packs against conserved residues provided by the pro-survival protein, whereas the solvent exposed aspartate forms an ionic interaction with a conserved arginine in the BH1 domain of the pro-survival protein. Further binding interactions arise from the three other hydrophobic residues Φ_1, Φ_3 and Φ_4 that project into pockets provided by the surface exposed hydrophobic groove on the pro-survival protein. All pro-survival proteins use the equivalent interface for their interactions with BH3-ligands and bind multiple BH3-ligands, but each BH3-only protein

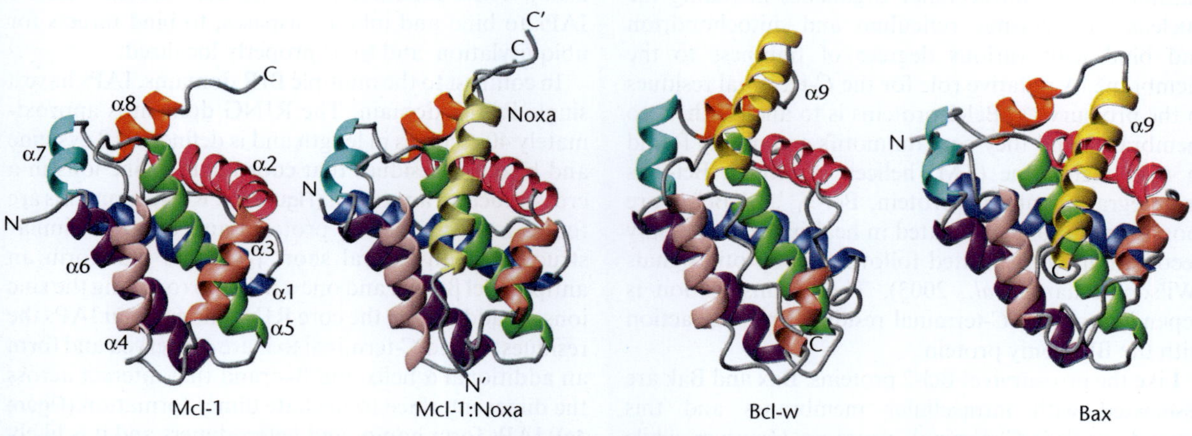

Figure 3 Structures of Bcl-2 proteins. Structures of *C*-terminally truncated Mcl-1 alone and in complex with the BH3 domain from Noxa are shown. Mcl-1 forms a small helical bundle with a binding groove formed by the helices α2–α5 and α8. The helices are labelled α1–α8 and the *N*- and *C*-termini for Mcl-1, N and C. In the Mcl-1:Noxa complex structure the N- and C-termini for Noxa are labelled N′ and C′. The structure of pro-survival Bcl-w and pro-apoptotic Bax are also shown and their termini are indicated. The ribbon diagram shows that the structures of pro-survival and the multidomain pro-apoptotic proteins have the same topology, with the *C*-terminal helix, α9, lying in the binding groove.

binds a distinct set of pro-survival proteins (Chen *et al.*, 2005). Some, like Bim, bind all pro-survival proteins tightly with nanomolar affinity, whereas others such as Noxa only bind to Mcl-1 and A1 with high affinity.

The interaction between BH3-only domains and pro-survival proteins is accompanied by conformational change in both binding partners. On their own many BH3-only proteins lack a defined structure, but the BH3 domain folds into a helix on binding to the hydrophobic groove of the pro-survival protein (Hinds *et al.*, 2007). The helices that constitute the binding groove on the pro-survival protein also adjust on binding the ligand and, although the surface area buried is approximately 1000 Å^2 for each interaction, the exact intermolecular interactions vary between different BH3-only:pro-survival complexes. The plasticity of the binding groove helps account for the ability of the pro-survival proteins to bind BH3 sequences that bear little sequence similarity. This plasticity also partly accounts for the observation that the BH3-only proteins are surprisingly tolerant of mutations, as are the pro-survival proteins, with only a few positions forming critical contacts that confer selectivity. Together these features constitute the basis for the promiscuous yet selective binding shown between the BH3-only and pro-survival Bcl-2 proteins.

Membrane Interactions and the Bcl-2 Family

Pro-survival Bcl-2 proteins are located on the cytoplasmic face of intracellular organelles including the nucleus, endoplasmic reticulum and mitochondrion and bind with various degrees of tightness to the membrane. A putative role for the *C*-terminal residues in the pro-survival Bcl-2 proteins is to anchor them to membranes and they contain motifs commonly found in *trans*-membrane (TM) helices. Although Bcl-2 is an integral membrane protein, Bcl-x_L and Bcl-w are loosely membrane associated in healthy cells and only become tightly associated following apoptotic signals (Wilson-Annan *et al.*, 2003). Their translocation is dependent on the *C*-terminal residues and interaction with the BH3-only protein.

Like the pro-survival Bcl-2 proteins, Bax and Bak are associated with intracellular membranes and this depends on their *C*-terminal sequences. However, while Bak is an integral membrane protein, in healthy cells Bax is largely a cytosolic protein and it only becomes tightly associated with membranes after initiation of apoptosis (Youle and Strasser, 2008). Apoptotic stimuli and inactivation of the pro-survival Bcl-2 proteins by the BH3-only proteins results in conformational changes to both Bak and Bax, and their homo-oligomerization to form large clusters on the mitochondrial outer membrane follows. Although the nature of the conformational changes remains uncertain, pro-survival Bcl-2 proteins can inhibit this and, therefore, promote membrane integrity. However, once the pro-survival proteins have been inactivated, conformational change of Bax and Bak promotes the formation of higher order oligomers, and these events lead to mitochondrial permeabilization and release of apoptogenic factors that activate the caspase cascade. Bax and Bak have a critical role and cells lacking them are highly resistant to multiple apoptotic stimuli (Wei *et al.*, 2001; Zong *et al.*, 2001).

Structural Basis of IAP Function

Not all IAPs regulate apoptosis, but those that do have both *N*-terminal baculoviral IAP repeat (BIR) domains and a *C*-terminal really interesting new gene 1 (RING) domain (**Figure 4a**). The BIR domain is approximately 80 residues in length and is defined by conserved histidine and cysteine residues that coordinate a zinc ion (Hinds *et al.*, 1999; Sun *et al.*, 1999). All BIR domains adopt a similar compact zinc-finger-like structure, with the zinc ion surrounded by three short β-strands and four α helices (**Figure 4b**). The BIR domain functions as a protein–protein interaction domain and many IAPs are able to interact with a number of different proteins by virtue of their multiple BIR domains (Vaux and Silke, 2005). These interactions of the BIR domains enable IAPs to bind and inhibit caspases, to bind targets for ubiquitylation, and to be properly localized.

In contrast to the multiple BIR domains, IAPs have a single RING domain. The RING domain is approximately 40 residues in length and is defined by 8 cysteine and histidine residues that coordinate 2 zinc ions in a cross-brace arrangement (**Figure 4c**). RING domains are found in a number of proteins and adopt a similar structure, with several short β-strands that form an antiparallel β sheet and one α helix surrounding the zinc ions. In addition to the core RING domain, in IAPs the residues *N*- and *C*-terminal to it are conserved and form an additional α helix and β-strand that interact across the dimer interface to mediate dimer formation (**Figure 4c**). IAPs form homo- and heterodimers and it is likely that these are structurally very similar (Mace *et al.*, 2008). In addition to mediating dimerization, the RING domains of IAPs are functional ubiquitin E3 ligases and this is important for their ability to regulate

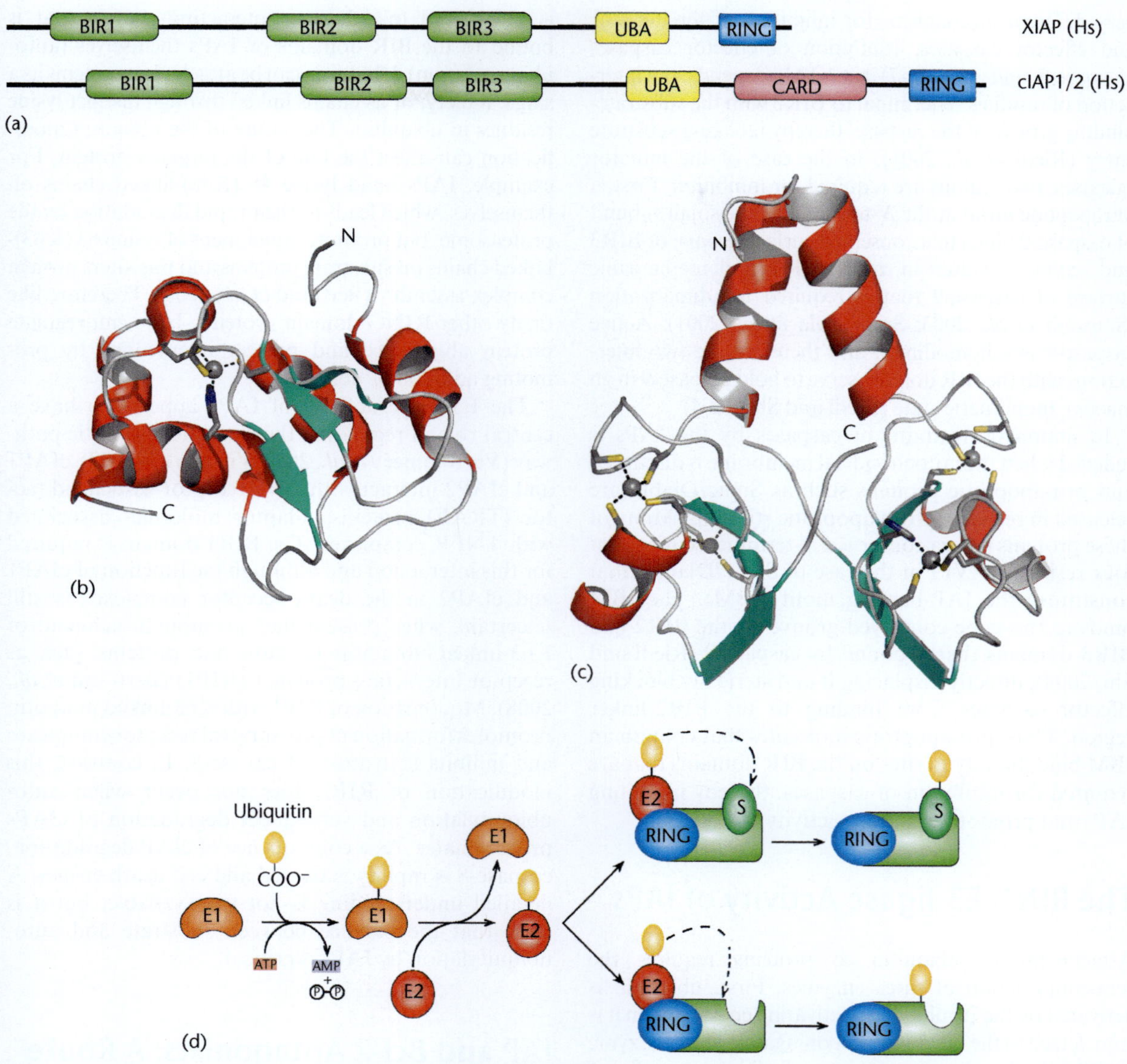

Figure 4 Structure and organization of IAPs. (a) Domain arrangement of selected mammalian IAPs. (b) The BIR domain has a compact structure centred on a single zinc ion (grey ball). The structure of BIR3 from XIAP is shown (PDB code 1g3f). (c) The *C*-terminal RING domain binds two zinc ions and has a conserved fold that is common to all RING domains. In IAPs residues *N*- and *C*-terminal to the core RING domain mediate dimer formation. The structure of the cIAP2 RING dimer is shown (PDB code 3eb5). (d) An enzyme cascade mediates ubiquitylation. Isopeptide bond formation between the *C*-terminus of ubiquitin and lysine residues in target proteins depends on ATP and E1, E2 and E3 enzymes. Substrate (S) (top) and autoubiquitylation (bottom) by RING E3 ligases is shown.

both protein abundance and protein complex formation. Some IAPs also contain a ubiquitin-associated (UBA) domain that binds to ubiquitin chains (Gyrd-Hansen *et al.*, 2008) and a caspase recruitment domain (CARD) of unknown function.

Caspase Regulation by IAPs

The BIR domains directly inhibit both initiator and effector caspases (Riedl and Shi, 2004), however, structures of caspases in complex with BIR domains showed very different mechanisms for inhibition of the initiator and effector caspases. Inhibition of effector caspases (caspase-3 and caspase-7) by XIAP depends on interaction of residues *N*-terminal to BIR2 with the substrate-binding groove of the caspase, thereby blocking substrate entry (Riedl *et al.*, 2001). In the case of the initiator caspases, two regions are required for inhibition. First, a tetrapeptide motif at the *N*-terminus of the small subunit of caspase-9 binds to a conserved surface groove of BIR3 and second, residues in $\alpha 4$ of BIR3 bind to the same surface of caspase-9 that is required for dimerization (Shiozaki *et al.*, 2003; Srinivasula *et al.*, 2001). Active caspase-9 is a homodimer, and therefore the two interactions with the BIR domain serve to hold caspase-9 in an inactive monomeric state (Riedl and Shi, 2004).

In mammals restraint of caspases by the IAPs is relieved when the mitochondrial membrane is disrupted and pro-apoptotic proteins such as Smac/Diablo are released in response to an apoptotic stimulus. Many of these proteins have a conserved *N*-terminal sequence of four residues (AVPI in the case of Smac/Diablo) that constitutes the IAP-binding motif (IBM). The IBM binds to the same conserved groove on the BIR2 and BIR3 domains that is bound by caspase-9 (Riedl and Shi, 2004), directly displacing it and sterically blocking effector caspases from binding to the BIR2-linker region. Thus, pro-apoptotic molecules that contain an IBM bind directly to sites on the BIR domains that are required for inhibition of caspases, thereby inhibiting IAPs and promoting caspase activity.

The RING E3-ligase Activity of IAPs

Attachment of ubiquitin to proteins requires the sequential action of three enzymes. First, ubiquitin is activated by the E1 ubiquitin-activating enzyme, then it is transferred to the E2 ubiquitin-conjugating (Ubc) enzyme and finally the ubiquitin E3-ligase facilitates transfer of ubiquitin from the E2 to a lysine sidechain in the substrate protein (**Figure 4d**). RING E3 ligases provide a scaffold that brings the E2 $\sim$ ubiquitin conjugate and substrate into proximity. The structure of the RING domain from cIAP2 bound to the E2, UbcH5b, shows that both RING domains of the homodimer interact with an E2 molecule (Mace *et al.*, 2008). Like other RING domains, the conserved hydrophobic residues at the centre of the cIAP2:RING interface are required for E2 recruitment. Nearby residues also make contacts and these are likely to contribute to the selective binding, as IAPs interact with a only subset of the available E2s. Once IAPs have bound the E2 $\sim$ ubiquitin conjugate, ubiquitin is transferred to either an interacting protein that is bound to the BIR domains or IAPs themselves (auto-ubiquitylation). Ubiquitin can be attached to proteins as a single moiety, or as chains linked through distinct lysine residues in ubiquitin. The nature of the ubiquitin modification can affect the fate of the targeted protein. For example, IAPs build lysine 48 (K48)-linked chains on themselves, which leads to their rapid degradation by the proteasome, but promote attachment of lysine 63 (K63)-linked chains on substrate proteins and this alters protein complex assembly (Bertrand *et al.*, 2008). Therefore like many other RING domain proteins, IAPs can regulate protein abundance and protein interactions by promoting addition of ubiquitin.

The E3-ligase activity of IAPs appears to have a central role in regulating the extrinsic apoptotic pathway (Varfolomeev *et al.*, 2007; Vince *et al.*, 2007). cIAP1 and cIAP2 interact with TNF receptor-associated factor (TRAF) proteins, adaptor molecules associated with TNFR complexes. The BIR1 domain is required for this interaction and although the function of cIAP1 and cIAP2 in the death receptor complexes is still uncertain, when present they promote attachment of K63-linked ubiquitin on substrate proteins such as receptor interacting protein 1 (RIP1) (Bertrand *et al.*, 2008). Modification of RIP1 with K63-linked ubiquitin promotes formation of pro-survival receptor complexes and inhibits activation of capase-8. In contrast, this modification of RIP1 does not occur when auto-ubiquitylation and subsequent degradation of cIAPs predominates. As a consequence of cIAP degradation, caspase-8 is rapidly activated and cell death ensues. A detailed understanding is not yet available, but it is clear that the balance between substrate and auto-ubiquitylation by IAPs is critical.

IAP and Bcl-2 Antagonists: A Route to New Cancer Therapies?

Defects in the regulation of apoptosis have been implicated as a cause, or contributing factor, in many diseases and impaired apoptosis is the hallmark of most, if not all, cancers (Hanahan and Weinberg, 2000). Oncogenic mutations often inactivate signalling pathways for cell death or activate pro-survival pathways. The most common oncogenic mutation in human malignancies inactivates the tumour suppressor p53, which is normally activated by deoxyribonucleic acid (DNA) damage and initiates DNA repair and cell cycle arrest, or triggers the destruction of damaged cells by apoptosis. Aberrant regulation of apoptosis is manifested not only as a fundamental mechanism underlying causality in cancer, but also makes many cancers refractory to treatment, as many cytotoxic therapies function by inducing apoptosis in cancer cells. These features have led to the search for small molecule antagonists for both the IAP and pro-survival Bcl-2 families as a way to mobilize the poised caspase machinery in cancer cells.

As a tetrapeptide based on the IBM motif could promote caspase activity and sensitize cancer cell lines to undergo apoptosis, several groups focused on developing molecules that could mimic this IAP-antagonist effect (Vucic and Fairbrother, 2007). Peptidomimetic compounds that bind to the IBM-binding pocket of BIR domains with high affinity have been developed. Structures of these compounds and peptide antagonists in complex with BIR domains show that very similar contacts are made with the IBM-binding site (**Figure 5a**). IBM-antagonist compounds have been shown to activate caspases, stimulate cell death in tumour cell lines and promote tumour regression in model systems (Vucic and Fairbrother, 2007). Although the compounds were designed to displace caspases from the IBM-binding pocket of IAPs, recent studies have suggested that the E3-ligase activity of the RING domain is critical for their activity (Varfolomeev *et al.*, 2007; Vince *et al.*, 2007). Notably, the IBM-antagonist compounds promote autoubiquitylation and degradation of cIAPs and this is required for their pro-apoptotic activity. Notwithstanding this finding, IAP antagonists show particular therapeutic promise and are now entering clinical trials.

To develop pro-survival Bcl-2 protein antagonists that mimic the BH3-only proteins a structure-based approach was utilized. Initially, libraries of small molecules were screened and then by linking low affinity hits that bound to adjacent pockets in the BH3-binding groove, optimized compounds were derived that bound Bcl-x_L, Bcl-2 and Bcl-w with high affinity (Oltersdorf *et al.*, 2005). The structure of one such compound, ABT-737, has been solved bound to Bcl-x_L and this ligand binds in the hydrophobic groove, but not in the same way as the BH3 domain ligand (**Figure 5b**). In contrast to IAPs, the plasticity of the binding grooves of the pro-survival Bcl-2 proteins makes predicting their contacts difficult. Binding of ABT-737 caused a rearrangement of the groove residues of Bcl-x_L resulting in a change in the pocket that accepts the residue at Φ_2 (see earlier discussion) (Lee *et al.*, 2007). Cell death initiated by ABT-737 is dependent on the presence of Bax and Bak indicating that this drug-like molecule activates the intrinsic apoptotic pathway. ABT-737 has the binding profile of the BH3-only protein Bad, which also binds Bcl-2, Bcl-x_L and Bcl-w with high affinity. As a consequence of this, cells that are protected from cell death by overexpression of other pro-survival proteins such as Mcl-1, that does not bind ABT-737, do not die. Like the IAP antagonists, molecules related to ABT-737 are currently in clinical trials.

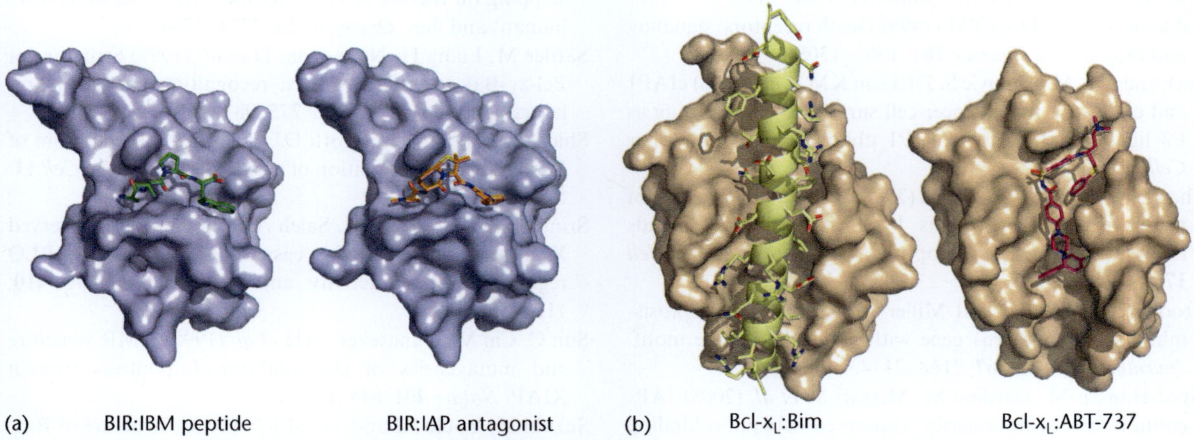

Figure 5 Structure-based development of drugs that antagonize IAPs and pro-survival Bcl-2 proteins. (a) Structure of a chimeric BIR3 domain showing the IBM-binding pocket with either a peptide or a small molecule antagonist present (PDB codes 2I3H and 2I3I). The BIR domain is shown as a surface and the bound ligands as stick representation. (b) Bcl-x_L is shown bound to the BH3 domain of Bim and to ABT-737, a small molecule anatagonist (PDB codes 1PQ1 and 2YXJ) that occupies the same binding groove. Bcl-x_L is shown as a surface, Bim as a ribbon and ABT-737 as a stick model. The IBM-binding site adopts a similar conformation when bound to both peptide and small molecule antagonist, whereas the binding groove of Bcl-x_L adopts a distinct structure when bound to ABT-737.

Summary

The IAP and Bcl-2 protein families are important regulators of programmed cell death, that act at different points in the apoptosis regulatory pathways, using entirely different mechanism, to prevent inappropriate cell death. In mammals, the Bcl-2 proteins govern the activation of initiator caspases by controlling the release of initiating factors through the mitochondrial outer membrane. In contrast, the IAPs can act on caspases directly by preventing the assembly of active dimers and by blocking the binding site, or they can ubiquitylate themselves and other proteins and this modulates pathways that lead to capsase activation. The subversion of apoptosis has important consequences and is an underlying pathophysiological basis for many diseases. The IAPs and pro-survival Bcl-2 proteins are frequently deregulated in cancer and are proving promising targets for pharmaceutical design. By mimicking the molecular mode of action of native inhibitors of the IAPs or the pro-survival Bcl-2 family members, it may be possible to initiate apoptosis in cancer cells and overcome their resistance to cell death. Understanding the molecular structures and interactions has been, and will be, key to this process.

References

Adams JM (2003) Ways of dying: multiple pathways to apoptosis. *Genes & Development* **17**: 2481–2495.

Ashkenazi A and Dixit VM (1998) Death receptors: signaling and modulation. *Science* **281**: 1305–1308.

Bertrand MJ, Milutinovic S, Dickson KM *et al.* (2008) cIAP1 and cIAP2 facilitate cancer cell survival by functioning as E3 ligases that promote RIP1 ubiquitination. *Molecular Cell* **30**: 689–700.

Chen L, Willis SN, Wei A *et al.* (2005) Differential targeting of prosurvival Bcl-2 proteins by their BH3-only ligands allows complementary apoptotic function. *Molecular Cell* **17**: 393–403.

Crook NE, Clem RJ and Miller LK (1993) An apoptosis-inhibiting baculovirus gene with a zinc finger-like motif. *Journal of Virology* **67**: 2168–2174.

Gyrd-Hansen M, Darding M, Miasari M *et al.* (2008) IAPs contain an evolutionarily conserved ubiquitin-binding domain that regulates NF-κB as well as cell survival and oncogenesis. *Nature Cell Biology* **10**: 1309–1317.

Hanahan D and Weinberg RA (2000) The hallmarks of cancer. *Cell* **100**: 57–70.

Hinds MG, Lackmann M, Skea GL *et al.* (2003) The structure of Bcl-w reveals a role for the *C*-terminal residues in modulating biological activity. *EMBO Journal* **22**: 1497–1507.

Hinds MG, Norton RS, Vaux DL and Day CL (1999) Solution structure of a baculoviral inhibitor of apoptosis (IAP) repeat. *Nature Structural Biology* **6**: 648–651.

Hinds MG, Smits C, Fredericks-Short R *et al.* (2007) Bim, Bad and Bmf: intrinsically unstructured BH3-only proteins that undergo a localized conformational change upon binding to prosurvival Bcl-2 targets. *Cell Death & Differentiation* **14**: 128–136.

Lee EF, Czabotar PE, Smith BJ *et al.* (2007) Crystal structure of ABT-737 complexed with Bcl-x_L: implications for selectivity of antagonists of the Bcl-2 family. *Cell Death & Differentiation* **14**: 1711–1713.

Mace PD, Linke K, Feltham R *et al.* (2008) Structures of the cIAP2 RING domain reveal conformational changes associated with ubiquitin-conjugating enzyme (E2) recruitment. *Journal of Biological Chemistry* **283**: 31633–31640.

Muchmore SW, Sattler M, Liang H *et al.* (1996) X-ray and NMR structure of human Bcl-x_L, an inhibitor of programmed cell death. *Nature* **381**: 335–341.

Oltersdorf T, Elmore SW, Shoemaker AR *et al.* (2005) An inhibitor of Bcl-2 family proteins induces regression of solid tumours. *Nature* **435**: 677–681.

Riedl SJ, Renatus M, Schwarzenbacher R *et al.* (2001) Structural basis for the inhibition of caspase-3 by XIAP. *Cell* **104**: 791–800.

Riedl SJ and Shi Y (2004) Molecular mechanisms of caspase regulation during apoptosis. *Nature Reviews. Molecular Cell Biology* **5**: 897–907.

Salvesen GS and Abrams JM (2004) Caspase activation – stepping on the gas or releasing the brakes? Lessons from humans and flies. *Oncogene* **23**: 2774–2784.

Sattler M, Liang H, Nettesheim D *et al.* (1997) Structure of Bcl-x_L-Bak peptide complex: recognition between regulators of apoptosis. *Science* **275**: 983–986.

Shiozaki EN, Chai J, Rigotti DJ *et al.* (2003) Mechanism of XIAP-mediated inhibition of caspase-9. *Molecular Cell* **11**: 519–527.

Srinivasula SM, Hegde R, Saleh A *et al.* (2001) A conserved XIAP-interaction motif in caspase-9 and Smac/DIABLO regulates caspase activity and apoptosis. *Nature* **410**: 112–116.

Sun C, Cai M, Gunasekera AH *et al.* (1999) NMR structure and mutagenesis of the inhibitor-of-apoptosis protein XIAP. *Nature* **401**: 818–822.

Suzuki M, Youle RJ and Tjandra N (2000) Structure of Bax: coregulation of dimer formation and intracellular localization. *Cell* **103**: 645–654.

Varfolomeev E, Blankenship JW, Wayson SM *et al.* (2007) IAP antagonists induce autoubiquitination of c-IAPs, NF-κB activation, and TNFα-dependent apoptosis. *Cell* **131**: 669–681.

Vaux DL and Silke J (2005) IAPs, RINGs and ubiquitylation. *Nature Reviews. Molecular Cell Biology* **6**: 287–297.

Vince JE, Wong WW, Khan N *et al.* (2007) IAP antagonists target cIAP1 to induce TNFα-dependent apoptosis. *Cell* **131**: 682–693.

Vucic D and Fairbrother WJ (2007) The inhibitor of apoptosis proteins as therapeutic targets in cancer. *Clinical Cancer Research* **13**: 5995–6000.

Wei MC, Zong WX, Cheng EH *et al.* (2001) Proapoptotic BAX and BAK: a requisite gateway to mitochondrial dysfunction and death. *Science* **292**: 727–730.

Wilson-Annan J, O'Reilly LA, Crawford SA *et al.* (2003) Proapoptotic BH3-only proteins trigger membrane integration of prosurvival Bcl-w and neutralize its activity. *Journal of Cell Biology* **162**: 877–887.

Youle RJ and Strasser A (2008) The BCL-2 protein family: opposing activities that mediate cell death. *Nature Reviews. Molecular Cell Biology* **9**: 47–59.

Zong WX, Lindsten T, Ross AJ *et al.* (2001) BH3-only proteins that bind pro-survival Bcl-2 family members fail to induce apoptosis in the absence of Bax and Bak. *Genes & Development* **15**: 1481–1486.

Further Reading

Adams JM and Cory S (2007) The Bcl-2 apoptotic switch in cancer development and therapy. *Oncogene* **26**: 1324–1337.

Danial NN and Korsmeyer SJ (2004) Cell death: critical control points. *Cell* **116**: 205–219.

Keskin O, Gursoy A, Ma B and Nussinov R (2008) Principles of protein–protein interactions: what are the preferred ways for proteins to interact? *Chemical Reviews* **108**: 1225–1244.

Lacasse E, Mahoney D, Cheung H *et al.* (2008) IAP-targeted therapies for cancer. *Oncogene* **27**: 6252–6275.

Lessene G, Czabotar PE and Colman PM (2008) BCL-2 family antagonists for cancer therapy. *Nature Reviews. Drug Discovery* **7**: 989–1000.

Pickart CM (2001) Mechanisms underlying ubiquitination. *Annual Review of Biochemistry* **70**: 503–533.

Reed JC, Doctor KS and Godzik A (2004) The domains of apoptosis: a genomics perspective. *Science's STKE* **2004**: re9.

Engulfment of Apoptotic Cells and its Physiological Roles

Rikinari Hanayama, *Department of Medical Chemistry, Kyoto University Graduate School of Medicine, Kyoto, Japan*

Masanori Miyanishi, *Department of Medical Chemistry, Kyoto University Graduate School of Medicine, Kyoto, Japan*

Hiroshi Yamaguchi, *Department of Medical Chemistry, Kyoto University Graduate School of Medicine, Kyoto, Japan*

Jun Suzuki, *Department of Medical Chemistry, Kyoto University Graduate School of Medicine, Kyoto, Japan*

Shigekazu Nagata, *Department of Medical Chemistry, Kyoto University Graduate School of Medicine, Kyoto, Japan*

Advanced article

Article Contents

- Introduction
- "Find-me" Signals
- "Eat-me" Signals
- Signaling Pathways for Engulfment
- Engulfment Mediated by Bridging Molecules
- Phosphatidylserine Receptors
- Immunosuppressive Roles of Engulfment
- Conclusions and Future Prospects

During development of animals, many cells undergo programmed cell death via apoptosis. In adult, senescent cells also undergo apoptosis and are replaced by newly generated cells. Apoptotic dying cells display an "eat-me" signal(s) on their surface, and are recognized by macrophages and immature dendritic cells for engulfment. Phosphatidylserine, which is exposed on the cell surface in a caspase-dependent manner, is the most likely candidate for the "eat-me" signal. Phagocytes recognize phosphatidylserine through specific

receptors or via bridging molecules that link them with the apoptotic cells. Phagocytes engulf the apoptotic cells in a Rac1-dependent manner probably through specific portals. When apoptotic cells are not engulfed efficiently, they undergo secondary necrosis and release their cellular contents, which leads to local inflammation, production of autoantibodies, and SLE (systemic lupus erythematosus)-type autoimmune diseases.

Introduction

In our bodies, billions of cells die by apoptosis every day and dead cells are swiftly removed by phagocytes to maintain the integrity and functioning of the surrounding tissues. The impaired clearance of apoptotic cells causes the release of their cellular components, which can act as self-antigens, and may induce inflammation and autoimmune diseases. The clearance of apoptotic cells is mediated by neighboring cells (fibroblasts and epithelial cells) and/or professional phagocytes (macrophages and immature dendritic cells). These phagocytes usually do not engulf healthy cells, which means that apoptotic cells display an "eat-me" signal(s) recognized by the phagocytes. Recently, several proteins have been proposed as molecules that mediate the engulfment of apoptotic cells. These include soluble proteins that act as a bridge between apoptotic cells and phagocytes, and surface receptors on phagocytes that directly recognize apoptotic cells (**Figure 1**). In this article, we discuss the current understanding of the molecular mechanisms underlying the engulfment of apoptotic cells. **See also**: Apoptosis: Regulatory Genes and Disease; The Siren's Song: This Death that Makes Life Live

"Find-me" Signals

Professional phagocytes are attracted to sites where apoptosis occurs, to engulf the dying cells. Therefore, during early apoptosis, apoptotic cells secrete one or more chemoattractants (i.e., "find-me" signals) to recruit the phagocytes. Lysophosphatidylcholine (LPC) is one of the molecules proposed to be such a chemoattractant signal; it is released from apoptotic cells in a caspase-3-dependent manner (Lauber *et al.*, 2003). The release of LPC is caused by the activation of calcium-independent phospholipase A2, which converts phosphatidylcholine into LPC. LPC binds to G2A, a G-protein coupled receptor expressed on phagocytes,

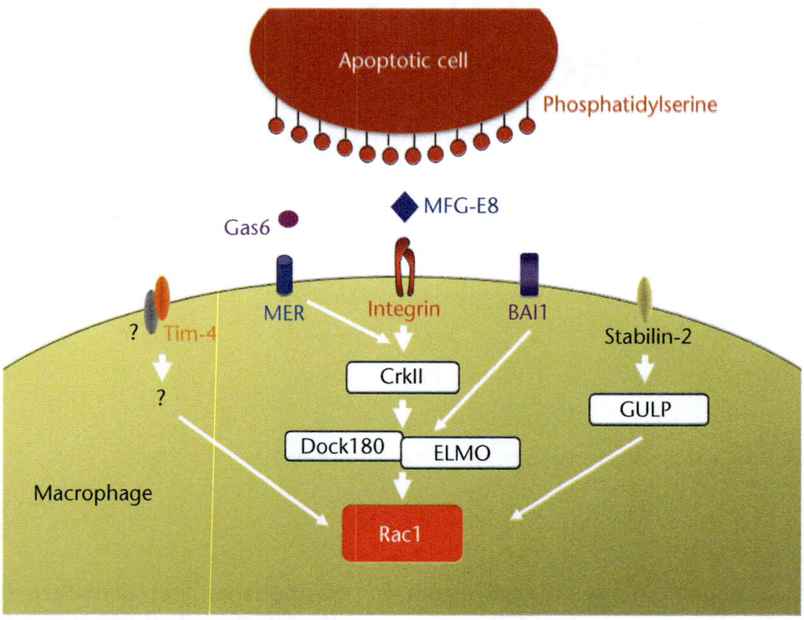

Figure 1 Signaling pathways for the engulfment of apoptotic cells. Apoptotic cells display phosphatidylserine on their surface as an "eat-me" signal. Phosphatidylserine can be bound by soluble bridging molecules such as MFG-E8 and Gas6, which are recognized, by integrin and MER on phagocytes, respectively. Phagocytes can directly recognize phosphatidylserine on apoptotic cells via surface receptors such as TIM4, BAI1, and stabilin2. The downstream signaling involves CrkII/Dock180/ELMO1 or GULP, which activates Rac1 to initiate the engulfment.

and triggers the chemotaxis of the phagocytes (Peter *et al.*, 2008). Although this model is attractive, the concentration of LPC required to cause the chemotaxis of phagocytes is rather high (20–30 μM), a level that is unlikely to be reached physiologically. Therefore, some other molecules may support the LPC-mediated recruitment of phagocytes.

A chemokine, CX3CL1/fractalkine was also proposed as a chemoattractant signal. It is rapidly processed and released as part of a caspase-dependent mechanism from apoptotic neurons in the central nervous system or from apoptotic B cells in the germinal centers of the spleen (Truman *et al.*, 2008). It binds to a specific receptor, CX3CR, on microglia and macrophages, induces their chemotaxis, and upregulates the expression of milk fat globule EGF factor 8 (MFG-E8, a factor that promotes engulfment; see below) (Fuller and Van Eldik, 2008). This finding indicates that when phagocytes sense "find-me" signals they can alter their gene expression profile in preparation for the engulfment of apoptotic cells.

"Eat-me" Signals

Phagocytes engulf apoptotic cells but not healthy cells, indicating that the dying cells display "eat-me" signals to phagocytes and that phagocytes recognize the signals using specific receptors. The best-studied and most likely "eat-me" signal is an aminophospholipid, phosphatidylserine (Fadok *et al.*, 1992). Phosphatidylserine is a component of the cell plasma membrane that is kept exclusively on the inner leaflet of the lipid bilayer in healthy cells by the action of ATP-dependent aminophospholipid translocases (Balasubramanian and Schroit, 2003). Two attractive models have been proposed to explain how phosphatidylserine is exposed on the outer plasma membrane leaflet of apoptotic cells. In one model, once the cells receive the signal to initiate apoptosis, the translocases are inactivated, causing the randomization of the membrane leaflets and the exposure of phosphatidylserine on the outer leaflet of the membrane (Balasubramanian and Schroit, 2003). The second model involves another possible player, the phospholipid scramblase protein family (Sahu *et al.*, 2007). Scramblases are thought to be activated by the increased intracellular calcium levels in dying cells, and to scramble the phospholipids in the plasma membranes, causing the exposure of phosphatidylserine.

Data supporting both of the above models have recently been obtained in studies on apoptosis in *Caenorhabditis elegans*, a nematode. Unlike in mammals, it has not been clear that phosphatidylserine is exposed during apoptosis in *C. elegans*. However, using ectopically expressed green fluorescent protein (GFP)-fused mouse MFG-E8, which specifically binds phosphatidylserine, Venegas and Zhou (2007) recently detected the specific exposure of phosphatidylserine on the surface of apoptotic cells in *C. elegans*. They showed that the *C. elegans* ortholog of phospholipid scramblase (PLSCR)1 promotes the exposure of phosphatidylserine on apoptotic germ cells but not somatic cells, while CED-7, a *C. elegans* ortholog of an ATP-binding cassette (ABC) transporter (see **Table 1**; Wu and Horvitz, 1998), is involved in the exposure of phosphatidylserine in the apoptotic somatic but not germ cells. In another study, Xue and colleagues reported that TAT-1, a *C. elegans* homolog of ATP-dependent aminophospholipid translocase, is critical for maintaining the cell surface asymmetry of the phosphatidylserine distribution (Darland-Ransom *et al.*, 2008). These reports apparently support the idea that the inactivation of aminophospholipid translocase, coupled with the activation of phospholipid scramblases or ABC transporters, induces the exposure of phosphatidylserine. However, Hengartner and colleagues did not find any change in phosphatidylserine exposure by knocking down PLSCR1 or the CED-7 ABC transporter (Zullig *et al.*, 2007). Rather, they observed that the TAT-1 aminophospholipid translocase functions to promote phosphatidylserine exposure in *C. elegans*. **See also**: Cell Death in *C. elegans*

Studies in mammals have also failed to support the model involving scramblases and translocases. There are four genes for phospholipid scramblases in humans (PLSCR1–4). All the PLSCRs have a nuclear localization signal, and PLSCR1 binds to the promoter of the inositol 1,4,5-triphosphate receptor type 1 gene to enhance its expression (Zhou *et al.*, 2005). $PLSCR1^{-/-}$ and $PLSCR3^{-/-}$ mice do not show abnormal phosphatidylserine exposure on apoptotic cells (Wiedmer *et al.*, 2004; Zhou *et al.*, 2002), suggesting that PLSCRs are not important players in the exposure of phosphatidylserine, at least in mammals. Thus, how phosphatidylserine is exposed to the surface of apoptotic cells remains a mystery.

Many other molecules, including oxidized lipids, sugars, glycoproteins, and the calcium-binding protein calreticulin, have been proposed as potential "eat-me" signals. However, how these molecules contribute to the recognition of the apoptotic cells by phagocytes is not clear. In contrast, the masking of phosphatidylserine on apoptotic cells inhibits their *in vitro* engulfment by various macrophages and causes systemic lupus

Table 1 *C. elegans* genes involved in engulfment of apoptotic cells and their mammalian counterparts

C. elegans	Mammalian counterpart	Properties
CED-1	MEGF10	A type I membrane protein (1111 aa) carrying multiple EGF-like sequences in the extracellular region
CED-6	GULP	A cytoplasmic protein (492 aa) carrying a phosphotyrosine-binding domain (PTB) and four SH3-binding motifs. It binds to CED-1 and functions upstream of CED-10
CED-7	ABC transporter	A huge protein (1691 aa) with two homologous repeats, each harboring six transmembrane segments and one ATP-binding site
CED-2	CrkII	A cytoplasmic protein (279 aa) carrying an SH2 and two SH3 domains, and works as an adaptor in the various signal transduction
CED-5	Dock180	A huge cytoplasmic protein (1781 aa) containing an SH3 domain. It associates with CrkII and ELMO, and activates Rac1 as a guanine nucleotide exchange factor
CED-10	Rac1	A small GTP-binding protein (191 aa) of the Ras superfamily
CED-12	ELMO	A cytoplasmic protein (731 aa). It associates with CrkII and Dock180, and activates Rac1
TAT-1	Aminophospholipid translocase	A large protein (1139 aa) containing 10 transmembrane region, and belongs to the P-type ATPase superfamily, ATP-dependent cation transporter

erythematosus (SLE)-type autoimmune disease *in vivo*, clearly implicating phosphatidylserine as a major player in the "eat-me" signal (Asano *et al*., 2004).

Signaling Pathways for Engulfment

Clearing away apoptotic cells is a fundamental process of programmed cell death and is conserved throughout species. In *C. elegans*, 131 of the 1090 somatic cells undergo programmed cell death during development. These dying cells are swiftly engulfed by their neighboring sister cells, and very few cell corpses can be recognized during late embryogenesis. By screening for *C. elegans* mutants defective in the engulfment of apoptotic cells, Horvitz and colleagues identified seven genes that mediate the recognition and engulfment of apoptotic cells (Reddien and Horvitz, 2004). Genetic analysis showed that these seven genes function in two parallel and partially redundant signaling pathways: the CED-1/-6/-7 and CED-2/-5/-10/-12 pathways.

CED-1 is a transmembrane receptor containing a typical epidermal growth factor (EGF)-like repeats and a cysteine-rich motif in its extracellular domain, and it shares high homology with multiple EGF-like-domains 10 (MEGF10) in mammals (see **Table 1**; Hamon *et al*., 2006). Some studies have suggested that CED-1 recognizes phosphatidylserine (Venegas and Zhou, 2007), but the direct binding of CED-1/MEGF10 to phosphatidylserine has not been demonstrated. CED-6 is an ortholog of PTB (phosphotyrosine-binding) domain-containing engulfment adaptor protein (GULP). It binds to the NPXY (Asn-Pro-X-Tyr) motif in the CED-1 intracellular domain and may transduce the signal for the engulfment of apoptotic cells (Su *et al*., 2002).

The CED-7 protein is homologous to the ABC transporters that actively transport a variety of substances across the plasma membrane. CED-7 was originally believed to promote the exposure of phosphatidylserine and/or other "eat-me" signals on the surface of apoptotic cells (Reddien and Horvitz, 2004). However, the direct interaction of ABCA1 transporter (CED-7) with MEGF10 (CED-1) on the phagocytes (Hamon *et al*., 2006) may require an alternative model for the involvement of CED-7 in the engulfment.

CED-2, -5, -10, and -12 of the second *C. elegans* phagocytic pathway correspond, respectively, to mammalian CrkII, Dock180, Rac1, and ELMO1. CrkII is an adaptor protein that associates with Dock180 and ELMO1. Dock180 is a guanine-nucleotide exchange factor for Rac1, and its activity is positively regulated by ELMO1 (Cote and Vuori, 2007). The CrkII/Dock180/Rac1/ELMO1 pathway for the engulfment of apoptotic cells seems to be evolutionarily well conserved from *C. elegans* to humans, except for its upstream signaling components. In mammals, integrin family members are upstream of the Rac1-activating signaling pathway, which sometimes involves activation of a different Rho family guanosine triphosphatase (GTPase), RhoG, as the intermediate (Katoh and Negishi, 2003). In contrast, none of the integrins seems to be involved in the engulfment of apoptotic cells in *C. elegans*, and the upstream receptor for this pathway remains to be determined.

By using a fluorescence resonance energy transfer (FRET) probe that detects active Rac1, Nakaya *et al.* (2008) monitored Rac1 activation during the engulfment of apoptotic cells, and found that phagocytes usually engulf apoptotic cells with their lamellipodia, where Rac1 is activated. The successive engulfment of different cells often occurs at the same lamellipodial site, suggesting that there might be "portal sites" for apoptotic cells. When phagocytes start to engulf apoptotic cells, activated Rac1 and integrin are recruited to the contact site to form phagocytic cups composed of actin patches. As soon as an apoptotic cell sinks into the phagocyte through the phagocytic cup, Rac1 appears to be inactivated and the actin depolymerized, indicating that the activation and deactivation of Rac1 must be regulated with specific timing for the efficient engulfment of apoptotic cells. Indeed, as proposed previously (Grimsley and Ravichandran, 2003), many molecules that positively and negatively regulate Rac1 may be components of engulfment synapses and function in the efficient engulfment of apoptotic cells.

Once the apoptotic cells are engulfed, they are transferred into lysosomes in which Rab5, another Rho family GTPase, plays an essential role (Kitano *et al.*, 2008). Interestingly, RhoA inhibits the engulfment of apoptotic cells (Nakaya *et al.*, 2006), indicating that their efficient engulfment requires the careful orchestration of the Rho family GTPases. Future studies will elucidate whether this Rho family-mediated signaling pathway for the engulfment of apoptotic cells is similar to that used for complement-mediated phagocytosis or for the engulfment of microorganisms.

Engulfment Mediated by Bridging Molecules

Several soluble proteins recognize phosphatidylserine on apoptotic cells and promote their engulfment. MFG-E8 is such a factor, and it is secreted by activated macrophages and immature dendritic cells (Hanayama *et al.*, 2002; Miyasaka *et al.*, 2004). MFG-E8 specifically binds to apoptotic cells by recognizing phosphatidylserine via its Factor-VIII-homologous domains in its C-terminal region. It also interacts with the $\alpha_v\beta_3$ or $\alpha_v\beta_5$ integrin on phagocytes through its RGD (Arg-Gly-Asp) motif, located in the second of two EGF domains in its N-terminal half (Yamaguchi *et al.*, 2008). The binding of MFG-E8 to the integrin recruits the CrkII/Dock180/ELMO complex to the plasma membrane at the nascent phagosome, which leads to the activation of Rac1 for the engulfment of apoptotic cells (Akakura *et al.*, 2004).

MFG-E8 is highly expressed in the germinal centers of the spleen and lymph nodes (Hanayama *et al.*, 2004), particularly by follicular dendritic cells and tingible-body macrophages in the germinal center (Kranich *et al.*, 2008). B lymphocytes are activated in the germinal centers, then they proliferated and differentiated into antibody-producing plasma cells. In this process, activated B lymphocytes that have low affinity for the antigen undergo apoptosis, and are swiftly removed by the tingible-body macrophages. In MFG-E8-deficient mice, many apoptotic B lymphocytes remain unengulfed on the tingible-body macrophages in the germinal centers (Hanayama *et al.*, 2004), indicating that MFG-E8 plays an essential role in the removal of the activated apoptotic B lymphocytes. On the other hand, the attachment of apoptotic lymphocytes to the MFG-E8-null tingible-body macrophages indicates that macrophages can recognize apoptotic cells without MFG-E8, supporting a two-step model: first tethering and then engulfment (Hoffmann *et al.*, 2001). MFG-E8 is involved in the engulfment step, and an unidentified ligand–receptor pair seems to work in the tethering step.

MFG-E8-deficient mice develop an age-dependent autoimmune disease resembling human SLE (Hanayama *et al.*, 2004). That is, MFG-E8-deficient mice, especially the females, produce high concentrations of anti-double-stranded deoxyribonucleic acid (DNA) and antinuclear antibody in their sera, and develop splenomegaly and glomerulonephritis by 40 weeks of age. In young mice, the autoantibodies are hardly detectable, but when these mice are immunized with keyhole limpet hemocyanin (KLH) to activate their B lymphocytes, they also produce antinuclear antibodies. These results confirm that if apoptotic cells, in this case B lymphocytes activated in the germinal centers, are not efficiently engulfed by phagocytes, the animals will develop autoimmune diseases (Baumann *et al.*, 2002).

The autoimmunity that develops in MFG-E8-null mice is probably caused by the intracellular components released from the dying cells. During apoptosis, plasma membranes are kept intact to prevent the release of cellular contents. However, when the apoptotic cells are left unengulfed, they undergo secondary necrosis by losing the integrity of their plasma membranes. Cellular components, such as nuclei, mitochondria, and DNA, thus released from the dead cells, stimulate quiescent autoreactive B lymphocytes to produce autoantibodies (**Figure 2**).

As expected from its function as a bridging molecule, MFG-E8 dose-dependently enhances the engulfment of apoptotic cells at low concentrations; however, it inhibits the engulfment at higher concentrations (Yamaguchi *et al.*, 2008). It seems that if MFG-E8 is present in excess, phosphatidylserine on apoptotic cells and integrins on phagocytes are occupied by different MFG-E8 molecules, preventing it from functioning as a bridge. Accordingly, the injection of recombinant MFG-E8 into mice causes SLE-type autoimmune diseases (Asano *et al.*, 2004). Thus, either a shortage or an excess of MFG-E8 leads to autoimmune disease, by preventing the engulfment of apoptotic cells. A sensitive

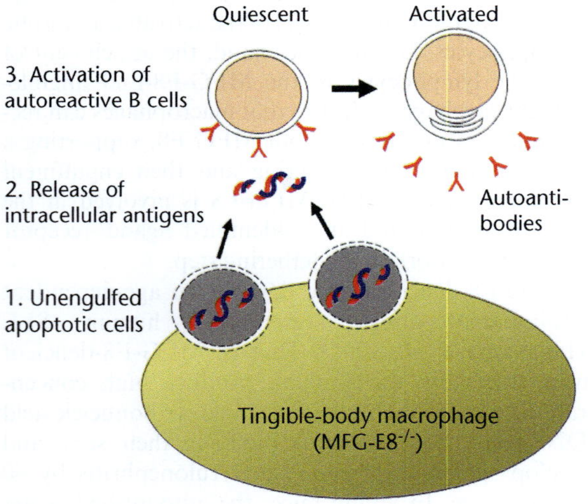

Figure 2 A model for the activation of autoimmune responses by the impaired engulfment of apoptotic cells. In germinal centers, apoptotic lymphocytes are swiftly engulfed by tingible-body macrophages which express MFG-E8. In MFG-E8-deficient mice, many apoptotic cells are left unengulfed, and undergo secondary necrosis, which allows the release of intracellular self-antigens such as DNA and chromatin. The released self-antigens may stimulate the proliferation and activation of autoreactive B lymphocytes, which are normally quiescent, to produce a large amount of autoantibodies. Reproduced from Hanayama R, Miyasaka K, Nakaya M and Nagata S (2006) MFG-E8-dependent clearance of apoptotic cells, and autoimmunity caused by its failure. *Current Directions in Autoimmunity* **9**: 162–172. Reproduced with permission from S. Karger AG, Basel.

enzyme-linked immunosorbent assay (ELISA) system for human MFG-E8 showed that some human SLE patients have a high serum level of MFG-E8 (Yamaguchi *et al.*, 2008). Although it is not clear whether the high level of MFG-E8 is the cause of the disease or a secondary effect of the disease, its high level would certainly make the disease worse.

In addition to the tingible-body macrophages in the spleen and thioglycollate-elicited peritoneal macrophages, MFG-E8 is expressed in various other tissues. Its expression in microglia is upregulated by CX3CL1/fractalkine secreted by apoptotic neurons, and may be involved in removing dying neurons from ischemic brain tissue (Fuller and Van Eldik, 2008). Langerhans cells, a subset of immature dendritic cells in the skin, also express a high level of MFG-E8, but its biological function in this context is unknown (Miyasaka *et al.*, 2004). MFG-E8 is expressed by macrophages that infiltrate into atherosclerotic lesions, and the deficiency of MFG-E8 accelerates atherosclerosis, suggesting a role of MFG-E8 in removing apoptotic cells generated in atherosclerotic lesions (Ait-Oufella *et al.*, 2007). As its name suggests, MFG-E8 is associated with milk fat globules (MFG), and it plays an essential role in the involution of mammary gland tissue (Hanayama and Nagata, 2005).

Other bridging molecules that interact with phosphatidylserine on apoptotic cells are two related plasma proteins, growth arrest-specific 6 (Gas6) and protein S. Both proteins contain a Gla domain in their N-terminal region, which has multiple glutamic acid residues that are γ-carboxylated in a vitamin K-dependent manner. Protein S is a vitamin K-dependent negative regulator in the blood coagulation system, and Gas6 may also work in the clotting system. In their C-terminal region, Gas6 and protein S have an SHBG (sex hormone-binding globulin) domain that is responsible for binding to TAM receptor family members (Tyro3, Axl, and Mer), which have a tyrosine kinase domain in their cytoplasmic region. Since Gas6 binds phosphatidylserine, and since mice expressing a kinase-dead mutant of Mer (Mer^{kD}) develop a SLE-like autoimmunity (Scott *et al.*, 2001), a role for the Gas6–TAM system in the engulfment of apoptotic cells, particularly in the testis and retina, has been proposed (Lemke and Rothlin, 2008). However, *Gas6*$^{-/-}$ mice show platelet dysfunction (Angelillo-Scherrer *et al.*, 2001), but do not develop autoimmune disease. A recent publication indicates that TAM receptors negatively regulate the innate immune reaction and that dendritic cells lacking TAM receptors overproduce various cytokines including interleukin (IL)-6, interferon α (IFN-α), and tumor necrosis factor α (TNFα) (Rothlin *et al.*, 2007). IFN-α and TNFα are known to regulate SLE-type autoimmune disease (Pascual *et al.*, 2008). Therefore, the SLE-like autoimmunity found in Mer-deficient mice (Scott *et al.*, 2001) may be explained by their deregulated expression of cytokines.

Most apoptotic cells are cleared away rapidly during early apoptosis. However, if apoptotic cells persist longer in tissues, either because of impaired engulfment or because the number of apoptotic cells overwhelm the phagocytotic process, the complement system may function to clear them (Trouw *et al.*, 2008). C1q is the most studied complement component that binds dead and necrotic cells in the later stages of apoptosis. Although the involvement of IgM in this process was reported (Ogden *et al.*, 2001), the ligand for C1q on dead cells is unknown. Notably, almost all individuals deficient in C1q develop severe SLE (Botto and Walport, 2002). In C1q-deficient mice, unengulfed dead cells persist in the glomeruli of the kidneys, and the mice develop a SLE-like phenotype; this is particularly evident in lupus-prone mouse strains such as MRL/MP, which have a defect in the clearance of apoptotic cells (Potter *et al.*, 2003). These findings suggest that Clq-mediated engulfment is a backup system for the clearance of dead cells. In this regard, it may be informative to cross to $C1q^{-/-}$ mice with MFG-E8-deficient mice, which have a defect in the engulfment of apoptotic cells.

Phosphatidylserine Receptors

Whether or not phagocytes directly recognize phosphatidylserine on apoptotic cells via surface receptors has been hard to ascertain. In 2000, Fadok *et al.* established a monoclonal antibody that inhibits the engulfment of apoptotic cells by human macrophages (Fadok *et al.*, 2000). Using a phage-display cloning technique, they then identified a complementary DNA (cDNA), carrying a sequence that could be recognized by the antibody, and designated it as the phosphatidylserine receptor (PSR). Subsequently, several groups reported that a "PSR" deficiency impairs the engulfment of apoptotic cells in mice and *C. elegans* (Kunisaki *et al.*, 2004; Li *et al.*, 2003; Wang *et al.*, 2003). However, it turned out that this molecule is located in the nucleus and contains a Jumonji C domain, which is known to have a role in chromatin remodeling. More careful analyses of PSR knockout mice showed that the monoclonal antibody used to identify this molecule doesn't recognize "PSR," and that the lethality of the PSR-null mutation in embryos is not due to a defect in the engulfment of apoptotic cells (Bose *et al.*, 2004). PSR was recently shown to be identical with JMJD6 (Jumonji domain-containing 6 protein), which bears a histone demethylase (Chang *et al.*, 2007). It seems that the increased number of "unengulfed TUNEL-positive apoptotic cells" in the PSR-null animals was due to increased cell death caused by the lack of JMJD6's chromatin remodeling function.

As described above, MFG-E8 is expressed by thioglycollate-elicited peritoneal macrophages, but not by the resident peritoneal macrophages, although resident peritoneal macrophages efficiently engulf apoptotic cells. By screening a library of monoclonal antibodies against mouse resident peritoneal macrophages, Miyanishi *et al.* (2007) identified an antibody (Kat5-18) that strongly inhibits the engulfment of apoptotic cells. Expression cloning of its antigen revealed that Kat5-18 recognizes a type I membrane protein called T-cell immunoglobulin- and mucin-domain-containing molecule 4 (Tim-4). Tim-4 consists of a signal sequence, immunoglobulin (IgV) domain, mucin-like domain, transmembrane domain, and cytoplasmic region.

Tim-4 is expressed by macrophages and dendritic cells in tissues that include the spleen, lymph nodes, thymus, and tonsils. Tim-4 specifically binds to phosphatidylserine on apoptotic cells with a dissociation constant (Kd) of about 2 nM via its extracellular IgV domain, and promotes the uptake of dead cells. Tim-4 belongs to the Tim family, which has eight members in mouse (Tim-1 through Tim-8). Among the Tim family members examined (Tim-1 through Tim-4), Tim-1 and Tim-4 bind phosphatidylserine. A crystal structure analysis revealed that both Tim-1 and Tim-4, but not Tim-2 or Tim-3, have a characteristic narrow cavity in the IgV domain where phosphatidylserine can interact in a metal ion-dependent manner (Santiago *et al.*, 2007). The short cytoplasmic region of Tim-4 has no obvious signaling motif, and is not necessary to enhance the engulfment of apoptotic cells (Park *et al.*, 2009), indicating that Tim-4 interacts with a signaling partner to perform its task. The inhibition of Tim-4 function in mice by the administration of Kat5-18 impairs the engulfment of apoptotic cells, leading to the production of autoantibodies. Whether this finding is related to the fact that the gene locus for the Tim family is within the region of susceptibility genes for the development of atopy and asthma remains to be studied.

The expression profile of Tim-4 is different from that of MFG-E8 (Miyanishi *et al.*, 2007). That is, resident peritoneal macrophages express Tim-4 but not MFG-E8, and thioglycollate-elicited peritoneal macrophages and bone marrow-derived dendritic cells express MFG-E8, but not Tim-4. It seems that Tim-4 is expressed in resident macrophages that function in the development and maintenance of tissues; in contrast, MFG-E8 is expressed in macrophages that function during inflammation. The heterogeneity of macrophages and the existence of distinct macrophage subsets have been recognized (Gordon and Taylor, 2005); the expression of MFG-E8 and Tim-4 by different macrophage populations supports this concept. It will be interesting to extend the study of which subsets of macrophages produce Tim-4 and MFG-E8, and what determines this difference in expression.

In addition to Tim-4/Tim-1, two other molecules, brain-specific angiogenesis inhibitor 1 (BAI1) and stabilin-2, have been identified as phosphatidylserine receptors. In a yeast two-hybrid screen with ELMO as the bait, Ravichandran and colleagues identified BAI1 as a potential phosphatidylserine receptor (Park *et al.*, 2007). BAI1 belongs to the G-protein coupled receptor family and has five thrombospondin type 1 repeats (TSRs) in its extracellular region, a seven-transmembrane region, and a 392-amino acid cytoplasmic tail that interacts with ELMO. It is believed to bind phosphatidylserine on apoptotic cells via its TSRs and to promote the engulfment of apoptotic cells. However, BAI1 also binds to cardiolipin and other phospholipids, and its ability to enhance the engulfment of apoptotic cells may not be very strong (Park *et al.*, 2009). To confirm the role of BAI1 in the engulfment of apoptotic cells, it will be necessary to establish a BAI1-deficient mouse line.

Stabilin-2 (Park *et al.*, 2008) is a multifunctional receptor containing a large extracellular domain that consists of seven FAS1 domains and four EGF-like repeats; it is also called HARE (hyaluronic acid receptor for endocytosis). It is expressed by the sinusoidal endothelial cells of the spleen, lymph nodes, and bone marrow. Stabilin-2 binds tightly to hyaluronic acids and heparin, and its role as a clearance receptor for hyaluronic acid and heparin is well established (Harris *et al.*, 2008). Stabilin-2 was reported to bind phosphatidylserine via its extracellular EGF-like repeats (particularly the second one) in a calcium-dependent manner. Whether it plays an essential role in the engulfment of apoptotic cells *in vivo* remains to be confirmed.

Immunosuppressive Roles of Engulfment

The binding or uptake of apoptotic cells by phagocytes induces the production of anti-inflammatory cytokines, such as transforming growth factor β (TGFβ) and IL-10, *in vitro* (Erwig and Henson, 2007; Fadok *et al.*, 1998). These anti-inflammatory cytokines inhibit the lipopolysaccharide-induced expression of TNFα. This TGFβ production can be blocked by masking the phosphatidylserine on apoptotic cells, indicating that the signal that upregulates the TGFβ expression is transduced by the receptors for phosphatidylserine. In the spleen of MFG-E8-deficient mice, the production of IL-10 decreases, and the IFN-γ level increases, suggesting a switch of the immune response to the Th1 proinflammatory phenotype in the absence of MFG-E8 (Ait-Oufella *et al.*, 2007). Learning how the engulfment of apoptotic cells leads to the production of anti-inflammatory cytokines such as IL-10 and TGFβ is a challenge for future research. **See also**: Caspases in Inflammation and Immunity

Conclusions and Future Prospects

Recent advances in this field suggest that phagocytes recognize the phosphatidylserine on apoptotic cells by multiple systems. The molecular mechanisms of apoptosis and the clearance of apoptotic cells have been extensively studied in *C. elegans*, where phosphatidylserine also seems to work as an "eat-me" signal. However, *C. elegans* does not encode an ortholog of any of the phosphatidylserine-binding molecules, and therefore, how phosphatidylserine is recognized in *C. elegans* remains unclear. Given that Tim-4 and MFG-E8 are expressed by macrophages and immature dendritic cells that are professional phagocytes, and that *C. elegans* has no professional phagocytes, this difference in cell clearance mechanism may explain why *C. elegans* has no ortholog for MFG-E8 or Tim-4.

Recently, several reports raised the possibility that phosphatidylserine serves as an "eat-me" signal not only for the engulfment of apoptotic cells, but also for other processes. During erythropoiesis, a large number of nuclei are expelled from erythroid precursor cells. The nuclei released from erythroblasts are surrounded by a plasma membrane, and the asymmetry of the plasma membrane is lost shortly after the separation from the reticulocyte, due to a lack of ATP, resulting in the exposure of phosphatidylserine (Yoshida *et al.*,

2005). These nuclei are recognized by unidentified receptor, and engulfed by macrophages in a phosphatidylserine-dependent manner. Another example is the axon pruning that takes place during the development of neural circuits. In this process, studied in *Drosophila*, unnecessary or extra axons degenerate, expose phosphatidylserine, and are engulfed by glia in a CED-6-dependent manner (Awasaki *et al.*, 2006). However, how phagocytes recognize phosphatidylserine on the pruned axons also remains to be elucidated.

The engulfment of dead cells is mainly to degrade their intracellular materials before the cellular contents are released and can activate the immune system. However, in some cases, phagocytes actively induce programmed cell death. For example, macrophages within the developing eye induce the programmed cell death of the vascular endothelial cells via the Wnt ligand (Lobov *et al.*, 2005), and the phagocyte-induced programmed cell death serves as a backup death-inducing system in *C. elegans* (Hoeppner *et al.*, 2001; Reddien *et al.*, 2001). How macrophages come to engulf apparently healthy cells is puzzling. The elucidation of this mechanism may help explain hemophagocytic syndrome in which activated macrophages engulf apparently normal red blood cells and other cells. In any event, future studies on the molecular mechanisms underlying the engulfment of apoptotic cells will help us understand the pathophysiology of various human diseases, especially autoimmune diseases, and will lead to the development of new therapeutic strategies for treating them.

References

Ait-Oufella H, Kinugawa K, Zoll J *et al.* (2007) Lactadherin deficiency leads to apoptotic cell accumulation and accelerated atherosclerosis in mice. *Circulation* **115**: 2168–2177.

Akakura S, Singh S, Spataro M *et al.* (2004) The opsonin MFG-E8 is a ligand for the alphavbeta5 integrin and triggers DOCK180-dependent Rac1 activation for the phagocytosis of apoptotic cells. *Experimental Cell Research* **292**: 403–416.

Angelillo-Scherrer A, de Frutos P, Aparicio C *et al.* (2001) Deficiency or inhibition of Gas6 causes platelet dysfunction and protects mice against thrombosis. *Nature Medicine* **7**: 215–221.

Asano K, Miwa M, Miwa K *et al.* (2004) Masking of phosphatidylserine inhibits apoptotic cell engulfment and induces autoantibody production in mice. *Journal of Experimental Medicine* **200**: 459–467.

Awasaki T, Tatsumi R, Takahashi K *et al.* (2006) Essential role of the apoptotic cell engulfment genes draper and ced-6 in programmed axon pruning during *Drosophila* metamorphosis. *Neuron* **50**: 855–867.

Balasubramanian K and Schroit AJ (2003) Aminophospholipid asymmetry: a matter of life and death. *Annual Review of Physiology* **65**: 701–734.

Baumann I, Kolowos W, Voll RE *et al.* (2002) Impaired uptake of apoptotic cells into tingible body macrophages in germinal centers of patients with systemic lupus erythematosus. *Arthritis and Rheumatism* **46**: 191–201.

Bose J, Gruber AD, Helming L *et al.* (2004) The phosphatidylserine receptor has essential functions during embryogenesis but not in apoptotic cell removal. *Journal of Biology* **3**: 15.

Botto M and Walport MJ (2002) C1q, autoimmunity and apoptosis. *Immunobiology* **205**: 395–406.

Chang B, Chen Y, Zhao Y and Bruick RK (2007) JMJD6 is a histone arginine demethylase. *Science* **318**: 444–447.

Cote JF and Vuori K (2007) GEF what? Dock180 and related proteins help Rac to polarize cells in new ways. *Trends in Cell Biology* **17**: 383–393.

Darland-Ransom M, Wang X, Sun CL *et al.* (2008) Role of *C. elegans* TAT-1 protein in maintaining plasma membrane phosphatidylserine asymmetry. *Science* **320**: 528–531.

Erwig LP and Henson PM (2007) Immunological consequences of apoptotic cell phagocytosis. *American Journal of Pathology* **171**: 2–8.

Fadok VA, Bratton DL, Konowal A *et al.* (1998) Macrophages that have ingested apoptotic cells *in vitro* inhibit proinflammatory cytokine production through autocrine/paracrine mechanisms involving TGF-beta, PGE2, and PAF. *Journal of Clinical Investigation* **101**: 890–898.

Fadok VA, Bratton DL, Rose DM *et al.* (2000) A receptor for phosphatidylserine-specific clearance of apoptotic cells. *Nature* **405**: 85–90.

Fadok VA, Voelker DR, Campbell PA *et al.* (1992) Exposure of phosphatidylserine on the surface of apoptotic lymphocytes triggers specific recognition and removal by macrophages. *Journal of Immunology* **148**: 2207–2216.

Fuller AD and Van Eldik LJ (2008) MFG-E8 regulates microglial phagocytosis of apoptotic neurons. *Journal of Neuroimmune Pharmacology* **3**: 246–256.

Gordon S and Taylor PR (2005) Monocyte and macrophage heterogeneity. *Nature Reviews. Immunology* **5**: 953–964.

Grimsley C and Ravichandran KS (2003) Cues for apoptotic cell engulfment: eat-me, don't eat-me and come-get-me signals. *Trends in Cell Biology* **13**: 648–656.

Hamon Y, Trompier D, Ma Z *et al.* (2006) Cooperation between engulfment receptors: the case of ABCA1 and MEGF10. *PLoS ONE* **1**: e120.

Hanayama R and Nagata S (2005) Impaired involution of mammary glands in the absence of milk fat globule EGF factor 8. *Proceedings of the National Academy of Sciences of the United States of America* **102**: 16886–16891.

Hanayama R, Tanaka M, Miwa K *et al.* (2002) Identification of a factor that links apoptotic cells to phagocytes. *Nature* **417**: 182–187.

Hanayama R, Tanaka M, Miyasaka K *et al.* (2004) Autoimmune disease and impaired uptake of apoptotic cells in MFG-E8-deficient mice. *Science* **304**: 1147–1150.

Harris EN, Weigel JA and Weigel PH (2008) The human hyaluronan receptor for endocytosis (HARE/Stabilin-2) is a systemic clearance receptor for heparin. *Journal of Biological Chemistry* **283**: 17341–17350.

Hoeppner DJ, Hengartner MO and Schnabel R (2001) Engulfment genes cooperate with ced-3 to promote cell death in *Caenorhabditis elegans*. *Nature* **412**: 202–206.

Hoffmann PR, deCathelineau AM, Ogden CA *et al.* (2001) Phosphatidylserine (PS) induces PS receptor-mediated macropinocytosis and promotes clearance of apoptotic cells. *Journal of Cell Biology* **155**: 649–659.

Katoh H and Negishi M (2003) RhoG activates Rac1 by direct interaction with the Dock180-binding protein Elmo. *Nature* **424**: 461–464.

Kitano M, Nakaya M, Nakamura T, Nagata S and Matsuda M (2008) Imaging of Rab5 activity identifies essential regulators for phagosome maturation. *Nature* **453**: 241–245.

Kranich J, Krautler NJ, Heinen E *et al.* (2008) Follicular dendritic cells control engulfment of apoptotic bodies by secreting Mfge8. *Journal of Experimental Medicine* **205**: 1293–1302.

Kunisaki Y, Masuko S, Noda M *et al.* (2004) Defective fetal liver erythropoiesis and T lymphopoiesis in mice lacking the phosphatidylserine receptor. *Blood* **103**: 3362–3364.

Lauber K, Bohn E, Krober SM *et al.* (2003) Apoptotic cells induce migration of phagocytes via caspase-3-mediated release of a lipid attraction signal. *Cell* **113**: 717–730.

Lemke G and Rothlin CV (2008) Immunobiology of the TAM receptors. *Nature Reviews. Immunology* **8**: 327–336.

Li MO, Sarkisian MR, Mehal WZ *et al.* (2003) Phosphatidylserine receptor is required for clearance of apoptotic cells. *Science* **302**: 1560–1563.

Lobov IB, Rao S, Carroll TJ *et al.* (2005) WNT7b mediates macrophage-induced programmed cell death in patterning of the vasculature. *Nature* **437**: 417–421.

Miyanishi M, Tada K, Koike M *et al.* (2007) Identification of Tim4 as a phosphatidylserine receptor. *Nature* **450**: 435–439.

Miyasaka K, Hanayama R, Tanaka M and Nagata S (2004) Expression of milk fat globule epidermal growth factor 8 in immature dendritic cells for engulfment of apoptotic cells. *European Journal of Immunology* **34**: 1414–1422.

Nakaya M, Kitano M, Matsuda M and Nagata S (2008) Spatiotemporal activation of Rac1 for engulfment of apoptotic cells. *Proceedings of the National Academy of Sciences of the United States of America* **105**: 9198–9203.

Nakaya M, Tanaka M, Okabe Y, Hanayama R and Nagata S (2006) Opposite effects of rho family GTPases on engulfment of apoptotic cells by macrophages. *Journal of Biological Chemistry* **281**: 8836–8842.

Ogden CA, deCathelineau A, Hoffmann PR *et al.* (2001) C1q and mannose binding lectin engagement of cell surface calreticulin and CD91 initiates macropinocytosis and uptake of apoptotic cells. *Journal of Experimental Medicine* **194**: 781–795.

Park D, Hochreiter-Hufford A and Ravichandran KS (2009) The phosphatidylserine receptor TIM-4 does not mediate direct signaling. *Current Biology* **19**: 346–351.

Park D, Tosello-Trampont AC, Elliott MR *et al.* (2007) BAI1 is an engulfment receptor for apoptotic cells upstream of the ELMO/Dock180/Rac module. *Nature* **450**: 430–434.

Park SY, Jung MY, Kim HJ *et al.* (2008) Rapid cell corpse clearance by stabilin-2, a membrane phosphatidylserine receptor. *Cell Death and Differentiation* **15**: 192–201.

Pascual V, Allantaz F, Patel P *et al.* (2008) How the study of children with rheumatic diseases identified interferon-alpha and interleukin-1 as novel therapeutic targets. *Immunological Reviews* **223**: 39–59.

Peter C, Waibel M, Radu CG *et al.* (2008) Migration to apoptotic "find-me" signals is mediated via the phagocyte receptor G2A. *Journal of Biological Chemistry* **283**: 5296–5305.

Potter PK, Cortes-Hernandez J, Quartier P, Botto M and Walport MJ (2003) Lupus-prone mice have an abnormal response to thioglycolate and an impaired clearance of apoptotic cells. *Journal of Immunology* **170**: 3223–3232.

Reddien PW, Cameron S and Horvitz HR (2001) Phagocytosis promotes programmed cell death in *C. elegans*. *Nature* **412**: 198–202.

Reddien PW and Horvitz HR (2004) The engulfment process of programmed cell death in *Caenorhabditis elegans*. *Annual Review of Cell and Development Biology* **20**: 193–221.

Rothlin C, Ghosh S, Zuniga E *et al.* (2007) TAM receptors are pleiotropic inhibitors of the innate immune response. *Cell* **131**: 1124–1136.

Sahu SK, Gummadi SN, Manoj N *et al.* (2007) Phospholipid scramblases: an overview. *Archives of Biochemistry and Biophysics* **462**: 103–114.

Santiago C, Ballesteros A, Martinez-Munoz L *et al.* (2007) Structures of T cell immunoglobulin mucin protein 4 show a metal-ion-dependent ligand binding site where phosphatidylserine binds. *Immunity* **27**: 941–951.

Scott RS, McMahon EJ, Pop SM *et al.* (2001) Phagocytosis and clearance of apoptotic cells is mediated by MER. *Nature* **411**: 207–211.

Su HP, Nakada-Tsukui K, Tosello-Trampont AC *et al.* (2002) Interaction of CED-6/GULP, an adapter protein involved in engulfment of apoptotic cells with CED-1 and CD91/low density lipoprotein receptor-related protein (LRP). *Journal of Biological Chemistry* **277**: 11772–11779.

Trouw LA, Blom AM and Gasque P (2008) Role of complement and complement regulators in the removal of apoptotic cells. *Molecular Immunology* **45**: 1199–1207.

Truman LA, Ford CA, Pasikowska M *et al.* (2008) CX3CL1/fractalkine is released from apoptotic lymphocytes to stimulate macrophage chemotaxis. *Blood* **112**: 5026–5036.

Venegas V and Zhou Z (2007) Two alternative mechanisms that regulate the presentation of apoptotic cell engulfment signal in *Caenorhabditis elegans*. *Molecular Biology of the Cell* **18**: 3180–3192.

Wang X, Wu YC, Fadok VA *et al.* (2003) Cell corpse engulfment mediated by *C. elegans* phosphatidylserine receptor through CED-5 and CED-12. *Science* **302**: 1563–1566.

Wiedmer T, Zhao J, Li L *et al.* (2004) Adiposity, dyslipidemia, and insulin resistance in mice with targeted deletion of phospholipid scramblase 3 (PLSCR3). *Proceedings of the National Academy of Sciences of the United States of America* **101**: 13296–13301.

Wu YC and Horvitz HR (1998) The *C. elegans* cell corpse engulfment gene ced-7 encodes a protein similar to ABC transporters. *Cell* **93**: 951–960.

Yamaguchi H, Takagi J, Miyamae T *et al.* (2008) Milk fat globule EGF factor 8 in the serum of human patients of systemic lupus erythematosus. *Journal of Leukocite Biology* **83**: 1300–1307.

Yoshida H, Kawane K, Koike M *et al.* (2005) Phosphatidylserine-dependent engulfment by macrophages of nuclei from erythroid precursor cells. *Nature* **437**: 754–758.

Zhou Q, Ben-Efraim I, Bigcas JL *et al.* (2005) Phospholipid scramblase 1 binds to the promoter region of the inositol 1,4,5-triphosphate receptor type 1 gene to enhance its expression. *Journal of Biological Chemistry* **280**: 35062–35068.

Zhou Q, Zhao J, Wiedmer T and Sims PJ (2002) Normal hemostasis but defective hematopoietic response to growth factors in mice deficient in phospholipid scramblase 1. *Blood* **99**: 4030–4038.

Zullig S, Neukomm LJ, Jovanovic M *et al.* (2007) Aminophospholipid translocase TAT-1 promotes phosphatidylserine exposure during *C. elegans* apoptosis. *Current Biology* **17**: 994–999.

Further Reading

Erwig LP and Henson PM (2008) Clearance of apoptotic cells by phagocytes. *Cell Death Differentiation* **15**: 243–250.

Gaipl US, Kuhn A, Sheriff A *et al.* (2006) Clearance of apoptotic cells in human SLE. *Current Directions in Autoimmunity* **9**: 173–187.

Hanayama R, Miyasaka K, Nakaya M and Nagata S (2006) MFG-E8-dependent clearance of apoptotic cells, and autoimmunity caused by its failure. *Current Directions in Autoimmunity* **9**: 162–172.

Nagata S (1997) Apoptosis by death factor. *Cell* **88**: 355–365.

Nagata S (2005) DNA degradation in development and programmed cell death. *Annual Review of Immunology* **23**: 853–875.

Nagata S (2007) Autoimmune diseases caused by defects in clearing dead cells and nuclei expelled from erythroid precursors. *Immunological Reviews* **220**: 237–250.

Ravichandran KS and Lorenz U (2007) Engulfment of apoptotic cells: signals for a good meal. *Nature Reviews. Immunology* **7**: 964–974.

Serhan CN and Savill J (2005) Resolution of inflammation: the beginning programs the end. *Nature Immunology* **6**: 1191–1197.

Autophagy

María Isabel Colombo, *IHEM-CONICET-Facultad de Ciencias Médicas, Universidad Nacional de Cuyo, Mendoza, Argentina*

Hans-Uwe Simon, *Institute of Pharmacology, University of Bern, Bern, Switzerland*

Advanced article

Article Contents

- General Overview: Introduction to Autophagy
- Molecular Dissection of Autophagy
- Regulation of the Autophagic Pathway
- Autophagy in Pathological Processes
- Autophagy and Cell Death
- Concluding Remarks

Autophagy is a conserved proteolytic mechanism that degrades cytoplasmic material including cell organelles. Although the importance of autophagy for cell homeostasis and survival has long been appreciated, our understanding of how autophagy is regulated at a molecular level just recently evolved. The importance of autophagy for the quality control of proteins is underscored by the fact that many neurodegenerative and myodegenerative diseases are characterized by an increased but still insufficient autophagic activity.

Similarly, if the cellular stress, leading to deoxyribonucleic acid (DNA) damage, mitochondrial damage and/or damaged proteins, does not result in sufficient autophagic repair mechanisms, cells seem to be prone to transform into tumour cells. Therefore, autophagy has multiple roles to play in the causation and prevention of human diseases.

General Overview: Introduction to Autophagy

As a cellular response to starvation and other stress situations autophagy is induced. This process delivers intracellular proteins and organelles sequestered by double-membrane vesicles, called autophagosomes, to lysosomes for degradation.

There is increased evidence for the key role of autophagy as a membrane trafficking system that delivers cytoplasmic components into the lysosomal compartment for bulk protein degradation not only as a survival mechanism but also as a protein quality-control mechanism that removes proteins with aberrant structures (Mizushima, 2005). In addition, autophagy has been involved in many physiological and pathological processes, such as antigen processing, innate immune defence against invading micro-organisms. Indeed autophagy is considered one of the most critical barriers of the intracellular host cell defence machinery that bacteria must face on cell invasion. Furthermore, impairment of autophagy has been implicated in the pathogenesis of neurodegenerative disorders, cancer and lysosomal storage diseases.

Although autophagy is considered a pro-survival mechanism, excess autophagy may induce the so-called type II programmed cell death, a type of programmed cell death that differs from apoptosis. Thus, the study of the molecular basis of autophagy would allow us to better understand the role of autophagy not only on survival but also in ageing and cell death, and would likely result in proposals for new therapeutic approaches for several diseases. Here, we will present a general overview of the molecular mechanism and regulation of autophagy, as well as the role of this process in physiopathological situations. **See also**: Apoptosis: Regulatory Genes and Disease; Autophagy in Non-Mammalian Systems; The Siren's Song: This Death that Makes Life Live

Molecular Dissection of Autophagy

Within the last decade a complete set of critical genes for autophagy has been discovered both in yeast and in mammalian cells. These findings in conjunction with the signalling pathways that regulate autophagy have greatly improved the molecular understanding and current interest in this fundamental process. Although we still do not completely understand the autophagic process at the molecular level, the role of several of these genes has been described in detail in specific events of this pathway. The family of genes or proteins called Atg (for autophagy related) (Klionsky *et al.*, 2003) constitutes the core of the molecular machinery of autophagy. They were originally identified in yeast, but now several Atg orthologues have been discovered in mammalian cells, suggesting that the basic machinery for autophagy is evolutionarily conserved. Numerous molecules have been identified to participate in the autophagic pathway. However, for brevity, we have not included in this article a comprehensive analysis of all the molecules involved. **See also**: Autophagy in Non-Mammalian Systems

Two ubiquitin-like conjugations systems

During autophagy, a preautophagosomal cup-shaped structure, known as phagophore or isolation membrane, which engulfs organelles and other cytoplasmic components, closes forming an autophagosome (Mizushima *et al.*, 2002). Subsequently, this compartment fuses with endo/lysosomal vesicles, leading to the proteolytic degradation of the sequestered material by lysosomal lytic enzymes (**Figure 1**; Eskelinen, 2005).

For the initial formation of the sequestering membrane, two ubiquitin-like modifications are required: the Atg12–Atg5 conjugation system (Mizushima *et al.*, 1998), in which the small protein Atg12 is covalently linked to Atg5, and the Atg8-phosphatidylethanolamine (PE) complex formation (Ichimura *et al.*, 2000). Even though Atg12 does not have an evident homology to ubiquitin, the conjugation reaction of Atg12 and Atg5 is similar to that of ubiquitination. Atg12 is first activated by Atg7 who works as an E1 enzyme leading to the formation of a thioester bond between both proteins. Subsequently, Atg12 is transferred to Atg10 (the E2-like enzyme) forming a new thioester bond and then, the Atg12 protein is attached via its *C*-terminal glycine with the amino group of a lysine in Atg5 (Shintani *et al.*, 1999) in an irreversible manner (please see inset in **Figure 1**). Atg16 (Atg16L in mammalian cells) binds to the Atg12–Atg5 complex via its

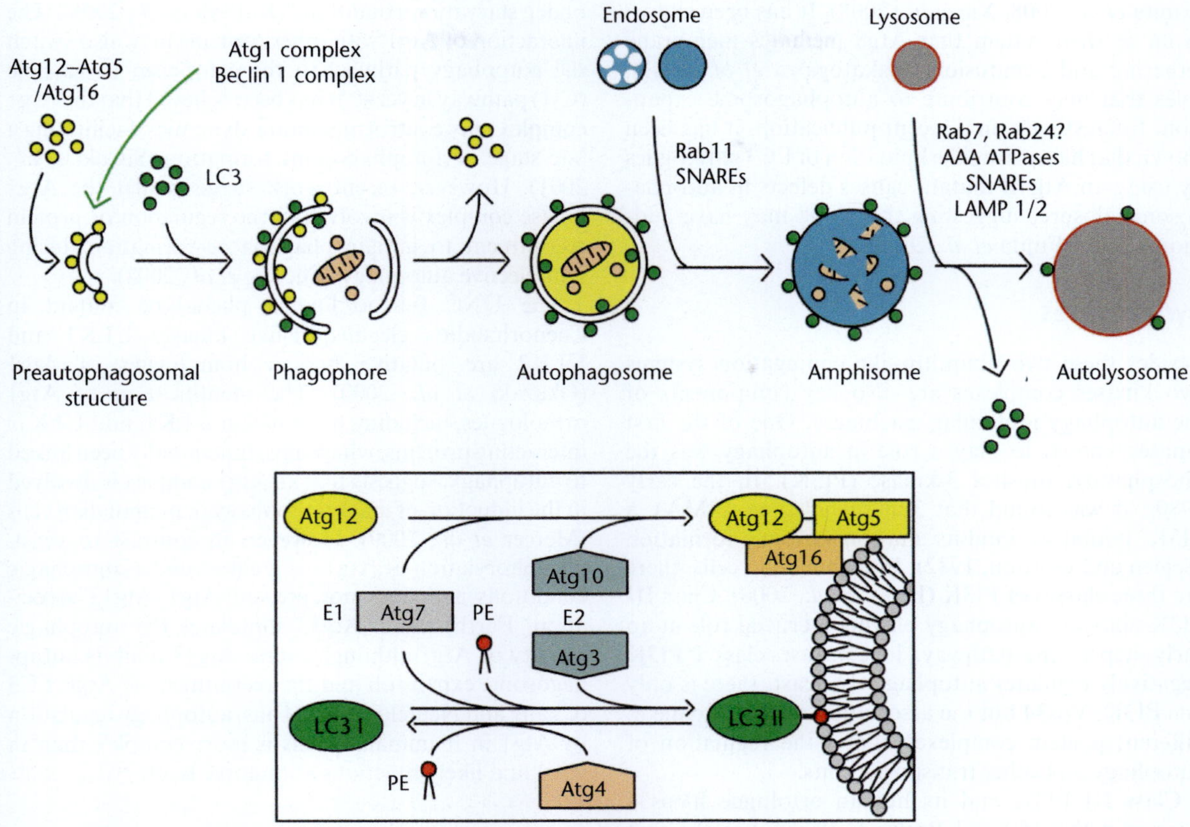

Figure 1 Major molecular components of the autophagy pathway. On activation a pre-autophagosomal structure is initially formed generating the phagophore or isolation membrane, which engulfs organelles and other cytoplasmic components. This structure growths and closes, forming the autophagosome, which, in turn, fuses with endocytic vesicles forming the so-called amphisome. Finally, fusion with the lysosomes takes place and the sequestered material is digested by the lysosomal enzymes. Key components of the molecular machinery involved in each step of the autophagy are depicted. Inset: the two ubiquitin-like conjugation systems that lead to the binding of LC3 to the autophagosomal membrane.

C-terminal coiled-coil domain and the whole complex transiently associates to the cup-shaped isolation membrane (Mizushima *et al.*, 2003). This complex is required for the association of the second conjugation system but once the autophagosome is formed Atg12–Atg5 conjugate dissociates.

As aforementioned, the second ubiquitin-like conjugation system is the Atg8/LC3–PE complex. LC3 (microtubule-associated protein 1 light chain 3) is the mammalian orthologue of the yeast Atg8 and the best-characterized marker for autophagosomes. Atg8 is processed by the cysteine proteinase Atg4 that cleaves the *C*-terminal Arg exposing a Gly which is activated by Atg7 and then Atg8 is transferred to the E2-lke enzyme Atg3 (Ichimura *et al.*, 2000). Finally, Atg8 is conjugated to the phospholipids PE that allows the attachment of the protein to the autophagosomal membrane. The same Atg4 enzyme deconjugates the Atg8–PE complex and Atg8 cycles back to the cytoplasm (see inset in **Figure 1**). Similarly, newly synthesized LC3 is cleaved by an Atg4 cysteine proteinase (i.e. autophagin) leading to the generation of the cytosolic LC3-I form (18 kDa). This form is subjected to a couple of ubiquitination-like reactions to produce the membrane bound protein LC3-II (16 kDa) that attaches to both the inner and outer sides of the forming autophagosome (Kabeya *et al.*, 2000).

It is important to mention that the Atg16L complex is required for the localization and recruitment of LC3 before its lipidation (Fujita *et al.*, 2008). This reaction is believed to take place at the surface of the isolation membrane. The precise function of the Atg8/LC3 remains unknown although evidence indicates that it might be related to determining the size of autophagosomes. Indeed, reduction of the expression levels of Atg8 leads to the formation of smaller autophagosomes

(Fujita *et al.*, 2008; Xie *et al.*, 2008). It has been shown in an *in vitro* system that Atg8 mediates membrane tethering and hemifusion (Nakatogawa *et al.*, 2007), roles that may contribute to autophagosome expansion. Interestingly, in a recent publication, it has been shown that hampering the lipidation of LC3 paralogues by using an Atg4B mutant causes defects in autophagosome closure, suggesting that Atg8 may have additional roles (Fujita *et al.*, 2008).

Two kinases

Besides these two ubiquitin-like conjugation systems two kinases complexes are also key components of the autophagy molecular machinery. One of the first kinases known to play a role in autophagy was the phosphatidyl inositol 3-kinase (PI3K). In the early 1980s, it was found that 3-methyladenine (3-MA), a PI3K inhibitor, inhibits autophagosome formation (Seglen and Gordon, 1982). In mammalian cells, there are three classes of PI3K (Petiot *et al.*, 2000). Class III PI3K activates autophagy playing a crucial role at an early step in the pathway. In contrast, class I PI3K negatively regulates autophagy. In yeast, there is only one PI3K, Vps34 but the association of this enzyme in different protein complexes allows the regulation of autophagy and other transport events.

Class III PI3K and its human ortologue hVps34 interact with p150 and Beclin 1. Beclin 1 is the first autophagy-related tumour suppressor gene which was originally isolated as a Bcl-2 interacting protein (Liang *et al.*, 1998). Similar to yeast, in mammalian cells it has been recently identified two distinct class III PI3K complexes, which likely function in different membrane trafficking pathways (Itakura *et al.*, 2008). Atg14 participates in one of the complexes, and in the other the so-called UV irradiation resistance-associated gene (UVRAG) is present. Though Atg14 is located on autophagic isolation membranes and is essential for autophagy, UVRAG is mainly present in endosomal structures.

The second key kinase is the serine/threonine kinase Atg1, which forms a complex with other regulatory proteins. It has been shown that the composition of this complex varies depending on nutrient conditions in yeast. Under starvation conditions, Atg13 is partially dephosphorylated and interacts with Atg1 leading to autophagosome formation. In contrast, under full-nutrient conditions, Atg13 is phosphorylated and its association to Atg1 is blocked and thus, autophagy is not activated. Another Atg protein, Atg17, associates with the Atg1–Atg13 complex, and this binding is enhanced under starvation conditions (Kabeya *et al.*, 2005). The interaction of Atg1 with other proteins may also switch the autophagy pathway to the cytoplasm to vacuole (Cvt) pathway in yeast. It has been believed that the Atg1 complex may control membrane dynamics, facilitating a late stage of autophagosome formation (Suzuki *et al.*, 2001). However, recent work suggests that the Atg1 kinase complex is involved in the regulation of protein recruitment to initiate phagophore formation during nonselective autophagy (Cheong *et al.*, 2008).

The UNC (uncoordinated phenotype mutant in Caenorhabditis elegans)-51-like kinases ULK1 and ULK2 are putative human homologues of Atg1 (Okazaki *et al.*, 2000). The identification of Atg1 orthologues, including mammalian ULK1, and ULK1-interacting proteins, which have functionally been linked to autophagy, suggests that an Atg1 complex is involved in the induction of macroautophagy in mammalian cells (Mercer *et al.*, 2009). However, in contrast to yeast, phosphorylation of Atg13 is greatest under autophagic conditions and does not prevent Atg1–Atg13 association. Furthermore, Atg13 stimulates the autophagic activity of Atg1 although excess Atg13 inhibits autophagosome expansion and the recruitment of Atg8/LC3 (Chang and Neufeld, 2009). Thus, autophagy regulation by Atg1 in mammalian cells is more complex than in yeast and likely functions at multiple levels.

Rabs, SNAREs and more

The molecular components involved in several vesicular transport events have been identified and extensively studied. Members of the small GTPases (guanosine triphsphatases) family Rabs and the SNARE proteins are key components of the molecular machinery involved in vesicle tethering and fusion. The Rab family is one of the largest families of small GTPases with more than 60 members in the human genome. The Rabs and their interacting proteins coordinate sequential steps in vesicular transport from vesicle formation, transport and tethering and fusion with the target compartment (Zerial and McBride, 2001). Rab proteins localize in specific intracellular compartments, where they recruit effector molecules leading to the approximation of the vesicles and facilitating the last fusion step, which is mediated by the formation of the trans-SNARE complexes (Martens and McMahon, 2008). NSF (*N*-ethylmaleimide sensitive factor) member of the AAA-ATPase (adenosine triphosphatase) family is a chaperone protein whose ATPase activity is required for the disassembly of the SNARE complex after fusion.

Both in yeast and mammalian cells, fusion of autophagosomes with endocytic compartments requires several of these factors. The Rab protein Ypt7/Rab7 is required for fusion of autophagosomes with the vacuole/lysosome (Gutierrez *et al.*, 2004a, b; Jager *et al.*, 2004). Overexpression of a Rab7 dominant negative mutant or siRNA (small interfering ribonucleic acid) depletion of Rab7 leads to the accumulation of immature autophagosomes and the concomitant decrease in the degradation of long-lived proteins. Another small GTPase Rab24 has been associated with the autophagy pathway although the precise role of this protein remains to be elucidated. The protein Rab24 changes its reticular distribution to a punctate pattern on autophagy induction colocalizing with the protein LC3 (Munafo and Colombo, 2002). In a recent publication, it has been shown that the small GTPase Rab11 is involved in fusion between autophagosomes and endocytic multivesicular bodies (MVBs) but was not required for fusion with lysosomes (Fader *et al.*, 2008). The AAA-ATPase SKD1 (suppressor-of-potassium-transport-growth-defect-1 protein mutant in Saccharomyces cerevisiae) is required for the maturation of the autophagic pathway. Overexpression of a mutant unable to hydrolyse ATP hampers endosome function and causes massive accumulation of early autophagosomes by inhibiting the formation of the autolysosome (Eskelinen, 2005).

In yeast, fusion of autophagosomes with the vacuole also requires the SNARE proteins Vam3, Vam7, Vti1 and Ykt6, as well the NSF and SNAP homologues Sec18 and Sec17. The mammalian homologue of Vti1 seems to have a role in fusion between autophagosomes and MVBs. Recent data indicates that other SNARE proteins are also required for fusion between autophagosomes and the endo/lysosomal compartment in mammalian cells. Also, alterations in the functioning of the ESCRT machinery (endosomal sorting complex required for transport) required for the formation of the MVBs inhibits the formation of autolysosomes. In addition, the maturation of autophagosomes into autolysosomes is prevented in cells deficient for the proteins LAMP-1 (lysosomal-associated membrane protein) and LAMP-2 (Eskelinen, 2005).

Regulation of the Autophagic Pathway

Both extracellular (amino acid deprivation, hormones, toxins and drug treatments) and intracellular signals (pathogens invasion, accumulation of misfolded proteins) are able to modulate the autophagic response. A complex set of signal transduction mechanisms are known to be involved in the regulation of autophagy both in a positive or negative manner, according to the stimuli (Meijer and Codogno, 2004).

For many years it has been known that amino acids inhibit the formation of autophagosomes. However, the signalling pathways involved have been just elucidated, in part, in the recent years. Among these pathways, PI3Ks and the protein kinase TOR (target of rapamycin) play central roles (**Figure 2**). As indicated earlier, PI3K class III (Vps34) forms a complex with Beclin 1 and other proteins. Amino acids negatively regulate autophagy by interfering with the activity of this class of PI3K. The Beclin 1 complex is involved in the initial step of autophagosome formation. Beclin 1 was originally identified as a Bcl-2 interacting protein and binding of Bcl-2 or Bcl-x_L to Beclin 1 reduces the capacity of Beclin 1 to activate autophagy (Pattingre *et al.*, 2005). Induction of autophagy under physiological conditions (i.e. starvation) stimulates the dissociation of the antiapoptotic protein Bcl-2, clearly showing the crosstalk between the core machineries regulating autophagy and apoptosis (**Figure 2**).

The kinase TOR is a serine/threonine kinase involved in several regulatory pathways activated in response to nutrient conditions. TOR negatively regulates autophagy. It is known that amino acids (mainly leucine) can stimulate TOR and as a consequence autophagy is inhibited (Meijer and Codogno, 2004). Conversely, inactivation of TOR by amino acid depletion or rapamycin stimulates autophagy (**Figure 2**). It has been shown that TOR modulates certain Atg proteins altering the formation of autophagosomes. Indeed, TOR physically interacts with Atg1 and both Atg1 and Atg13 are phosphorylated in a TOR- and Atg1 kinase-dependent manner. Furthermore, knockout of Atg1 or Atg13 results in defective autophagy in response to TOR inactivation.

TOR also controls translation and transcription. One of the TOR kinase targets is the eukaryotic initiation factor 4E-binding protein-1 (4E-BP1) which is directly phosphorylated by TOR. The phosphorylation of this factor leads to its dissociation from eukaryotic initiation factor 4E (eIF4E), which in turn binds to (Codogno and Meijer, 2005) the 5′-terminal cap of messenger RNA (mRNA) and promotes the translation. In addition, the p70S6 kinase (S6K) is also a substrate of TOR and S6K phosphorylates the ribosomal subunit of 40S known as S6, upregulating the translation of several mRNAs such as ribosomal protein, elongation factor (EFla and EF2) and polyA-binding protein.

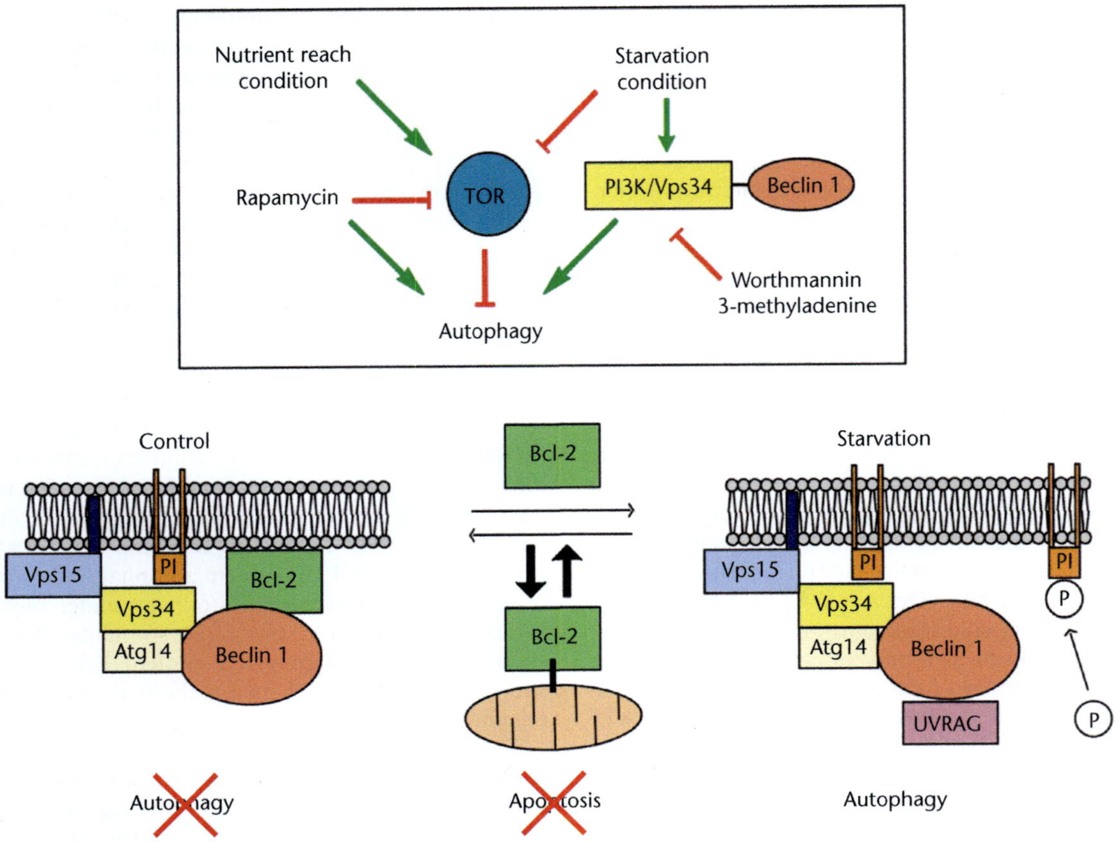

Figure 2 Regulation of autophagy. The PI3K class III and the protein kinase TOR play central roles in the signalling pathways involved in nutrient regulation of autophagy. The complex PI3K class III (Vps34)/Beclin 1 positively regulates autophagy whereas TOR inhibits this pathway. Binding of the antiapoptotic protein Bcl-2 to Beclin 1 decreases the capacity of Beclin 1 to activate autophagy. Starvation-dependent induction of autophagy stimulates the dissociation of Bcl-2 and Beclin 1. The crosstalk between the core machineries regulating autophagy and apoptosis is depicted.

TOR also phosphorylates the protein Tap42, leading to its association to the protein phosphatase 2A (PP2A) and inhibiting its activity. This phosphatase is involved in the dephosphorylation of the transcription factor Gnl3 promoting its dissociation from urease 2 (Ure2), which translocates to the nucleus activating the transcription of several genes (Beck and Hall, 1999). Genes of the autophagy machinery, such as Atg8 and Atg14, are upregulated on nitrogen starvation or rapamycin treatment in a Gnl3-dependent manner.

Hormones are known to be important regulators of autophagy. Insulin has an inhibitory effect on autophagy. The first part of this pathway, located upstream of the TOR, involves binding to the insulin receptor, which triggers the autophosphorylation of the tyrosine kinase receptor and the recruitment of P85, a regulatory subunit of the PI3K class I. The activated PI3K generates PIP_3 and PIP_2, leading to the activation of protein kinase B (AKT/PKB), followed by activation of TOR. In contrast, glucagon inhibits TOR by downregulating PI3K class I, leading to an increase in autophagy.

The ATP levels are also important in the regulation of autophagy. Increased levels of AMP promote autophagy via the activation of the AMP-dependent protein kinase (AMPK), which is involved in the control of TOR signalling (Meley *et al.*, 2006). Calcium is another key player in the autophagy pathway and depletion of sequestered Ca^{2+} ions from some intracellular storage compartments leads to autophagy inhibition.

Autophagy in Pathological Processes

As aforementioned, autophagy mostly represents a stress adaptation pathway that promotes cell survival.

Examples for stress include nutrient deprivation, growth factor depletion and hypoxia. Besides metabolic stress, autophagy is also induced during oxidative stress, infection or when protein aggregates accumulate in the cytoplasm of cells. Clearly such stress conditions are present under pathological conditions. It is, therefore, not surprising that pathologists noted ultrastructural features of autophagy associated with several human diseases already long time ago (De Duve and Wattiaux, 1966; Martinez-Vicente and Cuervo, 2007). Recently, however, it became clear that dysregulated autophagy can also be a causative factor, or at least a major player, in the pathogenesis of human diseases. In the following, we review our current understanding regarding the role of dysregulated autophagy in pathological processes.

Neurodegenerative diseases

In the brains of patients with neurodegenerative diseases, the accumulation of autophagosomes is often observed (Rubinsztein *et al.*, 2007; Williams *et al.*, 2006). But does the increased autophagic activity contribute to the pathogenesis in these disorders? Recent work in animals suggests that the accumulation of autophagosomes rather represents a beneficial response and not the cause of the disease (Rubinsztein *et al.*, 2007; Williams *et al.*, 2006). In these diseases, protein aggregates often occur that exert cellular toxicity causing neurodegeneration. Such protein aggregates often occur due to mutations and cannot be degraded by the proteasome (Williams *et al.*, 2006). As a consequence, autophagy is activated leading to a reduction of the protein aggregates. Therefore, pharmacological activation of autophagy represents a new strategy for the treatment of neurodegenerative diseases (Rubinsztein *et al.*, 2007).

That the increased autophagic activity in neurodegenerative diseases is not sufficient to prevent the development of clinical symptoms suggests that the capacity to clean up mutant proteins is somehow limited. The extent of the autophagic response may vary among different individuals due to genetic differences. In addition, an age-related decline of autophagy has been described (Shibata *et al.*, 2006) that could explain why these diseases are more frequent in older patients. Moreover, in Alzheimer disease, there is, besides accumulation of amyloid β protein, a clear defect in autophagolysomal maturation (Yu *et al.*, 2005). Taken together, neurodegenerative diseases are often caused by mutant proteins forming aggregates in the cytoplasm of neurons and autophagy can limit the accumulation of such toxic protein aggregates. A limited capacity of autophagy can contribute to the pathogenesis of these diseases and the pharmacological induction of autophagic activity represents a new strategy to prevent or delay neurodegeneration.

Myodegenerative diseases

Myodegenerative disease can, like neurodegenerative diseases, be caused by protein aggregations and/or limited capacity to activate autophagy. Also similar to neurodegenerative diseases, pharmacological upregulation of autophagy may protect or delay myodegeneration. Some genetic diseases strongly limit autophagy and are therefore the primary cause of a given myopathy. For instance, a failure of fusion between autophagosomes and lysosomes results in myodegeneration (e.g. due to LAMP-2 deficiency) (Nishino *et al.*, 2000). Moreover, deficiencies in lysosomal enzymes cause lysosomal defects. For instance, lack of acid glucosidase results in glycogen deposition inside and outside of lysosomes. The failure of glycogen digestion results in local starvation and consequent high autophagic activity leading to muscle wasting (Fukuda *et al.*, 2006).

Cardiac diseases

The rather rare genetic diseases resulting in myopathies also affect the heart. However, is there a role of autophagy in the more frequently occurring ischaemic heart diseases? Indeed, the accumulation of autophagosomes is a common phenomenon in these diseases (Terman and Brunk, 2005). Similar to the neurodegenerative diseases, it has been believed that autophagy contributes to cardiomyocyte degeneration under these pathological conditions. However, experimental models changed this view and autophagy is now seen as an adaptive response against the ischaemic stress (Hamacher-Brady *et al.*, 2006).

Cancer

A potential role of autophagy in the pathogenesis of cancer is suggested by the fact that several known tumour suppressor genes (p53, PTEN, TSC1 and TSC2) inhibit mTOR activation and consequently stimulate autophagy. In contrast, class I PI3K/Akt signalling pathway, which activates mTOR and therefore blocks autophagy, is often hyperactivated in cancer. Therefore, autophagy defects may contribute to tumourigenesis (Levine, 2007; Mathew *et al.*, 2007). Further support for this concept comes from the observation that Bcl-2 and Bcl-x_L, which are overexpressed in many cancers, inhibit autophagy by binding to Beclin 1 (Pattingre *et al.*, 2005; Maiuri *et al.*, 2007). Moreover, the death-associated protein kinase (DAPk) mediates increased autophagic activity and is often underexpressed in cancer (Gozuacik and Kimchi, 2006). The observation that Beclin 1 is deficient in many cancers provided the first direct link between a regulator of autophagy and cancer (Liang *et al.*, 1999; Miracco *et al.*, 2007). In addition, Atg5 demonstrated tumour suppressor effects in a mouse xenograft model (Yousefi *et al.*, 2006).

The molecular mechanisms of how autophagy blocks tumour development largely remain to be determined. However, increased deoxyribonucleic acid (DNA) damage, centrosome abnormalities, aneuploidy, chromosomal abnormalities and gene amplifications have been observed in mouse epithelial cell lines deficient in Beclin 1 or Atg5. Therefore, defective autophagy seems to promote genetic instability (Mathew *et al.*, 2007), although it remains unclear whether autophagy prevents genomic stability in normal cells (**Figure 3**).

Another possibility of how autophagy could block tumourigenesis could be by a direct inhibitory effect on cell proliferation. For instance, proliferation of tumour cell lines is reduced by enforced expression of Beclin 1 (Liang *et al.*, 1999; Koneri *et al.*, 2007), Atg5 (Yousefi *et al.*, 2006) or Atg1 (Scott *et al.*, 2007). The antiproliferative effect of Atg5 does not seem to depend on Atg5-induced increased autophagic activity (Yousefi *et al.*, 2006). Similarly, there is increasing evidence that tumour suppressor functions of certain autophagy genes are not dependent on both potential prodeath or prosurvival effects (Mathew *et al.*, 2007).

In the case of anticancer therapy, however, cell survival/death effects due to changes of autophagic activity are likely important. Many chemotherapeutic drugs increase the numbers of autophagosomes in cancer cells *in vitro*, a phenomenon, which is now believed to represent a cellular stress response in order to survive in the presence of cytotoxic agents (Maiuri *et al.*, 2007). Therefore, when a tumour is established, inhibition rather than stimulation of autophagy might be helpful in anticancer treatment (**Figure 4**). However, autophagy blocks angiogenesis and therefore tumour growth (Eisenberg-Lerner *et al.*, 2009). Clearly, inhibition of autophagy would here be the wrong strategy.

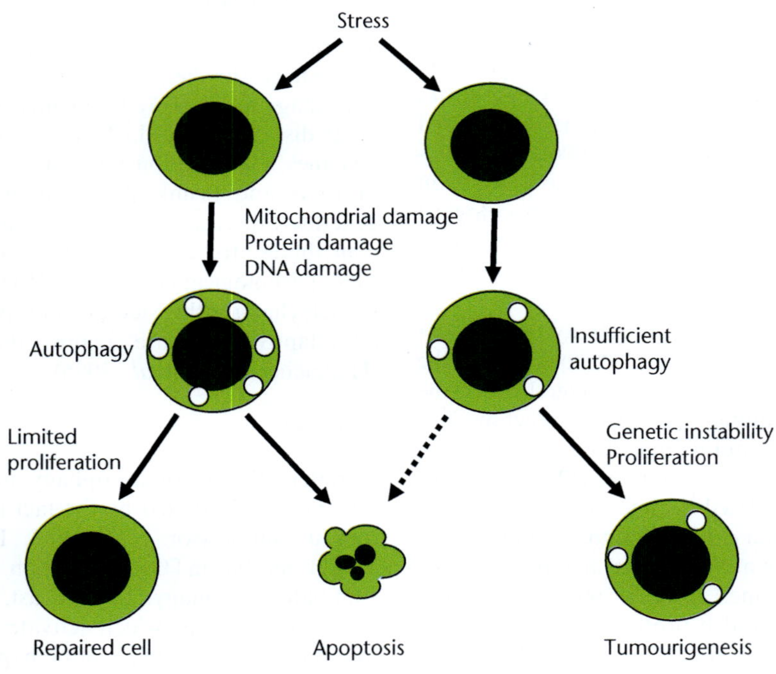

Figure 3 The potential role of defective autophagy in tumour development. Autophagy acts in tumour suppression by removing damaged organelles and proteins, and reduces genetic instability. It may also cooperate with apoptosis to kill damaged cells. Moreover, some autophagic genes have an inhibitory effect on cell proliferation.

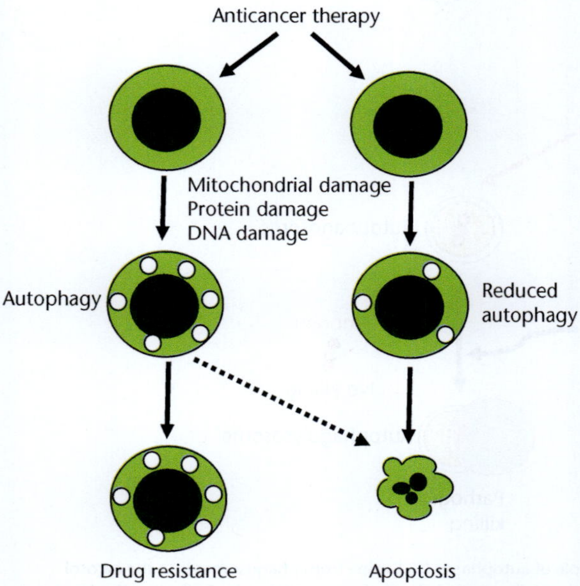

Figure 4 Autophagy acts as a cytoprotective mechanism in anticancer therapy leading to drug resistance. The specific reduction of autophagic activity in cancer cells might be beneficial under these circumstances.

Infectious and inflammatory diseases

Phagocytosis is a major mechanism of the innate immune system to eliminate bacteria, which are targeted into phagosomes and delivered to the lysosome for destruction (Dorn *et al.*, 2002). However, some bacteria are able to escape the phagosome and enter the cytosol to multiply (**Figure 5**). In other cases, bacteria modify the phagosome to prevent fusion with the lysosome (**Figure 5**) (Ogawa and Sasakawa, 2006). In these cases, autophagy is one mechanism that targets the bacteria into autophagosomes, which subsequently fuse with the lysosome (Nakagawa *et al.*, 2004; Gutierrez *et al.*, 2004a, b; Ogawa *et al.*, 2005; Birmingham *et al.*, 2006).

How this process of defence autophagy is triggered remains obscure. It is likely that such triggers are generated by bacteria themselves and it has been suggested that Toll-like receptors (TLRs) may play a role (Sanjuan *et al.*, 2007; Xu *et al.*, 2007; Delgado *et al.*, 2008). Since cytokines, such as TNFα (tumour necrosis factor α) (Subauste *et al.*, 2007), TRAIL (TNF-relatedapoptosis-inducing ligand) (Mills *et al.*, 2004) and IFNγ (interferon γ) (Gutierrez *et al.*, 2004a, b), are also able to increase autophagic activity, TLR-mediated targeting of intracellular pathogens might not be direct and could involve cytokines. CD40-mediated activation of macrophages also increased autophagic activity, which somehow helped to fuse *Toxoplasma gondii*-containing vacuoles with lysosomes resulting in killing of the pathogen (Andrade *et al.*, 2006). Interestingly, many pathogens developed strategies to escape killing mechanisms by blocking the fusion of lysosomes with the autophagosome (Sanjuan and Green, 2008), demonstrating the importance of autophagy induction for pathogen killing by monocytes/macrophages (**Figure 5**).

Autophagy has also been associated with innate immunity against pathogens in the absence of phagosome escape mechanisms. Here, autophagy seems to enhance the fusion of phagosomes with lysosomes, apparently without the formation of autophagosomes (Sanjuan *et al.*, 2007). As a result, rapid destruction of the engulfed pathogen occurs. Before fusion, the phagosome recruits LC3 and this process requires Atg5, Atg6, Atg7 and PI3K activity (Sanjuan *et al.*, 2007) as well as TLR signalling for recognition that the phagosome contains a pathogen (Sanjuan and Green, 2008).

Blocking autophagy in monocytes/macrophages has recently been demonstrated to enhance IL-1β (interleukin 1β) production (Saitoh *et al.*, 2008). Why the process of autophagy somehow limits the inflammatory response remains unclear. However, this mechanism might be important, since the induction of autophagy by pharmacological inhibition of mTOR may mediate anti-inflammatory activities, as shown in one patient with Crohn disease (Massey *et al.*, 2008). The cell type, however, in which the therapeutic induction of autophagy is important might differ from monocytes/macrophages. Recent work suggests that decreased autophagic activity of specialized intestinal epithelial cells results in inability to eliminate intestinal bacteria via autophagy in Crohn disease (Klionsky, 2009). Clearly, further experimental work, including clinical trials, is required to better understand the potential link between autophagy and the strength of innate immune responses.

Dendritic cells represent professional antigen-presenting cells, although other leukocytes such as B cells and macrophages are also able to present antigen. Apart from its role in innate immunity, autophagy might also alarm the adaptive immune system against pathogens that are delivered for lysosomal destruction by autophagy in these cells. In particular, lysosomal degradation products are presented via MHC (major histocompatibility complex) class II molecules to CD4+ T cells, and these helper T cells orchestrate the specific immune response (Münz, 2006). As

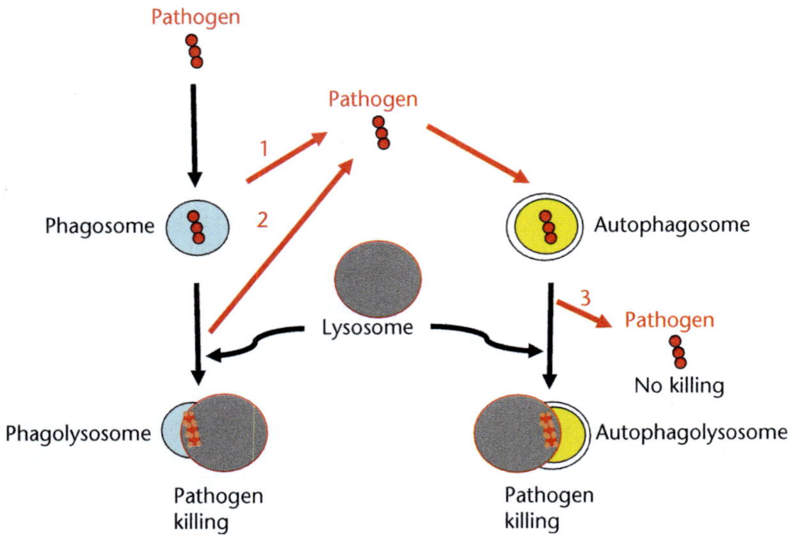

Figure 5 Three escape mechanisms of pathogens on phagocytosis and the role of autophagy. (1) Escape from phagosomes into the cytosol (e.g. *Listeria monocytogeneses*). (2) Prevention of fusion with lysosome (e.g. *Salmonella enterica*). Similarly, prevention of acidification has also been observed (e.g. *Mycobacterium tuberculosis*). (3) Pathogens in the cytosol can be trapped by autophagosomes and degraded via the autophagic machinery. However, some pathogens can also escape from this degradation system by preventing the fusion between autophagosomes and lysosomes (e.g. *Legionella pneumophila*).

described earlier, autophagy is used to target pathogens into autophagosomes, which subsequently fuse with the lysosome (Nakagawa *et al.*, 2004; Gutierrez *et al.*, 2004a, b; Ogawa *et al.*, 2005; Birmingham *et al.*, 2006). This way, epitopes of pathogens and intracellular antigens can be delivered to MHC class II molecules (Nimmerjahn *et al.*, 2003; Dengjel *et al.*, 2005; Paludan *et al.*, 2005; Zhou *et al.*, 2005). Interestingly, this pathway might be highly relevant for cells that express MHC class II molecules in the absence of high endocytic capacity, such as cortical thymic epithelial cells (Mizushima *et al.*, 2004).

Although we know now that autophagic pathways promote MHC class II antigen presentation, many questions still need to be answered. For instance, what is the contribution of autophagic delivery of pathogens to adaptive immunity? Clearly, to unravel the role of autophagic antigen processing and presentation not only has implications for infectious diseases, but also for cancer, autoimmunity, allergy and transplantation medicine.

Autophagy and Cell Death

As discussed earlier, high autophagic activity in muscle cells results in muscle wasting (Fukuda *et al.*, 2006). Earlier work suggested that lysosomal hydrolases are liberated into the sarcoplasm, leading to destruction of myofibrils and cell death (Beaulaton and Lockshin, 1977). Though autophagy can clearly contribute to cell death (Eisenberg-Lerner *et al.*, 2009), it is often not clear whether cell death truly occurs through induction of autophagy or whether cell death is just associated with autophagy (autophagy not required for cell death). Since 3-methyladenine (3-MA) inhibits autophagy (Seglen and Gordon, 1982), any cell death, which is blocked by 3-MA and demonstrates morphologic features of autophagy, should be considered as autophagic cell death (Bursch *et al.*, 1996). Morphologically, in cells undergoing autophagic cell death, it can be observed that the nucleus stays intact until the late phases of cell death and cellular fragmentation is not observed (**Table 1**). Unlike apoptosis, caspases are not activated in autophagic cell death. Moreover, autophagic cell death has often been described under experimental conditions, in which cells were exposed to a death trigger, but caspase activation was genetically or pharmacologically prevented.

Although autophagic cell death also occurs under physiologic conditions (e.g. hormone deprivation-induced regression of mammary gland or prostate) (Bursch, 2001), the concurrent occurrence of both autophagy and apoptosis is much more frequent,

Table 1 Morphological and biochemical characteristics of apoptosis and autophagic cell death

	Apoptosis	Autophagic cell death
Nucleus	Chromatin condensation	Partial chromatin condensation
	Pyknosis	Sometimes pyknosis
	Fragmentation	Intact until late stages
	DNA laddering	No DNA laddering
Cytoplasm	Condensation	Autophagosomes and autolysosomes
	Fragmentation to apoptotic bodies	No apoptotic bodies
	Possible lysosomal activity	Increased lysosomal activity
	Mitochondria often involved	Mitochondria might be involved
	Caspases activated	Caspase independent

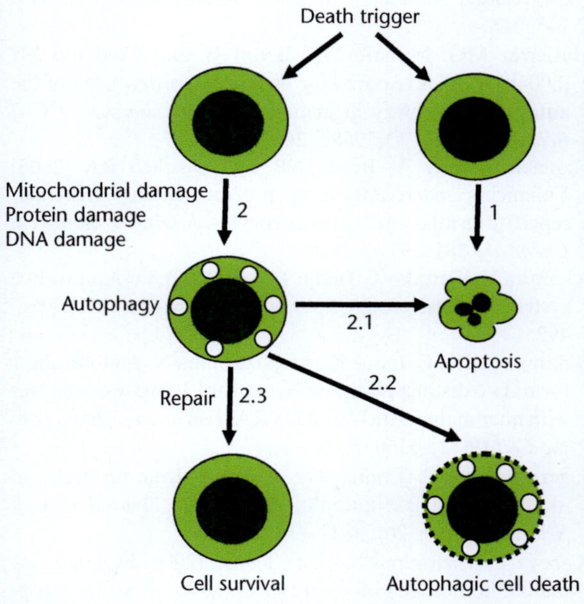

Figure 6 Schematic representation of the crosstalk between autophagy and apoptosis. (1) Following a death trigger the cell undergoes rapid apoptosis without induction of autophagy. (2) The death trigger induces autophagy, which is a prerequisite for apoptosis induction (2.1.). If apoptosis is blocked, the cell may undergo autophagic cell death (2.2.). Finally, the damage induced by the death trigger might be repaired and the cell survives due to autophagy (2.3.).

pointing to the possibility that both pathways may interplay and dependent on the circumstances, autophagy either promotes or blocks apoptosis (Eisenberg-Lerner *et al.*, 2009). There are multiple examples in the literature that suggest autophagy and apoptosis cooperate to lead to cell death. If one programme is blocked (by genetic defect or pharmacological inhibition), the other takes over. The goal is to efficiently kill the cell, even if one death programme does not function. In other scenarios, autophagy actually allows apoptosis to occur, for instance, if cells cannot provide sufficient levels of ATP. However, autophagy may attenuate apoptosis by creating a cellular milieu in which survival is favoured. Examples were discussed earlier in this article (e.g. removal of protein aggregates in neurodegenerative diseases). These different types of interplay between autophagy and apoptosis are displayed in **Figure 6** (Eisenberg-Lerner *et al.*, 2009).

Concluding Remarks

Autophagy is a key cellular catabolic pathway that serves to protect cells against stress, including lack of nutrients and growth factors, or due to hypoxia. It also functions to eliminate superfluous or damaged cell organelles, misfolded proteins and invading pathogens. Therefore, autophagy appears to be primarily an adaptive response. Only under certain circumstances, it is selectively responsible for cell death. The molecular understanding of the autophagic machinery was rapidly advanced within recent years. In particular, it became clear that the autophagosome formation involves two ubiquitin-like conjugation systems and signalling pathway(s) that initiate the induction of autophagy. Insufficient autophagy is directly involved in the pathogenesis of degenerative and infectious diseases, as well as cancer. Pharmacological activation of autophagy appears to be promising strategy to treat such pathological conditions.

References

Andrade RM, Wessendarp M, Gubbels MJ, Striepen B and Subauste CS (2006) CD40 induces macrophage anti-*Toxoplasma gondii* activity triggering autophagy-dependent fusion of pathogen-containing vacuoles and lysosomes. *Journal of Clinical Investigation* **116**: 2366–2377.

Beaulaton J and Lockshin RA (1977) Ultrastructural study of the normal degeneration of the intersegmental muscles of *Anthereae polyphemus* and *Manduca sexta* (Insecta, Lepidoptera) with particular reference of cellular autophagy. *Journal of Morphology* **154**: 39–57.

Beck T and Hall MN (1999) The TOR signalling pathway controls nuclear localization of nutrient-regulated transcription factors. *Nature* **402**(6762): 689–692.

Birmingham CL, Smith AC, Bakowski MA, Yoshimori T and Brumell JH (2006) Autophagy controls *Salmonella* infection in response to damage to the *Salmonella*-containing vacuole. *Journal of Biological Chemistry* **281**: 11374–11383.

Bursch W (2001) The autophagosomal-lysosomal compartment in programmed cell death. *Cell Death and Differentiation* **8**: 569–581.

Bursch W, Ellinger A, Kienzl H *et al.* (1996) Active cell death induced by the anti-estrogens tamoxifen and ICI 164 384 in human mammary carcinoma cells (MCF-7) in culture: the role of autophagy. *Carcinogenesis* **17**: 1595–1607.

Chang YY and Neufeld TP (2009) An Atg1/Atg13 complex with multiple roles in TOR-mediated autophagy regulation. *Molecular Biology of the Cell* **20**(7): 2004–2014.

Cheong H, Nair U, Geng J and Klionsky DJ (2008) The Atg1 kinase complex is involved in the regulation of protein recruitment to initiate sequestering vesicle formation for nonspecific autophagy in *Saccharomyces cerevisiae*. *Molecular Biology of the Cell* **19**(2): 668–681.

Codogno P and Meijer AJ (2005) Autophagy and signaling: their role in cell survival and cell death. *Cell Death and Differentiation* **2**: 1509–1518.

Delgado MA, Elmaoued RA, Davis AS, Kyei G and Deretic V (2008) Toll-like receptors control autophagy. *EMBO Journal* **27**: 1110–1121.

Dengjel J, Schoor O, Fischer R *et al.* (2005) Autophagy promotes MHC class II presentation of peptides from intracellular source proteins. *Proceedings of the National Academy of Sciences of the USA* **102**: 7922–7927.

Dorn BR, Dunn WA and Progulske-Fox A (2002) Bacterial interactions with the autophagic pathway. *Cell Microbiology* **4**: 1–10.

De Duve C and Wattiaux R (1966) Functions of lysosomes. *Annual Review of Physiology* **28**: 435–492.

Eisenberg-Lerner A, Bialik S, Simon HU and Kimchi A (2009) Life and death partners: apoptosis, autophagy and the cross-talk between them. *Cell Death and Differentiation* **16**: 966–975.

Eskelinen EL (2005) Maturation of autophagic vacuoles in mammalian cells. *Autophagy* **1**(1): 1–10.

Fader CM, Sanchez D, Furlan M and Colombo MI (2008) Induction of autophagy promotes fusion of multivesicular bodies with autophagic vacuoles in k562 cells. *Traffic* **9**(2): 230–250.

Fujita N, Itoh T, Omori H *et al.* (2008) The Atg16L complex specifies the site of LC3 lipidation for membrane biogenesis in autophagy. *Molecular Biology of the Cell* **19**(5): 2092–2100.

Fukuda T, Ahearn M, Roberts A *et al.* (2006) Autophagy and mistargeting of therapeutic enzyme in skeletal muscle in Pompe disease. *Molecular Therapy* **14**: 831–839.

Gozuacik D and Kimchi A (2006) DAPk protein family and cancer. *Autophagy* **2**: 74–79.

Liang XH, Jackson S, Seaman M *et al.* (1999) Induction of autophagy and inhibition of tumorigenesis by Beclin 1. *Nature* **402**: 672–676.

Gutierrez MG, Master SS, Singh SB *et al.* (2004a) Autophagy is a defense mechanism inhibiting BCG and *Mycobacterium tuberculosis* survival in infected macrophages. *Cell* **119**: 735–766.

Gutierrez MG, Munafo DB, Beron W and Colombo MI (2004b) Rab7 is required for the normal progression of the autophagic pathway in mammalian cells. *Journal of Cell Science* **117**(part 13): 2687–2697.

Hamacher-Brady A, Brady NR and Gottlieb RA (2006) Enhancing macroautophagy protects against ischemia/reperfusion injury in cardiac myocytes. *Journal of Biological Chemistry* **281**: 29776–29787.

Ichimura Y, Kirisako T, Takao T *et al.* (2000) A ubiquitin-like system mediates protein lipidation. *Nature* **408**(6811): 488–492.

Itakura E, Kishi C, Inoue K and Mizushima N (2008) Beclin 1 forms two distinct phosphatidylinositol 3-kinase complexes with mammalian Atg14 and UVRAG. *Molecular Biology of the Cell* **19**(12): 5360–5372.

Jager S, Bucci C, Tanida I *et al.* (2004) Role for Rab7 in maturation of late autophagic vacuoles. *Journal of Cell Science* **117**(part 20): 4837–4848.

Kabeya Y, Mizushima N, Ueno T *et al.* (2000) LC3, a mammalian homologue of yeast Apg8p, is localized in autophagosome membranes after processing. *EMBO Journal* **19**(21): 5720–5728.

Kabeya Y, Kamada Y, Baba M *et al.* (2005) Atg17 functions in cooperation with Atg1 and Atg13 in yeast autophagy. *Molecular Biology of the Cell* **16**(5): 2544–2553.

Klionsky DJ (2009) Crohn disease and autophagy. *Autophagy* **5**: 139.

Klionsky DJ, Cregg JM, Dunn WA Jr *et al* (2003) A unified nomenclature for yeast autophagy-related genes. *Developmental Cell* **5**(4): 539–545.

Koneri K, Goi T, Hirono Y, Katayama K and Yamaguchi A (2007) Beclin 1 gene inhibits tumor growth in colon cancer cell lines. *Anticancer Research* **27**: 1453–1457.

Levine B (2007) Cell biology: autophagy and cancer. *Nature* **446**: 745–747.

Liang XH, Kleeman LK, Jiang HH *et al.* (1998) Protection against fatal Sindbis virus encephalitis by Beclin, a novel Bcl-2-interacting protein. *Journal of Virology* **72**(11): 8586–8596.

Maiuri MC, Le Toumelin G, Criollo A *et al.* (2007) Functional and physical interaction between Bcl-X(L) and a BH3-like domain in Beclin-1. *EMBO Journal* **26**: 2527–2539.

Martens S and McMahon HT (2008) Mechanisms of membrane fusion: disparate players and common principles. *Nature Review of Molecular and Cellular Biology* **9**(7): 543–556.

Martinez-Vicente M and Cuervo AM (2007) Autophagy and neurodegeneration: when the cleaning crew goes on strike. *Lancet Neurology* **6**: 352–361.

Massey DCO, Bredin F and Parkes M (2008) Use of sirolimus (rapamycin) to treat refractory Crohn's disease. *Gut* **57**: 1294–1296.

Mathew R, Karantza-Wadsworth V and White E (2007) Role of autophagy in cancer. *Nature Review of Cancer* **7**: 961–967.

Meijer AJ and Codogno P (2004) Regulation and role of autophagy in mammalian cells. *International Journal of Biochemistry & Cell Biology* **36**(12): 2445–2462.

Meley D, Bauvy C, Houben-Weerts JH *et al.* (2006) AMP-activated protein kinase and the regulation of autophagic proteolysis. *Journal of Biological Chemistry* **281**(46): 34870–34879.

Mercer CA, Kaliappan A and Dennis PB (2009) A novel, human Atg13 binding protein, Atg101, interacts with ULK1 and is essential for macroautophagy. *Autophagy* **5**(5).

Mills KR, Reginato M, Debnath J, Queenan B and Brugge JS (2004) Tumor necrosis factor-related apoptosis-inducing ligand (TRAIL) is required for induction of autophagy during lumen formation *in vitro*. *Proceedings of the National Academy of Sciences of the USA* **101**: 3438–3443.

Miracco C, Cosci E, Oliveri G *et al.* (2007) Protein and mRNA expression of autophagy gene Beclin 1 in human brain tumours. *International Journal of Oncology* **30**: 429–436.

Mizushima N (2005) The pleiotropic role of autophagy: from protein metabolism to bactericide. *Cell Death and Differentiation* **12**(suppl. 2): 1535–1541.

Mizushima N, Noda T, Yoshimori T *et al.* (1998) A protein conjugation system essential for autophagy. *Nature* **395**(6700): 395–398.

Mizushima N, Ohsumi Y and Yoshimori T (2002) Autophagosome formation in mammalian cells. *Cell Structure and Function* **27**(6): 421–429.

Mizushima N, Yoshimori T and Ohsumi Y (2003) Role of the Apg12 conjugation system in mammalian autophagy. *International Journal of Biochemistry & Cell Biology* **35**(5): 553–561.

Mizushima N, Yamamoto A, Matsui M, Yoshimori T and Ohsumi Y (2004) *In vivo* analysis of autophagy in response to nutrient starvation using transgenic mice expressing a fluorescent autophagosome marker. *Molecular Biology of the Cell* **15**: 1101–1111.

Munafo DB and Colombo MI (2002) Induction of autophagy causes dramatic changes in the subcellular distribution of GFP-Rab24. *Traffic* **3**(7): 472–482.

Münz C (2006) Autophagy and antigen presentation. *Cell Microbiology* **8**: 891–898.

Nakagawa I, Amano A, Mizushima N *et al.* (2004) Autophagy defends cells against invading Group A *Streptococcus*. *Science* **306**: 1037–1040.

Nakatogawa H, Ichimura Y and Ohsumi Y (2007) Atg8, a ubiquitin-like protein required for autophagosome formation, mediates membrane tethering and hemifusion. *Cell* **130**(1): 165–178.

Nimmerjahn F, Milosevic S, Behrends U *et al.* (2003) Major histocompatibility complex class II-restricted presentation of a cytosolic antigen by autophagy. *European Journal of Immunology* **33**: 1250–1259.

Nishino I, Fu J, Tanji K *et al.* (2000) Primary LAMP-2 deficiency causes X-linked vacuolar cardiomyopathy and myopathy (Danon disease). *Nature* **406**: 906–910.

Ogawa M and Sasakawa C (2006) Bacterial evasion of the autophagic defense system. *Current Opinion in Microbiology* **9**: 62–68.

Ogawa M, Yoshimori T, Suzuki T *et al.* (2005) Escape of intracellular *Shigella* from autophagy. *Science* **307**: 727–731.

Okazaki N, Yan J, Yuasa S *et al.* (2000) Interaction of the Unc-51-like kinase and microtubule-associated protein light chain 3 related proteins in the brain: possible role of vesicular transport in axonal elongation. *Brain Research. Molecular Brain Research* **85**(1–2): 1–12.

Paludan C, Schmid D, Landthaler M *et al.* (2005) Endogenous MHC class II processing of a viral nuclear antigen after autophagy. *Science* **307**: 593–596.

Pattingre S, Tassa A, Qu X *et al.* (2005) Bcl-2 antiapoptotic proteins inhibit Beclin 1-dependent autophagy. *Cell* **122**: 927–939.

Petiot A, Ogier-Denis E, Blommaart EF, Meijer AJ and Codogno P (2000) Distinct classes of phosphatidylinositol 3′-kinases are involved in signaling pathways that control macroautophagy in HT-29 cells. *Journal of Biological Chemistry* **275**(2): 992–998.

Rubinsztein DC, Gestwicki JE, Murphy LO and Klionsky DJ (2007) Potential therapeutic applications of autophagy. *Nature Review of Drug Discovery* **6**: 304–312.

Saitoh T, Fujita N, Jang MH *et al.* (2008) Loss of the autophagy protein Atg16L1 enhances endotoxin-induced IL-1β production. *Nature* **456**: 264–268.

Sanjuan MA and Green DR (2008) Eating for good health. *Autophagy* **4**: 607–611.

Sanjuan MA, Dillon CP, Tait SWG *et al.* (2007) Toll-like receptor signalling in macrophages links the autophagy pathway to phagocytosis. *Nature* **450**: 1253–1257.

Scott RC, Juhász G and Neufeld TP (2007) Direct induction of autophagy by Atg1 inhibits cell growth and induces apoptotic cell death. *Current Biology* **17**: 1–11.

Seglen PO and Gordon PB (1982) 3-Methyladenine: specific inhibitor of autophagic/lysosomal protein degradation in isolated rat hepatocytes. *Proceedings of the National Academy of Sciences of the USA* **79**(6): 1889–1892.

Shibata M, Lu T, Furuya T *et al.* (2006) Regulation of intracellular accumulation of mutant Huntingtin by Beclin 1. *Journal of Biological Chemistry* **281**: 14474–14485.

Shintani T, Mizushima N, Ogawa Y *et al.* (1999) Apg10p, a novel protein-conjugating enzyme essential for autophagy in yeast. *EMBO Journal* **18**(19): 5234–5241.

Subauste CS, Andrade RM and Wessendarp M (2007) CD40-TRAF6 and autophagy-dependent anti-microbial activity in macrophages. *Autophagy* **3**: 245–248.

Suzuki K, Kirisako T, Kamada Y *et al.* (2001) The pre-autophagosomal structure organized by concerted functions of APG genes is essential for autophagosome formation. *EMBO Journal* **20**(21): 5971–5981.

Terman A and Brunk UT (2005) Autophagy in cardiac myocyte homeostasis, aging, and pathology. *Cardiovascular Research* **68**: 355–365.

Williams A, Jahreiss L, Sarkar S *et al.* (2006) Aggregate-prone proteins are cleared from the cytosol by autophagy: therapeutic implications. *Current Topics in Developmental Biology* **76**: 89–101.

Xie Z, Nair U and Klionsky DJ (2008) Atg8 controls phagophore expansion during autophagosome formation. *Molecular Biology of the Cell* **19**(8): 3290–3298.

Xu Y, Jagannath C, Liu XD *et al.* (2007) Toll-like receoptor 4 is a sensor for autophagy associated with innate immunity. *Immunity* **27**: 135–144.

Yousefi S, Perozzo R, Schmid I *et al.* (2006) Calpain-mediated cleavage of Atg5 switches autophagy to apoptosis. *Nature Cell Biology* **8**: 1124–1132.

Yu WH, Cuervo AM, Kumar A *et al.* (2005) Macroautophagy – a novel beta-amyloid peptide-generating pathway activated in Alzheimer's disease. *Journal of Cell Biology* **171**: 87–98.

Zerial M and McBride H (2001) Rab proteins as membrane organizers. *Nature Review of Molecualr and Cellular Biology* **2**(2): 107–117.

Zhou D, Li P, Lott JM *et al.* (2005) Lamp-2 facilitates MHC class II presentation of cytoplasmic antigens. *Immunity* **22**: 571–581.

Further Reading

Amano A, Nakagawa I and Yoshimori T (2006) Autophagy in innate immunity against intracellular bacteria. *Journal of Biochemistry* **140**: 161–166.

Cai H, Reinisch K and Ferro-Novick S (2007) Coats, tethers, Rabs, and SNAREs work together to mediate the intracellular destination of a transport vesicle. *Developmental Cell* **12**: 671–682.

Campoy E and Colombo MI (2009) Autophagy in intracellular bacterial infection. *Biochimica et Biophysica Acta* (in press).

Cardenas ME, Cutler NS, Lorenz MC, Di Como CJ and Heitman J (1999) The TOR signaling cascade regulates gene expression in response to nutrients. *Genes & Development* **13**: 3271–3279.

Fader CM and Colombo MI (2009) Autophagy and multivesicular bodies: two closely related partners. *Cell Death and Differentiation* **16**: 70–78.

Gozuacik D and Kimchi A (2004) Autophagy as a cell death and tumor suppressor mechanism. *Oncogene* **23**: 2891–2906.

Grosshans BL, Ortiz D and Novick P (2006) Rabs and their effectors: achieving specificity in membrane traffic. *Proceedings of the National Academy of Sciences of the USA* **103**: 11821–11827.

Klionsky DJ (2007) Autophagy: from phenomenology to molecular understanding in less than a decade. *Nature Review of Molecular and Cellular Biology* **8**: 931–937.

Kundu M and Thompson CB (2008) Autophagy: basic principles and relevance to disease. *Annual Review of Pathology: Mechanisms of Disease* **3**: 427–455.

Lerena C, Calligaris SD and Colombo MI (2008) Autophagy: for better or for worse, in good times or in bad times. *Current Molecular Medicine* **8**: 92–101.

Levine B and Deretic V (2007) Unveiling the roles of autophagy in innate and adaptive immunity. *Nature Review of Immunology* **7**: 767–777.

Levine B and Kroemer G (2008) Autophagy in the pathogenesis of disease. *Cell* **132**: 27–42.

Lünemann JD and Münz C (2009) Autophagy in CD4+ T-cell immunity and tolerance. *Cell Death and Differentiation* **16**: 79–86.

Meijer AJ (2008) Amino acid regulation of autophagosome formation. *Methods in Molecular Biology* **445**: 89–109.

Mizushima N (2007) Autophagy: process and function. *Genes & Development* **21**: 2861–2873.

Mizushima N, Levine B, Cuervo AM and Klionsky DJ (2008) Autophagy fights disease through cellular self-digestion. *Nature* **451**: 1069–1075.

Schmid D and Münz C (2007) Innate and adaptive immunity through autophagy. *Immunity* **27**: 11–21.

Xie Z and Klionsky DJ (2007) Autophagosome formation: core machinery and adaptations. *Nature Cell Biology* **9**: 1102–1109.

Yousefi S and Simon H-U (2009) Autophagy in cells of the blood. *Biochimica et Biophysica Acta*, doi:10.1016/j.bbamcr.2008.12.023.

Autophagy in Nonmammalian Systems

Jahda H Hill, *University of Maryland Biotechnology Institute, College Park, Maryland, USA*

Eric H Baehrecke, *University of Massachusetts Medical School, Worcester, Massachusetts, USA*

Advanced article

Article Contents

- Introduction
- Regulation of Autophagy
- Functions of Autophagy
- Disease Models
- Conclusion
- Acknowledgement

Autophagy, or 'self-eating', is a catabolic process that enables lysosome-mediated degradation of cytoplasmic contents and recycling of macromolecules to be used in essential cellular processes. This process enables cells to survive during nutrient restriction, participates in cell death during development, and functions in clearance of protein aggregates and intracellular pathogens. Autophagy has been widely studied in yeast, *Drosophila* and *Caenorhabditis elegans*, and studies in these organisms have revealed regulation by multiple cellular pathways. These include cell growth regulators, such as the Class I PI3 kinase and target of rapamycin (TOR), as well as cell death regulators, including caspases. Defects in autophagy are associated with cancer, neurodegeneration and reduced lifespan. Genetic experiments in *Drosophila* and *C. elegans* have been critical in determining how impairment of autophagy functions in these diseases.

Introduction

Autophagy is a lysosome-mediated process whereby long-lived proteins and organelles are degraded and their components are recycled for use in essential cellular processes. Diverse organisms ranging from yeast to mammals utilize autophagy for a number of physiological and developmental processes. The most widely studied function of autophagy is as a survival mechanism during nutrient restriction. This function was first characterized in yeast and is conserved in flies, nematodes and mammals. Autophagy also plays a developmental role during insect metamorphosis as the fruitfly *Drosophila melanogaster* uses autophagic cell death to degrade larval tissues. This form of cell death, also known as type II programmed cell death, differs from apoptosis in that it occurs in the absence of phagocytic cells. In addition to its pro-survival and developmental functions, recent studies in the fly support data from mammalian cell lines that revealed an immune function for autophagy. Intracellular pathogens such as listeria and mycobacteria can be engulfed and degraded through an autophagic pathway. Furthermore, several autophagy-associated disease models, including neurodegeneration, have emerged and are under investigation in the fruitfly, *D. melanogaster*, and the nematode, *Caenorhabditis elegans*. Because of the ease of their genetic manipulation, flies and worms are good systems for studying the function of autophagy in the context of an entire organism. **See also**: Autophagy

Regulation of Autophagy

Cell morphological changes associated with autophagy were first observed in the liver (Deter and de Duve, 1967), but it was not until a series of experiments in the 1990s in the budding yeast, *Saccharomyces cerevisiae*, that the mechanisms of autophagy were uncovered. The first of those showed that defects in vacuolar proteolysis led to accumulation of autophagic bodies under conditions of nutrient restriction (Takeshige *et al*., 1992). These autophagic bodies, now known as autophagosomes, are composed of a double membrane that surrounds cytosolic components targeted for degradation in the vacuole, the yeast equivalent of the lysosome. This study was followed by two independent genetic screens in yeast; one for mutants defective in degradation of autophagic bodies (Tsukada and Ohsumi, 1993), and the other for mutants that accumulate the cytosolic protein fatty acid synthase (Thumm *et al.*, 1994). These concurrent studies led to the discovery of the 14 *apg* genes and the 3 *aut* complementation groups, which are required for autophagy. Additionally, a screen for mutants in the cytoplasm to vacuole-targeting (CVT) pathway uncovered genetic overlap between this pathway and autophagy (Harding *et al.*, 1995, 1996). The CVT pathway functions in transport of the vacuolar protein aminopeptidase I (API). Thirty-one autophagy-related genes (now called *atg* genes) have been discovered (Klionsky *et al.*, 2003) and most are conserved in higher eukaryotes.

atg1, the first *atg* gene characterized in yeast, encodes a protein kinase. Mutant analysis showed that ATG1 kinase activity is required for autophagosome formation and viability during starvation (Matsuura *et al.*, 1997). ATG1 binds to ATG13 and this complex functions early in autophagosome formation. Regulation of the ATG1–ATG13 complex is under control of the serine/threonine kinase, target of rapamycin (Tor). Under nutrient-rich conditions, TOR hyperphosphorylates ATG13 reducing its binding affinity for ATG1. In response to nutrient starvation, TOR becomes inactivated, alleviating repression of ATG13 (Kamada *et al.*, 2000).

Regulation of autophagy by TOR is more complex in multicellular eukaryotes, where the class I phosphatidylinositol 3-kinase (PI3K) pathway signals upstream of TOR to induce cell growth and inhibit autophagy. The mechanisms of PI3K/TOR regulation of autophagy have been largely elucidated from genetic studies in the fruitfly, *D. melanogaster*. Mutations in either *tor* or *dp110*, the catalytic subunit of PI3K, lead to reduced cell growth and induction of autophagy in well-fed *Drosophila* larvae, whereas ectopic expression of either class I PI3K or TOR in *Drosophila* larvae inhibits starvation-induced autophagy (Scott *et al.*, 2004). In addition to its requirement for starvation-induced autophagy, ATG1 activity is also vital in *Drosophila* development, as animals lacking *ATG1* function die before completing metamorphosis (Scott *et al.*, 2004).

Further studies in yeast identified two ubiquitin-like conjugation pathways that function downstream of ATG1 activity (**Figure 1**). These conjugation pathways are composed of several ATG proteins, all of which are required for autophagosome formation. ATG7 was identified as an E1-like enzyme that activates ATG12 and ATG8 in two distinct conjugation pathways (Mizushima *et al.*, 1998; Tanida *et al.*, 1999; Ichimura *et al.*, 2000). ATG12 and ATG8 are then covalently conjugated to the E2-like proteins ATG10 and ATG3. This is followed by conjugation of ATG12 to ATG5 and lipidation of ATG8 with phosphatidylethanolamine (PE). The ATG5–ATG12 complex interacts with multimerized ATG16. These protein conjugation reactions occur in the membrane of the forming autophagosome, mediating membrane expansion around cytosolic components destined for degradation. Both conjugation pathways are conserved in higher eukaryotes and components of these pathways are required for autophagy in *Drosophila*, *C. elegans* and mammals.

A third group of autophagy-related genes recovered from yeast screens were first identified in a screen for mutants defective in vacuolar protein sorting (Bankaitis *et al.*, 1986). First identified as VPS30, ATG6 was found to interact in two distinct complexes, both of which include the class III PI3K, VPS34 and the protein kinase VPS15 (Kihara *et al.*, 2001). VPS15 activity is required for VPS34 function. The lone target of VPS34 is the lipid phosphatidylinositol (PI), which is converted to PI(3)P upon phosphorylation by VPS34. PI(3)P functions in recruitment of proteins to membranes, supporting a general role for VPS34 in intracellular membrane trafficking. A complex including ATG6, VPS34, VPS15 and VPS38 is required for vacuolar protein sorting, whereas a second complex containing ATG6, VPS34, VPS15 and ATG14 localizes to the preautophagosomal structure (PAS) and is required for autophagosome formation (Kihara *et al.*, 2001).

Similar VPS34 complexes have been described in mammalian cells, where several other proteins have been shown to interact with VPS34 and ATG6. The mammalian homologue of ATG6, known as BECLIN1, was identified in an yeast two-hybrid screen for proteins that interact with the antiapoptotic protein BCL-2 (Liang *et al.*, 1998). Biochemical experiments later confirmed that BECLIN1 also interacts with VPS34 and P150, the mammalian homologue of VPS15, and that this complex is required for starvation-induced autophagy. Additionally, the tumour suppressors UVRAG and BIF-1, and the previously uncharacterized AMBRA1, have been shown to interact biochemically with BECLIN1 (Liang *et al.*, 2006; Takahashi *et al.*, 2007; Fimia *et al.*, 2007). An ATG14 homologue was recently discovered and UVRAG was determined to be the homologue of VPS38 (Itakura *et al.*, 2008). Both interact with BECLIN1, but never in the same complex. These results suggest a mechanism similar to yeast where two distinct complexes function in autophagy and endocytosis.

Although there is not yet evidence of distinct ATG6/VPS34/VPS15 complexes in *Drosophila* and *C. elegans*, there is data to support that VPS34 is required for autophagy and endocytosis. *vps34* mutants have defects in starvation-induced autophagy

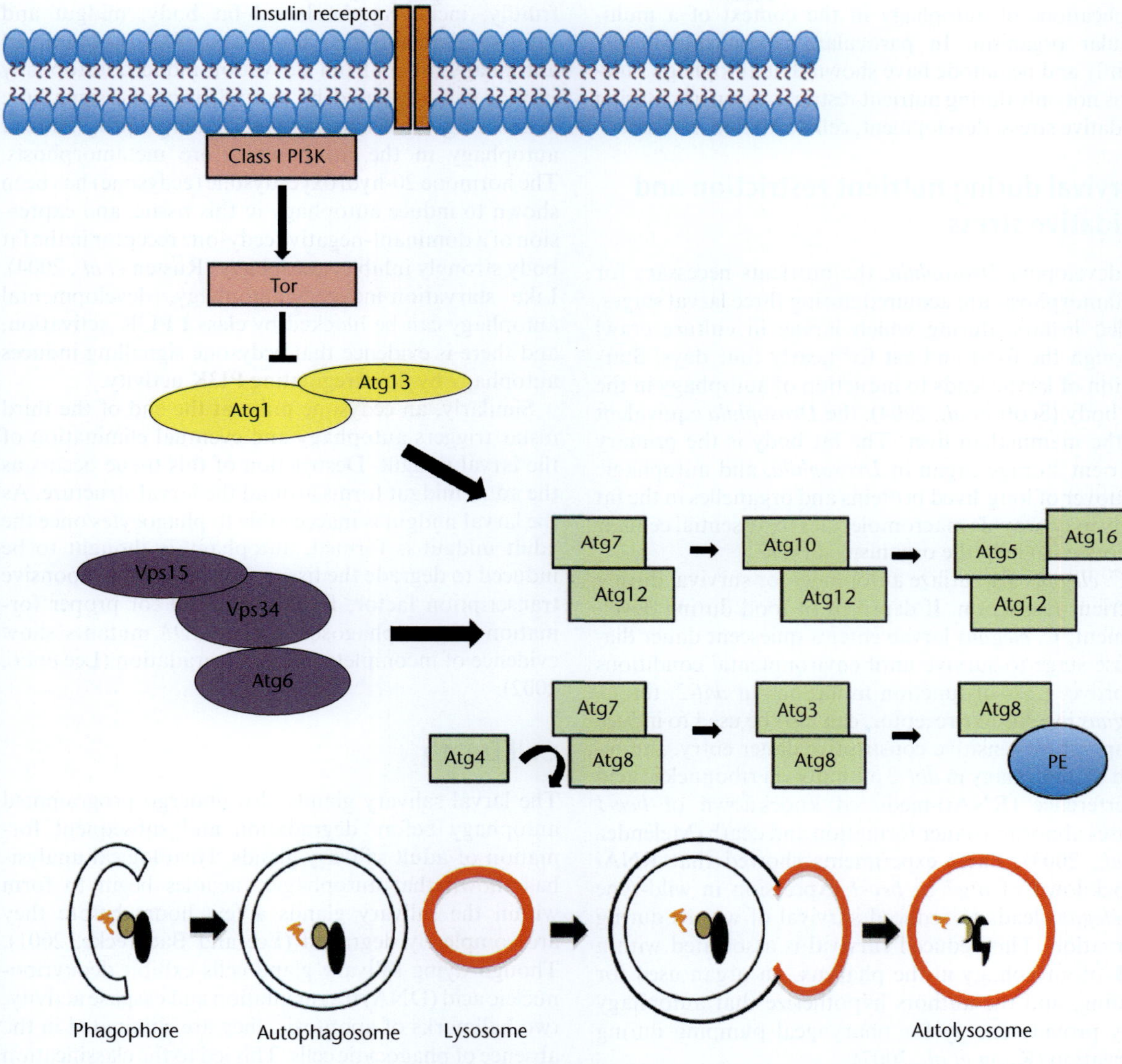

Figure 1 Regulation of autophagy. Class I PI3K negatively regulates autophagy via the kinase Tor. Inactivation of Tor by starvation or rapamycin treatment leads to induction of autophagy. Atg1 kinase activity is required for autophagy induction, and Atg1 binds to Atg13. Vps34, Vps15 and Atg6 regulate the formation of a double membrane phagophore around proteins and organelles to be degraded. Two ubiquitin-like conjugation pathways regulate the elongation of the autophagosome membrane. Atg8 is cleaved at the C-terminus by Atg4. Atg7 activates Atg12 and Atg8 in an adenosine triphosphate (ATP)-dependent manner. This is followed by covalent conjugation of Atg12 to Atg5 and Atg8 to the lipid phosphatidylinositol (PE). Atg5 binds noncovalently to Atg16, and Atg8-PE becomes anchored in the autophagosome membrane. The outer membrane of the autophagosome fuses with the lysosome, releasing the inner membrane and sequestered contents into the lysosome for degradation.

and fluid phase endocytosis and die before metamorphosis (Juhász *et al.*, 2008). Similarly, loss of *bec-1*, the *C. elegans* homologue of *atg6*, leads to a defect in PI(3)P localization to intracellular membranes, as well as embryonic lethality (Takacs-Vellai *et al.*, 2005).

Functions of Autophagy

Thanks to several studies in yeast, autophagy has been well characterized for its role in cell survival during nutrient deprivation. Studies in higher eukaryotes have revealed more details about the physiological implications of autophagy in the context of a multicellular organism. In particular, experiments in the fruitfly and nematode have shown that autophagy functions not only during nutrient restriction, but also during oxidative stress, development, cell death and immunity.

Survival during nutrient restriction and oxidative stress

In developing *Drosophila*, the nutrients necessary for metamorphosis are acquired during three larval stages, called instars, during which larvae in culture crawl through the food and eat for nearly four days. Starvation of larvae leads to induction of autophagy in the fat body (Scott *et al.*, 2004), the *Drosophila* equivalent of the mammalian liver. The fat body is the primary nutrient storage organ in *Drosophila*, and autophagic turnover of long-lived proteins and organelles in the fat body may provide macromolecules for essential cellular processes to help the organism survive.

C. elegans also utilize autophagy for survival during nutrient restriction. If deprived of food during development, *C. elegans* larvae enter a quiescent dauer diapause stage to survive until environmental conditions improve. Loss-of-function mutations in *daf-2*, the *C. elegans* insulin-like receptor, can also be used to induce temperature-sensitive constitutive dauer entry. Inhibition of autophagy in *daf-2* mutants via ribonucleic acid interference (RNAi)-mediated knockdown of *bec-1* causes abnormal dauer formation and death (Meléndez *et al.*, 2003). Later experiments showed that RNAi knockdown of *atg7* or *bec-1* expression in wild-type *C. elegans* leads to reduced survival of worms during starvation. This reduced survival is associated with a lack of autophagy in the pharynx, an organ used for feeding, and the authors hypothesize that autophagy may provide energy for pharyngeal pumping during starvation (Kang *et al.*, 2007).

In addition to its role in survival under starvation conditions, autophagy also functions to relieve oxidative stress by removing damaged organelles. Experiments in adult flies have shown that autophagy is required for survival in response to chemically induced oxidative stress. Loss-of-function *atg7* flies survive to adulthood but are hypersensitive to oxidative stress induced by treatment with paraquat, an herbicide that generates free radicals in the cell (Juhász *et al.*, 2007).

Development

Studies in *Drosophila* revealed a developmental function for autophagy. Several tissues in the developing fruitfly, including the larval fat body, midgut and salivary glands, utilize autophagy. This programmed autophagy differs from starvation-induced autophagy in that it is regulated by steroid hormone signalling. Wandering third instar larvae undergo programmed autophagy in the fat body before metamorphosis. The hormone 20-hydroxyecdysone (ecdysone) has been shown to induce autophagy in this tissue, and expression of a dominant-negative ecdysone receptor in the fat body strongly inhibits autophagy (Rusten *et al.*, 2004). Like starvation-induced autophagy, developmental autophagy can be blocked by class I PI3K activation, and there is evidence that ecdysone signalling induces autophagy by downregulating PI3K activity.

Similarly, an ecdysone pulse at the end of the third instar triggers autophagy and eventual elimination of the larval midgut. Destruction of this tissue occurs as the adult midgut forms around the larval structure. As the larval midgut is inaccessible to phagocytes once the adult midgut is formed, autophagy is thought to be induced to degrade the tissue. The ecdysone responsive transcription factor, E93, is required for proper formation of autophagosomes, and E93 mutants show evidence of incomplete midgut degradation (Lee *et al.*, 2002).

Cell Death

The larval salivary glands also undergo programmed autophagy before degradation and subsequent formation of adult salivary glands. Histological analysis has shown that autophagic vacuoles begin to form within the salivary glands a few hours before they are completely degraded (Lee and Baehrecke, 2001). Though dying salivary gland cells exhibit deoxyribonucleic acid (DNA) fragmentation and caspase activity, two hallmarks of apoptosis, they are eliminated in the absence of phagocytic cells. This led to the classification of salivary gland cell death as autophagic cell death.

Like midgut cell death, salivary gland cell death requires expression of ecdysone responsive genes. A developmentally regulated pulse of ecdysone induces expression of the transcription factor E93. Although direct targets of E93 have yet to be identified, microarray analysis of dying salivary glands showed that several *atg* genes and caspases are induced immediately before cell death with autophagy, and that the response of several of these genes was attenuated in salivary glands of *E93* mutants (Lee *et al.*, 2003). Genetic experiments showed that caspases and autophagy are required for salivary gland cell death, and that ectopic expression of ATG1 can induce early gland degradation independent of caspase activation (Martin and Baehrecke, 2004; Berry and Baehrecke, 2007). In the latter study, autophagy induction was linked to growth arrest through downregulation of class I PI3K signalling, and ectopic PI3K induction was sufficient to prevent normal gland degradation.

Caspases have also been shown to regulate autophagy and cell death in *Drosophila* ovaries, where starvation can trigger death of egg chambers before their normal developmental cell death during late stage oogenesis. An RNAi-based assay was used to identify known cell death genes that regulate starvation-induced autophagy in a larval blood cell line and the function of these genes in autophagy was tested *in vivo* (Hou *et al.*, 2008). Death caspase-1 (DCP-1) and BRUCE, an inhibitor of apoptosis, were among the known cell death regulators found to function in autophagy in ovaries. DCP-1 is required for induction of autophagy in degenerating egg chambers, whereas BRUCE plays an inhibitory role under nutrient-rich conditions. Consistent with caspase activity, degenerating egg chambers also show evidence of DNA fragmentation, and this requires autophagy, as knockout of *atg1* or *atg7* leads to reduced DNA fragmentation.

A recent model of neuronal cell death in *C. elegans* has linked autophagy to necrosis. Gain of function alleles of *mec-4(d)*, which encodes an ion channel subunit, is neurotoxic in early larval stages and this cell death is necrotic (Hall *et al.*, 1997). Neuronal cell death in *mec-4(d)* requires expression of the ATG1 orthologue, uncoordinated family member 51 (UNC-51), and a time course revealed that autophagy is highly induced in neurons during the early stages of necrosis, but is downregulated in later stages (Samara *et al.*, 2008). The authors hypothesize that high levels of autophagy trigger necrotic cell death and propose that inhibition of autophagy might protect neurons from cell death after ischaemic stroke.

Disease Models

Infectious disease

Autophagy shares multiple regulatory components, including PI(3)P and RAB GTPases (guanosine triphosphatase), with the phagocytosis lysosomal degradation pathway. Data from mammalian cell lines revealed that autophagy is able to eliminate some pathogens that impair host cell phagocytic machinery (reviewed in Deretic, 2006). Upon infection, intracellular bacteria are engulfed by phagosomes, which later fuse with lysosomes, leading to pathogen degradation. Some pathogens, like *Mycobacterium tuberculosis*, are able to evade the host's initial defence mechanisms by inhibiting phagosome maturation in macrophages. Induction of autophagy in macrophage cell lines infected with *M. tuberculosis*, by starvation or rapamycin treatment, suppresses the phagosome maturation defect and inhibits bacterial survival (Gutierrez *et al.*, 2004). Confocal microscopy analyses indicate that intracellular compartments containing *M. tuberculosis* also contain the autophagy proteins BECLIN1 and LC3, whereas electron microscopy data suggest that the phagosome maturation pathway merges with the autophagic pathway to rid cells of the pathogen via the lysosome.

Other pathogens, like *Listeria monocytogenes*, act by lysing host cell phagosomes. In response, autophagy is induced in the cell and autophagosomes engulf the bacteria for transport to the lysosome for degradation (Rich *et al.*, 2003). Recent experiments in *Drosophila* revealed the mechanisms behind the induction of autophagy in response to *L. monocytogenes* infection. The peptidoglycan recognition protein family member (PGRP)-LE is required for resistance to *L. monocytogenes*, as PGRP-LE mutants are significantly more susceptible to pathogen infection than wild-type flies. PGRP-LE recognizes peptidoglycan in the cell wall of *L. monocytogenes* that invades the cytoplasm following phagosome lysis. PGRP-LE is required for induction of autophagy in this context, as autophagosomes fail to form around invading bacteria in cells lacking PGRP-LE (Yano *et al.*, 2008). Furthermore, inhibition of autophagy by RNAi-mediated knockdown of *atg5* in haemocytes, the *Drosophila* blood cells, leads to increased susceptibility to infection, an indication that autophagy is essential for survival after *L. monocytogenes* infection.

Neurodegeneration

In recent years, several neurodegeneration models have been described in flies and worms, and autophagy defects can lead to aggregation of proteins in the nervous system. Ubiquitinated proteins accumulate in the brains of *atg7* mutant flies, and this leads to neuronal cell death (Juhász *et al.*, 2007). Similarly, deletion of *vps15* causes defects in autophagy and accumulation of ubiquitinated proteins in the fat body and gut (Lindmo *et al.*, 2008). These ubiquitinated protein aggregates also contain REF(2)P, the orthologue of mammalian P62, a protein involved in formation of inclusion bodies in cells deficient in autophagy (Komatsu *et al.*, 2007). Studies have shown that REF(2)P is required for protein aggregate formation in the brain of older adult flies. Significantly, inhibition of autophagy by mutation of ATG8 leads to accumulation of REF(2)P aggregates in the brains of young adult flies (Nezis *et al.*, 2008).

Huntington disease is a well-characterized human neurodegenerative condition that can be studied in flies and worms. It is caused by aggregation of mutant HUNTINGTON protein (HTT), which has a polyglutamine expansion in the *N*-terminus that prevents cleavage and degradation by the proteasome. Experiments in mammalian cell lines have shown that autophagy can degrade polyglutamine protein aggregates. *Drosophila* has been used as a model for Huntington's disease because the consequences of protein aggregation can be studied in the context of the whole organism. Cellular implications of impaired HTT degradation can be studied in flies by ectopically expressing mutant versions of the protein in the adult eye, which is made up of photoreceptor neurons that can be monitored by microscopy for cell death. Expression of mutant HTT in the *Drosophila* eye leads to degeneration and this can be rescued by induction of autophagy by the TOR inhibitor rapamycin (Ravikumar *et al.*, 2004). Autophagy was also found to be important in the *C. elegans* models of HTT disease. Expression of mutant human HTT in RNAi-treated *bec-1* or *atg7* mutant worms causes increased aggregate formation and degeneration of adult sensory neurons (Jia *et al.*, 2007).

Like Huntington disease, spinobulbar muscular atrophy (SBMA) is a neurodegenerative disorder caused by aggregation of a polyglutamine-containing protein in neurons. In the case of SBMA, the polyglutamine expansion occurs in the androgen receptor (AR). A fly model of SBMA was described in 2007, and this study revealed a link between proteasome impairment and autophagic degradation of protein aggregates (Pandey *et al.*, 2007). Expression of a polyglutamine AR mutant in the eye of flies treated with dihydrotestosterone (DHT), the ligand of AR, causes cell degeneration. Expression of a fluorescent proteasome reporter revealed that this degenerative phenotype is associated with proteasome impairment. The proteasome impairment and cell degeneration phenotypes were rescued by expression of histone deacetylase 6 (HDAC6), a microtubule-associated protein previously shown to interact with ubiquitinated proteins and facilitate their removal from the cell via dynein motors (Kawaguchi *et al.*, 2003). RNAi-mediated knock down of *atg12* and *atg6* showed that HDAC6 suppression of AR-induced degeneration requires autophagy. These results demonstrated that autophagy can compensate for proteasome impairment to rid the cell of harmful protein aggregates.

Ageing

The relationship between nutrition and lifespan extension has been well studied in *C. elegans*, but details about the possible role of autophagy in ageing have only recently been recognized. Downregulation of insulin receptor/TOR signalling has been associated with lifespan extension in worms. Worms with mutations in *daf-2*, the *C. elegans* insulin-like receptor, live twice as long as wild-type worms and this lifespan extension requires the autophagy gene *bec-1* (Kenyon *et al.*, 1993; Meléndez *et al.*, 2003). A mutation in the gene *eat-2* phenocopies dietary restriction due to an inability to feed, and this leads to lifespan extension. RNAi knock down of *bec-1* or *vps-34* in *eat-2* mutant adult worms shortens their lifespan by up to 30%, but does not affect lifespan in wild-type worms (Hansen *et al.*, 2008). Like yeast and fruitflies, autophagy is regulated by TOR in *C. elegans*, and RNAi experiments show that autophagy is required for lifespan extension in animals with reduced TOR signalling.

The relationship between autophagy and lifespan has also been studied in *Drosophila*. Several studies have shown that inhibition of autophagy by *atg* gene mutant analysis leads to the accumulation of ubiquitinated proteins in the brain, a phenotype usually associated with ageing. The *blue cheese* (*bchs*) gene is one such example, as mutation leads to the accumulation of ubiquitinated protein aggregates, neurodegeneration and decreased lifespan. *bchs* encodes a FYVE domain containing protein, and its human homologue, autophagy-linked FYVE protein (ALFY), co-localizes with autophagic machinery and ubiquitin-positive structures under proteasome impairment conditions (Simonsen *et al.*, 2004). A screen for genetic modifiers of a *bchs* gain of function phenotype in the eye revealed genes involved in lysosomal trafficking and autophagy (Simonsen *et al.*, 2007). Notably, a mutation in *atg8* resulted in decreased lifespan relative to wild-type flies. Subsequent gain of function experiments showed that ectopic expression of ATG8 in adult brains caused lifespan extension. Flies expressing ATG8 survive, on average, 56% longer than control flies (Simonsen *et al.*, 2008). These results are consistent with previous studies, which showed that the loss of *atg7* causes reduction of lifespan (Juhász *et al.*, 2007). Collectively, data from flies and worms provides compelling evidence that autophagy has an antiageing function, due in part, to its role in elimination of protein aggregates.

Conclusion

Studies in nonmammalian organisms have contributed immensely to the field of autophagy research. The budding yeast, *S. cerevisiae*, has been an invaluable tool for determining the molecular mechanisms of autophagosome formation. Originally discovered in yeast screens, the *atg* genes are conserved in worms, fruitflies and mammals, and are essential for autophagy function. Experiments in *C. elegans* have revealed details about the relationship between autophagy and cell death and the role of autophagy in lifespan extension induced by nutrient restriction. The relationship between autophagy and cell death has also been well characterized in *Drosophila* development. The short generation time and abundance of phenotypic markers allow for complex genetic experiments that cannot be easily done in mammalian systems. Disease models have also been developed in nematodes and flies, where autophagy has been studied for its function in neurodegeneration and infection. Further studies in these organisms will be necessary to complement the dearth of information from mammalian cell lines with *in vivo* data. Such studies will present a clearer picture of how autophagy functions in the context of a whole animal in regulating development, cell death and survival under stressful conditions.

Acknowledgement

We thank the NIH (GM59136 and GM079431) for support.

References

Bankaitis VA, Johnson LM and Emr SD (1986) Isolation of yeast mutants defective in protein targeting to the vacuole. *Proceedings of the National Academy of Sciences of the USA* **83**: 9075–9079.

Berry DL and Baehrecke EH (2007) Growth arrest and autophagy are required for salivary gland cell degradation in *Drosophila*. *Cell* **131**: 1137–1148.

Deretic V (2006) Autophagy as an immune defense mechanism. *Current Opinion in Immunology* **18**: 375–382.

Deter RL and de Duve C (1967) Influence of glucagon, an inducer of cellular autophagy, on some physical properties of rat liver lysosomes. *Journal of Cell Biology* **33**: 437–449.

Fimia GM, Stoykova A, Romagnoli A *et al.* (2007) Ambra1 regulates autophagy and development of the nervous system. *Nature* **447**: 1121–1125.

Gutierrez MG, Master SS, Singh SB *et al.* (2004) Autophagy is a defense mechanism inhibiting BCG and *Mycobacterium tuberculosis* survival in infected macrophages. *Cell* **119**: 753–766.

Hall DH, Gu G, García-Añoveros J *et al.* (1997) Neuropathology of degenerative cell death in *Caenorhabditis elegans*. *Journal of Neuroscience* **17**: 1033–1045.

Hansen M, Chandra A, Mitic LL *et al.* (2008) A role for autophagy in the extension of lifespan by dietary restriction in *C. elegans*. *Public Library of Science Genetics* **4**: e24.

Harding TM, Hefner-Gravink A, Thumm M and Klionsky DJ (1996) Genetic and phenotypic overlap between autophagy and the cytoplasm to vacuole protein targeting pathway. *Journal of Biological Chemistry* **271**: 17621–17624.

Harding TM, Morano KA, Scott SV and Klionsky DJ (1995) Isolation and characterization of yeast mutants in the cytoplasm to vacuole protein targeting pathway. *Journal of Cell Biology* **131**: 591–602.

Hou YC, Chittaranjan S, Barbosa SG, McCall K and Gorski SM (2008) Effector caspase Dcp-1 and IAP protein Bruce regulate starvation-induced autophagy during *Drosophila melanogaster* oogenesis. *Journal of Cell Biology* **182**: 1127–1139.

Ichimura Y, Kirisako T, Takao T *et al.* (2000) A ubiquitin-like system mediates protein lipidation. *Nature* **408**: 488–492.

Itakura E, Kishi C, Inoue K and Mizushima N (2008) Beclin 1 forms two distinct phosphatidylinositol 3-kinase complexes with mammalian Atg14 and UVRAG. *Molecular Biology of the Cell* **19**: 5360–5372.

Jia K, Hart AC and Levine B (2007) Autophagy genes protect against disease caused by polyglutamine expansion proteins in *Caenorhabditis elegans*. *Autophagy* **3**: 21–25.

Juhász G, Erdi B, Sass M and Neufeld TP (2007) Atg7-dependent autophagy promotes neuronal health, stress tolerance, and longevity but is dispensable for metamorphosis in *Drosophila*. *Genes & Development* **21**: 3061–3066.

Juhász G, Hill JH, Yan Y *et al.* (2008) The class III PI(3)K Vps34 promotes autophagy and endocytosis but not TOR signaling in *Drosophila*. *Journal of Cell Biology* **181**: 655–666.

Kamada Y, Funakoshi T, Shintani T *et al.* (2000) Tor-mediated induction of autophagy via an Apg1 protein kinase complex. *Journal of Cell Biology* **150**: 1507–1513.

Kang C, You YJ and Avery L (2007) Dual roles of autophagy in the survival of *Caenorhabditis elegans* during starvation. *Genes & Development* **21**: 2161–2171.

Kawaguchi Y, Kovacs JJ, McLaurin A *et al.* (2003) The deacetylase HDAC6 regulates aggresome formation and cell viability in response to misfolded protein stress. *Cell* **115**: 727–738.

Kenyon C, Chang J, Gensch E, Rudner A and Tabtiang R (1993) A *C. elegans* mutant that lives twice as long as wild type. *Nature* **366**: 461–464.

Kihara A, Kabeya Y, Ohsumi Y and Yoshimori T (2001) Beclin-phosphatidylinositol 3-kinase complex functions at the trans-Golgi network. *European Molecular Biology Organization Reports* **4**: 330–335.

Klionsky DJ, Cregg JM, Dunn WA Jr *et al.* (2003) A unified nomenclature for yeast autophagy-related genes. *Developmental Cell* **5**: 539–545.

Komatsu M, Waguri S, Koike M *et al.* (2007) Homeostatic levels of p62 control cytoplasmic inclusion body formation in autophagy-deficient mice. *Cell* **131**: 1149–1163.

Lee CY and Baehrecke EH (2001) Steroid regulation of autophagic programmed cell death during development. *Development* **128**: 1443–1455.

Lee CY, Clough EA, Yellon P *et al.* (2003) Genome-wide analyses of steroid- and radiation-triggered programmed cell death in *Drosophila*. *Current Biology* **13**: 350–357.

Lee CY, Cooksey BA and Baehrecke EH (2002) Steroid regulation of midgut cell death during *Drosophila* development. *Developmental Biology* **250**: 101–111.

Liang C, Feng P, Ku B *et al.* (2006) Autophagic and tumor suppressor activity of a novel Beclin 1-binding protein UVRAG. *Nature Cell Biology* **8**: 688–699.

Liang XH, Kleeman LK, Jiang HH *et al.* (1998) Protection against fatal *Sindbis virus* encephalitis by beclin, a novel Bcl-2-interacting protein. *Journal of Virology* **72l**: 8586–8596.

Lindmo K, Brech A, Finley KD *et al.* (2008) The PI 3-kinase regulator Vps15 is required for autophagic clearance of protein aggregates. *Autophagy* **4**: 500–506.

Martin DN and Baehrecke EH (2004) Caspases function in autophagic programmed cell death in *Drosophila*. *Development* **131**: 275–284.

Matsuura A, Tsukada M, Wada Y and Ohsumi Y (1997) Apg1p, a novel protein kinase required for the autophagic process in *Saccharomyces cerevisiae*. *Gene* **192**: 245–250.

Meléndez A, Talloczy Z, Seaman M *et al.* (2003) Autophagy genes are essential for dauer development and life-span extension in *C. elegans*. *Science* **301**: 1387–1391.

Mizushima N, Noda T, Yoshimori T *et al.* (1998) A protein conjugation system essential for autophagy. *Nature* **395**: 395–398.

Nezis IP, Simonsen A, Sagona AP *et al.* (2008) Ref(2)P, the *Drosophila melanogaster* homologue of mammalian p62, is required for the formation of protein aggregates in adult brain. *Journal of Cell Biology* **180**: 1065–1071.

Pandey UB, Nie Z, Batlevi Y *et al.* (2007) HDAC6 rescues neurodegeneration and provides an essential link between autophagy and the UPS. *Nature* **447**: 859–863.

Ravikumar B, Vacher C, Berger Z *et al.* (2004) Inhibition of mTOR induces autophagy and reduces toxicity of polyglutamine expansions in fly and mouse models of Huntington disease. *Nature Genetics* **36**: 585–595.

Rich KA, Burkett C and Webster P (2003) Cytoplasmic bacteria can be targets for autophagy. *Cell Microbiology* **5**: 455–468.

Rusten TE, Lindmo K, Juhász G *et al.* (2004) Programmed autophagy in the *Drosophila* fat body is induced by ecdysone through regulation of the PI3K pathway. *Developmental Cell* **7**: 179–192.

Samara C, Syntichaki P and Tavernarakis N (2008) Autophagy is required for necrotic cell death in *Caenorhabditis elegans*. *Cell Death and Differentiation* **15**: 105–112.

Scott RC, Schuldiner O and Neufeld TP (2004) Role and regulation of starvation-induced autophagy in the *Drosophila* fat body. *Developmental Cell* **7**: 167–178.

Simonsen A, Birkeland HC, Gillooly DJ *et al.* (2004) Alfy, a novel FYVE-domain-containing protein associated with protein granules and autophagic membranes. *Journal of Cell Science* **117**: 4239–4251.

Simonsen A, Cumming RC, Brech A *et al.* (2008) Promoting basal levels of autophagy in the nervous system enhances longevity and oxidant resistance in adult *Drosophila*. *Autophagy* **4**: 176–184.

Simonsen A, Cumming RC, Lindmo K *et al.* (2007) Genetic modifiers of the *Drosophila* blue cheese gene link defects in lysosomal transport with decreased life span and altered ubiquitinated-protein profiles. *Genetics* **176**: 1283–1297.

Takacs-Vellai K, Vellai T, Puoti A *et al.* (2005) Inactivation of the autophagy gene bec-1 triggers apoptotic cell death in *C. elegans*. *Current Biology* **15**: 1513–1517.

Takahashi Y, Coppola D, Matsushita N *et al.* (2007) Bif-1 interacts with Beclin 1 through UVRAG and regulates autophagy and tumorigenesis. *Nature Cell Biology* **9**: 1142–1151.

Takeshige K, Baba M, Tsuboi S, Noda T and Ohsumi Y (1992) Autophagy in yeast demonstrated with proteinase-deficient mutants and conditions for its induction. *Journal of Cell Biology* **119**: 301–311.

Tanida I, Mizushima N, Kiyooka M *et al.* (1999) Apg7p/Cvt2p: a novel protein-activating enzyme essential for autophagy. *Molecular Biology of the Cell* **10**: 1367–1379.

Thumm M, Egner R, Koch B *et al.* (1994) Isolation of autophagocytosis mutants of *Saccharomyces cerevisiae*. *FEBS Letters* **349**: 275–280.

Tsukada M and Ohsumi Y (1993) Isolation and characterization of autophagy-defective mutants of *Saccharomyces cerevisiae*. *FEBS Letters* **333**: 169–174.

Yano T, Mita S, Ohmori H *et al.* (2008) Autophagic control of listeria through intracellular innate immune recognition in *drosophila. Nature Immunology* **9**: 908–916.

Further Reading

Baehrecke EH (2002) How death shapes life during development. *Nature Reviews Molecular Cell Biology* **3**: 779–787.

Klionsky DJ (2007) Autophagy: from phenomenology to molecular understanding in less than a decade. *Nature Reviews Molecular Cell Biology* **8**: 931–937.

Levine B and Kroemer G (2007) Autophagy in the pathogenesis of disease. *Cell* **137**: 27–42.

Meléndez A and Neufeld TP (2008) The cell biology of autophagy in metazoans: a developing story. *Development* **135**: 2347–2360.

Xie Zhiping and Klionsky DJ (2007) Autophagosome formation: core machinery and adaptations. *Nature Cell Biology* **9**: 1102–1109.

Apoptosis: Regulatory Genes and Disease

James E Vince, *The University of Lausanne, Epalinges, Switzerland*
John Silke, *La Trobe University, Melbourne, Victoria, Australia*

Based in part on the previous version of this Encyclopedia of Life Sciences (ELS) article, Apoptosis: Regulatory Genes and Disease by Leonhard Müllauer, Sabine Wohlfart and Brigitte Scheucher.

Advanced article

Article Contents

- Introduction
- Dysregulation of the Intrinsic and Extrinsic Apoptotic Pathways in Disease
- Dysregulation and Disease in the Intrinsic Bcl-2 Blockable Apoptotic Pathway
- Perforin and Granzyme B
- Dysregulation and Disease in the Extrinsic Apoptosis Pathway

Apoptosis is essential for normal development and proceeds by the extrinsic death receptor pathway, or the intrinsic Bcl-2 blockable pathway. Both pathways activate a class of proteases termed caspases that cleave intracellular substrates resulting in cell death. Excessive apoptosis may excerbate neurological diseases and the damage caused by heart attacks and strokes. Apoptosis is used by metazoans as a defence against pathogens that have, therefore, evolved mechanisms to block the apoptotic machinery. The involvement of apoptosis in disease states has led to the rational design of compounds that target this pathway which are already successfully used in the clinic.

Introduction

The term apoptosis was coined by Kerr, Wyllie and Currie in 1972 to describe a form of programmed cell death that gives rise to dead cells with a defined morphology and is distinct from other modes of cell death such as necrosis. The apoptotic programme results in the activation of caspases, intracellular cysteine proteases that cleave cellular substrates after aspartate residues. Caspase mediated cleavage of cellular protein substrates results in the morphological changes that are the defining feature of apoptotic cell death and include the retraction of pseudopodes, membrane blebbing, chromatin condensation and internucleosomal deoxyribonucleic acid (DNA) fragmentation (Kroemer *et al.*, 2009). Apoptotic cells are efficiently targeted *in vivo* for phagocytosis by circulating immune cells and under most normal circumstances this elegant mode of cellular deconstruction and rapid clearance prevents an inappropriate immune response.

Although the key caspases involved in apoptosis have been well characterized (caspases-3, -6, -7, -8, -9 and -10), other caspases, termed inflammatory caspases (caspases-1, -4, -5, -11 and -12), are primarily involved in regulating inflammatory molecule production and immune responses induced by viral or microbial infection, and there is still no well defined physiological role for some caspases, such as caspase-2. Like most complex systems, caspases defy strict categorization and can play diverse, often opposing, functions depending on the cell type they are expressed in and the type and strength of the stimulus provided. For example, caspase-8, which is essential for death receptor apoptosis (see later discussion), is also required for antigen-driven T-cell expansion. **See also**: Caspases and Cell Death; Caspases in Inflammation and Immunity; Caspases, Substrates and Sequential Activation; Dismantling the Apoptotic Cell; Structure, Domains and functions in Cell Death (DD, DED, CARD, PYD); The Apoptosome: The Executioner of Mitochondria-mediated Apoptosis

There are two main pathways that result in caspase activation and apoptosis. These can be biochemically defined by the type of apoptotic stimuli, the initiator caspases involved, and the regulatory complexes formed to promote or inhibit the apoptotic cascade. They include the intrinsic mitochondrial, or Bcl-2 blockable apoptosis pathway, and the extrinsic or death receptor apoptosis pathway, which in most cases occurs independently of Bcl-2 and mitochondrial function. Both intrinsic and extrinsic apoptotic stimulation results in the activation of pathway-specific initiator (apical) caspases (caspases-2, -8, -9 and -10) after which the apoptotic pathways converge with activation of the effector, or executioner, caspases (caspases -3, -6 and -7), that are responsible for cellular demolition and the distinct apoptotic morphological phenotype. Two other death signalling pathways implicated in a number of human conditions, perforin–granzyme B-cell death and caspase-1-induced pyroptopsis, have been suggested as novel apoptosis pathways capable of operating independently from the intrinsic and extrinsic pathways, although their precise mechanisms are still being contested. **See also**: Death Receptors; The Bcl-2 Family Proteins – Key Regulators and Effectors of Apoptosis

Dysregulation of apoptosis is believed to contribute to cancer development, autoimmune diseases, defective immunity, viral and microbial pathenogenesis, developmental abnormalities and neurodegenerative disorders. This article briefly describes the intrinsic and extrinsic apoptotic pathways and then analyses their involvement in several disease states.

Intrinsic or Bcl-2 blockable apoptosis

Many intrinsic or cell internal stresses, including DNA damage, organelle damage, hypoxia and growth factor deprivation, induce apoptosis when the stimuli overwhelm the cells' ability to protect and repair itself. Intrinsic damaging agents include UV- and ionizing radiation, heat stress and toxins/chemicals that interfere with cellular metabolism. Similarly, receptor stimulation, or lack of, by glucocorticoids, cytokines, hormones and growth factors can also result in internal death responses, often through transcriptional gene regulation of Bcl-2 family members. Protein folding diseases (ie. Alzheimer's disease) also cause intracellular stress and can result in excessive apoptosis, as do genetic lesions that increase or decrease a protein's activity (i.e. Huntington disease). How cells decide whether to try to resist and repair damage or commit suicide has yet to be determined but is likely to involve the delicate equilibrium that exists between the levels of pro- and antiapoptotic Bcl-2 family members. If levels of the antiapoptotic Bcl-2 family members (including Bcl-2 itself) are high enough, or the pro-apoptotic Bcl-2 family members are absent, a cell is extremely resistant to death caused by intrinsic cellular stress. Diseases, such as cancer, that result from insufficient cell death are frequently caused by ectopic overexpression of pro-survival Bcl-2 family members, or alternatively, by defects in the pro-apoptotic arm. **See also**: BH3-Only Proteins; Structure and Function of IAP and Bcl-2 Proteins; The Bcl-2 Family Proteins – Key Regulators and Effectors of Apoptosis

The Bcl-2 family can be subdivided into three groups on the basis of conservation of Bcl-2 homology (BH) domains. The first group contains the founding member of the family, Bcl-2 and other multidomain pro-survival proteins, Bcl-x_L, Mcl-1, A1 and Bcl-w. The second group is the multidomain pro-apoptotic effectors Bak, Bax and Bok/Mtd1. They are highly related to the pro-survival proteins and structurally are almost indistinguishable despite having the opposite function to their protective counterparts. The last group, of eight BH3-only proteins, including Bim, Bid, Puma, Noxa and Bad are regulated transcriptionally or post-translationally in response to stress and initiate the apoptotic response.

A widely accepted, but still incomplete, model for intrinsic cell death is that the pro-apoptotic effectors Bax and Bak are kept in an inactive state by the pro-survival Bcl-2 family members and that BH3-only proteins once activated or upregulated compete the pro-survival proteins away from Bax and Bak and allow these proteins to oligomerize and form pores in the outer mitochondrial membrane (**Figure 1**). Mitochondria with damaged outer membranes are no longer able to generate energy by oxidative phosphorylation and also leak apoptogenic factors such as cytochrome *c* and Smac/DIABLO (direct IAP binding protein with low pI) from the intermembrane space. Cytochrome *c* release activates the platform protein Apaf-1 so that it recruits and activates the initiator caspase, caspase-9. Smac/DIABLO binds to a member of the inhibitor of

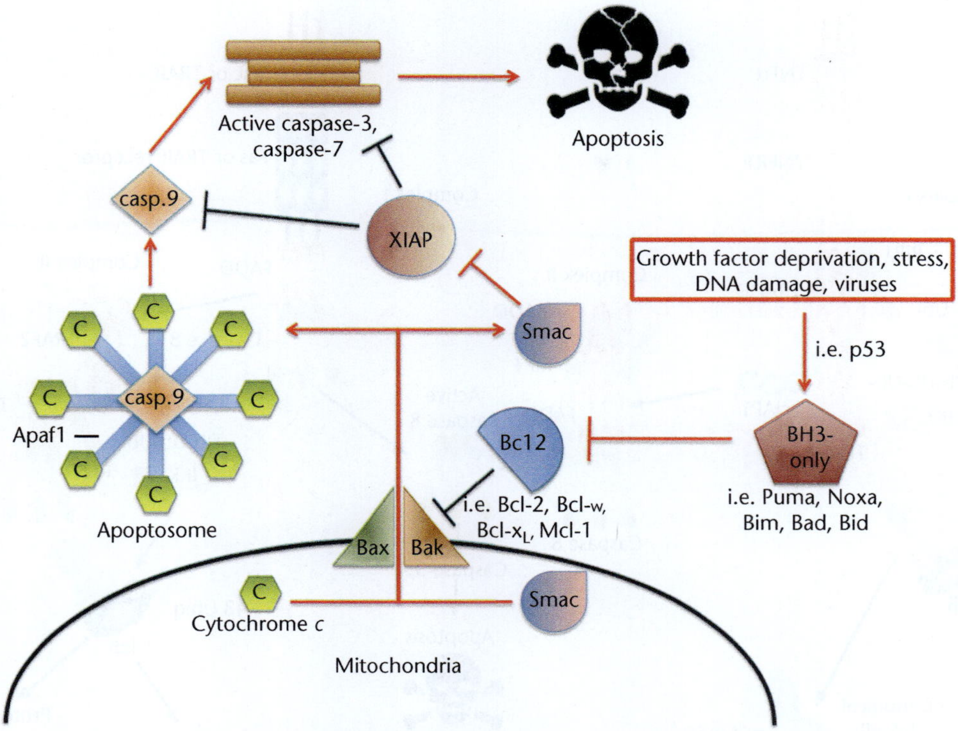

Figure 1 Model for the intrinsic, or Bcl-2 blockable, apoptosis pathway. Cell stresses, such as DNA damage induced by UV irradation, results in the transcriptional or posttranslational activation of pro-apoptotic BH3-only proteins. Their subsequent sequestration of the pro-survival Bcl-2 family members relieves Bcl-2 mediated inhibition of Bax and Bak and leads to Bax- and Bak-induced loss in mitochondrial membrane integrity. Cytochrome *c* is released into the cytosol, and together with Apaf-1, forms the caspase-9 activating apoptosome complex. IAP antagonists, such Smac, are released from the mitochondria concurrently with cytochrome *c* to antagonize XIAP inhibition of caspase activity. Caspase-9-mediated activation of the effector caspases, caspases-3 and caspases-7, results in apoptosis.

apoptosis family, XIAP (X-linked inhibitor of apoptosis), and prevents XIAP binding to and inhibiting caspase-9. Caspase activation ensures rapid apoptosis, which may be important in cases such as viral infection. However, disruption of the cells energy production alone probably delivers the mortal blow because, even if caspase activity is inhibited, the majority of cells with permeabilized mitochondrial outer membranes are destined to die. **See also**: Inhibitor of Apoptosis (IAP) and BIR-containing Proteins; Mitochondria Fusion and Fission; Mitochondrial Outer Membrane Permeabilization

Extrinsic death receptor apoptosis pathways

In addition to cells deciding for themselves whether they should kill themselves, they are also able to take lethal instructions from other cells, particularly those of the immune system. In nearly all cases, extrinsic apoptosis is initiated by death receptors, which belong to a subgroup of the tumour necrosis factor (TNF) receptor super family (TNFRSF). The classic model for signalling from these receptors is provided by the TRAIL (tumor necrosis factor-related apoptosis-inducing ligand) (DR4/TRAILR1 and DR5/TRAILR2), Fas (APO-1 and CD95) and TNFR1 death receptors. Ligand-induced oligomerization of the Fas, and TRAIL receptors results in the recruitment of the death domain containing adaptor protein FADD (FAS-associated death domain), which then binds to the initiator caspase, caspase 8, through homeotypic death effector domain interactions. This complex, termed the DISC (death-inducing signalling complex), results in oligomerization-induced activation and autocatalytic proteolytic processing of caspase-8 leading to downstream effector caspase activity and apoptosis (**Figure 2**). **See also**: Death Receptors; Death Receptors at the Molecular Level: Therapeutic Implications; Death Receptor-Induced Necroptosis

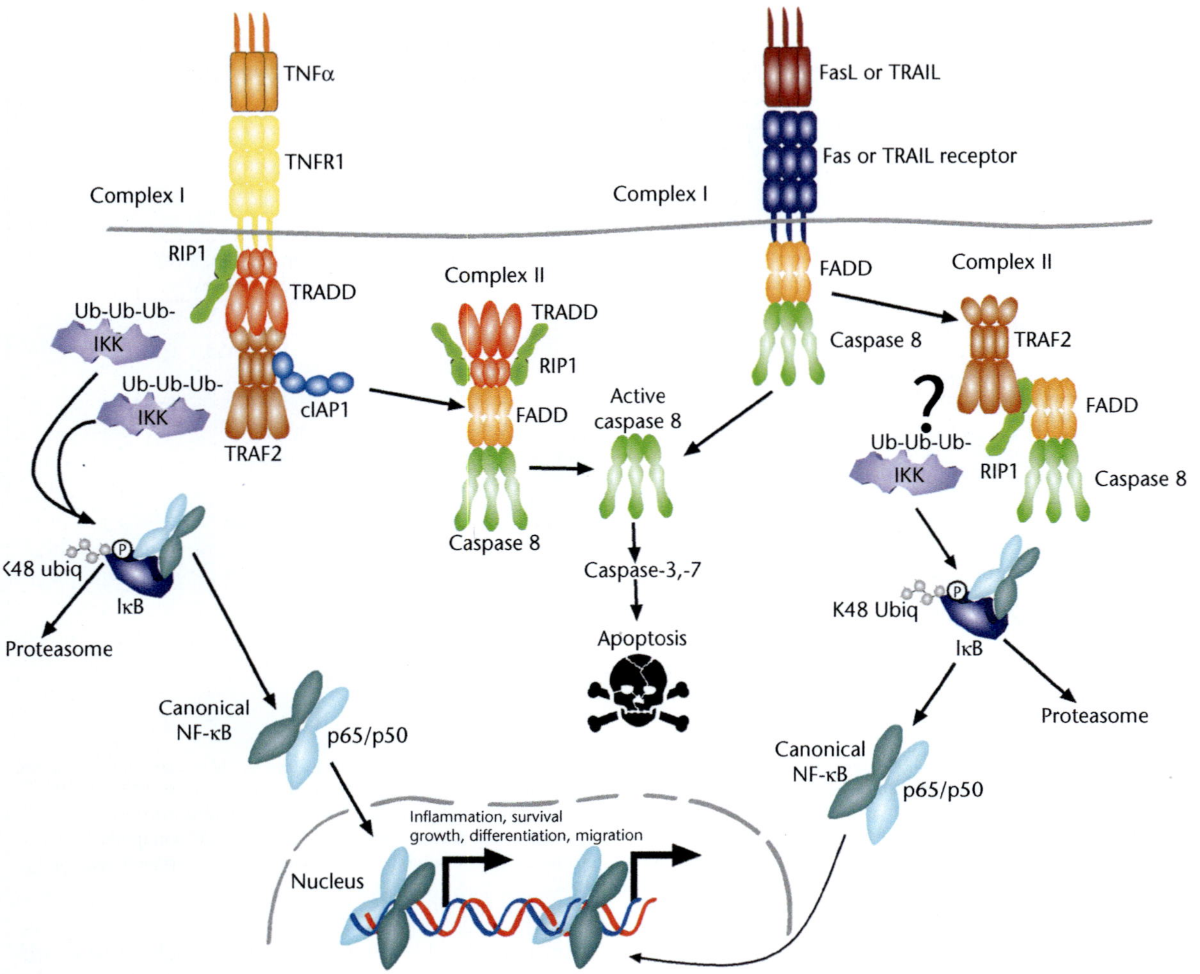

Figure 2 Models for death receptor signalling. TNFR1 activation results in the formation of an NFκB inducing complex (complex I) at the cell surface. TRAF2 and/or cIAP1 and cIAP2 ubiquitinate RIP1, and together with TRAF2 ubiquitination, this leads to the recruitment and activation of an IKK kinase complex. IKK-mediated phosphorylation of IκB results in its degradation by the proteasome, allowing the NFκB heterodimers, p65/p50, to translocate to the nucleus and initiate gene transcription. Following TRADD dissociation from TNFR1, FADD and caspase-8 are recruited to form a secondary complex, complex II, which results in cell death in the absence of NFκB function (see text). In contrast to TNFR1, Fas and TRAIL receptors initiate cell death complexes (complex I) at the plasma membrane. Although, TRAIL-induced NFκB has been reported to occur following FADD and caspase-8 dissociation from TRAILR1 and recruitment of RIP1, TRAF2 and IKK to form a cytoplasmic complex II, how Fas signals NFκB remains uncertain.

In some cell types there is cross talk between the extrinsic and Bcl-2 blockable pathways. In what has been called ‘type I’ cells, Fas and TRAIL induce cell death directly, in the manner described earlier. In contrast, Fas killing of ‘type II’ cells, such as hepatocytes, involves activation of the Bcl-2 blockable pathway (Yin *et al.*, 1999). Fas-induced caspase-8 activity results in processing of the BH3-only protein Bid, to tBid. Active tBid displaces Bcl-2 proteins (i.e. Bcl-x_L and Mcl-1) from Bax/Bak, leading to MOM (mitochondrial outermembrane) permeabilization and apoptosome formation. TNFα has been shown to kill hepatocytes by a similar mechanism, although Bim activation through jun *N*-terminal kinase (JNK) signalling, as well as Bid processing, is required for efficient TNFα killing of type II hepatocyte cells.

Less well understood is death receptor killing that has been reported to occur independently of caspase activation and has now been termed necroptosis (Kroemer *et al.*, 2009). On caspase inhibition it has been demonstrated that Fas, TRAIL and TNFα can initiate a necrotic form of cell death involving RIP1 (receptor interacting protein 1) (Holler *et al.*, 2000). Although the molecular mechanism of necroptosis remains to be properly elucidated, physiologically necroptosis may play a role in the regulation of T-cell proliferation and elimination (Holler *et al.*, 2000). As many pathogens express caspase inhibitory proteins to block the apoptotic process, necroptosis may also provide an alternate means of host cell suicide to help limit pathogen propagation. **See also**: Death Receptor-Induced Necroptosis

Death receptor activation of pro-survival signalling

Apart from signalling apoptosis, the death receptors can also simultaneously induce a pro-surivival nuclear factor-κB (NFκB) transcription factor response. TNFR1-induced NFκB signalling has been extensively studied and is initiated following the recruitment of TRADD (TNF receptor-associated death domain), RIP1, TRAF2 (TNF receptor associated factor 2) and cIAP1 (cellular inhibitor of apoptosis 1) and cIAP2 (**Figure 2**). It has been proposed that the ubiquitin E3 ligase activity of either the TRAF2 and/or cIAP1/cIAP2 RING (really interesting new gene) domains results in the modification of RIP1 with polyubiquitin chains. Both ubiquitin-modified RIP1 and/or TRAF2 have been proposed to bind and activate IKK complexes resulting in phosphorylation, ubiquitination and 26S proteasome-mediated degradation of the inhibitor of NFκB, IκB. The loss of IκB, which normally retains the canonical NFκB subunit p65 in the cytoplasm, liberates p65/p50 heterodimers allowing them to translocate into the nucleus and initiate canonical NFκB-mediated gene transcription. TNFR1 stimulation of NFκB activity induces transcription of a number of inflammatory mediators and pro-survival genes, including cFLIP (cellular FLICE-like inhibitory protein), which is a catalytically inactive homologue of caspase-8, and is able to compete with caspase-8 binding to FADD and thereby prevent TNFR1-induced apoptotic signalling.

TNFR1-induced activation of caspase-8 occurs in a secondary complex that does not contain TNFR1. Following TNFα binding and activation of NFκB in complex I, TRADD and RIP1 disscoiate from TNFR1 (**Figure 2**). The death domains of activated TRADD are now free to bind FADD/caspase-8 and the resulting oligomerized caspase-8 thereby becomes activated in what has been termed complex II. Although Fas and TRAIL receptors also activate NFκB, a major difference in these pathways is that the NFκB induced by TNFα is sufficient to block the apoptotic activity of the secondary complex. The dominant physiological function of TNFR1 is, therefore, not apoptosis but the induction of inflammatory cytokines that are required for immune cell organization and to combat infection, and TNFR1 only appears to induce apoptosis if NFκB signalling is disrupted. Consistent with this, TNFR1 knockout mice are resistant to endotoxic shock but succumb more easily to infection. Far from killing cells, in several carcinomas, TNFα-induced NFκB promotes tumour cell growth, increases vascular permeability and enhances immune cell recruitment to tumour nodules. Insidiously, metastatic lung carcinomas secrete a protein, versican, that is able to induce macrophage TNFα production, which in turn enables efficient metastasis of murine lung cancer (Kim *et al.*, 2009).

Although nearly always associated with apoptosis, it has been reported that Fas and TRAIL receptor activity, like TNFR1, can be associated with cell activation, differentiation, survival and growth depending on the cell and tissue type and context of stimulation. Although Fas signalling induces the destruction of liver cells during viral hepatitis, liver cirrhosis or Wilson disease, it has been reported that both Fas-deficient mice and mice where caspase-8 has been deleted from hepatocytes display impaired liver regeneration following partial hepatectomy, and that Fas agonist antibodies can expedite liver repair in wild-type mice (Peter *et al.*, 2007). Although the precise mechanisms for Fas-induced cell proliferation and survival remain unclear, Fas and TRAIL receptors can interact with the NFκB modulator RIP1 and both have been shown to induce NFκB and MAP (mitogen-activated protein) kinase signalling pathways. Opposite to the spatial organization of TNFR1 signalling, it has been proposed that TRAIL receptors signal death from a surface receptor complex (complex I) and that subsequently an internalized complex (complex II) is responsible for NFκB activation (Varfolomeev *et al.*, 2005; **Figure 2**). This has been proposed to occur through the recruitment of the signalling components TRAF2, RIP1 and IKK that are also responsible for NFκB induction of TNFR1.

Although it remains unclear whether Fas induction of NFκB is mediated in a similar manner to TRAIL receptor activity, MAP kinase activation following Fas stimulation of sensory neurons promotes neurite outgrowth through ERK (Extracellular signal-regulated kinase)-mediated induction of p35 and has also been shown to contribute to neuronal regeneration and branching. Consistent with this pro-life role in neurons, nonapoptotic Fas signalling can stimulate glioma tumour cell growth, and it has been suggested that glioma cell invasion is regulated through Fas recruitment and activation of phosphatidylinositol-3 kinase activity and downstream glycogen synthase kinase 3β-induced transcription of matrix metalloproteinase genes (Kleber *et al.*, 2008). The role of Fas in the CNS may extend to protecting from Parkinson disease, as Fas-deficient mice are highly susceptible to neurotoxin-induced Parkinson disease when compared with wild-type mice and Fas is required for protection against neurotoxin toxicity (Landau *et al.*, 2005). Collectively, these studies suggest that Fas signalling may be important for protection against neurodegeneration through its pro-survival signalling functions and that its dysregulation may contribute to disorders of the central nervous system (CNS).

Downstream apoptotic signalling components of Fas and TRAIL receptors have also been shown to harbour critical pro-life functions. Both caspase-8 and FADD have been reported to be required for T-lymphocyte proliferation in response to antigenic or mitogenic stimulation and toll-like receptor (TLR)-induced proliferation of B cells, whereas FADD and TRADD are also required for innate immune responses resulting in interferon and cytokine production. Notably, gene knockout mice for caspase-8, TRADD and FADD results in embryonic lethality associated with defects in vascular development and haematopoiesis, whereas deletion of the death receptors for FasL, TNFα or TRAIL, or the ligands themselves, results in viable mice. Currently it is unknown how the components required for intracellular death receptor signalling function to influence essential developmental programmes, but it has been shown that these adaptor molecules are shared among other signalling modules in addition to death receptors.

The findings outlined earlier, among many others, have made it apparent that death receptors have an inappropriate name that fails to describe their ability to signal other responses in addition to death, especially at physiological or low levels of receptor stimulation.

Perforin and granzyme B-induced cell death

Cytotoxic T lymphocytes and natural killer (NK) cells initiate the destruction of virally infected cells and also contribute to tumour cell surveillance and removal by perforin-mediated delivery of serine protease granzymes into the cytoplasm of target cells. Perforin is a membrane pore forming protein secreted by NK cells and cytotoxic lymphocytes into a tightly enclosed extracellular junction formed between the killer and target cell. Although granzymes can enter cells in the absence of Perforin, it is clear that perforin assists the entry of granzymes into the target cell and both molecules are required for efficient NK and T cell-induced target cell death. It remains unclear as to how perforin assists in granzyme delivery to the target cell. At one time it was thought that perforin formed channels at the target cell membrane through which granzymes could pass, but the pores of 16 nm diameter are now considered too small to work in this direct fashion. A number of different models have been put forward and then contested in recent years and researchers in the field have yet to agree on the mechanism by which perforin mediates granzyme delivery (Cullen and Martin, 2008). Nevertheless, the lack of efficient granzyme delivery in perforin null mice contributes to impaired NK and T cell-induced cytotoxicity.

One of the most abundant and studied granzymes, and the main cytotoxic granzyme that induces apoptosis in the target cell, is granzyme B. Granzyme B, although a serine protease rather than an aspartate protease, has a similar substrate specificity to caspase-8 and caspase-9 and similar to these initiator caspases, is able to process caspase-3 and caspase-7. Granzyme B can also activate Bid and cleave the pro-survival Bcl-2 member Mcl-1 and several essential cellular housekeeping and structural components such as Lamin B, tubulin, ICAD (inhibitor of caspase-3-activated DNase) and topoisomerase I (Cullen and Martin, 2008). It may therefore induce apoptosis by directly cleaving and activating caspases, such as caspase-3, and also induce the Bcl-2 blockable apoptotic pathway by activating the pro-apoptotic Bcl-2 family member Bid, leading to Bax/Bak activation and mitochondrial permeabilization. Consistent with this, deletion of Bid or overexpresssion of Bcl-2 is able to protect against granzyme B-induced cell death. Although, granzyme B-mediated mitochondrial depolarization has been reported to occur independently of Bax/Bak, its prevention does not block granzyme B killing. Therefore, killing by perforin–granzyme B is likely to occur via the intrinsic apoptosis pathway and/or direct caspase cleavage, although Bcl-2 independent mechanisms of granzyme B-induced cell death involving ROS (reactive oxygen species) production have been proposed. On this note, it is feasible that two granzyme B-cell death pathways exist to ensure destruction of tumours or viral inhabited cells whose intrinsic cell death pathway has been blocked. The coexistence of two independent granzyme B death signalling pathways is reminiscent of death receptor-induced apoptosis, which can occur via caspase activation, or if caspases are inhibited, by caspase independent necroptosis. **See also**: Immunity, Granzymes and Cell Killing

Dysregulation of the Intrinsic and Extrinsic Apoptotic Pathways in Disease

Inappropriate apoptosis is the cause or a contributing factor to many human diseases (**Table 1**). Excessive apoptosis may contribute to neurological diseases, and acute situations involving ischaemia reperfusion, such as heart attacks and strokes. However, it is frequently unclear whether apoptosis is really a cause of a disease rather than a response, and as a corollary whether inhibiting apoptosis would really cure the disease. Encouragingly from a drug development perspective, there are some cases where caspase inhibition has been shown to limit apoptosis and disease (Pockros *et al.*, 2007). **See also**: Apoptosis: Inherited Disorders

Insufficient apoptosis is likely to be a key contributor to autoimmune diseases and tumourigenesis. Experimental proof for this statement has been provided in gene knockout mice studies and genetic mapping of disease causing genes in humans. For example in mice that coexpress an oncogene such as Eμ-myc and an antiapoptotic protein such as Bcl-2, an experimental situation that mimics certain human tumours such as follicular lymphoma, the mice succumb much more rapidly to tumours than expression of the oncogene alone (Strasser *et al.*, 1990). Furthermore, the presence of Bcl-2 can be required for tumour maintenance, as on its elimination leukaemia cells undergo apoptosis and animal survival is prolonged. One of the problems when considering whether a particular apoptotic gene is really causal or contributory to tumour formation is that tumours are rapidly evolving and many genes are differentially regulated. It has not always been rigorously demonstrated that a particular apoptotic gene makes a decisive contribution to the disease and therefore the later discussion focuses on cases where there is compelling data.

Dysregulation and Disease in the Intrinsic Bcl-2 Blockable Apoptotic Pathway

p53

The transcription factor p53 is known as the guardian of the genome and is a key molecule in the defence against cellular transformation due to DNA damage. Induction of apoptosis by p53 proceeds by the intrinsic pathway, and is mostly due to transcriptional induction of pro-apoptotic genes, although non-transcriptional mechanisms, such as direct binding to Bcl-2 family members, have been reported. Mutation of p53 is the most common occurring genetic aberration observed in cancer, and is found in approximately 50% of all human tumours. Individuals with Li Fraumeni syndrome, who have germline heterozygous mutations in p53, are highly prone to diverse cancers at a young age and mutant mice homozygous or heterozygous for p53 deletion are highly likely to develop tumours early in life. Amazingly, given the genetic complexity and instability of cancer, restoration of wild-type p53 expression alone has been shown as sufficient to induce tumour regression in several mouse cancer models and this depended on the ability of p53 to induce apoptosis and/or cellular senescence. **See also**: P53 and Cell Death

p53 is negatively regulated by the ubiquitin E3 ligase MDM2 (murine double minute 2), which targets it for proteasome-mediated degradation, whereas MDM2 is in turn regulated by p19ARF (ADP-ribosylation factor). DNA damage activates p19ARF, which inhibits MDM2 resulting in increased levels of p53. Much recent research has focused on developing therapeutic strategies to target this arm of the p53 pathway, as apart from p53 mutation in many cancers, MDM2 overexpression is also frequently observed, and this is sufficient to prevent even wild-type p53 responses. Preclinical studies are focusing on creating small-molecule compounds that disrupt the MDM2–p53 interaction or MDM2 function to allow p53 stabilization and apoptosis. Others are looking at restoring p53 mutant activity; for example, in China, treatment has been approved for adenovirus-mediated cancer cell restoration of wild-type p53. Clinical trials are also ongoing for several compounds, such as R-roscovitine, shown to both activate p53 and inhibit NFκB signalling, thereby targeting simultaneously two of the most common pathways dysregulated in cancer.

Table 1 Apoptotic genes and their role in human disease

Gene	Function	Disease association	Knockout/Mutant Mouse
Extrinsic apoptosis			
TNFRSF6 (Fas/CD95)	Death receptor	ALPS I: autoimmune	lpr mice: lymphoproliferation i.e. develop autoimmune disease
		Lymphoproliferative syndrome	Resistant to Fas ligand
TNFSF6 (FasL)	Death ligand	ALPS 1B, systemic lupus erythematosus	gld mice: general lymphoproliferative disease, that is, develop autoimmune disease
TNFRSF1A (TNF-R1)	Death receptor	TRAPS: TNF receptor-associated periodic fever syndrome	Low-dose LPS resistance, susceptible to infection by some pathogens
TNFα	Death ligand	Implicated in Psoriasis, ankylosing spondylitis, rheumatoid arthritis. SNPs associated with susceptibility to a number of pathogens, autoimmune diseases and septic shock	Viable, normal development. Resistant to septic shock. Susceptible to infection by a number of pathogens. Protected against obesity induced insulin resistance
TNFRSF10A (TRAILR1/ DR4)	Death receptor		Viable, normal development. Increased tumour metastasis and increased inflammation on pathogen infection
TNFSF10 (TRAIL)	Death ligand		Viable, normal development. Increased tumour metastasis and spontaneous lymphoma development. Susceptible to collagen induced arthritis
FADD (FAS-associated death domain)	Caspase-activating adaptor		Embryonic lethal, Die E10, heart defect impaired activation induced cell death
CASP8 (Caspase-8)	Apoptosis initiator protease	ALPS II, suppression implicated in neuroblastoma metastasis	Embryonic lethal, heart defect. Required for lymphocyte homeostasis
CASP10 (Caspase-10)	Apoptosis initiator protease	ALPS II, familial non-Hodgkin lymphoma, gastric cancers, simultaneous TNFRSF1A mutation	(No caspase-10 gene in mouse)
BIRC2 (cIAP1) BIRC3 (cIAP2)	Antiapoptotic IAP family members	Hepatocellular carcinoma, pancreatic carcinoma, lung cancer (amplification), multiple myeloma (deletion). cIAP2 translocation in MALT lymphoma	Single knockout of either gene viable, developmentally normal
Intrinsic apoptosis			
TP53 (p53)	Transcription factor	Multiple tumour types, Li Fraumeni syndrome	Early lethality, spontaneous tumourigenesis, decreased apoptosis in response to genotoxic stress
Bcl-2	Antiapoptotic Bcl-2 family member	Non-Hodgkin lymphoma	Growth retardation, increased lymphoid cell and melanocyte apoptosis, polycystic kidney disease
Bax	Pro-apoptotic Bcl2 family member	Colon cancer	Viable, thymocyte and B-cell hyperplasia, deficient spermatogenesis

(*Continued*)

Table 1 Continued

Gene	Function	Disease association	Knockout/Mutant Mouse
Bak	Pro-apoptotic Bcl2 family member	Gastric and colorectal cancer	Viable, increased platelet lifespan
Apaf1	Caspase-activating adaptor	Melanoma and colon cancer	Brain hyperplasia
CASP9 (Caspase-9)	Apoptosis initiator protease		Brain hyperplasia
BIRC4 (XIAP)	Antiapoptotic IAP family member	X-linked lymphoproliferative disease	Viable, developmentally normal
Perforin–granzyme killing			
PRF1 (Perforin)	Cytotoxic facilitates granzyme entry	Familial haemophagocytic lymphohistiocytosis type 2 (FHL)	Susceptibility to infection by some viruses. Develop B-cell lymphoma

In response to cellular damage, elevated p53 activates several transcriptional pathways that can lead to cell cycle arrest, DNA repair, senescence and apoptosis. These responses allow dividing cells more time to repair DNA damage, or if the insult is too great, activate apoptosis to dispose of the damaged or potential harmful cell. Work on mouse models of lymphoma demonstrated that the ability of p53 to induce apoptosis, as opposed to cell cycle arrest, is probably key in its prevention of lymphomagenesis, as overexpression of Bcl-2 or a dominant negative caspase alleviated the selective pressure for p53 mutation in Eμ-myc-induced lymphoma. p53 probably exerts much of its apoptotic effects by induction of the BH3 proteins Puma and Noxa because loss of Puma in many cell types and Noxa in fibroblasts and keratinocytes results in almost complete loss of p53-dependant apoptosis. Several, but not all, cell types from Puma Noxa double knockout mice, which are born in a Mendelian ratio and develop normally, are as resistant to gamma irradiation as p53 knockout mice are (Michalak *et al.*, 2009). Remarkably though, unlike p53 knockout mice, Puma Noxa double knockout mice do not develop tumours either spontaneously or when challenged with whole body radiation (Michalak *et al.*, 2009). These results suggest that under some circumstances the ability of p53 to activate the apoptotic pathway alone is not crucial for its ability to suppress tumour formation. Alternatively, other pro-apoptotic p53 target genes may play important roles in suppressing the spontaneous tumourigenesis observed in p53 mutant mice.

Bcl-2

Bcl-2 derives its name from B-cell lymphoma 2, initially described as a reciprocal gene translocation in chromosomes 14 and 18 in follicular lymphomas. The chromosome rearrangement results in overexpression of Bcl-2 protein, which protects lymphoma cells from apoptosis. Both elevated levels of Bcl-2 and other pro-survival Bcl-2 family members have since been implicated in a range of malignancies including various lymphoid and haematopoietic cancers and their overexpression often correlates with resistance to chemotherapeutic treatments which invariably induce cell death via Bax/Bak activity and mitochondrial permeabilization. **See also**: BH3-Only Proteins; Structure and Function of IAP and Bcl-2 Proteins; The Bcl-2 Family Proteins – Key Regulators and Effectors of Apoptosis

The role of Bcl-2 and other pro-survival Bcl-2 family members in antagonizing Bax/Bak function has lead to the development of small molecule BH3 mimetic compounds designed to mimic pro-apoptotic BH3 only proteins in their antagonism of Bcl-2 family member function and thereby induce apoptosis through Bax/Bak activation and mitochondrial permeabilization (**Figure 1**). However, one study has shown that several small molecule BH3 mimetics still induce cell death in Bax/Bak double knock-out cells. This demonstrates conclusively that their toxicity is not due to their ability to antagonise pro-survival Bcl-2 family members (van Delft *et al.*, 2006). Perhaps the best BH3 mimetic described to date is ABT-737 and its derivatives, which cannot kill in the absence of Bax and Bak, and has been shown to selectively target Bcl-2, Bcl-x_L and Bcl-w. Unlike other BH3 mimetics ABT-737 was rationally designed using nuclear magnetic resonance (NMR)-based structure screening and medicinal chemistry and has been shown to act as a single agent to kill primary patient tumour cells, and in mouse models induce tumour regression and often lead to a complete cure (Oltersdorf *et al.*, 2005). It has also been established that ABT-737 synergizes with a large range of current chemotherapeutics to efficiently destroy cancer cells. As ABT-737 is unable to efficiently bind Mcl-1, cells expressing high levels of Mcl-1 remain refractory against treatment, and it is most efficacious against cancer cells expressing low levels of Mcl-1, or where Mcl-1 is concurrently downregulated with ABT-737 treatment. Consistent with this, in the absence of the Eμ-Bcl-2 transgene, elevated levels of Mcl-1 appeared to prevent ABT-737 efficacy in treating transplanted murine Eμ-myc lymphoma, although cotreatment with cylcophosphamide, a DNA crosslinking agent currently used to treat lymphoma, resulted in complete remission. This work fits the current frame of thinking that for most carcinomas no magic bullet exists, and that tailor made combination therapies offer the best chance for successful treatment.

Bcl-2 is required for the survival of mature T- and B lymphocytes. Similarly, genetic knockout of other Bcl-2 family members has demonstrated that they play important roles in the survival of platelets, erythroid progenitors and neuronal cells (Bcl-x_L), haematopoietic stem cells and lymphoid progenitors (Mcl-1) and sperm cells (Bcl-w). Therefore, despite the amazing success of ABT-737 in cancer cell lines and mouse models, and promising early clinical data, it remains to be determined how well it and other BH3 mimetics, currently in clinical trials, will be tolerated in humans.

Bcl-x_L

Platelets are anuclear cytoplasmic fragments essential for blood clotting and may also regulate cytokine production and inflammatory responses. Recently, it has become apparent that the defined lifespan of platelets is determined by levels of the pro-survival Bcl-x_L protein that limits the pro-apoptotic activity of Bak. As platelets are anuclear they cannot replace Bcl-x_L, which degrades at a quicker rate than Bak. Bcl-x_L mutant mice show markedly reduced platelet half-life and platelets from Bak-deficient mice live longer than normal (Mason *et al.*, 2007). These results raise the possibility that some inherited thrombocytopenias will occur due to mutations in this apoptotic pathway.

XIAP

XIAP is an inhibitor of apoptosis protein that has been shown to inhibit caspase-3, caspase-7 and caspase-9 with a Ki_i in the nanomolar range. XIAP also contains a RING finger, a ubiquitin E3 ligase domain, that has been shown to be capable of promoting caspase ubiquitylation, although the physiological significance of this is not clear, because there have been conflicting claims. Despite this ability to inhibit these caspases at physiological levels, and potentially to mark them for destruction by the proteasome, XIAP expression rarely appears to provide long-term survival to cells *in vivo*. This is probably due to the fact that mitochondria, once breached by Bax or Bak, leak IAP antagonists that prevent XIAP binding and XIAP inhibition of caspases, and also that mitochondria that have breached outer mitochondrial membranes are no longer able to generate energy. One model is that XIAP functions as a safety catch that prevents and disposes of inappropriately activated caspases but which at physiological levels is unable to contain full activation of the apoptotic pathway. **See also**: Inhibitor of Apoptosis (IAP) and BIR-containing Proteins

XIAP knockout mice are born in a normal Mendelian ratio, develop normally and show normal numbers of NKT (natural killer T) cells in haematopoetic organs. It has been suggested that XIAP plays a different role in humans as XIAP mutations leading to the loss of XIAP expression has been linked with an X-linked lymphoproliferative (XLP) syndrome, which is associated with a marked deficiency in NKT cell numbers (Rigaud *et al.*, 2006). XLP is a rare immunodeficiency characterized by abnormal immune responses, hypogammaglobulinaemia and lymphomas which usually develops in males following infection with *Epstein–Barr virus* (EBV). Mutations in the signalling lymphocyte activation molecule-associated protein (SAP) are associated with most cases of familial XLP. Lymphocytes from XIAP-deficient patients were more susceptible to extrinsic apoptotic stimuli and could be rescued by reconstitution with wild-type XIAP, suggesting that it is the antiapoptotic potential of XIAP that is required to prevent the disease. However, XIAP impinges on several other receptor signalling pathways and it is still possible that XIAP regulation of these pathways also contributes to the disease.

Perforin and Granzyme B

Perforin plays a central role in NK and T-cell cytotoxic functions and is essential for the control of some, but not all, viral infections (Russell and Ley, 2002). Although this has mostly been attributed to perforin–granzyme B target cell killing, its inhibition of viral virulence may also result from nonapoptotic mechanisms. For example, granzyme B targeted at *Herpes simplex virus* type-1 (HSV-1) infected neuronal cells does not induce target cell apoptosis but maintains HSV-1 latency by inducing cleavage of a protein required for transcription of early and late viral genes. This function is reminiscent of caspase-1, which can induce cell death in addition to mediating the processing and maturation of cytokines required for the innate immune response. **See also**: Immunity, Granzymes and Cell Killing

Mutation of perforin is observed in 20–40% of people suffering from familial haemophagocytic lymphohistiocytosis (FHL), an autosomal recessive disorder that is fatal if left untreated. FHL is characterized by an accumulation of activated lymphocytes, monocytes and histiocytes, with the latter exhibiting phagocytosis of blood cells and defective NK cell function. It is a rare disorder of young children presenting with fever, often associated with viral infection, hepatosplenomegaly and destructive organ infiltration and severe neurological symptoms. The disease involves either a gene locus on chromosome 9q21.3–22 or chromosome 10q21–22. In a study of 8 unrelated 10q21–22 linked FHL patients' nine different mutations in the perforin gene were detected. NK and T cells from these patients showed a greatly reduced capacity to lyse target cells. Although perforin deletion in mice does not induce FHL by itself, challenging these animals with pathogens does induce an FHL-like disease, suggesting environmental factors may also contribute to FHL in humans.

The cancer immune surveillance hypothesis proposed by Burnet has some experimental support and the immune system may be a target for the development of cancer immunotherapies. The perforin–granzyme cell death pathway appears to play an important role in tumour cell surveillance and removal as perforin mutant mice display an increased susceptibility to tumour formation, with around half spontaneously developing B-cell lymphomas. Consistent with its role in NK- and T-cell cytotoxicity, perforin deficiency does not appear to directly cause lymphoma but is indirectly involved, as B-cell lymphomas derived from perforin deficient mice are controlled on transplantation into wild-type mice by T-lymphocyte activity. Perforin null mice are 1000-fold more susceptible to transplanted lymphomas compared to wild-type mice and perforin loss cooperates with oncogene activation and/or tumour suppressor loss to increase lymphoma development and severity. In humans, perforin mutations have been reported in patients suffering with lymphoma, and this correlated with the loss of perforin expression and NK cell function.

Dysregulation and Disease in the Extrinsic Apoptosis Pathway

Targeting death receptors in cancer

Targeting the extrinsic pathway in cancer has become attractive as many mutations implicated in cancer affect the intrinsic apoptotic pathway (i.e. p53), limiting the utility of compounds seeking to exploit this mode of cell death. Most cancer types express death receptors, thus enabling intrinsic mutations to be effectively bypassed by activating the extrinsic apoptotic pathway. In addition, oncogenic transformation, such as that driven by the Myc and Ras oncogenes, has been observed to prime tumour cells for death receptor killing, although why this is the case has yet to be clearly established. Conversely, death receptor stimulation can also synergize with traditional therapeutics to initiate tumour cell death, thereby adding to the armament of potential tailor made and/or combinatorial therapeutics.

TNFα and Fas ligand treatment alone currently have limited therapeutic potential in treating cancer as Fas is a potent hepatocyte killer and TNFα treatment results in systemic toxicity and in some circumstances can promote tumour cell survival. However, isolated limb perfusion of TNFα is currently successful in the treatment of some soft tissue sarcomas, and TNFα sensitizing agents, such as IAP antagonist compounds, may lower the therapeutic window of TNFα levels required to kill cancer cells and thereby increase its therapeutic uses in the future.

Current evidence suggests that TRAIL is the best death receptor ligand for initiating effective and specific tumour cell killing. Most cancer types express one or both of the TRAIL receptors, DR4 and DR5, and TRAIL treatment preferentially kills cancer cell lines over normal cells. Several pharmaceutical companies have focused on making TRAIL receptor agonist antibodies or purified recombinant TRAIL protein.

Several phase I clinical trials have reported the overall safety of these agents and provided some encouraging preliminary therapeutic data, and phase II trials are currently underway to more accurately assess their potential in a range of carcinomas (Ashkenazi, 2008). However, it has been noted that primary tumour cells, as opposed to tumour cell lines, are often resistant to TRAIL-induced death and TRAIL must therefore be combined with other chemotherapeutic agents to initiate efficient cell death. Furthermore, some TRAIL resistant tumour cell lines can respond to TRAIL treatment with increased growth and invasiveness.

Physiologically, TRAIL plays an important role in suppressing innate immunity as TRAIL receptor knockout mice (there is only one TRAIL death receptor in the mouse) display increased innate immune cytokine production and enhanced CMV (cauliflower mosaic virus) viral clearance. TRAIL has also been reported to play an important role in preventing tumour growth and metastasis. TRAIL-deficient mice show increased spontaneous development of lymphoma and sarcoma on both a p53 wild-type and heterozygous background and although primary tumour growth is not affected, the loss of TRAIL in mice increases the incidence of tumour metastasis. The death receptor initiator caspase, caspase-8, has also been implicated in metastasis, and its suppression or deletion is observed in childhood neuroblastoma. Neuroblastoma cells advancing into the surrounding stroma select for loss of caspase-8, and when caspase-8 is reintroduced these cells undergo integrin-dependent apoptosis and tumour metastasis is perturbed.

Fas and disease

Although Fas-induced apoptosis has previously been suggested to be essential for the deletion of autoreactive T lymphocytes and the shut down of acute immune responses by initiating the death of antigen activated T- and B cells, this has since been proven to be mediated through the intrinsic cell death pathway and the upregulation of the pro-apoptotic BH3 only protein Bim. However, together with perforin, Fas ligand (FasL) accounts for nearly all the cytotoxic activity of NK and T cells and it has been shown that Fas-induced apoptosis does mediate T-cell receptor stimulation-induced cell death of normal cycling T lymphocytes and together with the intrinsic cell death pathway plays a critical role in the deletion of activated B cells. Recently, using gene knockout mice, it has been demonstrated that Fas, together with Bim, is also required for T-cell killing and immune shut down during chronic infection and for the prevention of systemic autoimmunity. Therefore, Fas death receptor-induced apoptosis is critical for T-cell termination during chronic T-cell activation by both pathogen and self-derived antigens. The involvement of Fas-induced apoptosis in immune function extends to other cell types, including the removal of epithelial cells infected with bacteria. It has been demonstrated that lung fibroblasts undergo Fas-induced apoptosis rapidly following infection with *Pseudomonas aeruginosa*, a causative agent of pneumonia and sepsis. This protective autodestruct mechanism is absent in Fas or FasL-deficient mice, which unlike wild-type mice succumb to sepsis and death following *P. aeruginosa* infection.

Consistent with the earlier findings patients with autoimmune lymphoproliferative syndrome (ALPS) have mutations in the Fas–FasL pathway and mice that are deficient for either Fas (lpr: lymphoproliferation) or FasL (gld: generalized lymphoproliferative disorder) develop autoimmune disease characterized by hypergammaglobulinaemia, glomerulonephritis, lymphadenopathy and expansion of otherwise rare lymphocytes. ALPS results from dysregulation of lymphocyte apoptosis whereby the inability of lymphocytes to die in response to normal developmental signals leads to lymphadenopathy, hypersplenism and autoimmune symptoms including thrombocytopenia, neutropenia and haemolytic anaemia. Both ALPS patients and Fas signalling deficient mice are also predisposed to developing lymphoma, demonstrating that Fas-induced apoptotic signalling and continual immune surveillance is important for the removal of potentially harmful cells. **See also**: Apoptosis: Inherited Disorders

The human Fas gene is located on chromosome 10q24.1 and the majority of patients with ALPS type Ia are heterozygous for a mutation in this gene (Rao and Straus, 2006). Over 70 different Fas gene mutations have now been described in ALPS patients; two-thirds are in the intracellular domain, usually in the death domain itself, and a third in the extracellular domains. Notably, mutations in the intracellular domain of Fas have shown to result in a more severe phenotype, with ALPS-related morbidity occurring in over 40% of patients, whereas extracellular Fas mutations do not generally result in death. Many of the Fas mutations act dominant negatively, and yet siblings with the same mutations may not present with the disease. This clearly indicates that there are other factors, genetic or environmental, that may contribute to the disease.

Patients with homozygous mutations in caspase-8 have defective lymphocyte apoptosis and homeostasis, like other ALPS patients, and caspase-10 has been shown to be associated with a small number of people suffering from type II ALPS. In contrast to mice, where caspase-8 deficiency is embryonic lethal, the earlier results demonstrate that caspase-8 loss in humans is compatible with normal development and shows that caspase-8 has a postnatal role in immune activation of naive lymphocytes. This difference in the requirement of caspase-8 for mouse and human development is accounted for by the fact that humans express a caspase-8 homologue, caspase-10, which is not present in the mouse genome.

Apart from neuroblastoma (see earlier discussion), caspase-8 mutation has also been associated with several other types of cancer. A 6 nucleotide deletion in the caspase-8 promoter region resulting in decreased caspase-8 expression has been linked to decreased T-lymphocyte death and a decreased risk of developing several types of cancer, including that of the lung, breast, cervical, colorectal and oesophageal. In contrast, another study demonstrated that a single nucleotide polymorphism (SNP) (D320 H) in caspase-8 correlated with an increased breast cancer risk. Although the functional consequence of this mutation has not been determined, these earlier results demonstrate the potential deleterious or beneficial effects that the alteration of one gene can have.

TNF receptor signalling and disease

More than 30 years since its discovery, new critical TNF receptor signalling components are still being identified and the mechanisms of TNFα signalling still being dissected. TNFα was first identified in 1975 and initially gained much attention for its ability to kill xenografted tumours in mouse cancer models. Since then, although its inflammatory properties have limited its use in cancer treatment, it has been implicated in a large number of inflammatory conditions. These include important roles in psoriasis, rheumatoid arthritis (RA), ankylosing spondylitis and polycystic kidney disease. Variation in the TNFα promoter region has also been linked to the susceptibility to a number of pathogens including malaria, leishmaniasis, leprosy and autoimmune conditions, such as inflammatory bowel disease, asthma and septic shock. **See also**: Death Receptors; Death Receptors at the Molecular Level: Therapeutic Implications; Death Receptor-Induced Necroptosis

TNFα does not usually induce apoptosis but rather promotes secretion of a range of inflammatory cytokines and is itself pyrogenic. The importance of regulating the TNFα pathway is demonstrated by a large number of knockout mice including A20, RIP1 and TRAF2 that die due to hyperactive TNFα signalling before or soon after birth. Based on the phenotypes of these knockouts, the lethal effects of TNFα overproduction are most likely due to the systemic inflammatory properties of TNFα rather than its cytotoxic activity on particular cell types. An inherited periodic fever syndrome has been associated with TNF-R1 mutations that appear to reduce the clearance of TNF-R1 from the plasma membrane and results in increased free circulating TNFα in the plasma. However, there was no evidence that this mutation causes an apoptosis defect. Similarly, in mice, the administration of TNFα mimics endotoxin and bacterially induced septic shock and polymorphisms within the TNFα promoter in humans have been associated with a significantly increased severity and chance of death due to sepsis. Despite these findings and the observation that neutralizing TNFα activity in mice prevents endotoxin and bacterially induced septic shock, several clinical trials found that TNFα neutralization had no significant effect on septic shock morbidity in humans, indicating that more efficient TNFα neutralization, or additional therapeutic targets, are required for septic shock treatment (Abraham, 1999).

Transgenic mice developed to overexpress TNFα develop synovial inflammation and display a remarkably similar pathology to that observed in RA. Likewise, inhibitors of TNFα function and signalling reduce or abrogate inflammation and have revolutionized the treatment of RA, which previously relied on antirheumatic drugs that were only partially effective and were associated with significant toxicity. Notably TRAIL knockout mice are prone to collagen-induced arthritis, although why this is the case remains uncertain. However, it has been shown that thymocyte negative selection is impaired on TRAIL loss and the predisposition of TRAIL knockout mice to developing arthritis may, therefore, lead to increased lymphocytes within arthritic synovial membranes and increased inflammation.

The effectiveness of TNFα inhibitors in stemming RA is due to the ability of TNFα to induce cartilage and bone destruction, activate endothelial cells, attract leukocytes and mediate the production of inflammatory cytokines and chemokines such as IL-1, IL-6, IL-8, RANTES (regulated on activation normal T cells expressed and secreted), MCP-1 (monocyte chemoattractant protein 1) GM-CSF (granulocyte–macrophage colony-stimulating factor), all of which contribute to and/or propagate the chronic inflammation associated with RA. Current TNFα-targeted RA therapeutics include TNFα neutralizing antibodies (infliximab and adalimumab) and soluble TNF receptor fused to the Fc region of IgG1 (etanercept). A number of other antibodies and decoy receptor fusion proteins targeting other cytokines and TNFSF members associated with RA are currently in phase II clinical trials as despite the success of TNFα targeted therapies, immunological tolerance and therefore RA cure has yet to be achieved.

In addition to RA, the TNFα neutralizing treatments discussed earlier have so far been used to successfully treat other inflammatory conditions including the inflammatory skin disorder, psoriasis and ankylosing spondylitis. One of the more surprising findings has been the relationship between TNFα and obesity. Levels of TNFα are increased in the adipose tissue of overweight animals and people and it has been shown to negatively affect the ability of insulin to stimulate glucose uptake. TNFα knockout mice become obese at a similar rate as wild-type mice when fed a high calorie diet. However, insulin levels, insulin receptor activity and insulin-mediated glucose acquisition are significantly improved, and circulating free fatty acids are lower in obese TNFα knockout mice compared to wild-type animals (Uysal *et al.*, 1997). Hence, TNFα may contribute to the development of insulin resistance and obesity-related disorders and consistent with this, clinical studies have demonstrated that insulin sensitivity is increased in humans following treatment with neutralizing TNFα antibodies.

cIAP1 and cIAP2 and disease

The cIAPs belong to the same family as XIAP and the defining feature of this family is the presence of the bacuoloviral IAP repeat domain (BIR). Although cIAP1 and cIAP2 are able to bind caspases they do not inhibit them. Their antiapoptotic role appears to be restricted to limit signalling from death receptors such as TNFR1 and TRAILR1 because removal of IAPs sensitizes many cells to the cytotoxic activity of these receptors (Vince *et al.*, 2007; Li *et al.*, 2004). In humans and mice cIAP1 and cIAP2 are found in the same genetic locus and only are separated by approximately 20 kb. Both cIAP1 and cIAP2, or the cIAP1 and cIAP2 locus (which also contains the Yap oncogene), have been shown to be amplified in many types of cancer including squamous cell carcinoma and pancreatic, liver, lung and oesophageal cancer. A recent study also showed in an experimental mouse model of hepatocellular carcinoma driven by c-Myc overexpression that this amplification, resulting in cIAP1 and cIAP2 overexpression, was required to sustain rapid growth of these tumours (Zender *et al.*, 2006). **See also**: Inhibitor of Apoptosis (IAP) and BIR-containing Proteins

cIAPs are also implicated in two other tumour types; MALT (mucosa-associated lymphoid tissue) lymphoma and multiple myeloma. A *t*(11;18)(q21;q21) chromosomal translocation results in an AP12–MALT1 fusion gene (Dierlamm *et al.*, 1999) and this translocation is specific to and the most common translocation observed in MALT lymphomas. This translocation results in activation of NFκB, as do two other genetic alterations, *t*(1;14)(p22;q32) and *t*(14;18)(q32;q21) which deregulate BCL10 and MALT1 genes, respectively.

Similarly, in multiple myeloma, there is frequently a loss of regulation of the NFκB inducing kinase, NIK1, that results in high levels of noncanonical NFκB which is thought to drive cancer growth and survival (Keats *et al.*, 2007; Annunziata *et al.*, 2007). Both cIAP1 and cIAP2 are recruited to NIK1 by TRAF2 where they act as E3 ligases for NIK1 to keep NIK1 to undetectable levels in unstimulated cells. Loss of either cIAP1, cIAP2 or TRAF2, among other genes in this pathway is therefore frequently observed in multiple myeloma (Keats *et al.*, 2007; Annunziata *et al.*, 2007). Therefore, although inhibitor of apoptosis proteins are involved in lymphoma and myeloma, it is not their antiapoptotic function that contributes to the disease but rather their separate ability to regulate NFκB signalling pathways.

References

Abraham E (1999) Why immunomodulatory therapies have not worked in sepsis. *Intensive Care Medicine* **25**: 556–566.

Annunziata CM, Davis RE, Demchenko Y *et al.* (2007) Frequent engagement of the classical and alternative NF-kappaB pathways by diverse genetic abnormalities in multiple myeloma. *Cancer Cell* **12**: 115–130.

Ashkenazi A (2008) Directing cancer cells to self-destruct with pro-apoptotic receptor agonists. *Nature Reviews. Drug Discovery* **7**: 1001–1012.

Cullen SP and Martin SJ (2008) Mechanisms of granule-dependent killing. *Cell Death Differentiation* **15**: 251–262.

van Delft MF, Wei AH, Mason KD *et al.* (2006) The BH3 mimetic ABT-737 targets selective Bcl-2 proteins and efficiently induces apoptosis via Bak/Bax if Mcl-1 is neutralized. *Cancer Cell* **10**: 389–399.

Dierlamm J, Baens M, Wlodarska I *et al.* (1999) The apoptosis inhibitor gene API2 and a novel 18q gene, MLT, are recurrently rearranged in the t(11;18)(q21;q21) associated with mucosa-associated lymphoid tissue lymphomas. *Blood* **93**: 3601–3609.

Holler N, Zaru R, Micheau O *et al.* (2000) Fas triggers an alternative, caspase-8-independent cell death pathway using the kinase RIP as effector molecule. *Nature Immunology* **1**: 489–495.

Keats JJ, Fonseca R, Chesi M *et al.* (2007) Promiscuous mutations activate the noncanonical NF-kappaB pathway in multiple myeloma. *Cancer Cell* **12**: 131–144.

Kim S, Takahashi H, Lin WW *et al.* (2009) Carcinoma-produced factors activate myeloid cells through TLR2 to stimulate metastasis. *Nature* **457**: 102–106.

Kleber S, Sancho-Martinez I, Wiestler B *et al.* (2008) Yes and PI3K bind CD95 to signal invasion of glioblastoma. *Cancer Cell* **13**: 235–248.

Kroemer G, Galluzzi L, Vandenabeele P *et al.* (2009) Classification of cell death: recommendations of the Nomenclature Committee on Cell Death 2009. *Cell Death Differentiation* **16**: 3–11.

Landau AM, Luk KC, Jones ML *et al.* (2005) Defective Fas expression exacerbates neurotoxicity in a model of Parkinson's disease. *Journal of Experimental Medicine* **202**: 575–581.

Li L, Thomas RM, Suzuki H *et al.* (2004) A small molecule Smac mimic potentiates TRAIL- and TNFalpha-mediated cell death. *Science* **305**: 1471–1474.

Mason KD, Carpinelli MR, Fletcher JI *et al.* (2007) Programmed anuclear cell death delimits platelet life span. *Cell* **128**: 1173–1186.

Michalak EM, Jansen ES, Happo L *et al.* (2009) Puma and to a lesser extent Noxa are suppressors of Myc-induced lymphomagenesis. *Cell Death Differentiation* **16**: 16.

Oltersdorf T, Elmore SW, Shoemaker AR *et al.* (2005) An inhibitor of Bcl-2 family proteins induces regression of solid tumours. *Nature* **435**: 677–681.

Peter ME, Budd RC, Desbarats J *et al.* (2007) The CD95 receptor: apoptosis revisited. *Cell* **129**: 447–450.

Pockros PJ, Schiff ER, Shiffman ML *et al.* (2007) Oral IDN-6556, an antiapoptotic caspase inhibitor, may lower aminotransferase activity in patients with chronic hepatitis C. *Hepatology* **46**: 324–329.

Rao VK and Straus SE (2006) Causes and consequences of the autoimmune lymphoproliferative syndrome. *Hematology* **11**: 15–23.

Rigaud S, Fondaneche MC, Lambert N *et al.* (2006) XIAP deficiency in humans causes an X-linked lymphoproliferative syndrome. *Nature* **444**: 110–114.

Russell JH and Ley TJ (2002) Lymphocyte-mediated cytotoxicity. *Annual Review of Immunology* **20**: 323–370.

Strasser A, Harris AW, Bath ML and Cory S (1990) Novel primitive lymphoid tumours induced in transgenic mice by cooperation between myc and bcl-2. *Nature* **348**: 331–333.

Uysal KT, Wiesbrock SM, Marino MW and Hotamisligil GS (1997) Protection from obesity-induced insulin resistance in mice lacking TNF-alpha function. *Nature* **389**: 610–614.

Varfolomeev E, Maecker H, Sharp D *et al.* (2005) Molecular determinants of kinase pathway activation by Apo2 ligand/tumor necrosis factor-related apoptosis-inducing ligand. *Journal of Biological Chemistry* **280**: 40599–40608.

Vince JE, Wong WW, Khan N *et al.* (2007) IAP antagonists target cIAP1 to induce TNFalpha-dependent apoptosis. *Cell* **131**: 682–693.

Yin XM, Wang K, Gross A *et al.* (1999) Bid-deficient mice are resistant to Fas-induced hepatocellular apoptosis. *Nature* **400**: 886–891.

Zender L, Spector MS, Xue W *et al.* (2006) Identification and validation of oncogenes in liver cancer using an integrative oncogenomic approach. *Cell* **125**: 1253–1267.

Further Reading

Adams JM and Cory S (2007) Bcl-2-regulated apoptosis: mechanism and therapeutic potential. *Current Opinion in Immunology* **19**: 488–496.

van Delft MF and Huang DC (2006) How the Bcl-2 family of proteins interact to regulate apoptosis. *Cell Research* **16**: 203–213.

Huang DC and Strasser A (2000) BH3-only proteins-essential initiators of apoptotic cell death. *Cell* **103**: 839–842.

Lozano G (2007) The oncogenic roles of p53 mutants in mouse models. *Current Opinion in Genetic Development* **17**: 66–70.

Nichols KE, Ma CS, Cannons JL, Schwartzberg PL and Tangye SG (2005) Molecular and cellular pathogenesis of X-linked lymphoproliferative disease. *Immunological Review* **203**: 180–199.

Stagg J, Johnstone RW and Smyth MJ (2007) From cancer immunosurveillance to cancer immunotherapy. *Immunological Review* **220**: 82–101.

Strasser A, Jost PJ and Nagata S (2009) The many roles of FAS receptor signaling in the immune system. *Immunity* **30**: 180–192.

Vazquez A, Bond EE, Levine AJ and Bond GL (2008) The genetics of the p53 pathway, apoptosis and cancer therapy. *Nature Reviews. Drug Discovery* **7**: 979–987.

Caspases in Inflammation and Immunity

Philippe M LeBlanc, *McGill University, Montreal, Canada*
Maya Saleh, *McGill University, Montreal, Canada*

Advanced article

Article Contents

- Preface
- Immune Response to 'Danger'
- Caspases are Essential Effectors of Innate Immunity and Inflammation
- Caspases in Inflammatory Disorders
- Caspases Tailor the Adaptive Immune Response
- Conclusion

Two of the main challenges that eukaryotic multicellular organisms faced during evolution were to cope with invading microorganisms and to eliminate infected cells. The immune system evolved to handle both tasks. Intertwined with immunity is programmed cell death that tailors the immune response and provides an effective way to remove infected cells. Caspases, effectors of inflammation and programmed cell death cascades, are perfectly suited to regulate the host response to invaders and injury. Their activity as inflammatory and killer proteases is regulated by direct binding to sensors of pathogens or danger, they can be modified following activation of signalling cascades by feedback mechanisms to terminate the inflammatory response and control cell survival, and they associate with coactivators and corepressors whose expression is closely linked to the needs of the cell.

Preface

The innate immune system provides first line defences against pathogens through constitutive and inducible means. Among constitutive defences are the physical barriers of the skin and internal epithelial layers, the acidic nature of the stomach, the mechanical defences of the mucous and cilia and the commensal bacteria that populate the intestine and ward off pathogens through competition. When pathogens breach anatomical barriers and enter otherwise sterile tissues, they are 'sensed' by evolutionarily conserved germline-encoded receptors known as pattern recognition receptors (PRRs). Through various effector molecules, most notably the inflammatory caspases, the sensing of pathogens by PRRs activates inducible responses such as the production of antimicrobial substances and inflammation that alerts the organism of the presence of danger. Innate immunity is also a sentinel that primes adaptive immunity, which eliminates any remaining pathogen and builds memory against rechallenge. Intertwined with immunity is programmed cell death that tailors the immune response and provides an effective way to remove infected cells. Indeed, it seems that innate immune mechanisms were essentially grafted to this response over evolutionary time. Through duplication and divergence of pattern recognition receptors and adaptor molecules, cell death versus infection-resistance mechanisms became somewhat but not completely separate.

Immune Response to 'Danger'

Our knowledge of how and why the body reacts to pathogens and other aberrant physiological conditions has changed greatly in the last half century. For a long time, it was thought that the immune system reacts by discriminating between self (what we are born with) and nonself (what came later) and attacking nonself while maintaining tolerance to self. However, this early model has gone through many adaptations since the late 1950s, firstly by Burnet and Medawar, whose work earned them the Nobel Prize in 1960. Their self–nonself (SNS) model described that each lymphocyte (B cell) expresses on its surface many copies of a single receptor (antibody) capable of recognizing a specific foreign entity (antigen), and that signalling through this antibody stimulates the immune response. It also addressed the obvious problem of autoreactive lymphocytes by indicating that these were deleted early in development. Since then, the model has seen the addition of a second cell type, the T-helper cell (T cell), by Bretscher and Cohn in 1969, who demonstrated the requirement of a 'helper signal' from T cells to keep B cells from dying following antigen–antibody recognition. Lafferty and

Cunningham added the final cell type to the currently accepted model in 1975 by proposing that a T cell also requires a secondary signal called 'costimulation' from an antigen-presenting cell (APC). APCs were to capture antigens from the extracellular space and present them to T cells, thus initiating the helper signal. In 1989, Janeway postulated that APCs are quiescent until activated through stimulation via PRRs that recognize evolutionarily conserved infectious nonself 'patterns' termed pathogen-associated molecular patterns (PAMPs). This model was termed the 'infectious nonself' (INS) model since PRRs endow APCs with the ability to trigger immune responses discriminately towards infectious nonself while tolerating the non-infectious self. Polly Matzinger later proposed an adaptation of this theory termed the 'danger model' in 1994 suggesting that APCs were also activated by 'danger' signals, specifically by danger-associated molecular patterns (DAMPs) and 'alarmins'. DAMPs, such as elevated adenosine triphospahte (ATP) levels, K^+ efflux, or the deposition of crystalline structures, have been recently shown to activate APCs and induce inflammation through inflammatory caspase activation within the inflammasome (McIntire *et al.*, 2009). Similarly, 'alarmins' have the ability to chemoattract cells to the danger site and modulate inflammation, acting as endogenous immunoenhancing adjuvants. Most known alarmins are host proteins that perform specific cellular roles under normal conditions but can be rapidly released to the extracellular milieu following infection or tissue injury to trigger inflammatory responses. These include various antimicrobial peptides such as defensins and cathelicidins, heat shock proteins, the antiviral eosinophil-derived neurotoxin (EDN), the chromatin-associated nuclear protein high mobility group box 1 (HMGB1) and products released from extracellular matrix degradation like hyaluronan (Oppenheim and Yang, 2005).

Caspases are Essential Effectors of Innate Immunity and Inflammation

Immunity in *Drosophila*: the caspase DREDD activates RELISH and is required for resistance to infection by Gram-negative bacteria

Devoid of an adaptive immune system, *Drosophila*'s host defence relies on innate immunity, which comprises cellular and humoral reactions. Although the cellular response is best illustrated by the strong phagocytic activity of the predominant blood cells, the plasmatocytes, its inducible arsenal against invading pathogens relies on humoral immunity in the fat body (analogous to the liver), specifically on two cellular signalling pathways, TOLL and IMD (for immune deficiency) (reviewed in Lemaitre and Hoffmann, 2007). Fungi, yeasts and Gram-positive bacteria activate the TOLL pathway whereas the IMD pathway is triggered by Gram-negative bacterial infection (**Figure 1**). Of greater simplicity than the mammalian innate immune system, gradual advances in the understanding of these two pathways and their signalling cascades have helped elucidate inducible innate immune mechanisms in other more complex species, including our own. **See also**: Death Receptors; Death Receptors at the Molecular Level: Therapeutic Implications; Death Receptor-Induced Necroptosis

The TOLL pathway consists of a signalling complex scaffolded by the protein TOLL, a single-pass transmembrane receptor with an ectodomain marked by leucine-rich repeat (LRR) motifs and a cytoplasmic TOLL/interleukin 1 (IL-1) receptor (TIR) domain. This structural organization is very similar to that of the mammalian Toll-like receptors (TLRs), which were subsequently discovered and so named (Medzhitov *et al.*, 1997). The *Toll* gene was initially identified for its essential role in development as the determinant of dorsoventral polarity in the fruitfly embryo. Subsequently, Toll was found to have a role in innate immunity. Although the *Drosophila* genome encodes a total of nine Tolls (TOLL or TOLL-1, 18-WHEELER or TOLL-2 and TOLL receptors 3–9), only TOLL mediates systemic antifungal immunity. The expression patterns observed for the other Tolls during embryogenesis and metamorphosis suggest that they have developmental roles. Unlike mammalian TLRs, Toll is not activated by a microbial product thus is not a PRR. The cascade of events that lead to its activation is initiated in the haemolymph. Following proteolytic processing and dimerization, the cytokine protein SPÄTZLE, a growth factor of the cysteine knot family with structural similarity to mammalian neurotrophins, binds to Toll and initiates its dimerization at the cell surface. The crosslinking of TOLL in turn leads to an intracellular cascade of activation culminating in the nuclear translocation of the nuclear factor-κB (NFκB) transcription factors, DORSAL and DIF, and expression of their target genes, particularly antimicrobial peptides (AMPs) such as the antifungal Drosomycins, and the Defensins with activity against Gram-positive

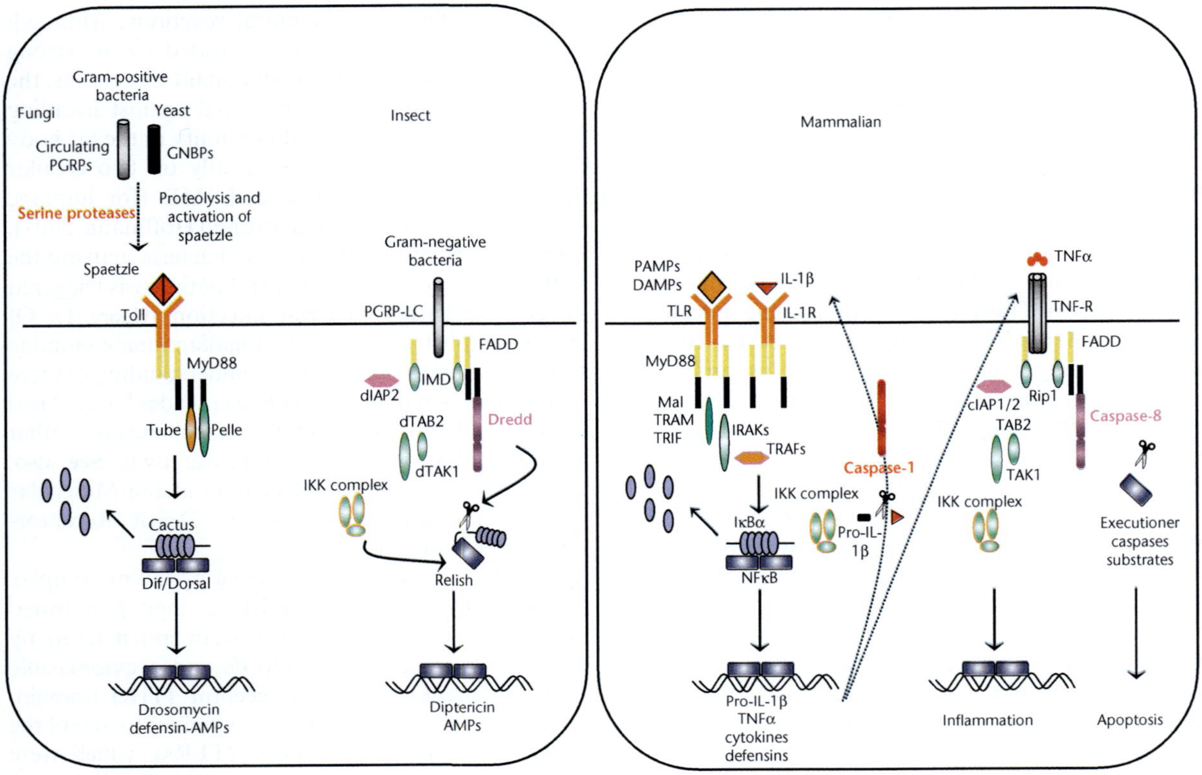

Figure 1 Insect and mammalian innate immune pathways. Antimicrobial peptides (AMPs) and cytokines are the main effectors of insect and mammalian immunity, respectively. The pathways leading to their production are evolutionarily conserved. Mammalian TLR and TNF-R signalling components are related to those operating in insect Toll and Imd pathways. Few modifications to the pathways and expansion of adaptors/signalling molecules occurred through evolution: (1) caspase-8 does not process NFκB proteins in mammalian cells nonetheless its function in modulating NFκB activity downstream of the TNF receptor is conserved. (2) Toll is not a PRR and the insect path to AMP production requires the proteolytic processing of the cytokine spaetzle. Although TLRs are PRRs and could trigger cytokine and AMP production directly, caspase-1 processing of pro-IL-1β in response to pathogen sensing results in signal transduction to AMP production via the IL-1 receptor, which shares with TOLL and TLRs the TIR domain and downstream activation cascade. (3) A PRR (PGRP-LC) initiates the Imd pathway. In contrast, TNFR signalling is activated following a TNFα autocrine loop. (4) Apoptosis does not occur following IMD-DREDD activation while it is induced by TNFα.

bacteria (**Figure 1**, left panel). An interesting parallel to the SPÄTZLE-AMP pathway has been reported in the human epidermis, whereby caspase-1 activation in Langerhans cells and macrophages results in the maturation and release of IL-1β, which binds to its receptor on keratinocytes and signals the production of human β-defensin-2 (HBD-2) (**Figure 1**, right panel).

The IMD pathway also leads to the expression of AMPs. The *imd* gene product is an intracellular death domain containing protein related to RIP1 of the TNF-R (tumour necrosis factor receptor) pathway in mammals. The IMD pathway senses the presence of Gram-negative PAMPs, mostly peptidoglycans (PGNs), through a number of PGN recognition proteins (PGRPs). In this case, immunity is dependent on a member of the caspase family, the caspase-8 homologue DREDD (death-related CED-3/NEDD2-like protein), as it is required for the activation of the NFκB transcription factor RELISH (**Figure 1**). Unlike DORSAL and DIF, RELISH is not inhibited by CACTUS, the *Drosophila* IκB homologue, but carries in its *C*-terminus an autoinhibitory ankyrin repeat domain. In response to infection, DREDD is activated and executes the proteolytic cleavage of RELISH, freeing its translocation to the nucleus (**Figure 1**). RELISH induces the expression of AMPs involved in the defence against Gram-negative bacteria, notably the Diptericins, Attacins, Cecropins and Drosocin. The requisite role of the IMD–FADD (Fas-associated death domain)–DREDD–RELISH pathway in fly immunity to Gram-negative bacteria is illustrated by the low inducibility of all AMPs and a high susceptibility to Gram-negative bacterial infections in *imd*, *fadd*, *dredd* and *relish* mutant flies (reviewed in Lemaitre and Hoffmann, 2007). The DREDD findings represent early evidence that caspases are not only merely effectors of apoptosis but also function in other physiological processes.

Caspases bind to mammalian PRRs and regulate immune signalling

TLRs

The identification of TOLL in *Drosophila* led to the discovery of the first mammalian TLR, classified as a PRR capable of activating NFκB signalling (Medzhitov *et al.*, 1997). To date, there are 12 known mammalian TLRs, structurally similar to TOLL with extracellular LRRs for ligand sensing and intracellular TIR domain to initiate signalling. TLR mutant or knockout mouse models have shed light on some of the functions of these receptors in microbial sensing. The spectrum of microbial agonists recognized by mammalian TLRs encompasses PGN and lipopeptides (TLR-2), lipopolysaccharide (LPS) (TLR-4), flagellin (TLR-5) and nucleic acids (TLR-3, TLR-7 and TLR-9). Following TLR sensing, the signalling cascade that ultimately leads to the activation of host defences is transduced through the recruitment of TIR-containing adaptor proteins by the cytoplasmic TLR TIR domains. Known adaptors include MyD88, TIRAP (TIR-domain-containing adaptor protein)/MAL (MyD88 adaptor-like), TRIF (TIR-domain-containing adapter-inducing interferon β) and TRAM (TRIF-related adaptor molecule), the recruitment of which tailors a specific immune response by activating the NFκB, mitogen-activated protein kinase (MAPK) and interferon regulatory factor (IRF) pathways (Kawai and Akira, 2006). Caspase-8 modulates NFκB activity downstream of TNF-R1 and has been recently shown to be required for antigen receptor and TLR signalling (**Figure 1**). Thus it appears, that most of the elements of fly innate immunity, including the role of caspases in the activation of antimicrobial responses, are conserved in mammals.

NLRs

The cell's ability to sense PAMPs is not restricted to the membrane-bound TLRs. Cytosolic receptors, believed to have evolved from resistance or 'R' proteins in plants, called nod-like receptors (NLRs), are also often involved in the recognition of PAMPs and DAMPs (DeYoung and Innes, 2006). Like TLRs, NLRs oligomerize following agonist sensing and activate pro-inflammatory effectors, most notably the inflammatory caspases. Although TLR signalling has been extensively characterized, very little is known about the regulation of NLRs or their signal transduction mechanisms. Nonetheless, it is becoming increasingly apparent that the NLR pathways are evolutionary related to apoptosis cell death pathways. NOD1, the first mammalian NLR to be discovered, was initially identified as an APAF-1-related protein. Both APAF-1 and NOD1 share a tripartite organization consisting of a caspase recruitment domain (CARD) at the *N*-terminus, a central nucleotide binding and oligomerization domain (Nod) and a *C*-terminal agonist/ligand-binding domain. In APAF-1, this domain is a WD-40 repeat motif that binds cytosolic cytochrome *c* following its release from the mitochondria during apoptosis, whereas in NOD1 it is an LRR domain that recognizes the bacterial PGN derivative diaminopimelic acid (DAP) (**Figure 2**). Twenty-two NLRs have been identified so far, including 14 NALP proteins (NALP1–14), 5 NODS, IPAF, NAIP5 and CIITA (class II transcription activator). Although the structural and functional similarities allow these NLRs to be grouped together, they do nevertheless possess some striking differences, most notably the variability in their *N*-terminal effector domains. NALP proteins have an *N*-terminal pyrin domain (PYD), NAIP5 has three baculovirus IAP repeats (BIR) domains, whereas NOD proteins and IPAF contain CARD domains (One in NOD1 and IPAF, two in NOD2) (**Figure 2a**). **See also**: The Apoptosome: The Executioner of Mitochondria-mediated Apoptosis

The Nod signalosome

The CARDs in NOD1 and NOD2 allow them to interact with a CARD-containing kinase termed RIP2. Once bound to RIP2, they form a multiprotein complex, the 'Nod signalosome', which recruits TRAF proteins and cellular inhibitor of apoptosis proteins (cIAP1 and 2) that polyubiquitinate and activate RIP2. It has been recently shown that the pathways activated downstream of NOD1 and NOD2 are closely related to the TNF-R pathway. NOD receptors and TNF-R1 use structurally and functionally similar components and are subjected to common regulatory mechanisms and feed-forward interactions. NOD1 and NOD2 proteins act as cytosolic PRRs for bacterial PGN. On NOD1 or NOD2 stimulation, RIP2 is conjugated with K63-ubiquitin chains, in a manner analogous to that of RIP1 ubiquitination at the TNFR complex. Polyubiquitinated RIP proteins then serve as a scaffold for the binding of TAK1

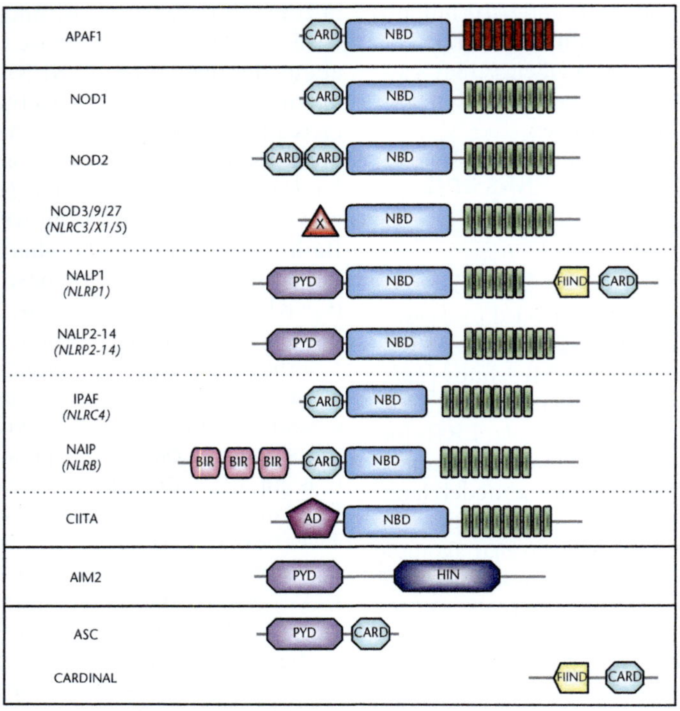
APAF1
CARD
NBD
NOD1
NOD2
NOD3/9/27
(NLRC3/X1/5)
X
NALP1
(NLRP1)
PYD
FIIND
NALP2-14
(NLRP2-14)
IPAF
(NLRC4)
NAIP
(NLRB)
BIR
CIITA
AD
AIM2
HIN
ASC
CARDINAL

(a)

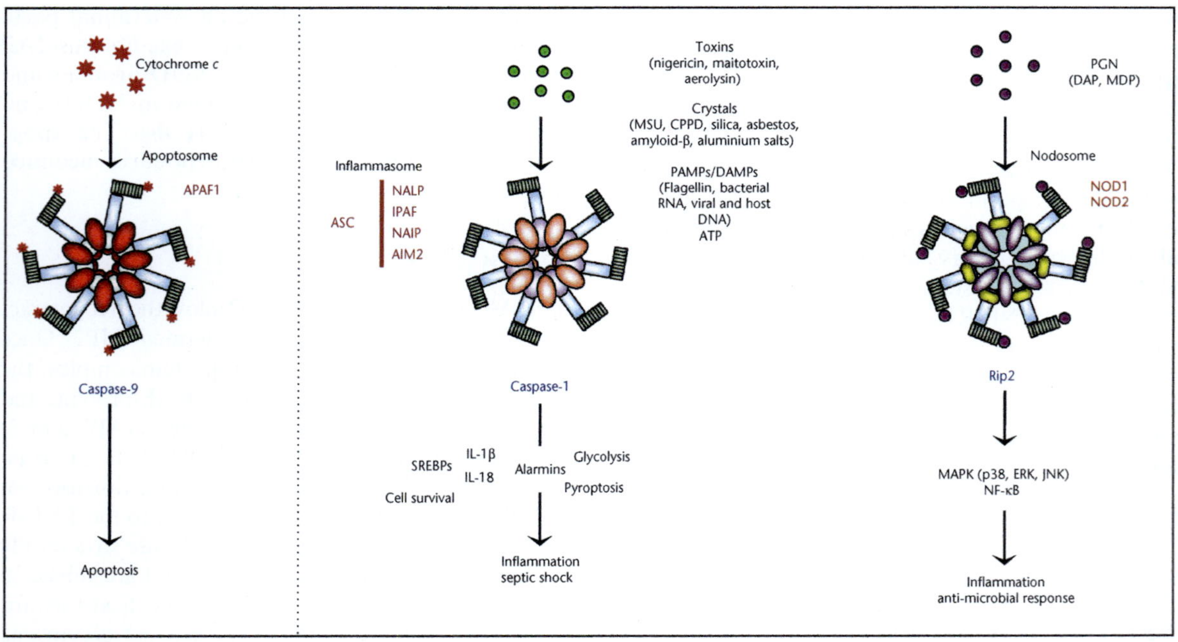
Cytochrome c
Apoptosome
APAF1
Caspase-9
Apoptosis
Inflammasome
ASC
NALP
IPAF
NAIP
AIM2
Toxins
(nigericin, maitotoxin, aerolysin)
Crystals
(MSU, CPPD, silica, asbestos, amyloid-β, aluminium salts)
PAMPs/DAMPs
(Flagellin, bacterial RNA, viral and host DNA)
ATP
Caspase-1
SREBPs
IL-1β
IL-18
Cell survival
Alarmins
Glycolysis
Pyroptosis
Inflammation
septic shock
PGN
(DAP, MDP)
Nodosome
NOD1
NOD2
Rip2
MAPK (p38, ERK, JNK)
NF-κB
Inflammation
anti-microbial response

(b)

(transforming growth factor-β-activated kinase 1) and TAK1-binding protein 2 (TAB2), and for the activation of the MAPK and NFκB pathways leading to cytokine, chemokine and antimicrobial peptide production (**Figure 2b**). Although NOD1 and NOD2 do not signal through inflammatory caspases, recent evidence indicates interplay among these proteins. Notably, the NOD signalosome appears to interact indirectly with caspase-1 on sensing of the PGN derivative muramyl dipeptide (MDP) (Hsu *et al.*, 2008). Additionally, caspase-12 dampens NOD signalling by displacing TRAF6 from the NOD signalosome and blocking RIP2 polyubiquitination (LeBlanc *et al.*, 2008). Other NLRs, the NALPs, IPAF and NAIP, do not assemble Nod signalosomes. As discussed a little further, these assemble other multiprotein complexes called 'inflammasomes' that recruit and activate caspase-1 to induce inflammation and cell death.

The inflammatory caspases

Group I inflammatory caspases, all classified as initiators, include caspases-1, 4, 5 and 12 in humans and caspases-1, 11 and 12 in rodents (reviewed in McIntire *et al.*, 2009). Mouse caspase-11 is the orthologue of human caspases-4 and 5, which are believed to have emerged in humans by tandem duplication. From telomere to centromere, inflammatory caspases are arranged as caspases-1, 5, 4 and 12 on chromosome 11q22.2-q22.3 in humans and as caspases-1, 11 and 12 in a syntenic region of chromosome 9A1 in the mouse (**Figure 3a**). This narrow clustering supports the standing hypothesis that these inflammatory caspases all originated from the same ancestral gene (**Figure 3b**). Their activity in the inflammatory response mainly regulates cytokine maturation, although it is now appreciated that their activation can also lead to a type of inflammatory cell death called pyroptosis (Labbé and Saleh, 2008). Their activity is controlled by various cellular mechanisms such as gene induction by inflammatory stimuli, alternative splicing and tissue-specific distribution. **See also**: Caspases and Cell Death

Of all the inflammatory caspases, caspase-1 has thus far been the most studied. This has conferred valuable insights into activation mechanisms, which is central to the induction of the inflammatory response. Originally discovered as the IL-1β-converting enzyme (ICE) during attempts to identify the enzyme responsible for processing of pro-IL-1β into its mature biologically active form (Thornberry *et al.*, 1992), it is now known to process a vast number of substrates (**Figure 2b**). Among these, members of the IL-1 family, which include pro-IL-1β, pro-IL-18, also known as interferon γ-inducing factor (IGIF), pro-IL-33 and IL-1F7b, which renders it an important effector in innate immunity and host resistance to pathogens. The release of these mature cytokines exerts various effects on different tissues, resulting, for example, in the induction of anorexia, fatigue, fat catabolism, fever, as well as the secretion of acute phase proteins and activation of immune cells leading to the release of other cytokines and chemokines (Dinarello, 1996). However, its functions extend beyond processing of IL-1 family cytokines. Caspase-1 executes an inflammatory form of programmed cell death, distinct from the immunologically silent apoptosis, termed pyroptosis (Labbé and Saleh, 2008). Paradoxically, it is also required for cell survival in response to pore-forming toxins, by regulating lipid biogenesis and membrane repair, which is thought to give the cell a survival window to try and repair itself before committing cell death. Recently, caspase-1 was shown to be necessary for unconventional protein secretion, of various factors including cytokines and possible alarmins (Keller *et al.*, 2008). Therefore, it seems that its activity in innate immunity extends beyond IL-1 maturation and that the threshold or mechanisms of its activation determines its cellular effects. Simplistically, a low level of active caspase-1 may process cytokines, release danger signals and repair membranes without killing, as inhibitory mechanisms have sufficient time to counteract the potentially deadly effects of the caspase unless the latter exceeds some threshold. We and others have recently identified inhibitors of caspase-1 (e.g. caspase-12, ICEBERG)

Figure 2 A parallel between apoptosis and innate immunity. (a) Nod-like receptors (NLRs) share with APAF-1 a tripartite structure that mediates ligand sensing, nucleotide binding and oligomerization and recruitment of caspases. The NLR family consists of the NOD, NALP and IPAF subfamilies. In addition, it includes the MHC II gene transactivator CIITA. AIM2 is a novel cytosolic deoxyribonucleic acid (DNA) sensor that contains a *N*-terminal pyrin domain (PYD) and assembles an inflammasome. ASC and Cardinal are inflammasome adaptors.
(b) The inflammasome is reminiscent of the apoptosome. The apoptosome is scaffolded by the protein APAF-1 and assembles in response to cytochrome *c* release from the mitochondria to activate caspase-9 and apoptosis cascades. NLR proteins scaffold the inflammasome, which is activated by a wide spectrum of triggers including PAMPs (pathogen nucleic acids, flagellin) and DAMPs (ATP, ionic perturbations, crystals and protein aggregates). The inflammasomes recruit and activate caspase-1 to induce inflammation, cell survival or pyroptosis. The Nodosome, which is related to the TNFR signalosome, is linked to caspases indirectly. It is activated by PGN derivatives and signals through the kinase RIP2 to induce inflammation and antimicrobial responses.

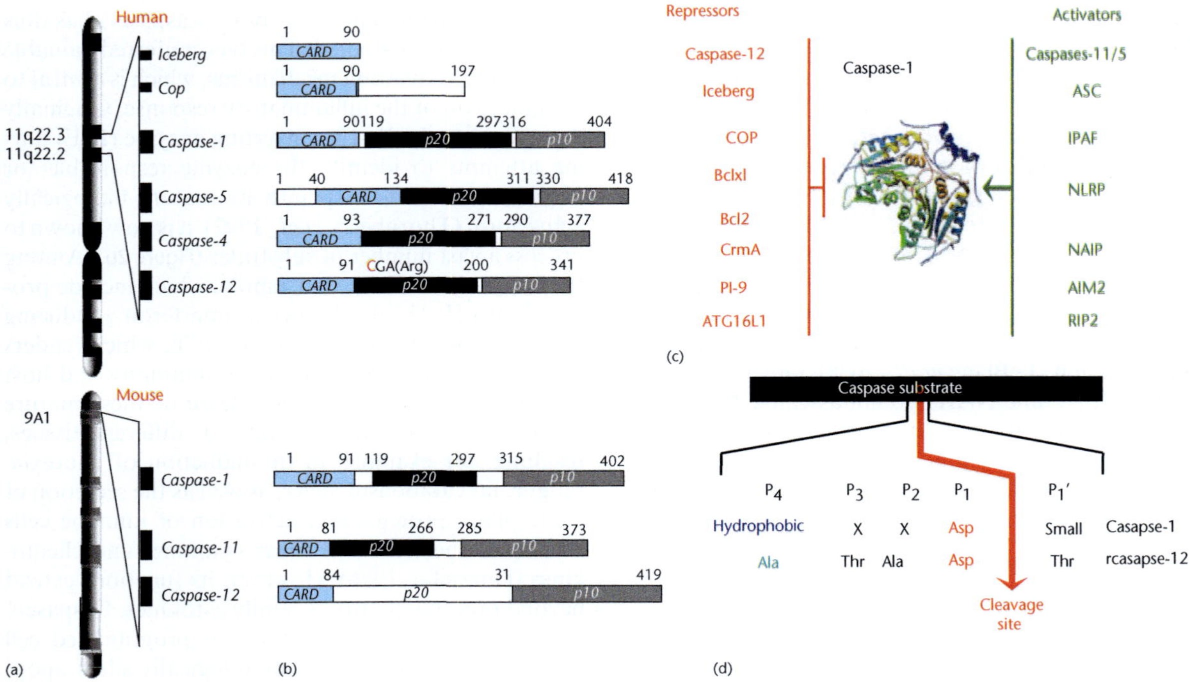

Figure 3 Characteristics of the inflammatory caspases. (a) Chromosomal arrangement of the inflammatory caspase genes on human chromosome 11 and on a syntenic region on mouse chromosome 9. Genes encoding the CARD-only proteins COP and ICEBERG are present only in humans. (b) Domain organization of the inflammatory caspases. CARD, caspase-recruitment domain; p20, large subunit (20 kDa) and p10, small subunit (10 kDa). (c) The crystal structure of caspase-1 is shown. Caspase-1 inhibitors are depicted in red and caspase-1 activators in green. (d) Inflammatory caspases preference for the substrate cleavage site. With the exception of caspase-12, the inflammatory caspases prefer a bulky hydrophobic residue at position P4 C-terminus of the scissile bond that invariably occurs at an aspartate (P1).

(Saleh *et al.*, 2006), however, little is known about how these and other cellular factors engage and regulate the inflammatory caspases (**Figure 3c**). These additional functions of caspase-1, which are quite distinct from its cytokine converting role, begin to illustrate the multi-faceted contributions of this enzyme in immunity and host defence.

The active form of caspase-1 is that of a heterotetramer composed of two large (20 kDa) and two small (10 kDa) subunits derived from a 45-kDa precursor. Its tertiary structure was resolved by X-ray crystallography to be of the form $(\alpha\beta)_2$ (**Figure 3c**). The active enzymatic site spans residues from both subunits, with the p20 containing the catalytic cysteine (Cys^{285}) at its *C*-terminus, whereas the p10 containing residues involved in substrate specificity (Thornberry *et al.*, 1992). The substrate specificity of caspase-1 has been extensively studied in *in vitro* cleavage assays of synthetic peptides. The enzyme requires peptide chains of a minimum of 5 amino acids, P4-P3-P2-P1-P1′, with the catalysis occurring between the P1 and P1′ positions. Although caspase-1 absolutely requires an Asp at P1 and a small hydrophobic residue at P1′ (either Gly or Ala), conservative substitutions are tolerated in the P2 and P3 positions but a large hydrophobic residue is preferred at P4 (**Figure 3d**). The activation of caspase-1 depends on oligomerization, and the stabilization of this active conformation is achieved through an initial processing between the small and large subunits and subsequent cleavage of the prodomain (Boatright *et al.*, 2003). With the exception of caspase-12, the other inflammatory caspases possess similar properties to those of caspase-1, being activated by proximity and preferring a large hydrophobic residue at the P4 position of the substrate recognition sequence. In contrast, and despite having key features necessary for proteolysis, caspase-12 catalytic activity appears to be confined to autoprocessing. In addition, the sequence at this site is relatively featureless indicating, that unlike other caspases, caspase-12's active site cleft might not tolerate more elaborate side chains on the residues upstream of the P1 Asp (Roy *et al.*, 2008; **Figure 3d**).

The inflammasome

The activation of the inflammatory caspases relies on mechanistic triggers that result in the formation of macromolecular complexes, termed 'inflammasomes'. Inflammasomes are assembled by certain NLRs on ligand sensing, as introduced earlier in this article. These are NALP1–14, IPAF and NAIP, and their activation results predominantly in the recruitment and activation of caspase-1. The analogy between NLR and apoptosis signalling is not restricted to the NOD–TNFR pathways, as the inflammasome appears to be reminiscent of the apoptosome. Indeed, the first described mammalian NLR was identified in a screen for APAF-1-related proteins and as described earlier, NLRs share with APAF-1 a common tripartite domain organization. In addition, structurally, the inflammasome resembles the apoptosome, as determined by cryoelectron microscopy; both complexes form oligomeric 'wheels' that recruit and activate caspases by proximity. Interestingly, a recent report extended the parallel between the two pathways by uncovering a role of the antiapoptotic proteins BCL-2 (B-cell lymphoma 2) and BCL-x_L, in modulating inflammasome activation through direct association with Nalp proteins (Bruey *et al.*, 2007; **Figure 2b** and **Figure 3c**).

The recruitment of caspases to the NLRs occurs through CARD–CARD homotypic interactions. This can be either direct or through the help of adaptor proteins. Direct NLR–caspase interactions include the binding of caspase-1 by IPAF and that of caspase-5 by NALP1. NALP2–14 do not contain CARD domains, and thus require adaptors to engage caspases. Two inflammasome adaptors have been identified, and these are *a*poptosis *s*peck-like protein containing a *C*ARD (ASC, a PYD-CARD-containing molecule also known as PYCARD) and CARD inhibitor of NFκB-activating ligands (Cardinal, structurally similar to the *C*-terminal part of NALP1) (**Figure 2a**). Human caspase-5 and its murine ortholog caspase-11 are believed to be recruited to certain inflammasomes and to act as coactivators of caspase-1. Inversely, recent studies have identified caspase-12 as a negative regulator of caspase-1 activity (Saleh *et al.*, 2004; Saleh, 2006). It is hypothesized that caspase-12 acts through a dominant negative, independently of its catalytic activity, analogous to cFLIP's (cellular caspase-8 (FLICE)-like inhibitory proteins) inhibition of caspase-8.

Caspases in Inflammatory Disorders

Although the activity of caspase-1 is important for host defence against pathogens, it has also been linked to a number of inflammatory conditions. Among the most severe are systemic inflammatory response syndrome (SIRS) and septic shock. Caspase-1 is hyperactivated during septic shock, and its activity mediates the disease as *casp1*$^{-/-}$ deficient mice are resistant to shock. Although, the function of caspase-1 in the activation of the IL-1 pathway contributes to the cytokine 'storm' that occurs in septic shock, hyperproduction of these cytokines alone does not account for the totality of its effects in this condition as *casp1*$^{-/-}$ mice, but not IL-1β/IL-18 double knockout mice, are protected from septic shock. To understand the role of caspase-1 in pyroptosis and septic shock, we sought to characterize its cellular substrates. Using the diagonal gel proteomic approach, we have recently identified 44 novel caspase-1 targets (Shao *et al.*, 2007). These included glycolysis enzymes, chaperones, translation machinery proteins as well as proteins involved in immunity. We have demonstrated that the glycolysis pathway is specifically targeted by caspase-1 during infection and septic shock. This provided new insights into how caspase-1 action during intracellular infection and sepsis may lead to cell death and tissue dysfunction. Consistent with the essential role of caspase-1 in bacterial clearance, deficiency in its antagonist, caspase-12, confers resistance to infection and severe sepsis (Saleh *et al.*, 2006). Interestingly, a single nucleotide polymorphism (SNP) in the *caspase-12* gene that results in the expression of caspase-12 in a proportion of African descendents is associated with severe sepsis and sepsis mortality in the clinic (Saleh *et al.*, 2004). Other inflammatory conditions associated with increased caspase-1 levels or activities include acute renal failure (ARF), metastatic melanoma, cutaneous T-cell lymphoma (CTCL), multiple sclerosis (MS), arthritis and asthma (reviewed in McIntire *et al.*, 2009). Studies pertaining to the implications of caspase-1 in these conditions have suggested that caspase-1 levels might constitute a reliable physiological marker of ongoing immune-inflammatory responses. Of note, mutations in the gene encoding NALP3 are associated with autoinflammatory diseases. Gain-of-function (GOF) mutations in the *Nalp3* gene, at various positions throughout the NACHT (neuronal apoptosis inhibitory protein, CIITA, HET-E and TP1) domain, have been shown to induce constitutive receptor activation and IL-1β release. These mutations have been linked to three different autosomal dominant inherited autoinflammatory diseases, jointly classified as familial periodic fever syndromes (reviewed in Kastner, 2005). Among these, familial cold urticaria (FCU), characterized by episodes of rash, fever and joint pain following exposure to cold, Muckle–Wells syndrome (MWS), resulting in chronic urticaria, periodic arthritis, sensorineural deafness and amyloidosis and finally neonatal-onset multisystem inflammatory disease (NOMID), associated with various signs of uncontrolled inflammation such as fever and joint pain, often leading to persistent neurological damage. *Nod2* mutations are linked to Crohn disease (CD) (Hugot *et al.*, 2001) and Blau syndrome (Miceli-Richard *et al.*, 2001), two inflammatory conditions, the first being a lifelong inflammatory disorder of the digestive tract and the second causing a constitutive state of inflammation. In the case of CD, mutations in the LRR result in a loss-of-function (LOF) phenotype, blunting bacterial sensing by NOD2. Recently, a regulatory region downstream of the *Nalp3* gene was found to be associated with Crohn disease susceptibility in individuals from European descent. SNPs in this region were associated with decreased NALP3 expression and dampened IL-1β levels in response to LPS (Villani *et al.*, 2009). These findings are in support of the CD immunophathogenesis paradigm suggesting that impaired innate immune responses leads to defective clearance of microbes which in turns leads to chronic intestinal inflammation.

The local production of host-derived danger signals can also lead to chronic inflammation. The metabolic disorders gout and pesudogout are inflammatory crystalline arthritic conditions caused by the excessive activation of the NALP3 inflammasome (Martinon *et al.*, 2006). Unlike the familial periodic fever syndromes discussed earlier, this aberrant NALP3 activation is not mediated by genetic factors but from the accumulation of monosodium urate monohydrate (MSU) and calcium pyrophosphate dihydrate (CPPD) crystals, in gout and pseudogout, respectively. The accumulation of these crystals in the synovial fluids results in a chronic sterile inflammation of the joints. In many cases, diseases associated with elevated IL-1β production such as those listed earlier can be effectively treated with targeted anti-IL-1β therapies. The recombinant IL-1 receptor antagonist (IL-1RA), anakinra, which blocks activation of the IL-1R by competitive binding, has proven beneficial for patients with hereditary periodic fevers and gout (Hoffman *et al.*, 2004).

Interestingly, the implication of caspases in inflammatory disease is not limited to the inflammatory caspases. Notably, sepsis-associated mortality results from an impaired immune response due to widespread immune effector cell death by apoptosis (Hotchkiss and Nicholson, 2006). This state abrogates the body's ability to clear the initial infection and engenders the development of secondary infections. Consistent with a role of apoptotic caspases in mediating severe sepsis, selective caspase-3 inhibitors or the transgenic expression of the antiapoptotic factor BCL-2 have proven protective in experimental sepsis and rescued animals from sepsis mortality.

Caspases Tailor the Adaptive Immune Response

Although cell death plays important roles in the host–pathogen battle during the innate phase of the immune response, it is also crucial for adaptive immunity. Particularly, apoptosis is required for the elimination of unwanted cells. The deletion of autoreactive immune cells is central to the immunological tolerance of self and because it relies on the cellular apoptotic machinery, constitutes an elegant example of the involvement of caspases in adaptive immunity. Apoptotic caspases play an essential role in maintaining homeostasis and preventing the appearance of reactive T cells during T-lymphocyte development in the thymus (Opferman, 2008). This process begins with the migration of a common lymphoid progenitor from the bone marrow to the thymus where it develops into a double-negative thymocyte. At this stage, the survival of thymocytes is dependent on stimulatory signals including the presence of IL-7 (Lee and Surh, 2005). The absence of the required cytokines or growth factors results in the triggering of intrinsic apoptosis in these cells, a process mediated by a core dependence on the BH3-only protein BIM and inhibited by the antiapoptotic BCL-2 protein MCL-1. The developing lymphocytes then rearrange their antigen receptor genes and acquire pre-T-cell receptors (pre-TCRs). These pre-TCRs may be responsive to self-antigens, which would be detrimental to the host, and must be eliminated through a process known as negative selection. For this, CD4 and CD8 double-positive T cells that strongly recognize self-major histocompatibility complex (MHC) molecules are eliminated by apoptosis in a Bim-dependent manner. This contribution is highlighted by the fact that $bim^{-/-}$ mice develop autoimmune disease through defective deletion of autoreactive thymocytes. Positively selected cells then migrate to peripheral lymphoid tissues and differentiate into CD4 or CD8 T cells to join the pool of mature circulating cells. Apoptotic caspases also play an essential role in the course of an immune response, where excessive clonal expansion is restricted by survival factor deprivation and activation-induced cell death (AICD)

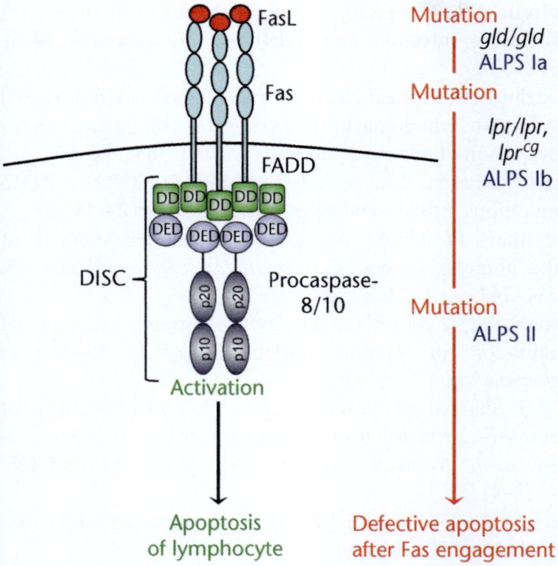

Figure 4 Mutations in the initiator caspase-8 and caspase-10 result in autoimmune lymphoproliferation syndromes (ALPS). Caspase-8 and caspase-10 are recruited to the death-inducing signalling complex (DISC) on binding of the trimeric FAS ligand (FASL) to its receptor FAS. Caspase recruitment and activation occurs via the adaptor protein FADD (Fas-associated death domain-containing protein), and is essential for lymphocyte apoptosis. Autosomal recessive mutations in FASL and FAS are found in the *gld/gld* and *lpr/lpr* mice, respectively. In humans, mutations in FASL, FAS and caspase-8 and caspase-10 are present in ALPS patients.

(Krammer *et al.*, 2007). In the periphery, potentially self-reactive clones that have escaped thymic deletion can still be removed by peripheral deletion. Recent studies into the apoptotic pathways involved in peripheral deletion have revealed that unlike the predominant role of mitochondrial apoptosis during thymic T-lymphocyte development, both extrinsic and intrinsic pathways are necessary for peripheral deletion. This is illustrated by the accumulation of lymphoid cells and expanded lymphoid organs observed in mice deficient in both FAS function and BIM. BIM-mediated apoptosis is activated in response to lack of support in the microenvironment by survival factors whereas Fas-mediated apoptosis, is induced following FASL recognition. Antigen-stimulated T cells express FASL leading to killing of neighbouring T cells and APCs. It has been suggested that this elevated FASL production is sufficient to account for most of the T-cell apoptosis or clonal contraction observed during acute infections, whereas Bim-mediated apoptosis becomes crucial during chronic infections. **See also**: Immunity, Granzymes and Cell Killing

Mutations in the FAS extrinsic apoptosis pathway have been linked to immunodeficient and autoimmune states. Autosomal recessive mutations in the FAS transmembrane receptor and in FASL, found in naturally arising mouse strains, *lpr*/*lpr* (lpr) and *gld*/*gld* (gld), respectively, cause profound lymphoproliferation, hypergammaglobulinaemia, autoantibodies, systemic autoimmunity and lymphomas as a result of defective lymphocyte apoptosis (Cohen and Eisenberg, 1991). In humans, autoimmune lymphoproliferative syndrome (ALPS) is triggered when mutations in *Casp-8*, *Casp-10*, *Fas* or *FasL* disrupt the FAS signalling pathway, leading to a similar defective FAS-induced apoptosis in lymphocytes and subsequent autoimmunity (reviewed in Bidere *et al.*, 2006; **Figure 4**). **See also**: Apoptosis: Inherited Disorders

Conclusion

The investigation of the molecular mechanisms linking the sensing of a pathogen or danger signal to the induction of an inflammatory and antimicrobial response has witnessed a renaissance in the past few years. This was initiated by the identification of PRRs, including TLRs and, recently, cytosolic NLRs, that brought innate immunity to centre stage and opened the field to the study of signal transduction pathways linked to PRRs. This led to the characterization of the inflammasome, a macromolecular complex, scaffolded by NLRs, that recruits and activates inflammatory caspases, central effectors of innate immunity: they regulate the processing and maturation of pro-inflammatory cytokines, nonconventional protein secretion of alarmins and cytokines, glycolysis and lipid biogenesis and the execution of an inflammatory form of cell death termed 'pyroptosis'. Ablation of inflammatory caspase activity is associated with susceptibility to infections and severe sepsis. However, misdirected activity is detrimental and leads to numerous diseases, notably inflammatory disorders and septic shock. In addition to the intricate involvement of the inflammatory caspases in innate immunity, apoptotic caspases tailor adaptive immune responses, maintain immune homeostasis and protect the host from autoimmunity.

References

Bidere N, Su HC and Lenardo MJ (2006) Genetic disorders of programmed cell death in the immune system. *Annual Review of Immunology* **24**: 321–352.

Boatright KM, Renatus M, Scott FL *et al.* (2003) A unified model for apical caspase activation. *Molecular Cell* **11**: 529–541.

Bruey JM, Bruey-Sedano N, Luciano F *et al.* (2007) Bcl-2 and Bcl-XL regulate proinflammatory caspase-1 activation by interaction with NALP1. *Cell* **129**: 45–56.

Cohen PL and Eisenberg RA (1991) Lpr and gld: single gene models of systemic autoimmunity and lymphoproliferative disease. *Annual Review of Immunology* **9**: 243–269.

DeYoung BJ and Innes RW (2006) Plant NBS-LRR proteins in pathogen sensing and host defense. *Nature Immunology* **7**: 1243–1249.

Dinarello CA (1996) Biologic basis for interleukin-1 in disease. *Blood* **87**: 2095–2147.

Hoffman HM, Rosengren S, Boyle DL *et al.* (2004) Prevention of cold-associated acute inflammation in familial cold autoinflammatory syndrome by interleukin-1 receptor antagonist. *Lancet* **364**: 1779–1785.

Hotchkiss RS and Nicholson DW (2006) Apoptosis and caspases regulate death and inflammation in sepsis. *Nature Review of Immunology* **6**: 813–822.

Hsu LC, Ali SR, McGillivray S *et al.* (2008) A NOD2-NALP1 complex mediates caspase-1-dependent IL-1beta secretion in response to *Bacillus anthracis* infection and muramyl dipeptide. *Proceedings of the National Academy of Sciences of the USA* **105**: 7803–7808.

Hugot JP, Chamaillard M, Zouali H *et al.* (2001) Association of NOD2 leucine-rich repeat variants with susceptibility to Crohn's disease. *Nature* **411**: 599–603.

Kastner DL (2005) Hereditary periodic Fever syndromes. Hematology. American Society of Hematology. Education Program 74–81.

Kawai T and Akira S (2006) TLR signaling. *Cell Death and Differentiation* **13**: 816–825.

Keller M, Ruegg A, Werner S and Beer HD (2008) Active caspase-1 is a regulator of unconventional protein secretion. *Cell* **132**: 818–831.

Krammer PH, Arnold R and Lavrik IN (2007) Life and death in peripheral T cells. *Nature Review of Immunology* **7**: 532–542.

Labbé K and Saleh M (2008) Cell death in the host response to pathogens. *Cell Death and Differentiation* **15**: 1339–1349.

LeBlanc PY, Rutherford G, Doiron N *et al.* (2008) Caspase-12 modulates NOD signaling and regulates antimicrobial peptide production and mucosal immunity. *Cell Host & Microbe* **3**: 146–157.

Lee SK and Surh CD (2005) Role of interleukin-7 in bone and T-cell homeostasis. *Immunological Review* **208**: 169–180.

Lemaitre B and Hoffmann J (2007) The host defense of *Drosophila melanogaster*. *Annual Review of Immunology* **25**: 697–743.

Martinon F, Pétrilli V, Mayor A, Tardivel A and Tschopp J (2006) Gout-associated uric acid crystals activate the NALP3 inflammasome. *Nature* **440**(7081): 237–241.

McIntire CR, Yeretssian G and Saleh M (2009) Inflammasomes in infection and inflammation. *Apoptosis* **14**(4): 522–535.

Medzhitov R, Preston-Hurlburt P and Janeway CA Jr (1997) A human homologue of the *Drosophila* Toll protein signals activation of adaptive immunity. *Nature* **388**: 394–397.

Miceli-Richard C, Lesage S, Rybojad M *et al.* (2001) CARD15 mutations in Blau syndrome. *Nature Genetics* **29**: 19–20.

Opferman JT (2008) Apoptosis in the development of the immune system. *Cell Death and Differentiation* **15**: 234–242.

Oppenheim JJ and Yang D (2005) Alarmins: chemotactic activators of immune responses. *Current Opinion in Immunology* **17**: 359–365.

Roy S, Sharom JR, Houde C *et al.* (2008) Confinement of caspase-12 proteolytic activity to autoprocessing. *Proceedings of the National Academy of Sciences of the USA* **105**: 4133–4138.

Saleh M (2006) Caspase-1 builds a new barrier to infection. *Cell* **126**: 1028–1030.

Saleh M, Mathison JC, Wolinski MK *et al.* (2006) Enhanced bacterial clearance and sepsis resistance in caspase-12 deficient mice. *Nature* **440**: 1064–1068.

Saleh M, Vaillancourt JP, Graham RK *et al.* (2004) Differential modulation of endotoxin responsiveness by human caspase-12 polymorphisms. *Nature* **429**: 75–79.

Shao W, Yeretssian G, Doiron K, Hussain SN and Saleh M (2007) The caspase-1 digestome identifies the glycolysis pathway as a target during infection and septic shock. *Journal of Biological Chemistry* **282**: 36321–36329.

Thornberry NA, Bull HG, Calaycay JR *et al.* (1992) A novel heterodimeric cysteine protease is required for interleukin-1 beta processing in monocytes. *Nature* **356**(6372): 768–774.

Villani AC, Lemire M, Fortin G *et al.* (2009) Common variants in the NLRP3 region contribute to Crohn's disease susceptibility. *Nature Genetics* **41**: 71–76.

Further Reading

Ferrandon D, Imler JL, Hetru C and Hoffmann JA (2007) The *Drosophila* systemic immune response: sensing and signalling during bacterial and fungal infections. *Nature Review of Immunology* **7**(11): 862–874.

Green DR (2008) Fas Bim boom! *Immunity* **28**: 141–143.

Matzinger P (2002) The danger model: a renewed sense of self. *Science* **296**: 301–305.

Scott AM and Saleh M (2007) The inflammatory caspases: guardians against infections and sepsis. *Cell Death and Differentiation* **14**: 23–31.

Yeretssian G, Labbe K and Saleh M (2008) Molecular regulation of inflammation and cell death. *Cytokine* **43**: 380–390.

Immunity, Granzymes and Cell Killing

Nigel J Waterhouse, *Apoptosis and Natural Toxicity Laboratory, Peter MacCallum Cancer Centre, Melbourne, Victoria, Australia*

Olivia Susanto, *Apoptosis and Natural Toxicity Laboratory, Peter MacCallum Cancer Centre, Melbourne, Victoria, Australia*

Karin A Sedelies, *Apoptosis and Natural Toxicity Laboratory, Peter MacCallum Cancer Centre, Melbourne, Victoria, Australia*

Joseph A Trapani, *Cancer Cell Death Laboratory, Peter MacCallum Cancer Centre, Melbourne, Victoria, Australia*

Advanced article

Article Contents

- Introduction
- Granule Exocytosis-induced Death
- Granzyme B and Apoptosis
- Granule-mediated Caspase-independent Cell Death
- Cell Death Induced by Granzymes Other than Granzyme B
- Conclusion

Cytotoxic lymphocytes protect us from viral infection and cancer by directly killing tumour cells, or cells harbouring a virus. One crucial mechanism they use to kill their targets is known as the 'granule exocytosis' pathway. This involves secretion of a potent mix of toxins, resulting in transfer of granule proteases (granzymes) from the killer cell into the target cell, where they cleave various intracellular substrates to activate diverse signalling pathways to cell death. Access to the target cell cytosol depends on a pore-forming protein toxin, perforin. Because granule exocytosis is a crucial part of the body's natural defence against such dangerous cells, understanding how granzymes kill their targets may yield novel strategies and identify new molecular targets for anticancer or antiviral therapies.

Introduction

To remain functional and contribute to a healthy organism, the structural and functional integrity of cells must be maintained, and they must also signal their well-being to their neighbours and to cells of the immune system (Matzinger, 2002). Cells constantly monitor and react to changes in their biochemical status to determine whether their integrity has been compromised. If problems are detected, the cell may either repair any damage, or alternatively initiate cell death by apoptosis. Although one or more cells are lost through this process, the organism as a whole is protected, either from spread of the virus, or continued proliferation of malignant cell clones. This protective response may be defective or blocked in cancer, if the pathways that regulate cell death are inhibited, or during infection if the virus encodes proteins that block cell death. Under such circumstances, target cells must undergo death enforced by the immune system. Removal of these dangerous cells is the central function of cytotoxic lymphocytes, cells of the immune system that can distinguish between healthy cells and those that pose a risk (Trapani and Smyth, 2002). **See also**: Apoptosis: Regulatory Genes and Disease; Caspases in Inflammation and Immunity

There are two common types of cytotoxic lymphocyte, cytotoxic T lymphocytes (CTL) and natural killer (NK) cells. Cytotoxic lymphocytes are produced in vast numbers in the spleen, and circulate around the body in the blood and lymphatics, searching for dangerous cells by systematically crawling and scanning through and over tissue. They recognize dangerous cells because of antigen presented on the outside of that cell (CTL), or by diminished presentation of molecules that are intrinsic to healthy cells (NK cells). Regardless of the method of detection, NK cells and CTL generally utilize highly conserved mechanisms to kill their targets, one of which is ligation of death receptors (reviewed in Chavez-Galan *et al.*, 2009), the other being granule exocytosis. Because cytotoxic lymphocytes are motile and kill their targets by direct contact-dependent mechanisms, they are generally very efficient at picking out and specifically eliminating dangerous cells while limiting damage to the surrounding tissue ('bystander cell death'), as could occur more readily in a systemic response. **See also**: Cell Death in Inflammation and Immunity

Granule Exocytosis-induced Death

When a lymphocyte detects its target, it forms a tight junction (the immunological synapse), and mobilizes its cytotoxic granules along the microtubular apparatus so that the granules cluster close to the target cell. These granules fuse with the plasma membrane and the granule contents are released into the immunological synapse (Stinchcombe *et al.*, 2006). Some of the granule components enter the target cell where they activate specific cell death pathways. Cytotoxic lymphocytes have been observed to kill their targets through lysis *in vitro*, resulting in death by lysis/necrosis, so it is also plausible that target cell death also occurs through this process (Russell *et al.*, 1980). However, it has not been formally demonstrated that this process can occur *in vivo*, and it is very likely that CL preferentially kill their targets by nonlytic mechanisms, as discussed later.

Cytotoxic granules also contain serine proteases (granzymes) that can trigger nonlytic death (Chowdhury and Lieberman, 2008). In particular, it is now emerging that cytotoxic lymphocytes typically induce apoptosis in their target cell. Apoptosis is a form of cell death commonly encountered in a wide variety of developmental processes, in response to many noxious stimuli and following the withdrawal of cellular growth factors. Perforin applied alone to cells is incapable of inducing apoptosis, but can trigger apoptotic death when used in combination with granzyme B (Waterhouse *et al.*, 2006; Shi *et al.*, 1992). It is therefore most likely that perforin is essential for cytotoxic lymphocyte-induced death by granule exocytosis because it is essential for correct delivery of granzymes. In support of this notion, granzymes can be taken up by a target cell's natural sampling processes such as endocytosis or pinocytosis, but the cell remains healthy in the absence of perforin (Bird *et al.*, 2005). The essential role for perforin in the delivery of granzymes is clear from studies showing that CTL from mice genetically deficient in perforin cannot kill their targets by granule exocytosis, but mice that lack certain granzymes or even combinations of granzymes can (Kagi *et al.*, 1994; Smyth *et al.*, 2003) (for a table on phenotype of mice deficient in components of the granule exocutosis pathway, refer to Trapani and Smyth, 2002).

It is not known how perforin exerts its function, but it is clear from experiments performed *in vitro* with purified perforin that it is capable of forming holes in the plasma membrane which allows granzymes direct access to the target cell cytoplasm (Baran *et al.*, 2009). This function is dependent on the presence of calcium ions, which are required for the binding of perforin monomers to the target cell membrane (Ishiura *et al.*, 1990). Subsequently, perforin monomers coalesce into polymeric transmembrane structures that are more than large enough to allow the passive diffusion of granzymes. However, at best this mechanism is an oversimplification of the true physiological function of perforin, as the pores appear to exert a level of specificity for the granzymes. Thus, dextran molecules considerably smaller than granzymes remain excluded from cells treated with concentrations of perforin that can deliver granzymes but do not cause significant cell damage when applied alone (such concentrations of perforin are commonly referred to as 'sublytic') (Trapani *et al.*, 1998). An alternate plausible hypothesis is that granzymes and perforin are taken up together by cell sampling mechanisms and that granzymes are released into the target cell cytoplasm following perforin-mediated lysis of endosomal vesicles (Froelich *et al.*, 1996). Alternatively, granzyme delivery may be as part of the membrane repair process that accompanies perforin-mediated osmotic damage to the target cell (Keefe *et al.*, 2005). The exact mechanism of granzyme delivery is yet to be fully elucidated, and remains a topic of intense interest in the field. Many other questions also remain about perforin function. In particular, how do cytotoxic lymphocytes escape from the toxic effects of perforin, during its biosynthesis, trafficking and storage in the lymphocyte, and especially following its release into the immunological synapse?

Granzyme B and Apoptosis

Apoptosis is a specific form of cell death and a sequence of morphological changes that includes cell shrinkage, plasma membrane blebbing, chromatin condensation, fragmentation of deoxyribonucleic acid (DNA) into oligonucleosomal (approximately 200 base pair) fragments, loss of plasma membrane asymmetry and eventual phagocytosis of the dead cell (reviewed in Fadeel and Orrenius, 2005). The timing of these discrete events may vary between cell types, but apoptosis invariably requires activation of caspases, ubiquitous proteases that are present in their zymogenic form, and once activated in response to various stimuli cleave their targets after the *C*-terminus of certain aspartic acid residues. Caspase activation and the term 'apoptosis' have therefore become inextricably linked, so that nonapoptotic cell death is often referred to as caspase-independent cell death. To elucidate the discussions of granule-induced death in this article, we will use this biochemical definition of caspase-dependent and caspase-independent death to differentiate apoptosis from all other forms of death.

Executioner caspases, including caspases-3 and -7, are generally activated by cleavage after two specific aspartic acid residues. Granzyme B is a serine protease that shares with the caspases (cysteine proteases) the capacity to cleave substrates at the *C*-terminus of aspartic acid residues, and can therefore participate in caspase activation by direct cleavage. For this reason, granzyme B is the most potent pro-apoptotic granzyme. Granzyme B cannot cleave both sites required for full activation of caspases-3 and -7, and cleavage of the second site requires autocatalytic proteolysis by the partially activated caspases themselves (Sutton *et al.*, 2003). This process is regulated by inhibitor of apoptosis proteins (IAPs), endogenous proteins that directly bind and inhibit caspase activity, suggesting that direct activation of caspases by granzyme B is a regulated process (Hunter *et al.*, 2007).

Granzyme B can also cleave Bid, a Bcl-2 family member which releases soluble proteins from the mitochondrial intermembrane space by inducing mitochondrial outer membrane permeabilization (MOMP) (**Figure 1**; Barry *et al.*, 2000). One of these proteins, direct IAP-binding protein with low pI (Diablo/SMAC), can displace IAPs and thus prevent IAPs from inhibiting caspase activity, and may permit full activation of granzyme B-cleaved caspases. MOMP also releases cytochrome *c* from the mitochondrial intermembrane space. In turn, this promotes activation of caspase-9 by forming an oligomeric complex with apoptotic protease-activating factor 1 (Apaf-1). Caspase-9 then activates executioner caspases by direct cleavage. This cytochrome *c*-dependent process is the major caspase-activating pathway utilized during stress- or drug-induced apoptosis, and does not appear to require release of SMAC/Diablo or pre-cleavage of caspases. Evidence to support this view is that cells deficient in SMAC/Diablo can still activate executioner caspases during drug-induced apoptosis, but cells deficient in cytochrome *c*, Apaf-1 or caspase-9 do not. These findings suggest that cleavage of Bid by granzyme B is sufficient for caspase activation through the mitochondrial pathway, even if direct activation by granzyme B has not taken place (Sutton *et al.*, 2003).

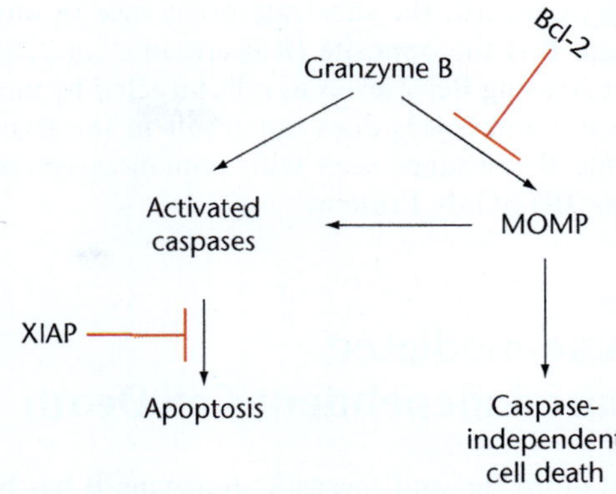

Figure 1 Cytotoxic lymphocytes kill their targets through granule exocytosis, and granzyme B initiates apoptosis via two main pathways. Granzyme B cleaves Bid to induce MOMP and subsequent release of cytochrome *c* from the mitochondria results in caspase activation and 'classic' apoptosis. However, granzyme B may also activate caspases directly (particularly in the mouse) and can also cleave many other substrates directly, potentially leading to caspase-independent death. Granzyme B also shows species variation in substrate specificity, with human granzyme B preferentially cleaving Bid, and mouse granzyme B favouring direct caspases cleavage.

Granzyme B can therefore trigger caspase-dependent apoptosis via one of the two mechanisms: either directly by pro-caspase cleavage or indirectly, via MOMP. Blocking MOMP, which prevents release of SMAC/Diablo, does not prevent CL-induced activation of caspases demonstrating that granzyme B can directly activate caspases in cells that express endogenous levels of IAPs. Further, overexpression of IAPs is not sufficient to block the mitochondrial pathway because SMAC/Diablo can deregulate IAPs, and cytochrome *c* can promote caspase activation independent of SMAC/Diablo. Blocking either pathway alone is therefore not sufficient to prevent cytotoxic lymphocyte-induced apoptosis, but blocking both pathways independently can prevent cytotoxic lymphocyte-induced apoptosis, at least for human cytotoxic lymphocytes (Sedelies *et al.*, 2008). Which pathway is utilized by granzyme B in individual cells may therefore be determined by (1) the concentration of granzyme B delivered to the target cell, (2) the relative level of Bcl-2 family members and (3) the relative level of caspases and IAPs expressed in the target cell. Interestingly, there are some notable species differences in the substrate preference of granzyme B. For instance, human granzyme B cleaves Bid with far greater efficiency than pro-caspases; consequently, human granzyme B-mediated cell death is very efficiently blocked in a target cell that overexpresses

Bcl-2. By contrast, the substrate preference of mouse granzyme B is the opposite (Kaiserman *et al.*, 2006). Thus, increasing Bcl-2 levels in cells targeted by mouse cytotoxic lymphocytes does not result in the level of granzyme B-resistance seen with human target cells. **See also**: BH3-Only Proteins

Granule-mediated Caspase-independent Cell Death

In addition to Bid and caspases, granzyme B has been shown to cleave at least 25 other substrates, many of which are also cleaved by caspases during apoptosis. It has, therefore, been postulated that granzyme B may bypass caspases and orchestrate caspase-independent cell death with features similar to apoptosis. This mechanism is probably not particularly prevalent in the human immune system, as blocking MOMP and caspase activity (e.g. by Bcl-2 overexpression) protected the vast majority of target cells from granzyme B-induced death, but it may be more prevalent in mice. As mouse granzyme B shows a slightly different substrate specificity to human granzyme B (see earlier), blocking both MOMP and caspase activity was not sufficient to block death induced by mouse CTL or mouse granzyme B (Pardo *et al.*, 2008). Using a mouse system of cytotoxic lymphocyte-induced death, we found that although mouse CTL could trigger MOMP in their targets, blocking caspase activity prevented target cells from rounding up (**Figure 2**). Therefore, if mouse granzyme B can bypass caspases, it does not appear to trigger caspase-independent cell death by a process similar to apoptosis. How cleavage of these additional granzyme B substrates results in death is therefore unclear.

Cell Death Induced by Granzymes Other than Granzyme B

Other granzymes, including granzyme A, C, H, K and M, have also been shown to trigger death when delivered to cultured cells by purified perforin (Beresford *et al.*, 1999; Johnson *et al.*, 2003; Fellows *et al.*, 2007; Zhao *et al.*, 2007; Kelly *et al.*, 2004), and CTL from mice deficient in granzyme B can still kill

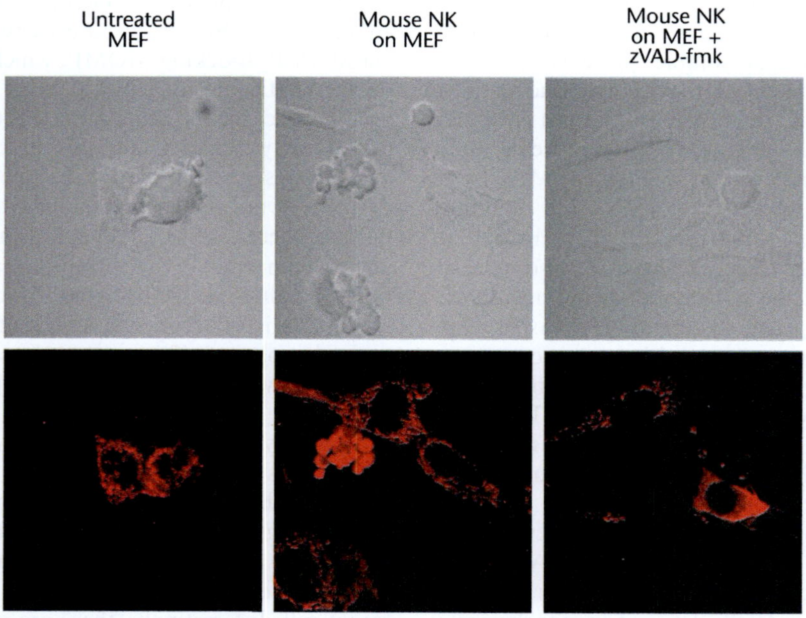

Figure 2 Mouse NK cells kill their targets via diverse pathways. Mouse NK cells were incubated with plastic-adherent mouse embryonic fibroblasts (MEF) at a 2:1 ratio for 2 h. The NK cells were then washed off, and the target cells fixed, permeabilized and stained for cytochrome *c*. The cells were then visualized by confocal microscopy for morphological changes and cytochrome *c* localization. MEF cells showed rounding and blebbing consistent with apoptosis, and released cytochrome *c* from the mitochondria indicating that MOMP has occurred. Although the apoptotic phenotype was being blocked by the caspase inhibitor zVAD-fmk, clonogenic survival was not rescued (data not shown) indicating that caspase-independent death was taking place.

their targets by a novel, nonapoptotic process. The molecular mechanism behind these deaths is less well defined, but they usually do not require caspase activity, or fragmentation of DNA into oligonucleosomal fragments. Accordingly, DNA fragmentation is greatly inhibited and markedly slowed in target cells killed by granzyme $B^{-/-}$ CTL (Heusel *et al.*, 1994). Clearly then, these granzymes must kill cells in such a way that the cell does not activate its intrinsic stress response, because this would result in caspase-mediated fragmentation of DNA (**Figure 3**).

Granzyme A has been shown to induce caspase-independent cell death by a mechanism that is similar to apoptosis, because it involves cell rounding, blebbing and loss of plasma asymmetry before loss of plasma membrane integrity (Beresford *et al.*, 1999). A number of granzyme A substrates have been identified, including poly (ADPribose) polymerase (PARP), lamins and histones, which are also targeted by caspases. This is consistent with the possibility that granzyme A could orchestrate an apoptosis-like death by a process that does not require caspase activation. However, cleavage of these proteins is most likely to impair nuclear function, and this is likely to result in activation of apoptosis pathways. In support of this, PARP inhibitors are well known to induce apoptotic cell death. Similarly, cleavage of novel granzyme A substrates, such as Ape-1, Ku70 and proteins in the endoplasmic reticulum associated complex (SET) complex, is likely to impair the cells' ability to initiate or detect DNA damage, and impairing DNA repair is one classic process that can greatly augment death by apoptosis.

Recently, mitochondria have been implicated in granzyme A-induced death because subunit 3 of NADH dehydrogenase ubiquitone flavoprotein (NDUFS3), a subunit of complex I of the electron transport chain, was identified as a substrate for granzyme A (Martinvalet *et al.*, 2008). The electron transport chain is required for production of adenosine triphosphate (ATP) by respiration, and blocking this process would appear to be a

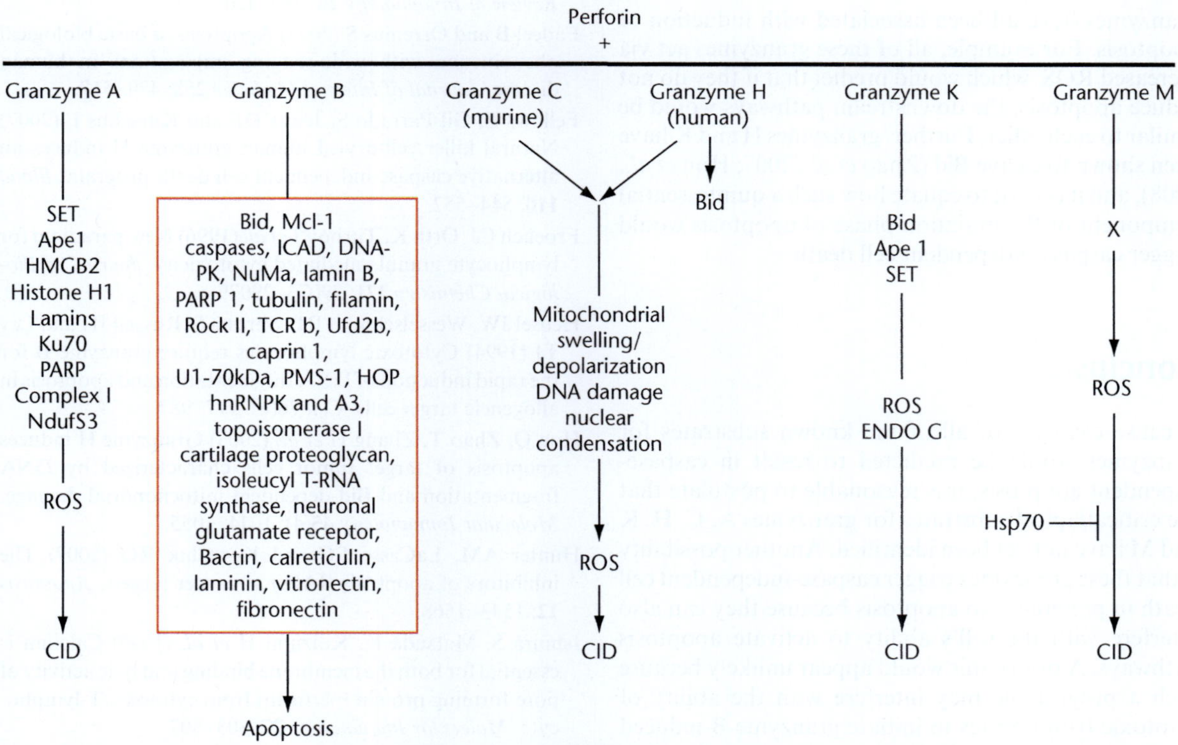

Figure 3 Granzymes cleave many substrates to initiate diverse pathways to cell death. Of the granzyme family, granzyme B is well characterized and initiates apoptosis of target cells (**Figure 1**). The other granzymes have also been shown to induce target cell death when applied directly to target cells *in vitro*; however, the morphology of cell death is not classically apoptotic and does not involve caspase activation. Although these other granzymes cleave a diverse range of substrates, granzymes A, C, K and M also lead to reactive oxygen species generation as a key feature. Many of the noncaspase substrates include critical cytoskeletal proteins and DNA repair proteins, cleavage of which is implicated in cell death.

bona fide mechanism to induce death. However, rotenone, a poison which also blocks mitochondrial complex I, is well known to induce caspase-dependent apoptosis (Lee *et al.*, 2008). Cleavage of NDUFS3 was also shown to promote elevated generation of reactive oxygen species (ROS), and ROS scavengers blocked granzyme A-induced death (Martinvalet *et al.*, 2008). Again, ROS is also a well known pro-apoptotic stimulus. It is still not clear how granzyme A is capable of accessing NDUFS3, which is located on the mitochondrial inner membrane, because the mitochondrial outer membrane is impermeable to all except the smallest molecules, and protein import into the mitochondria is a tightly regulated process which involves mitochondrial targeting, unfolding, refolding and removal of leader sequences in the protein.

Less studied, but equally important, granzymes, C, H, K and M have also been shown to trigger caspase-independent cell death when applied to cells with membrane-disruptive agents, but the mechanism of death is less well studied. Similar to granzyme A, the substrates and pathways identified for each of these granzymes have all been associated with induction of apoptosis. For example, all of these granzymes act via increased ROS, which would predict that if they do not induce apoptosis, the downstream pathways would be similar to each other. Further, granzymes H and K have been shown to cleave Bid (Zhao *et al.*, 2007; Hou *et al.*, 2008), and it is hard to equate how such a quintessential component of the initiation phase of apoptosis would trigger caspase-independent cell death.

Conclusion

Because cleavage of all of the known substrates for granzymes would be predicted to result in caspase-dependent apoptosis, it is reasonable to postulate that the critical/apical substrates for granzymes A, C, H, K and M have not yet been identified. Another possibility is that these granzymes trigger caspase-independent cell death in preference to apoptosis because they can also interfere with the cell's ability to activate apoptosis pathways. A priori, this would appear unlikely because such a programme may interfere with the ability of cytotoxic lymphocytes to initiate granzyme B-induced apoptosis. However, it is also tempting to speculate that death induced by these granzymes is the consequence of cleavage of so many substrates that cumulatively apoptotic morphology is effectively bypassed before it can be fully engaged through caspase activation.

References

Baran K, Dunstone M, Chia J *et al.* (2009) The molecular basis for perforin oligomerization and transmembrane pore assembly. *Immunity* **30**: 684–695.

Barry M, Heibein JA, Pinkoski MJ *et al.* (2000) Granzyme B short-circuits the need for caspase 8 activity during granule-mediated cytotoxic T-lymphocyte killing by directly cleaving Bid. *Molecular and Cellular Biology* **20**: 3781–3794.

Beresford PJ, Xia Z, Greenberg AH and Lieberman J (1999) Granzyme A loading induces rapid cytolysis and a novel form of DNA damage independently of caspase activation. *Immunity* **10**: 585–594.

Bird CH, Sun J, Ung K *et al.* (2005) Cationic sites on granyzme B contribute to cytotoxicity by promoting its uptake into target cells. *Molecular and Cellular Biology* **25**: 7854–7867.

Chavez-Galan L, Arenas-Del Angel MC, Zenteno E, Chavez R and Lascurain R (2009) Cell death mechanisms induced by cytotoxic lymphocytes. *Cellular & Molecular Immunology* **6**(1): 15–25.

Chowdhury D and Lieberman J (2008) Death by a thousand cuts: granzyme pathways of programmed cell death. *Annual Review of Immunology* **26**: 389–420.

Fadeel B and Orrenius S (2005) Apoptosis: a basic biological phenomenon with wide-ranging implications in human disease. *Journal of Internal Medicine* **258**: 479–517.

Fellows E, Gil-Parrado S, Jenne DE and Kurschus F (2007) Natural killer cell-drvied human granzyme H induces an alternative caspase-independent cell death program. *Blood* **110**: 544–552.

Froelich CJ, Orth K, Turbov J *et al.* (1996) New paradigm for lymphocyte granule-mediated cytotoxicity. *Journal of Biological Chemistry* **271**: 29073–29079.

Heusel JW, Wesselschmidt RL, Shresta S, Russell JH and Ley TJ (1994) Cytotoxic lymphocytes require granzyme B for the rapid induction of DNA fragmentation and apoptosis in allogeneic target cells. *Cell* **76**(6): 977–987.

Hou Q, Zhao T, Zhang H *et al.* (2008) Granzyme H induces apoptosis of target tumor cells characterized by DNA fragmentation and Bid-dependent mitochondrial damage. *Molecular Immunology* **45**(4): 1044–1055.

Hunter AM, LaCasse EC and Korneluk RG (2007) The inhibitors of apoptosis (IAPs) as cancer targets. *Apoptosis* **12**: 1543–1568.

Ishiura S, Matsuda K, Koizumi H *et al.* (1990) Calcium is essential for both the membrane binding and lytic activity of pore-forming protein (perforin) from cytotoxic T-lymphocyte. *Molecular Immunology* **27**: 803–807.

Johnson H, Scorrano L, Korsmeyer SJ and Ley TJ (2003) Cell death induced by granzyme C. *Blood* **101**(8): 3093–3101.

Kagi D, Ledermann B, Burki K *et al.* (1994) Cytotoxicity mediated by T cells and natural killer cells is greatly impaired in perforin-deficient mice. *Nature* **369**: 31–37.

Kaiserman D, Bird CH, Sun J *et al.* (2006) The major human and mouse granzymes are structurally and functionally divergent. *Journal of Cell Biology* **175**: 619–630.

Keefe D, Shi L, Feske S *et al.* (2005) Perforin triggers a plasma membrane-repair response that facilitates CTL induction of apoptosis. *Immunity* **23**: 249–262.

Kelly JM, Waterhouse NJ, Cretney E *et al.* (2004) Granzyme M mediates a novel form of perforin-dependent cell death. *Journal of Biological Chemistry* **279**(21): 22236–22242.

Lee J, Huang M-S, Yang I-C *et al.* (2008) Essential roles of caspases and their upstream regulators in rotenone-induced apoptosis. *Biochemical and Biophysical Research Communications* **371**: 33–38.

Martinvalet D, Dykxhoom DM, Ferrini R and Lieberman J (2008) Granzyme A cleaves a mitochondrial complex I protein to initiate caspase-independent cell death. *Cell* **133**: 681–692.

Matzinger P (2002) The danger model: a renewed sense of self. *Science* **296**: 301–305.

Pardo J, Wallich R, Martin P *et al.* (2008) Granzyme B-induced cell death exerted by ex vivo CTL: discriminating requirements for cell death and some of its signs. *Cell Death Differentiation* **15**: 567–579.

Russell JH, Masakowski VR and Dobos CB (1980) Mechanisms of immune lysis. I. Physiological distinction between target cell death mediated by cytotoxic T lymphocytes and antibody plus complement. *Journal of Immunology* **124**: 1100–1105.

Sedelies KA, Ciccone A, Clarke CJP *et al.* (2008) Blocking granule-mediated death by primary human NK cells requires both protection of mitochondria and inhibition of caspase activity. *Cell Death and Differentiation* **15**: 708–717.

Shi L, Kraut RP, Aebersold R and Greenberg AH (1992) A natural killer cell granule protein that induces DNA fragmentation and apoptosis. *Journal of Experimental Medicine* **175**: 553–566.

Smyth MJ, Street SEA and Trapani JA (2003) Granzymes A and B are not essential for perforin-mediated tumor rejection. *Journal of Immunology* **171**: 515–518.

Stinchcombe JC, Majorovits E, Bossi G, Fuller S and Griffiths GM (2006) Centrosome polarization delivers secretory granules to the immunological synapse. *Nature* **443**: 462–465.

Sutton VR, Wowk ME, Cancilla M and Trapani JA (2003) Caspase activation by granzyme B is indirect, and caspase autoprocessing requires the release of proapoptotic mitochondrial factors. *Immunity* **18**: 319–329.

Trapani JA and Smyth MJ (2002) Functional significance of the perforin/granzyme cell death pathway. *Nature Reviews. Immunology* **2**: 735–747.

Trapani JA, Jans P, Smyth MJ *et al.* (1998) Perforin-dependent nuclear entry of granzyme B precedes apoptosis, and is not a consequence of nuclear membrane dysfunction. *Cell Death and Differentiation* **5**: 488–496.

Waterhouse NJ, Sutton VR, Sedelies KA *et al.* (2006) Cytotoxic T lymphocyte-induced killing in the absence of granzymes A and B is unique and distinct from both apoptosis and perforin-dependent lysis. *Journal of Cell Biology* **173**(1): 133–144.

Zhao T, Zhang H, Guo Y and Fan Z (2007) Granzyme K directly processes Bid to release cytochrome *c* and endonuclease G leading to mitochondria-dependent cell death. *Journal of Biological Chemistry* **282**(16): 12104–12111.

Further Reading

Bolitho P, Voskoboinik I, Trapani JA and Smyth MJ (2007) Apoptosis induced by the lymphocyte effector molecule perforin. *Current Opinion in Immunology* **19**: 339–347.

Browne KA, Blink E, Sutton VR *et al.* (1999) Cytosolic deliver of granzyme B by bacterial toxins: evidence that endosomal disruption, in addition to transmembrane pore formation, is an important function of perforin. *Molecular and Cellular Biology* **19**: 8604–8615.

Cullen SP, Adrain C, Luthi AU, Duriez PJ and Martin SJ (2007) Human and murine granzyme B exhibit divergent substrate preferences. *Journal of Cell Biology* **176**(4): 435–444.

Shi L, Mai S, Israels S *et al.* (1997) Granzyme B (GraB) autonomously corsses the cell membrane and perforin initiates apoptosis and GraB nuclear localization. *Journal of Experimental Medicine* **185**: 855–866.

Smyth MJ, Cretney E, Kelly JM *et al.* (2005) Activation of NK cell cytotoxicity. *Molecular Immunology* **42**: 501–510.

Sutton VR, Davis JE, Cancilla M *et al.* (2000) Inititation of apoptosis by granzyme B requires direct cleavage of Bid, but not granzyme B-mediated caspase activation. *Journal of Experimental Medicine* **192**: 1403–1413.

Trapani JA, Sutton VR and Smyth MJ (1999) CTL granules: evolution of vesicles essential for combating virus infections. *Immunology Today* **20**: 351–356.

Uellner R, Zvelebil MJ, Hopkins J *et al.* (1997) Perforin is activated by a proteolytic cleavage during biosynthesis which reveals a phospholipid-binding C2 domain. *EMBO Journal* **16**: 7287–7296.

P53 and Cell Death

Advanced article

Kamil Wolyniec, *The Peter MacCallum Cancer Centre, Melbourne, Victoria, Australia*

Sue Haupt, *The Peter MacCallum Cancer Centre, Melbourne, Victoria, Australia*

Ygal Haupt, *Lautenberg Center for General and Tumor Immunology, The Hebrew University Hadassah Medical School, Jerusalem, Israel*

Article Contents

- Introduction
- Multiple Roles of p53 in Apoptosis
- P53 Family Members and Apoptosis
- Reactivation of p53 and Tumour Suppression
- Cancer Therapy and p53
- Conclusion and Future Perspective

The p53 transcription factor emerges not only as the most important tumour suppressor, but also as an intriguing scientific puzzle characterized by immense functional complexity. P53-mediated apoptosis is a well-established tumour suppression mechanism forming a critical barrier to tumourigenesis. Intense research into the mechanism of p53-induced apoptosis revealed an involvement of p53 in the extrinsic and intrinsic cell death pathways, reactive oxygen species signalling and antisurvival responses. Notably, p53 exerts its effects by various direct and indirect mechanisms engaging transcriptional activation or repression of target genes as well as transcriptional independent modes of action. Importantly, more than 20 years of intense research has paved the way for a rational design of selective anticancer therapies aimed at restoration of p53 functionality in cancer cells.

Introduction

Cancer may be defined as a disease of uncontrolled cell proliferation. Complex organisms with renewable tissues have evolved mechanisms to constrain the development of cancer. In this regard, p53 plays a fundamental role as a tumour suppressor which can either ensure irreversible cell cycle arrest (senescence) or trigger programmed cell death (apoptosis); both of which are processes that suppress oncogenic transformation (Vousden and Prives, 2009). P53 is a transcription factor that is regarded as an essential guardian of the genome. P53 responds to multiple stress signals and integrates these into appropriate growth inhibitory outcomes, thereby maintaining homeostasis and preventing the proliferation of aberrant cells (**Figure 1**).

Since the initial observation that p53 induces programmed cell death (apoptosis) (Yonish-Rouach *et al.*, 1991), over 20 years of diligent scientific research has revealed detailed information concerning the regulation and mechanism of p53-mediated apoptosis. Although most studies of p53 have been linked to cancer, it is worth mentioning that hyperactivation of p53 has been linked to neurodegenerative diseases and has also been recently proposed to play roles in glycolysis, autophagy, invasion, motility and angiogenesis, differentiation and bone remodelling (Vousden and Lane, 2007; Vousden and Prives, 2009). Therefore, p53 may emerge as a pleiotropic transcriptional factor affecting multiple cellular functions and cell types in the whole organism.

A plethora of stress signals, such as deoxyribonucleic acid (DNA) damage, oncogenic activation, hypoxia, nucleotide depletion and nutrient deprivation, activate p53. In turn, activated p53 promotes cell cycle arrest, cellular senescence or apoptosis (**Figure 1**). The outcome of p53 activation depends on various factors, such as the nature and intensity of the cellular stress, cell type and genetic background, as well as microenvironment. If these vital p53 functions become disrupted, consequent aberrations may occur in checkpoint responses and genomic instability, leading to enhanced survival; which in turn allows for the proliferation of damaged cells, cellular immortalization and eventually the onset of neoplasia (**Figure 1**).

In light of the fundamental restraint yielded by p53, it is not surprising then, that it is commonly inactivated in human cancer. Approximately 50% of sporadic human cancers bear *P53* gene mutations. In the majority of the remaining cancer cases, the integrity of the p53 signalling pathways is compromised indirectly by various mechanisms, such as the upregulation of key negative regulators of p53, Mdm2 or Mdm4 (Mdmx). Likewise, downregulation of the p53 activator, ARF, leads to

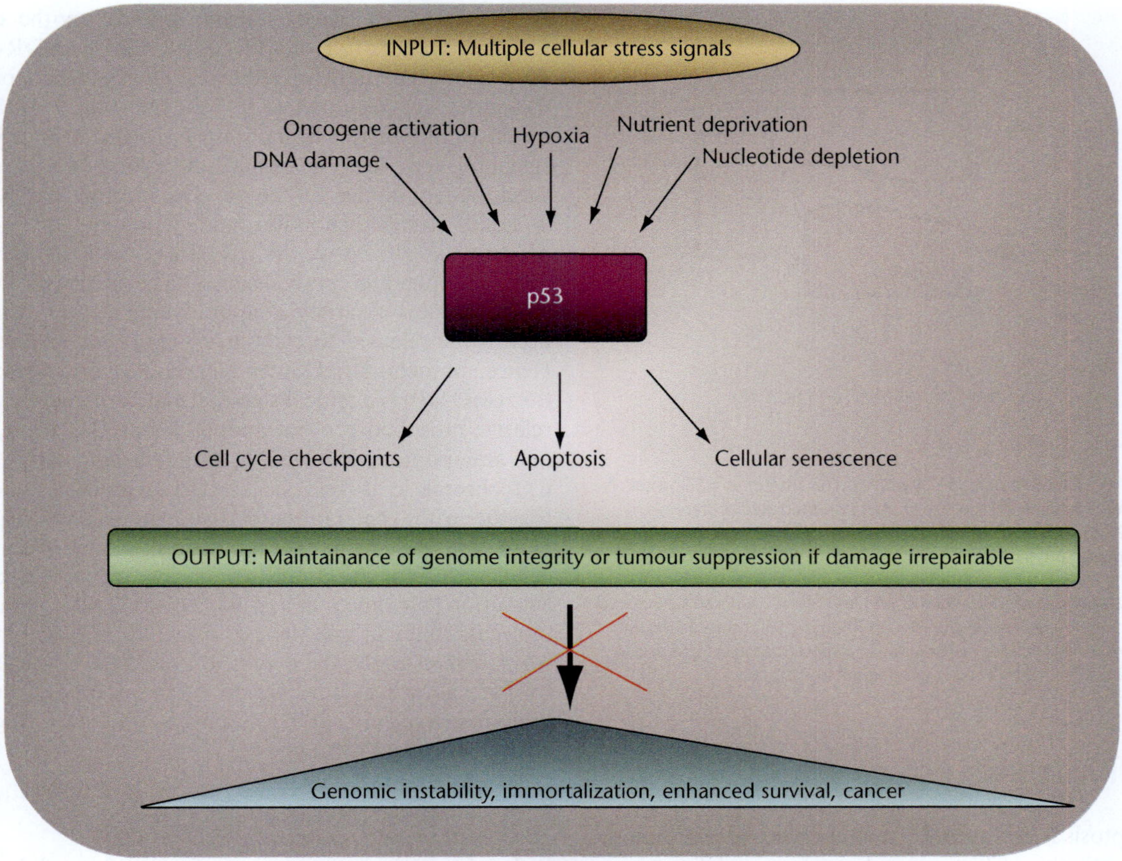

Figure 1 Multiple signals activate p53 to induce various growth inhibitory responses. P53 tumour suppressor protein responds to diverse intercellular (e.g. oncogene activation) and extracellular stimuli (e.g. nutrients deprivation) through activation of relevant downstream mechanisms (e.g. apoptosis) that act to protect cells from oncogenic transformation.

mouse tumourigenesis, reminiscent of p53 deficiency (Royds and Iacopetta, 2006; Sharpless, 2005). In Li-Fraumeni syndrome patients, germline mutations in p53 results in early onset of multiple cancer types (Royds and Iacopetta, 2006). A knockin of these mutations in mouse models further confirms the oncogenic ability of mutant p53 (Lang *et al.*, 2004; Olive *et al.*, 2004).

The activation of p53 entails posttranslational stabilization, increased DNA binding affinity and enhanced transactivation potential, and is often associated with extensive posttranslational modifications (Lavin and Gueven, 2006). In contrast, under nonstress conditions, multiple mechanisms operate to negatively regulate p53 activity, stability and nuclear localization to prevent potential detrimental effects of chronic p53 signalling in healthy tissues. Hence, under basal conditions p53 has a short half-life and low transcriptional activity. This essential role is largely attributed to Mdm2 (an E3 ubiquitin ligase, also known as Hdm2 in humans), which targets p53 for proteasomal degradation, and Mdm4, which inhibits p53 transcriptional acitivity (**Figure 2**; Marine *et al.*, 2007). Interestingly, Mdm2 is a transcriptional target of p53, and therefore Mdm2-p53 regulation operates within a negative feedback loop (Toledo and Wahl, 2006). Given the profound tumour suppressor properties of p53 and its relevance to human cancer, efforts endeavouring to restore p53 activity in established tumours represents a promising approach for the treatment of cancer (Haupt and Haupt, 2006). Further, understanding the regulation of tumour suppression by p53 may impact on cancer diagnosis and prognosis.

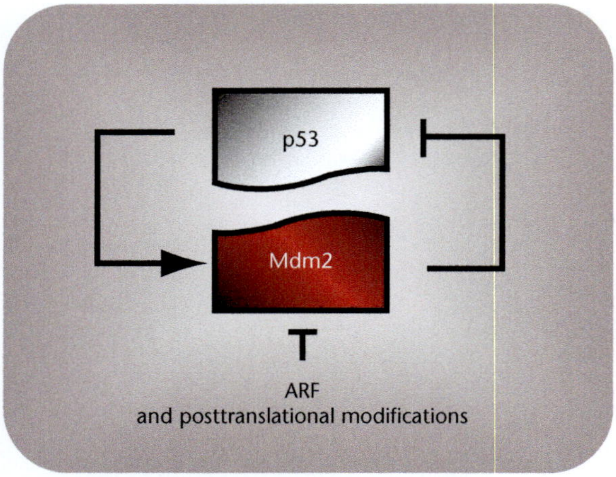

Figure 2 Regulation of p53 by Mdm2. The amount of p53 protein is determined mainly by the rate of ubiquitin-mediated proteolysis. Mdm2 is the E3 ligase of p53, inhibiting its activity and promoting it for degradation. In response to stress, the negative loop between p53 and Mdm2 is interrupted by proteins such as ARF, and by posttranslational modifications of p53 and Mdm2. ARF can be induced by oncogenes such as Myc or E1A resulting in p53 accumulation and apoptosis. Also, various posttranslational modifications of p53 and Mdm2 prevent Mdm2-mediated degradation of p53.

Multiple Roles of p53 in Apoptosis

Apoptosis is undoubtedly a vital aspect of p53 biology. Interestingly, p53 sequence homology and function is evolutionarily conserved in *Drosophila melanogaster* and *Caenorhabditis elegans* (Jin *et al.*, 2000; Schumacher *et al.*, 2001) emphasizing a physiological relevance of p53 in stress responses. P53-dependent apoptosis is not only critical for tumour suppression per se, but also determines chemosensitivity and radiosensitivity of at least certain types of cancer cells (El-Deiry, 2003). P53 is a nuclear transcription factor, which binds to its gene targets in a sequence-specific manner. The majority of p53-pro-apoptotic effects are dependent on transcriptional induction of apoptotic target genes (Vousden and Prives, 2009). Interestingly, during tumourigenesis there is a strong selective pressure to abrogate transcriptional activity of p53 as evidenced by the majority of tumour-derived p53 mutants which have lost the ability to bind specific DNA sequences and are unable to induce apoptosis (Bullock and Fersht, 2001). It is well established now that p53 is implicated in two main networks of apoptosis (extrinsic and intrinsic) and induces these pathways at multiple steps (**Figure 3**). The extrinsic pathway involves binding of various ligands (e.g. TNFα (tumour necrosis factor α), FasL and TRAIL (TNF-related apoptosis-inducing ligand)) to the cell surface receptors (e.g. TNFR, Fas/CD95 and DR-5). After ligation of the death receptor, cytoplasmic adaptor molecules are recruited (e.g. TRADD and FADD), which results in the formation of the death-inducing signalling complex (DISC) and activation of Caspase-8 to ultimately induce cell death. The intrinsic pathway is linked to mitochondria and represents a death programme responsive to cell stress, such as DNA damage, injury and survival factors. The intermembrane space contains a variety of apoptogenic factors, which need to be released to elicit a pro-apoptotic response. Hence, mitochondrial outer membrane permeabilization (MOMP) is a critical step that is determined by the relative ratio between pro-apoptotic Bax/Bak proteins and antiapoptotic Bcl-2/Bcl-x_L proteins. After MOMP, Cytochrome c is released to the cytoplasm where together with Apaf-1 (apoptosis-activating factor 1) and Caspase-9 they form the apoptosom, which executes cellular disintegration through a cascade of caspase activation (Meulmeester and Jochemsen, 2008). Bid is a molecule that can link the extrinsic and intrinsic pathways. Interestingly, the expression of Bid is controlled by p53, allowing coordination of the apoptotic programmes (Sax *et al.*, 2002). **See also**: Apoptosis: Regulatory Genes and Disease

The extrinsic pathway and p53

P53 induces the transcription of a number of death receptors (Meulmeester and Jochemsen, 2008). This includes the induction of the transmembrane death receptor Fas/CD95 in response to irradiation in a tissue-specific manner (Haupt *et al.*, 2003; Meulmeester and Jochemsen, 2008). P53 can also ameliorate Fas function by facilitating its trafficking from Golgi to QJ;the membrane (Haupt *et al.*, 2003). Another direct transcriptional target of p53 associated with the extrinsic apoptotic pathway is DR-5, the death-domain-containing receptor that is activated by TRAIL. The increased p53-mediated expression of DR-5 occurs on activation of p53 by genotoxic stress, e.g. by treatment with doxorubicin or etoposide, or by ectopic expression of p53 in cancer cells and whole body γ-irradiation of mice results in elevation of DR-5 in spleen, thymus and small intestine (Haupt *et al.*, 2003). Interestingly, not only death receptors but the TNFSF10 (TRAIL) death ligand itself as well as Fas ligand TNFSF6 (FasL), were also identified as p53-transcriptional targets that mediate p53-dependent cell death (Maecker *et al.*, 2000; Kuribayashi *et al.*, 2008). The combined induction of receptors and their ligand

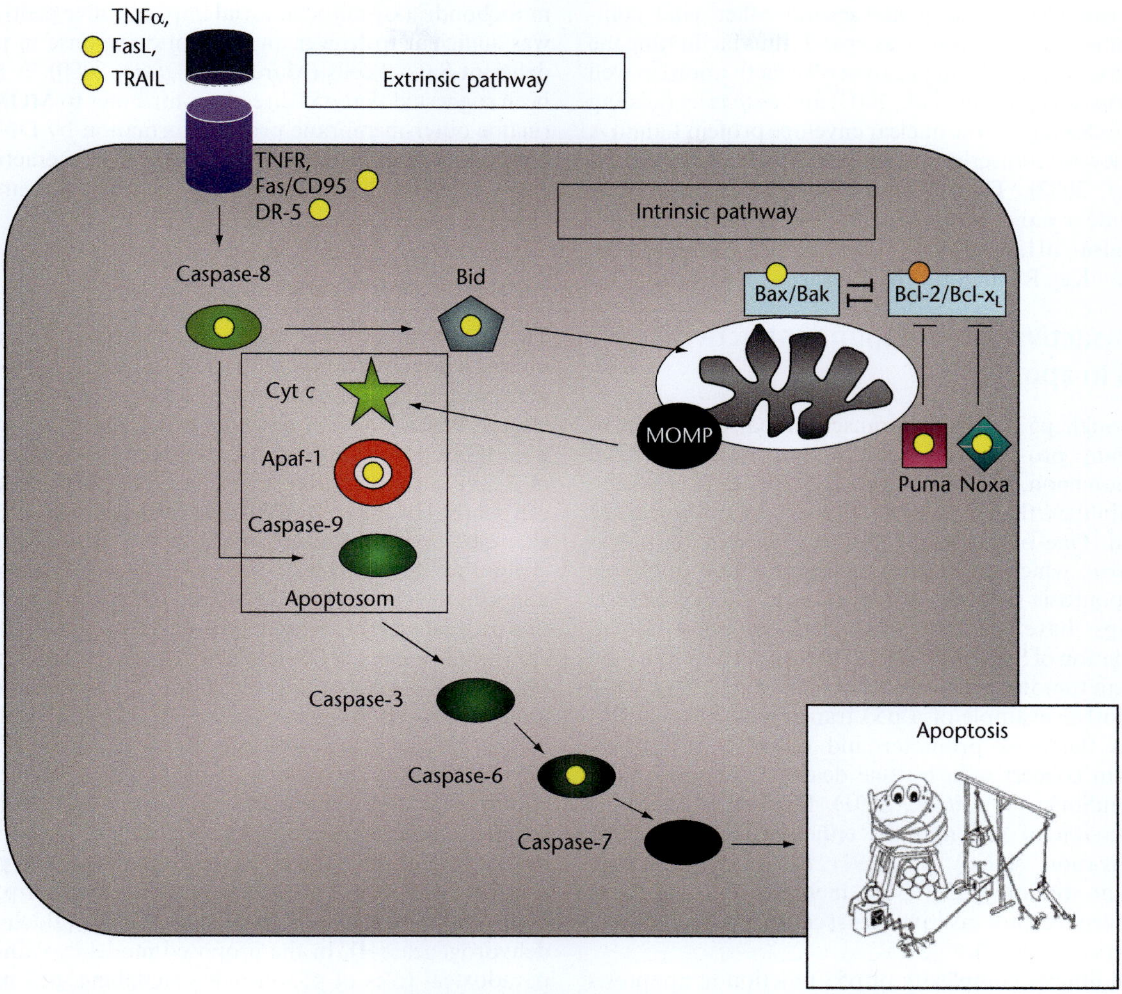

Figure 3 P53 activates the extrinsic and intrinsic cell death pathways. P53 activates the extrinsic pathway by directly inducing the expression of death receptors, such as Fas/CD95 or DR-5 as well as caspases 8 and 6. P53 also activates the intrinsic pathway where p53 act at multiple levels mainly through transactivation of multiple genes associated with the mitochondria as well as downstream factors such as: Apaf-1. Small yellow circles mark p53 target genes, whereas an orange small circle indicates p53-repressed gene.

amplifies the apoptotic signalling via these death receptors. **See also**: Death Receptors

The intrinsic pathway and p53

Stabilized p53 accumulates in the nucleus to regulate the expression of a number of the pro-apototic members of the Bcl-2 family that are implicated in the intrinsic mitochondrial pathway. These include *bax* and 'BH3-only' members *puma* and *noxa* (Haupt *et al.*, 2003). Importantly, ablation of *puma*, and to a lesser extent *noxa*, abroagates p53-induced apoptosis in response to genotoxic stress, at least in certain cell types (Villunger *et al.*, 2003). Therefore, p53 is able to shift the balance towards pro-apoptotic Bcl-2 proteins and cause the release of apoptogenic proteins such as Cytochrome c. The requirement of Bax in p53-mediated apoptosis appears to be cell-type and context dependent. Bax is required for the apoptotic response of the developing nervous system to irradiation and *bax*-deficient, oncogene-overexpressing MEFs (mouse embryonic fibroblasts) are refractory to chemotherapy-induced apoptosis (Haupt *et al.*, 2003). However, Bax was found to be redundant in colonic epithelial cells undergoing irradiation-induced cell death (Pritchard *et al.*, 1999). A growing body of evidence indicates that p53 can also transactivate several other vital components of apoptosis such as *apaf-1* (thus facilitating the caspase cascade through caspase 9 coactivation) as well as *caspase-8* (Haupt *et al.*, 2003) and *caspase-6* (leading to the cleavage of the nuclear envelope protein Lamin A and several transcription factors) (MacLachlan and El-Deiry, 2002). The induction of these caspases may provide a rapid apoptotic response to p53 activation. **See also**: BH3-Only Proteins; The Bcl-2 Family Proteins – Key Regulators and Effectors of Apoptosis

Transactivation-independent activities of p53 in apoptosis

Although p53-mediated transcriptional activation of relevant pro-apoptoic targets is a well-established phenomenon, p53 also seems to have transrepression capabilities that may play roles in programmed cell death. One of the genes that is repressed by p53 is *survivin*, which encodes antiapoptotic IAP (inhibitor of apoptosis protein) (Hoffman *et al.*, 2002). Several groups have demonstrated that antisense down-regulation of Survivin is sufficient to induce apoptosis in human tumour cell lines (Ambrosini *et al.*, 1997). Bcl-2 is another example of a p53 transrepressed gene. P53 binds the *bcl-2* promoter and represses its expression in concert with histone deacetylase-1 (HDAC1) and mSin3a (Wu *et al.*, 2001). Interestingly, ectopic expression of p53 mutants either lacking a nuclear localization signal (p53Δ291) or a DNA-binding domain mutant (p53(173L)), increased Bcl-2 levels in estrogen receptor positive breast cancer cell line MCF-7 (Pratt *et al.*, 2007).

Additional complexity of p53 function in apoptosis stems from the fact that there are other modes of p53 action operating independently of up- and down-regulation of transcription (Green and Kroemer, 2009). The initial suggestion for a nontranscriptional apoptotic activity of p53 was based on the observation that p53 induced apoptosis in cells treated with transcription- or translation-blocking drugs (Green and Kroemer, 2009). Further studies have demonstrated it more directly by using p53 transcription-defective mutants that retained the ability to induce apoptosis in certain contexts (Haupt *et al.*, 1995). Recently, the transcription-independent apoptotic activity of p53 has been linked to the intrinsic mitochondrial pathway (Green and Kroemer, 2009). It has been demonstrated that stress-induced redistribution of p53 to mitochondria precedes Cytochrome c release and caspase activation. In addition, this occurs only during p53-dependent cell death. Relocating p53 to the mitochondria by mitochondrial import leader peptides was sufficient to trigger an apoptotic response in p53 deficient Saos-2 cells (Marchenko *et al.*, 2000). It has been suggested that p53 directly contributes to MOMP (major outer-membrane protein) formation by DNA-binding domain mediated protein–protein interaction with protective Bcl-x_L and Bcl-2 (Green and Kroemer, 2009).

Role of p53 in redox signalling

The activation of p53 has been also linked to an increase in ROS (reactive oxygen species) levels before programmed cell death. Although the exact mechanism is largely unknown, p53 is capable of upregulating a number of genes implicated in redox metabolism, e.g. *pig3*, and certain antioxidants suppress p53-dependent cell death (Hwang *et al.*, 2001; Polyak *et al.*, 1997). For example, treatment of colon carcinoma cells with the chemotherapeutic agent 5-fluorouracil (5-FU) leads to a specific upregulation of ferredoxin reductase (FDXR) in a p53-dependent manner, whereas disruption of this enzyme decreases ROS levels and reduces apoptosis (Hwang *et al.*, 2001). Furthermore, it has been suggested that oxidoreductases are involved in p53 stabilization (Asher *et al.*, 2001; Chang *et al.*, 2003). One possibility is that activation of p53 leads to ROS elevation, which in turn interferes with mitochondrial function and integrity, and consequently to apoptotic death. Importantly, under certain circumstances p53 can also attenuate ROS levels by engagement of p53-regulated genes such as *TIGAR* and *ALDH4* (aldehyde dehydrogenease-4). In the proposed model explaining paradoxical roles of p53 in redox signalling, p53 may have a dual role depending on the extent of the stress signals. In response to low DNA damage signals p53 inhibits ROS production and promotes survival of cells, allowing DNA repair. However, in response to severe damage, p53 promotes apoptosis by inducing ROS levels (Vousden and Lane, 2007).

Inhibition of pro-survival signals by p53

As described so far p53 induces a range of genes with obvious apoptosis promoting activities. Intriguingly, p53 can also induce genes that in turn abrogate anti-apoptotic pro-survival signals, and therefore indirectly stimulate cell death. The best example is the ability of p53 to transactivate *PTEN* (tumour suppressor gene encoding a phosphatase), a negative regulator of the PI3K (phosphoinositide kinase) pathway. PI3K is implicated in the transduction of signals from cell surface receptors, e.g. EGFR (epidermal growth factor receptor) or oncogenes, such as Ras and Her2/Neu, and activation of signalling pathways that promote proliferation and viability of cells. AKT (serine/threonine kinase) plays a fundamental downstream role in this process as it phosphorylates a number of important targets (e.g. Akt phosphorylates and activates Mdm2-mediated p53 inhibition) to promote cell survival. Disruption of PTEN activity can compromise p53-mediated apoptosis in some cell types. In addition to its role in PTEN regulation, p53 also initiates caspase-mediated degradation of AKT itself (Mayo and Donner, 2002).

P53 Family Members and Apoptosis

P53 is part of a larger gene family including p63 and p73. Both p63 and p73 share homology with p53 in crucial functional domains; however, their genetic locus organization and developmental roles seem to be much more complex (**Figure 4**). The *p63* gene has been shown to be required for stem cell maintenance in epithelium and p63 knockout mice live only for a few hours after birth. The p73 protein exerts its functions mainly in the central nervous system. Interestingly, while the precise role of p63 and p73 is currently being studied, both have been linked to apoptosis and tumour suppression. In contrast to p53 null animals, mice deficient for p73 are not susceptible to spontaneous tumours (note that due to early lethality of p63 null mice, tumourigenesis studies have been impossible) and mutations in p63 or p73 have not been identified so far (McKeon and Melino, 2007). However, recent studies have revealed that p73 and p53 cooperate in tumour suppression as evidenced by analyses of mice hemizygous for p53, p63 and p73 in various combinations (McKeon and Melino, 2007). Importantly, mice lacking pro-apoptotic TAp73 isoforms develop spontaneous tumours, particularly lung adenocarcinomas and exhibit an enhanced rate of carcinogen-induced tumours (Tomasini *et al.*, 2008). In addition, transactivation domain-deficient isoforms of p63 and p73 have been shown to be overexpressed in some human tumours and were suggested to act in a dominant-negative fashion with respect to wild-type p53 (Benard *et al.*, 2003). Conversely, mutant p53 was reported to bind and inactivate p73, hence interfering

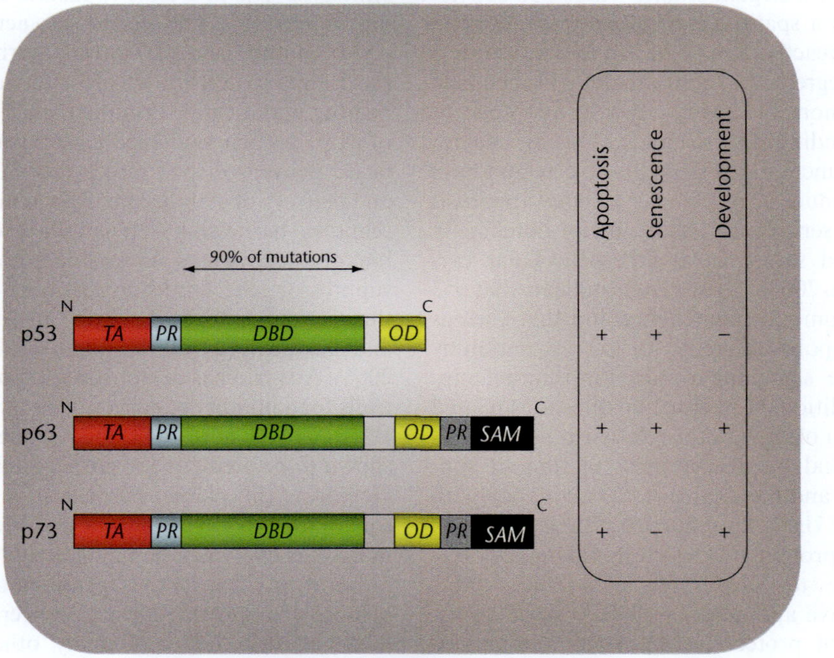

Figure 4 Structure of p53 family members and their roles in growth suppression and development. The p53 family consists of p53, p63 and p73, which have similar structure and share homology in major domains such as transactivation domain (TA), DNA-binding domain (DBD), proline-rich domain (PR) and oligomerization domain (OD). P63 and p73 also contain additional proline-rich domain (PR) and sterile alpha motif (SAM). The involvement of each member in growth suppression and development is indicated.

with chemotherapy-induced apoptosis (Haupt and Haupt, 2006). Intriguingly, a polymorphism encoding either arginine (72R) or proline (72P) at codon 72 of mutant p53 determines the suppression of p73 and the clinical response to cisplatin-based chemotherapy for head and neck cancer patients (Bergamaschi *et al.*, 2003). It is very likely that p63 and p73 are a part of cell death machinery operating synergistically with p53. Notably, p63 and p73 are required for efficient induction of apoptotic genes such as: *perp*, *bax* and *noxa* in response to DNA damage in MEFs (Flores *et al.*, 2002). This may explain why many p53-activating insults drive apoptosis in p53-deficient cells, although to a lesser extent.

Reactivation of p53 and Tumour Suppression

Although the sequential inactivation of tumour suppressors defines the onset of many tumours, a vital question related to cancer therapy has been established whether tumours display a dependency on sustained 'tumour suppressor loss'. This issue has been recently addressed in several elegant studies where p53 activity was regulated in a spatio-temporal manner. As demonstrated, brief reactivation of p53 in mouse tumours lead to cancer regression. Interestingly, p53-mediated tumour suppression was linked either to apoptosis or senescence depending on the context. In this regard, restoration of functional p53 in Eμ-myc lymphomas and radiation-induced lymphomas lead to apoptosis whereas cellular senescence was the major outcome in radiation-induced sarcomas and hepatocellular carcinomas (Kastan, 2007). These results indicate that p53 is required for tumour maintenance and that various tumour types respond differently to p53 restoration by engaging diverse signalling routes for tumour suppression. An additional fundamental questions related to the restoration of p53 are the relative importance of DNA damage and oncogenic stress in tumour suppression by p53, and the timing of p53 activation with respect to stress. Using an inducible p53ER (estrogen receptor) fusion protein, p53 was restored immediately before exposure to whole-body irradiation. This resulted in massive apoptosis in all radiosensitive tissues, but did not protect the mice from radiation-induced cancer. However, delayed restoration of p53 after exposure to DNA damage suppressed tumourignesis and increased survival (Kastan, 2007). The conclusion of this study was that p53 exerts its anticancer effects not at the time of radiation-induced DNA damage, despite induction of massive cell death, but rather at a later phase when oncogenic-stress pathways have been elicited and the development of cancer is in progress.

Cancer Therapy and p53

The concept of extrapolating from natural host initiatives to fight cancer has prompted the pursuit of p53 as a diagnostic and prognostic marker, and as a druggable therapeutic target. Extensive analysis of breast cancers has demonstrated that mutant p53 sequence analyses have strong prognostic significance. Further, specific p53 mutations and polymorphisms (notable codon72 polymorphism, Arg/Pro) appear to determine chemotherapeutic drug sensitivities (Bertheau *et al.*, 2008). The understanding of the importance of p53 in tumour suppression has been driving scientists to utilize p53 as a druggable therapeutic target. Priming cancer cells with p53, in contexts where it has become disabled, has been conceived as either an alternative or a supplement to conventional genotoxic anticancer treatments (Haupt and Haupt, 2006). Essentially, there are three major strategies. The first aims to instate wild-type (wt) p53 into cancer cells. The second is to activate endogenous p53 in tumour cells harbouring 'dormant' wt p53. The third aims to restore wt p53 function in cancer cells bearing mutant p53. Priming the antitumour activities of wt p53 is being pursued both through gene therapy-based delivery of p53 into tumour cells using viruses and delivery of stabilized p53 proteins. P53-based gene therapy, harnessing replication-incompetent recombinant adenoviruses containing a normal p53 tumour suppressor gene, has led to a number of clinical trials of the main prototypes Advexin (an American initiative and Gendicine (a Chinese initiative); Haupt and Haupt, 2006). Advexin has demonstrated some effect in limited trails for a number of cancer types, and is now in phase III trials. Of note, p53 gene sequence analysis was shown to be predictive of efficacy (Olivier *et al.*, 2009). Gendicine (injectable) launched for the treatment of squamous cell carcinoma of the head and neck was more effective when combined with chemotherapy or radiotherapy and its therapeutic potential is currently being tested for other types of cancers such as ovarian, nonsmall cell lung and many other solid tumours (Haupt and Haupt, 2006). Intriguingly, modifications mimicking phosphorylation of p53 at Thr18 and Ser20, which are essential for inhibiting p53 interaction with its negative p53 regulator Mdm2, significantly enhanced the level of apoptosis induced by this approach (Nakamizo *et al.*, 2008). Over the last couple of years there have also been efforts to utilize p53 in transduction therapy. This method is based on the delivery of synthesized recombinant p53 to a tumour. Certain modifications of p53, such as the substitution of *C*-terminal lysines with arginines, confering resistance to Mdm2-mediated ubiquitination, and proteasomal degradation have been employed to increase the half-life, and therefore the efficacy of p53 in tumours (Haupt and Haupt, 2006). Pharmacological intervention to interfere with the interaction between wt p53 and Mdm2 has become a vibrant field of research, yielding the very promising drugs, the Nutlins. These synthetic *cis*-imidazoline analogues have high affinity for binding to the p53 pocket of Mdm2 resulting in dis-localization of p53 (**Figure 5**). Nutlins are currently in advanced clinical trials. Reactivation of p53 and induction to tumour cell apoptosis (RITA) is a drug also aimed towards inhibiting p53–Mdm2 interaction; however, its mode of action is distinct from Nutlins, apparently primarily binding to p53. It has shown efficacy in mouse models (Issaeva *et al.*, 2004). A potentially critical perspective on restricting the application of these compounds to a wt p53 context, however, is implied from recent studies which indicates that mutant p53 can also be held in check by Mdm2 (Terzian *et al.*, 2008). One could therefore conceive of situations where systemic delivery of Nutlins to target a specific malignancy could actually result in unfortunate triggering of mutant p53 activity at a secondary location. Such scenarios may be relevant for combinatorial Nutlin therapy with genotoxic regimes. Promoting wt p53 consensus DNA binding by mutant p53 has been an aim of extensive high throughput screens of drug libraries. A small molecule, styrylquinazoline, CP-31398, has been proposed to act on mutant p53 through allosteric effects to impose appropriate conformation and stabilize it (Foster *et al.*, 1999). Interestingly this activation appears independent of p53–Mdm2 disruption and appears to involve protection of p53 from ubiquitination (Wang *et al.*, 2003). Two other important prototype small molecular weight molecules, with similar potential and modus operandi, are PRIMA (p53 reactivation and induction of massive apoptosis) and MIRA (propoxy-methyl meleimide). The precise

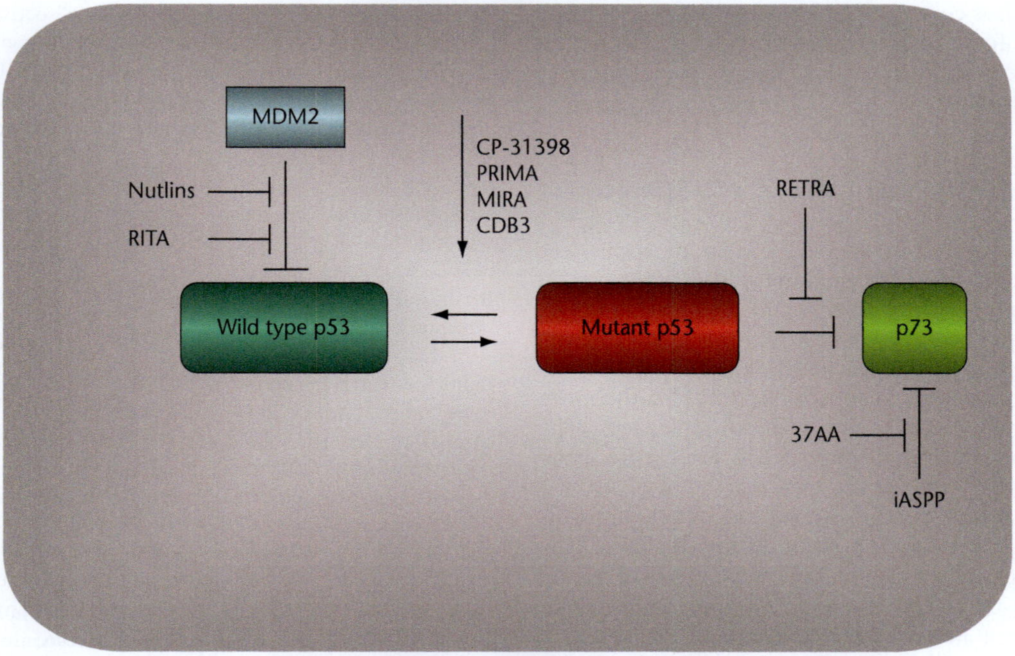

Figure 5 Strategies for activation of p53 in cancer cells. Wild-type p53 can be activated by inhibition of the interaction between p53 and Mdm2 using Nutlins and RITA. Mutant p53 can be reactivated using small molecules such as CP-31398, PRIMA-1 or MIRA. CDB3 is able to shift the equilibrium existing between wild-type p53 and muatant p53 towards functional p53. P73 chemosensitivity can be restored by exposure to RETRA or 37AA, if its activities are inhibited through sequestration by mutant p53 or is APP, respectively.

mechanisms underlying the actions of these drugs are still under investigation (**Figure 5**). An alternative approach is aimed at shifting thermodynamic equilibrium that exists between wt p53 and structural p53 mutants towards the correctly folded functional protein using core domain bindng 3 (CDB3). This molecule represents a peptide fragment naturally interacting with p53, the p53-binding protein 2 (53BP2), and was suggested to act as a chaperone capable of thermostabilization of p53 mutants, allowing for appropriate folding and DNA binding. CBD was shown to confer proper (native) conformation on p53R175H and p53R273H mutants leading to induction of p53-dependent gene expression and apoptosis (Haupt and Haupt, 2006). The p53 family member p73 is also an important determinant of chemosensitivity. In cancer cells, p73 may be inactivated through sequestration. Mutant p53 may impose resistance to chemosensitivity by restraining p73 and 'reactivation of transcriptional reporter activity' (RETRA) is a prototype synthetic small molecule aimed towards relieving p73 from mutant p53 (Kravchenko *et al.*, 2008). P73 may also be inhibited by interaction with iASPP (inhibitor of apoptosis-stimulating protein of p53), and a p53-derived peptide '37AA' has the capacity to disrupt this interaction and induce cell death (Bell *et al.*, 2007; **Figure 5**). It is conceivable that in the future anticancer therapies for a diverse range of malignancies will be tailored on an individual genetic basis for optimal patient response.

Conclusion and Future Perspective

The extensive study of p53-induced apoptosis has lead to the undisputed conclusion that p53 is crucial for the cellular apoptotic response to a variety of stress signals. Intriguingly, p53 activates the extrinsic death receptor signalling and the mitochondrial intrinsic pathways, as well as multiple downstream effectors. Furthermore, p53 promotes these pathways in multiple steps. This is achieved by transcriptional activation of a panel of apoptotic target genes, transrepression of survival genes, and transcriptional-independent functions, including an effect of p53 at the mitochondrial level. This involvement of p53 at multiple steps of the apoptotic pathways may achieve one or more of the following scenarios: a need for a rapid as well as sustained apoptotic signalling; allowing apoptosis under diverse conditions, including distinct cell types, differentiation stage and multiple stress signals; a need to amplify the apoptotic signals. The contribution of each pathway, in particular the direct effect of p53 at the mitochondria, to the overall apoptotic response of p53 is yet to be established. To what extent p53 isoforms, and p53 family members, contribute to the regulation of p53-mediated apoptosis is important to ascertain. Though the induction of apoptosis by p53 is important for its tumour suppression, it is clear that other functions of p53 are also critical. The induction of senescence emerges as an important antitumour function of p53 in various nonhaematopoietic malignancies. What determines the cellular choice between p53 outcomes is only partially understood. Recent evidence revealed that restoration of p53 leads to tumour regression and hence provides a sound basis for anticancer treatment via activation of p53.

References

Ambrosini G, Adida C and Altieri DC (1997) A novel anti-apoptosis gene, survivin, expressed in cancer and lymphoma. *Nature Medicine* **3**: 917–921.

Asher G, Lotem J, Cohen B, Sachs L and Shaul Y (2001) Regulation of p53 stability and p53-dependent apoptosis by NADH quinone oxidoreductase 1. *Proceedings of the National Academy of Sciences of the USA* **98**: 1188–1193.

Bell HS, Dufes C, O'prey J *et al.* (2007) A p53-derived apoptotic peptide derepresses p73 to cause tumor regression *in vivo*. *Journal of Clinical Investigation* **117**: 1008–1018.

Benard J, Douc-Rasy S and Ahomadegbe JC (2003) TP53 family members and human cancers. *Human Mutation* **21**: 182–191.

Bergamaschi D, Gasco M, Hiller L *et al.* (2003) p53 polymorphism influences response in cancer chemotherapy via modulation of p73-dependent apoptosis. *Cancer Cell* **3**: 387–402.

Bertheau P, Espie M, Turpin E *et al.* (2008) TP53 status and response to chemotherapy in breast cancer. *Pathobiology* **75**: 132–139.

Bullock AN and Fersht AR (2001) Rescuing the function of mutant p53. *Nature Reviews of Cancer* **1**: 68–76.

Chang NS, Doherty J and Ensign A (2003) JNK1 physically interacts with WW domain-containing oxidoreductase (WOX1) and inhibits WOX1-mediated apoptosis. *Journal of Biological Chemistry* **278**: 9195–9202.

El-Deiry WS (2003) The role of p53 in chemosensitivity and radiosensitivity. *Oncogene* **22**: 7486–7495.

Flores ER, Tsai KY, Crowley D *et al.* (2002) p63 and p73 are required for p53-dependent apoptosis in response to DNA damage. *Nature* **416**: 560–564.

Foster BA, Coffey HA, Morin MJ and Rastinejad F (1999) Pharmacological rescue of mutant p53 conformation and function. *Science* **286**: 2507–2510.

Green DR and Kroemer G (2009) Cytoplasmic functions of the tumour suppressor p53. *Nature* **458**: 1127–1130.

Haupt S and Haupt Y (2006) Importance of p53 for cancer onset and therapy. *Anticancer Drugs* **17**: 725–732.

Haupt S, Berger M, Goldberg Z and Haupt Y (2003) Apoptosis: the p53 network. *Journal of Cell Science* **116**: 4077–4085.

Haupt Y, Rowan S, Shaulian E, Vousden KH and Oren M (1995) Induction of apoptosis in HeLa cells by trans-activation-deficient p53. *Genes & Development* **9**: 2170–2183.

Hoffman WH, Biade S, Zilfou JT, Chen J and Murphy M (2002) Transcriptional repression of the anti-apoptotic survivin gene by wild type p53. *Journal of Biological Chemistry* **277**: 3247–3257.

Hwang PM, Bunz F, Yu J *et al.* (2001) Ferredoxin reductase affects p53-dependent, 5-fluorouracil-induced apoptosis in colorectal cancer cells. *Nature Medicine* **7**: 1111–1117.

Issaeva N, Bozko P, Enge M *et al.* (2004) Small molecule RITA binds to p53, blocks p53-HDM-2 interaction and activates p53 function in tumors. *Nature Medicine* **10**: 1321–1328.

Jin S, Martinek S, Joo WS *et al.* (2000) Identification and characterization of a p53 homologue in *Drosophila melanogaster*. *Proceedings of the National Academy of Sciences of the USA* **97**: 7301–7306.

Kastan MB (2007) Wild-type p53: tumors can't stand it. *Cell* **128**: 837–840.

Kravchenko JE, Ilyinskaya GV, Komarov PG *et al.* (2008) Small-molecule RETRA suppresses mutant p53-bearing cancer cells through a p73-dependent salvage pathway. *Proceedings of the National Academy of Sciences of the USA* **105**: 6302–6307.

Kuribayashi K, Krigsfeld G, Wang W *et al.* (2008) TNFSF10 (TRAIL), a p53 target gene that mediates p53-dependent cell death. *Cancer Biology & Therapy* **7**: 2034–2038.

Lang GA, Iwakuma T, Suh YA *et al.* (2004) Gain of function of a p53 hot spot mutation in a mouse model of Li-Fraumeni syndrome. *Cell* **119**: 861–872.

Lavin MF and Gueven N (2006) The complexity of p53 stabilization and activation. *Cell Death and Differentiation* **13**: 941–950.

Maclachlan TK and El-Deiry WS (2002) Apoptotic threshold is lowered by p53 transactivation of caspase-6. *Proceedings of the National Academy of Sciences of the USA* **99**: 9492–9497.

Maecker HL, Koumenis C and Giaccia AJ (2000) p53 promotes selection for Fas-mediated apoptotic resistance. *Cancer Research* **60**: 4638–4644.

Marchenko ND, Zaika A and Moll UM (2000) Death signal-induced localization of p53 protein to mitochondria. A potential role in apoptotic signaling. *Journal of Biological Chemistry* **275**: 16202–16212.

Marine JC, Dyer MA and Jochemsen AG (2007) MDMX: from bench to bedside. *Journal of Cell Science* **120**: 371–378.

Mayo LD and Donner DB (2002) The PTEN, Mdm2, p53 tumor suppressor-oncoprotein network. *Trends n Biochemical Science* **27**: 462–467.

Mckeon F and Melino G (2007) Fog of war: the emerging p53 family. *Cell Cycle* **6**: 229–232.

Meulmeester E and Jochemsen AG (2008) p53: a guide to apoptosis. *Current Cancer Drug Targets* **8**: 87–97.

Nakamizo A, Amano T, Zhang W *et al.* (2008) Phosphorylation of Thr18 and Ser20 of p53 in Ad-p53-induced apoptosis. *Neuro-Oncology* **10**(3): 275–291.

Olive KP, Tuveson DA, Ruhe ZC *et al.* (2004) Mutant p53 gain of function in two mouse models of Li-Fraumeni syndrome. *Cell* **119**: 847–860.

Olivier M, Petitjean A, Marcel V *et al.* (2009) Recent advances in p53 research: an interdisciplinary perspective. *Cancer Gene Therapy* **16**: 1–12.

Polyak K, Xia Y, Zweier JL, Kinzler KW and Vogelstein B (1997) A model for p53-induced apoptosis. *Nature* **389**: 300–305.

Pratt MA, White D, Kushwaha N, Tibbo E and Niu MY (2007) Cytoplasmic mutant p53 increases Bcl-2 expression in estrogen receptor-positive breast cancer cells. *Apoptosis* **12**: 657–669.

Pritchard DM, Potten CS, Korsmeyer SJ, Roberts S and Hickman JA (1999) Damage-induced apoptosis in intestinal epithelia from bcl-2-null and bax-null mice: investigations of the mechanistic determinants of epithelial apoptosis *in vivo*. *Oncogene* **18**: 7287–7293.

Royds JA and Iacopetta B (2006) p53 and disease: when the guardian angel fails. *Cell Death and Differentiation* **13**: 1017–1026.

Sax JK, Fei P, Murphy ME *et al.* (2002) BID regulation by p53 contributes to chemosensitivity. *Nature Cell Biology* **4**: 842–849.

Schumacher B, Hofmann K, Boulton S and Gartner A (2001) The *C. elegans* homolog of the p53 tumor suppressor is required for DNA damage-induced apoptosis. *Current Biology* **11**: 1722–1727.

Sharpless NE (2005) INK4a/ARF: a multifunctional tumor suppressor locus. *Mutation Research* **576**: 22–38.

Terzian T, Suh YA, Iwakuma T *et al.* (2008) The inherent instability of mutant p53 is alleviated by Mdm2 or p16INK4a loss. *Genes & Development* **22**: 1337–1344.

Toledo F and Wahl GM (2006) Regulating the p53 pathway: *in vitro* hypotheses, *in vivo* veritas. *Nature Reviews of Cancer* **6**: 909–923.

Tomasini R, Tsuchihara K, Wilhelm M *et al.* (2008) TAp73 knockout shows genomic instability with infertility and tumor suppressor functions. *Genes & Development* **22**: 2677–2691.

Villunger A, Michalak EM, Coultas L *et al.* (2003) p53- and drug-induced apoptotic responses mediated by BH3-only proteins puma and noxa. *Science* **302**: 1036–1038.

Vousden KH and Lane DP (2007) p53 in health and disease. *Nature Reviews. Molecular Cellular Biology* **8**: 275–283.

Vousden KH and Prives C (2009) Blinded by the light: the growing complexity of p53. *Cell* **137**: 413–431.

Wang W, Takimoto R, Rastinejad F and El-Deiry WS (2003) Stabilization of p53 by CP-31398 inhibits ubiquitination without altering phosphorylation at serine 15 or 20 or MDM2 binding. *Molecular and Cellular Biology* **23**: 2171–2181.

Wu Y, Mehew JW, Heckman CA, Arcinas M and Boxer LM (2001) Negative regulation of bcl-2 expression by p53 in hematopoietic cells. *Oncogene* **20**: 240–251.

Yonish-Rouach E, Resnitzky D, Lotem J *et al.* (1991) Wild-type p53 induces apoptosis of myeloid leukaemic cells that is inhibited by interleukin-6. *Nature* **352**: 345–347.

Further Reading

Campisi J and D'adda Di Fagagna F (2007) Cellular senescence: when bad things happen to good cells. *Nature Reviews. Molecular Cellular Biology* **8**(9): 729–740.

Michalak E, Villunger A, Erlacher M and Strasser A (2005) Death squads enlisted by the tumour suppressor p53. *Biochemical and Biophysical Research Communications* **331**(3): 786–798.

Peters G and Vousden KH (eds) (1997) *Oncogenes and Tumour Suppressors*. Oxford: Oxford University Press.

Riley T, Sonatg E, Chen P and Levine A (2008) Transcriptional control of p53-regulated genes. *Nature Reviews of Molecualr and Cellular Biology* **9**(5): 402–412.

Weinberg R (2007) *Biology of cancer*. New York, London: Garland Science Press.

Cornification of the Skin: A Non-apoptotic Cell Death Mechanism

Advanced article

Article Contents

- Epidermal Differentiation and Cornified Cell Envelope Formation
- Skin Transglutaminases
- Transglutaminase Substrates
- Lipid Envelope
- Cell Adhesion and Desquamation
- Conclusions

Eleonora Candi, *University of Tor Vergata, Department of Experimental Medicine, Rome, Italy*

Richard A Knight, *University College London, London, UK*

Gerry Melino, *University of Tor Vergata, Department of Experimental Medicine, Rome, Italy*

The most important function of the epidermis is to form a barrier against the environment by means of several layers of terminally differentiated, dead keratinocytes, the cornified envelope (CE). CEs consist of keratins enclosed within an insoluble amalgam of proteins and lipids. Transglutaminase enzymes catalyse the formation of characteristic crosslinks between structural proteins to form the protein part of the CE. Another form of cell death, that is, apoptosis, which has a completely different molecular mechanism and physiological significance, also occurs in the skin. Defects of apoptosis are related to the development of cancer, whereas CE abnormalities are associated with barrier abnormalities and ichthyosis.

Epidermal Differentiation and Cornified Cell Envelope Formation

The biology of mammalian skin involves two distinct models of cell death: cornification and apoptosis. The epidermis, the uppermost multilayered compartment of the skin, has evolved to provide a physical and a permeability barrier, both of which are essential for survival as an adaptation to terrestrial life. The barrier is provided and continuously regenerated by terminally differentiating keratinocytes. The process of keratinocyte differentiation (cornification) is a highly organized process, both in space and time, in which keratinocytes transit from proliferative cells in the basal cell layer through the granular layer where the cornified envelope

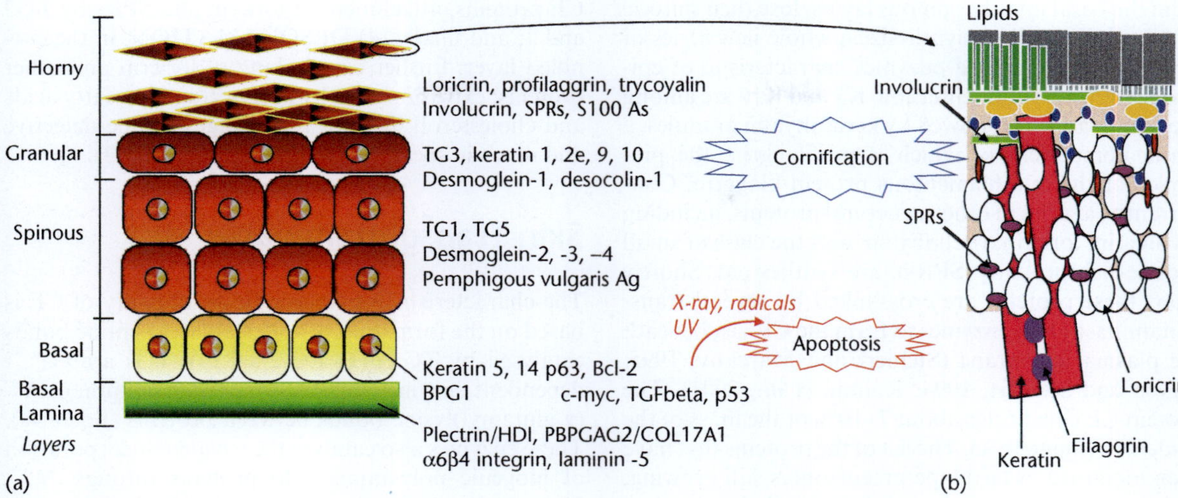

(c)

	G	S	K	Q	P	Epidermis	Forestomach
CE	50.3	21.8	3.4	5.9	5.1 (%)	Epidermis	Forestomach
Loricrin	59.4	24.1	2.0	3.4	1.8	82%	65% (w/w)
Keratins	23.8	13.9	4.8	12.7	0.8	-	10%
Filaggrin	15.7	19.8	0.0	21.4	2.8	6%	5%
Involucrin	5.0	3.2	10.0	41.3	9.1	-	-
SPRs	0.4	3.8	12.4	21.6	43.6	8%	18%

Figure 1 Apoptosis and terminal differentiation in the epidermis. (a) The genes expressed during skin differentiation and apoptosis are summarized. Apoptosis is restricted only to the lower levels as shown by TUNEL staining and expression of apoptosis-related genes. (b) A theoretical model of the CE of human foreskin epidermis. The proteins expressed in the epidermis are indicated according to their relative localization. The proteins involved in apoptosis are indicated in red; those related to cornification are indicated in light blue. Death of keratinocytes (CEs) occurs in the upper layers and its defects results in ichthyosis. Death by apoptosis occurs in the basal layer (e.g. after sunburn) and its defects are related to cancer development. (c) Comparison of the amino acid composition of the CE with the amino acid composition of the CE precursors (loricrin, keratins, filaggrin, involucrin, SPRs). As indicated, the most abundant amino acids in the CE are G, S, K, Q and P; these amino acids are also highly represented in loricrin and the SPRs, the two major protein components of the CE (82% and 8% of the epidermal CE, respectively). Adapted from Candi E, Schmidt R and Melino G (2005) The cornified envelope: a model of cell death in the skin. *Natural Review. Molecular Cell Biology* **6**(4): 328–340.

(CE) is formed, to an association of flattened dead cells in the uppermost layer of the skin, the stratum corneum (**Figure 1a–b**). The formation begins with the synthesis of an immature type of envelope beneath the plasma membrane. This envelop matures by the covalent attachment of preformed dedicated molecules to form a rigid structure that confers physical and dehydration resistance. **See also**: The Siren's Song: This Death that Makes Life Live

Besides terminal differentiation, another type of cell death, called apoptosis, occurs in the skin. Apoptosis is a developmental remodelling programme and a defensive, organized self-destruction of the cell in reaction to severe damage. Unlike necrosis, apoptosis occurs inside an intact plasma membrane, thus preventing the release of intracellular contents and inflammation. Apoptosis is the major but not the sole mechanism of physiological cell death in mammalian cells. Despite the fact that epidermal terminal differentiation and apoptosis share some common features, they are two profoundly distinct processes with two distinct endpoints occurring at different locations within the epidermis. Here, we focus on cornification.

In normal epidermis the proliferation rate in the basal layer is precisely balanced by desquamation of the CE at the skin surface (see below). This epidermal homeostasis constantly rejuvenates the epidermis. Epidermal proliferation and differentiation occurs in stages and is characterized by the sequential expression of distinct proteins (**Figure 1a–b**). Keratins K5 and K14 are the main structural proteins in proliferating basal keratinocytes (Fuchs and Cleveland, 1998; Porter and Lane, 2003; Strelkov *et al.*, 2003). Upon signals that are not yet well understood, certain basal keratinocytes migrate from the basal into the spinous layers, lose their mitotic activity, and begin to synthesize a whole new series of structural proteins and enzymes characteristic of epidermal terminal differentiation. K1 and K10 are among the first to appear followed by keratohyalin granules, a major component of which is profilaggrin, the precursor of the interfilamentous protein filaggrin. Concurrently, a series of other structural proteins, including involucrin, loricrin, trichohyalin and the class of small proline-rich proteins (SPRs), are synthesized. Shortly after, these proteins are crosslinked by several transglutaminase (TG) enzymes to form the CE just beneath the plasma membrane (Steinert and Marekov, 1995; Steven and Steinert, 1994; Kalinin *et al.*, 2001). The protein CE constitutes about 7–10% of the mass of the epidermis (**Figure 1b–c**). The list of the proteins that have been identified as envelope precursors is still growing, and some of these proteins are discussed below in more detail. Besides the proteins, a complex series of lipids, ceramides, are synthesized, some of which become covalently attached to the protein CE, and most of which form intercellular lamellae to form a complete barrier. Thus, the stratum corneum is composed of terminally differentiated dead, cornified, flattened cells called corneocytes or CE with mostly KIF embedded into a filaggrin matrix surrounded by the insoluble and highly resistant CE. CEs are tightly attached to each other by corneodesmosomes, modified desmosomal structures, which are proteolytically degraded in the uppermost layers of the stratum corneum to allow desquamation (Serre *et al.*, 1991).

Similar structures are formed in many other types of stratified squamous epithelia, including hair, nail cuticles, and the alimentary canal and urogenital tract, by using the same and/or different structural proteins in varying amounts. The relative composition allows different properties (elasticity, mechanical resistance, water impermeability) for distinct epithelia (Kartasova *et al.*, 1996).

The formation of the CE is better understood than the formation of apoptotic bodies with each step being attributable to specific proteins, and specific diseases resulting from their abnormalities according to the model indicated in **Figure 2**. Thus, the formation of the CE can be divided schematically into distinct stages: (1) INITIATION, in the spinous layer; synthesis of the structural proteins; synthesis and extrusion into the intercellular space of specific lipids; (2) FORMATION OF THE LIPID ENVELOPE, in the granular layer; crosslinking of proteins (envoplakin, periplakin, involucrin) by TG5 and 1; (3) REINFORCEMENT, in the granular layer; covalent attachment of some lipids to the CE proteins, attachment of loricrin and SPRs by TG3 and 1; and finally (4) DESQUAMATION, in the cornified layer; further crosslinking of loricrin and other proteins; extrusion of ω-hydroxy-ceramides, fatty acids and cholesterol. Each of these stages can be defective and result in specific skin diseases (see **Figure 3**).

Skin Transglutaminases

The characteristic resistance and insolubility of CE is based on the formation of very stable isopeptide bonds catalysed by TGs (TG EC 2.3.2.13). TGs are Ca^{2+}-dependent enzymes that catalyse the formation of N^{ε}-(γ-glutamyl)lysine bonds between proteins (**Figure 4a**). These enzymes also catalyse the covalent incorporation of biogenic polyamines into proteins through *N*,*N*-bis(γ-glutamyl) bonds, which act as bridges between molecules. In mammals, nine distinct types of TGs (**Figure 4b**) have been characterized and at least four of these (TG1, 2, 3 and 5) are expressed during terminal differentiation of the human epidermis. Many *in vitro* and *in vivo* data have documented the involvement of TG1, 3 and 5 in CE formation (Lorand and Graham, 2003).

TG1 is expressed primarily in keratinocytes. In proliferating or basal keratinocytes a very low specific activity form is almost entirely bound to membranes through multiple myristate adducts. During terminal differentiation, TG1 is proteolytically processed at specific sites (residues 93 and 573) to form a complex of 10, 33 and 67 kDa chains that remain held together by secondary bonding interactions and attached to membranes through palmitate linkages. This last membrane-anchored form has a very high specific activity and is responsible for most of the TG1 activity in epithelial cells. In addition, some of the full-length and processed forms lose the lipid adducts and are liberated into the cytosol where they possess significant activity. The exact role of these different forms of TG1 in CE assembly is not yet known, although $TG1^{-\backslash-}$ mice die at birth to the impaired skin barrier function and human TG1 mutations result in the genetic disease, lamellar ichthyosis (Matsuki *et al.*, 1998). Some of their known substrate CE structural proteins include involucrin and loricrin. Moreover, *in vitro* data have shown that the membrane-anchored TG1 enzyme can covalently esterify ω-hydroxy-ceramides with long-chain fatty acids (>C30) onto glutamine residues of the scaffold proteins (i.e. involucrin). This reaction creates a direct link between the protein (inner) and lipid (outer) component of the CE.

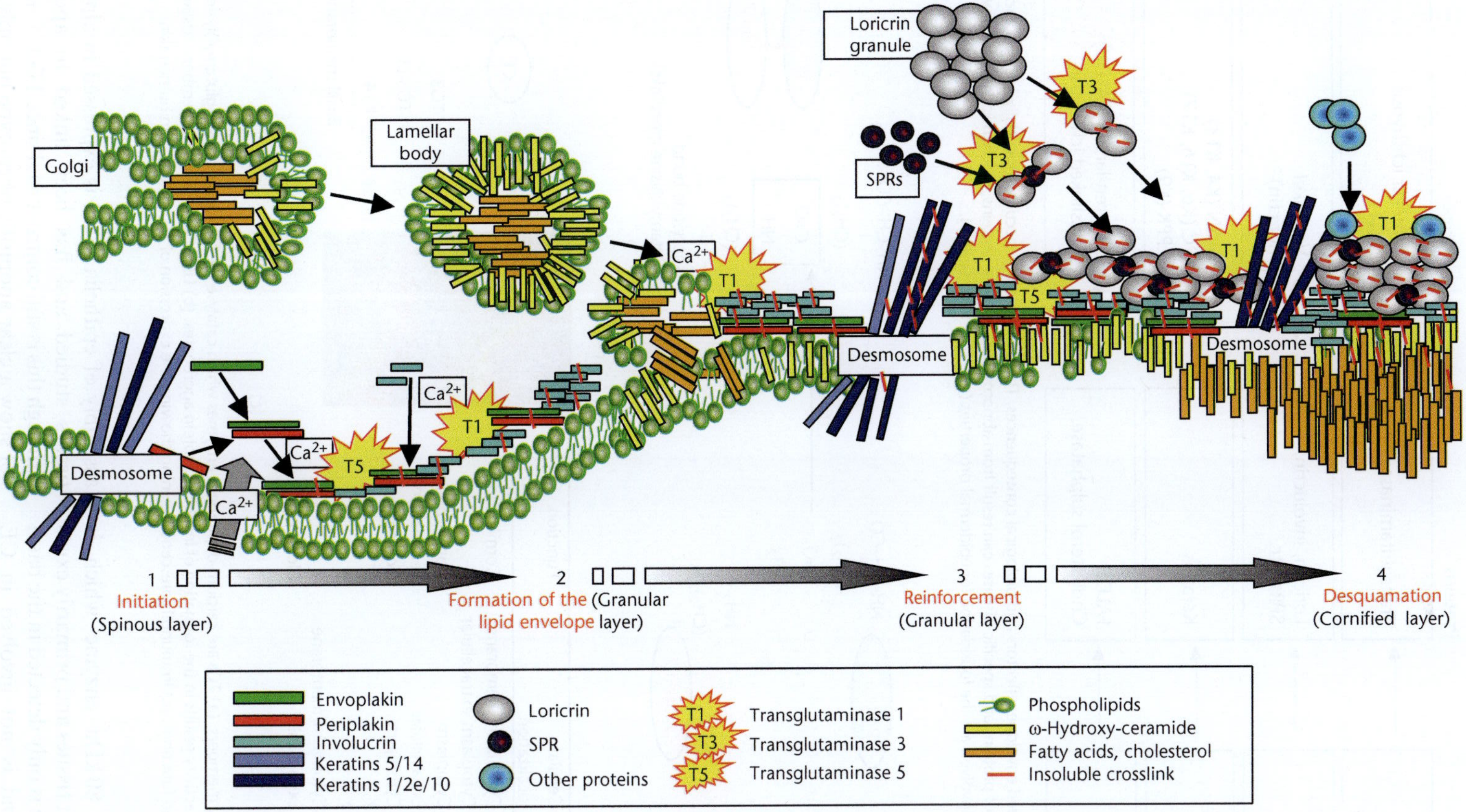

Figure 2 Progressive steps in the formation of CE. (a) Schematic sequence of the different phases, with the related proteins involved, the biological effects and the related diseases. (b) Molecular description of the progressive steps: (1) initiation, (2) formation of the lipid envelope, (3) reinforcement and (4) desquamation. In the initiation phase (1), TG1 and 5 crosslink envoplakin and periplakin under the cell membrane, anchoring them to the desmosome. In the formation of the lipid envelope (2), lipids from the lamellar body, derived from the Golgi, are attached to the already crosslinked proteins and exposed outside the membrane. In the reinforcement phase (3), heavy crosslinking occurs on the desmosome, using keratins, loricrin and SPRs as substrates for TG1 and 3. In the desquamation phase (4), the final crosslinking is catalysed by TG1 on the protein scaffold in addition to lipid deposition. The physical properties of the CE depend on the nature of the substrates and the crosslinks. Adapted from Candi E, Schmidt R and Melino G (2005) The cornified envelope: a model of cell death in the skin. *Natural Review. Molecular Cell Biology* **6**(4): 328–340.

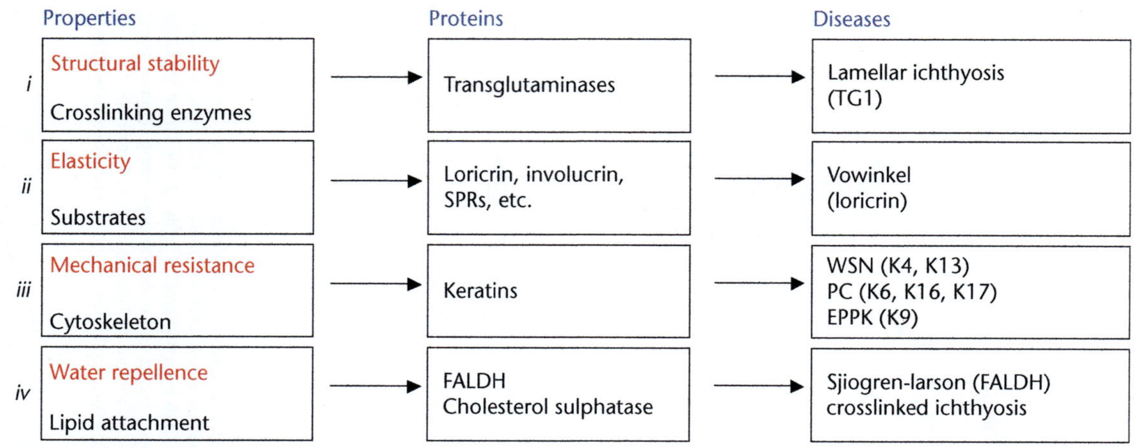

Figure 3 Formation of the cornified envelope: effectors and biological consequences. The differentiation occurs sequentially in steps. Each process is characterized by the expression of proteins and specific diseases can result from abnormalities in these proteins. Biochemical and biophysical properties of the specific proteins involved are the foundation for epidermal properties (e.g. structure stability).

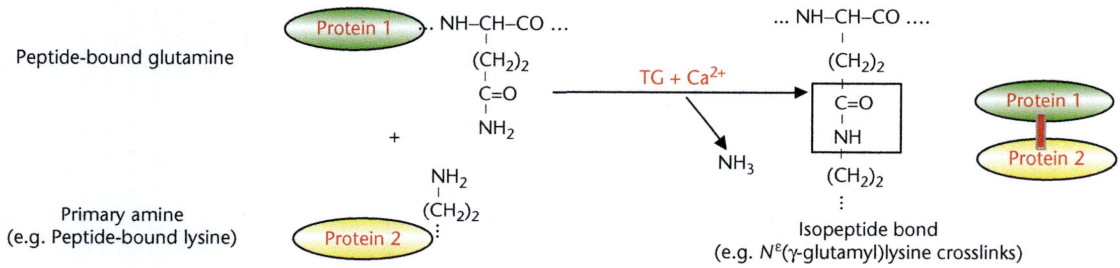

(a)

	Tissue distribution	Cellular localization	Function
TG1	Skin, brain	Cytoplasm/membrane	Cornification
TG2	Ubiquitous	Cytoplasm/extracellular	
TG3	Skin/hair	Cytoplasm	Cornification
TG4	Prostate	Extracellular	Seminal clot
TG5	Skin	Cytoplasm	Cornification
TG6	?	?	?
TG7	?	?	?
Band4.2	Red cells	Cytoplasm/membrane	?
fXIIIa	Plasma	Extracellular	Blood clot

(b)

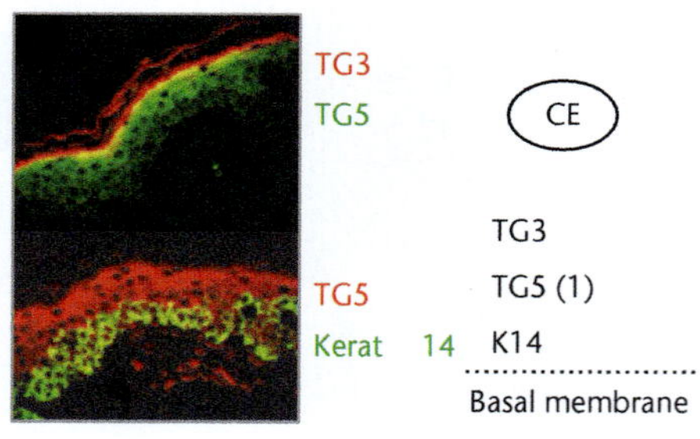

(c)

Figure 4 Transglutaminase family members. (a) TGs are calcium-dependent enzymes which catalyse the formation of *N*-(γ-glutamyl)lysine bonds between proteins. TG crosslinking activity results in the formation of insoluble protein aggregates. (b) List of the TG family members, tissue distribution, cellular localization and functions. (c) Immunofluorescence staining showing the expression of TG3 and TG5 in human skin.

TG2 is a monomeric 80 kDa enzyme which is expressed in many different tissues and primarily exists in the cytosol. This enzyme is only detected in the basal layer of the epidermis and is not involved in CE assembly of epithelia. TG2 is expressed in almost all tissues studied and has been linked to apoptosis, although this is still controversial since $TG2^{-/-}$ mice do not show a clear apoptotic phenotype but only a mild erythrocyte fragility, abnormal insulin secretion and defects in phagocytosis.

TG3, first identified in hair follicles, is expressed in terminally differentiating epithelial cells (**Figure 4c**). It is initially expressed as an inactive zymogen of about 77 kDa, but it may be activated into a very high specific activity form following proteolytic cleavage at or about residue 469 into 50 and 27 kDa fragments that remain bound together by secondary forces. Based on *in vitro* assays and immunocytochemistry, the activated TG3 enzyme seems to be primarily responsible for the crosslinking of proteins such as loricrin, trichohyalin and SPRs.

TG5 also plays a role in CE assembly. Indeed, TG5 is detected in the upper layers of normal human epidermis (**Figure 4c**) following a gradient of expression stretching from the spinous layer to the horny layer, and it is induced during the early stages of keratinocyte differentiation *in vitro*. Moreover, kinetic constant (Km, Kcat/Km) showed high affinity for TG5 in using loricrin, involucrin and SPRs as substrates.

In addition to the fine regulation exerted by calcium ions, TG2, 3 and 5 are also regulated by guanosine triphosphate (GTP) as well as by nitric oxide. While the biochemical basis of this regulation is clear, its biological relevance is less evident, especially in the context of skin crosslinking.

Transglutaminase Substrates

Several protein substrates of TGs are present in the CE. The major substrates crosslinked by TGs are discussed below.

Involucrin

Involucrin (Rice and Green, 1977), the first CE structural protein identified, is a classic if not obligatory component of the CE. Furthermore, an involucrin-like protein is a principal component of the apoptotic bodies of liver cells, suggesting that the processes of epithelial differentiation may be a recent elaboration of the more ancient apoptosis. Human involucrin (70 kDa) is extraordinarily enriched in G and D residues. Like several other CE structural proteins, it is largely built from repeating peptide units, in this case of 10 residues, which have undergone extensive rearrangements during recent evolutionary times. Involucrin is highly α-helical, and TG1 preferentially uses Q495 and 496 for crosslinking (Simon and Green, 1988). Since many other Q and K are crosslinked, multiple TGs are involved in this process. This is consistent with the idea that involucrin is an early component in the assembly of CEs providing a scaffold to which other proteins subsequently become crosslinked. Within the CE structure, involucrin is adjacent to the cell membrane (which is subsequently replaced) and may therefore be the preferred substrate to which the externalized lipids, mainly ceramides, are covalently attached to form the exterior surface of the CE. Surprisingly, in transgenic mice overexpressing involucrin no abnormalities are detected in the skin.

Loricrin

Human loricrin (26 kDa) is expressed in the granular layer during epidermal differentiation. It is unusually enriched in G (47% human, 55% mouse), S (22.8%) and C (6%). The loricrin gene is located on chromosome 1q21.1 (the so-called epidermal differentiation complex) in which several other genes expressed in late epidermal differentiation are also included, such as involucrin, SPRs and profilaggrin genes. Models for the secondary structures of mouse and human loricrins have been proposed based on computer modelling of the deduced amino acid sequences, and can be subdivided into discrete domains (**Figure 5a**) (Hohl *et al.*, 1991). There are three Glycine-Serine (GC)-rich domains which are interspersed with short regions enriched in Q, and these are flanked at the *N*- and *C*-termini by highly conserved sequences rich in Q and K. The G residues are configured in tandem inexact peptide repeats and may fold into a unique protein conformation known as the 'glycine loop'. The G residues are interspersed by occasional long-chain aliphatic or aromatic residues which may associate themselves by hydrophobic interactions, thereby displacing the glycine sequences into an Ω-loop-like configuration (**Figure 5a**).

Loricrin is the major component of the epidermal CE, comprising 70–85% of total protein mass (**Figure 1c**), depending on the body site of epidermis. Limited proteolysis experiments have been performed on foreskin epidermal CEs to isolate crosslinked peptides for amino acid sequencing; most of the peptides consisted of loricrin–loricrin sequences crosslinked together via isopeptide bonds, as well as smaller amounts of loricrin crosslinked to SPR1 and 2, which seem to serve as crossbridging proteins between the loricrin molecules (see **Table 1**). Loricrin appears to function as a major reinforcement protein for the CE on its cytoplasmic face (**Figure 2**). Loricrin is a substrate for TG1, 2 and 3 *in vitro*, although these produce different types of crosslinking (Candi *et al.*, 1995) (**Figure 5a**). Together, the *in vitro* and *in vivo* data suggest that accretion of loricrin

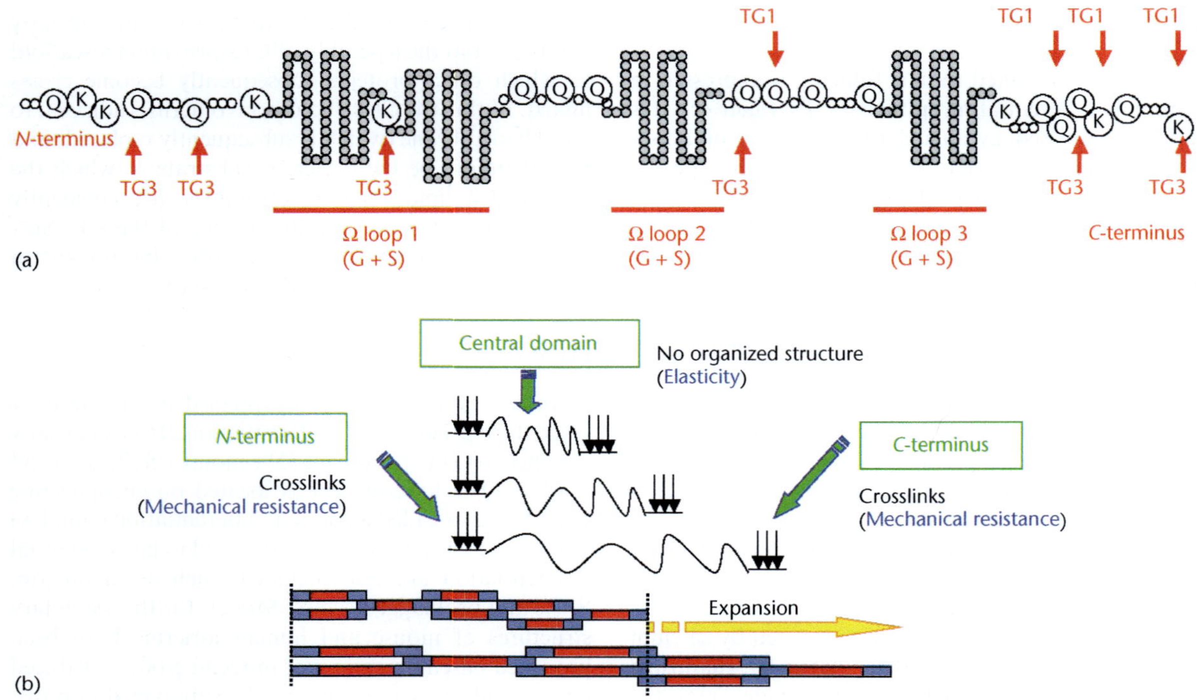

Figure 5 Lack of ordered structure in the substrates of skin TGs: loricrin. (a) Schematic diagram of loricrin, showing the three Ω loops, formed by GS residues aligned by the recurrent hydrophobic residues (mostly valine, isoleucine, phenylalanine and tyrosine); Q and K are highlighted. The quantitatively major TG crosslinking sites in loricrin are indicated by the arrows: (1) crosslinks by TG1, (3) crosslinks by TG3. These crosslinks may be intra- or interchain and form a highly insoluble structure that is essential for the barrier function of the epidermis. (b) Schematic representation of the loricrin/SPR structural model upon crosslinking by TGs. Due to their peculiar amino acid composition, loricrin/SPR structure is predicted to be 'spring-like': the *N*- and *C*-terminal end domains contain most of the Q and K that can be used by TGs as substrates (mechanical resistance). In contrast, the central domain forms a very flexible and non-ordered structure responsible for the expansion and elasticity of the CE.

by crosslinking onto the CE *in vivo* may involve at least two steps: an initial attachment by TG1 (and perhaps TG5) which oligomerize loricrin by way of interchain crosslinks, and a reinforcement process involving compaction by the TG3 enzyme (see **Figure 2**). **Figure 5b–C** shows a model in which the lack of ordered structure (glycine loop) is responsible for the elasticity of the CE, whereas the TG crosslinks are responsible for the mechanical resistance of the protein net. Gene disruption studies in the mouse have shown a delay in barrier development in the embryo and a reduction in stratum corneum stability in the newborn mice. However, this shortcoming is compensated by an increase in expression of the other envelope precursor proteins of the SPR family.

Small Proline-Rich Proteins

Small proline-rich proteins consist of a family of related small (6–18 kDa) proline-rich (up to 55%) proteins subdivided into SPR1 (two genes in mouse and human), SPR2 (at least 10 genes) and SPR3 (one gene) classes. They contain head and tail domains rich in glutamines and lysines, as well as a central domain consisting of variable numbers (2–20) of repeating peptide units of eight (SPR1 and SPR3) or nine (SPR2) residues that are extraordinarily enriched in prolines (Steinert and Marekov, 1995; Broome *et al.*, 2003).

In *in vitro* crosslinking assays, also confirmed in *in vivo* data, the SPRs are TG substrates (although TG3 is greatly preferred), and Q and K located on the head and tail domains are utilized for crosslinking. This indicates that the model suggested for loricrin, where the disorganized structure is responsible for elasticity and the TG crosslinks for mechanical resistance, also applies to SPRs (**Figure 5b–c**).

The exclusive utilization of head and tail domain sequences for crosslinking and the nature of the crosslinked peptides recovered from CEs strongly suggest that the SPRs function as cross-bridging proteins among the more abundant loricrins of the CE (**Table 1**). Expression

Table 1 SPRs and loricrin function as cross-bridging proteins in CE

loricrin-spr1-loricrin
loricrin-spr2-loricrin
loricrin-spr3-loricrin
spr1-loricrin-spr1-loricrin
spr1-loricrin-spr2-loricrin
spr2-loricrin-spr1-loricrin
spr3-loricrin-spr1-loricrin
spr1-loricrin-spr1-loricrin-spr1
spr1-loricrin-spr3-loricrin-spr1
loricrin-spr1-loricrin-spr1-loricrin-spr1
loricrin-spr1-loricrin-loricrin
loricrin-spr2-loricrin-loricrin
spr2-loricrin-spr1-loricrin-loricrin
loricrin-spr1-loricrin-envoplakin
loricrin-spr1-envoplakin
loricrin-spr1-envoplakin-desmoplakin
loricrin-spr1-involucrin-envoplakin
loricrin-spr1-involucrin
envoplakin-spr1-involucrin
loricrin-loricrin-spr1-keratin1
loricrin-spr1-keratin1
loricrin-spr1-loricrin-keratin1
loricrin-spr2-loricrin-keratin1
loricrin-spr1-loricrin-spr1-keratin1
spr1-loricrin-spr1-loricrin-keratin1
spr2-loricrin-spr1-loricrin-keratin1
spr3-loricrin-spr1-loricrin-keratin1
spr1-spr1-loricrin-spr2-loricrin
loricrin-spr1-spr1-loricrin
trichohyalin-spr1-trichohyalin-spr1
loricrin-trichohyalin-spr3-trichohyalin
trichohyalin-envoplakin-spr2-loricrin

of mouse SPR1 correlates well with increases in epidermal thickness and/or a requirement for extreme flexibility. The content of SPRs can markedly alter the physical properties of the CE with significant consequences for the epidermal structure and barrier function. On the basis of the wide variations in the amount of SPRs in the CEs of different epithelia, we conclude that the mechanical properties of the CE and respective epithelia vary depending on the type and amount of SPRs expressed in the tissue, and thereby render the tissues uniquely adapted to function.

Keratin Intermediate Filaments

The keratin intermediate filaments (KIFs) are structures of 10 nm forming the cytoskeleton of all cells, including epithelia. In the epidermis, expression of the K5/K14 pair of chains in the basal layer is shut down following cessation of cell division and the commitment to terminal differentiation, whereupon K1/K10 are produced in large amounts, and presumably incorporated simply by interchange of the new proteins into the preexisting K5/K14 KIF network. Specialized locations express significant amounts of other keratins such as K9 in palms/soles, and K2e in these and other thickened sites. Each of the three type II keratin chains (K1, K2e and K5) becomes crosslinked to the CE structure during terminal differentiation by use of a specific lysine residue located in a conserved region of the V1 subdomain of the head domain. This results in coordination and stabilization of the KIF–filaggrin network within the cornified cell and the CE at the periphery. Loss of this K by mutation results in gross shape malformations, which are manifest especially as thickening of especially the palms and soles. Similarly, abnormal differentiation of skin in benign and malignant tumours correlates well with an aberrant expression of the differentiation-specific K1/K10 pair 51.

Profilaggrin

Human profilaggrin, also located in the chromosomal 1q21 cluster, is a very large (about 500 kDa) and complex protein. Its product, filaggrin, becomes crosslinked to the epidermal CE and participates in this way in coordinating the structure of the cornified cells (Steinert and Marekov, 1995). The bulk of protein sequences consist of a tandem array of repeating filaggrin units of about 35 kDa. Each filaggrin unit is separated by a short 7–10 amino acid residue 'linker' peptide. Profilaggrin, initially highly phosphorylated and accumulated in keratohyalin granules during terminal differentiation, is dephosphorylated and cleaved by specific proteases that selectively recognize and excise the linker regions to release individual filaggrin molecules. Based on *in vitro* assays, filaggrin has the remarkable property of aggregating KIF and other types of intermediate filaments into tight bundles in which the individual filaments are closely aligned in regular arrays (Steinert *et al.*, 1981). Filaggrin functions *in vivo* in the same way by organizing the KIF into bundles in the cells. The effect of this would result in collapse of the KIF cytoskeleton, which in turn would promote a major change in shape from an ellipsoid to a flat cell in which the KIFs are aligned parallel to the outer surface of the epidermis. The half-life of filaggrin is about 6 h, so it is clear that these events occur rapidly. In the cornified cell, it is mostly degraded into free amino acids. The high concentration of hydrophilic amino acids (in the 100 mM range) is essential for the retention of water and contributes to the osmolarity, and hence the flexibility, of the stratum corneum. Thus, the profilaggrin system plays at least three critical roles in epidermal differentiation: (1) the alignment of KIF, (2) major changes in cell shape, and subsequently (3) the maintenance of epidermal texture.

Lipid Envelope

The CE is embedded into the lipid envelope. Skin lipids are essential for barrier function of the epidermis to avoid transepidermal water loss. The majority of these lipids are ceramides, cholesterol, fatty acids and cholesterol esters that form lamellae between the corneocytes of the stratum corneum. Intercellular skin lipids are synthesized and stored in granules known as lamellar granules (or Odland bodies), which are extruded (probably by fusion) into the extracellular space in the upper granular layer during cornification (**Figure 2**) (Bouwstra *et al.*, 2003; Madison, 2003). The lamellar bodies are composed of lipid lamellae which contain glucosylceramides, phospholipids and cholesterol, which are the precursors of the intercellular skin lipids. Complex changes in lipid composition occur after the extrusion of the lamellar granules' contents by the enzymatic action of a group of acid hydrolases, which are also secreted into the extracellular space. Enzymes such as β-glucocerebroside (involved in Gaucher's disease), acid sphingomyelinase and phospholipase A2 are involved in metabolic modification of the extracellular lipids. Even though several animal models show a defined role for lipids in the epidermal barrier, the physical structural organization of the extracellular lipids of the CE is not known at the molecular level. Electron micrographic studies show that the lipids are organized as stacked membrane sheets which occupy all the extracellular space. Biophysical studies indicate that the free fatty acids and the amide-linked fatty acid chains in the ceramides are tightly packed to form gel-phase membrane domains, which coexist with liquid crystalline domains. The experimental difficulties in studying these lipids have led to the formulation of different models: the 'domain mosaic' model and the more recent 'sandwich' and the 'single gel phase' models.

An important though still unresolved area is the mechanism of interaction of the extracellular lipids and the CE. Long-chain ω-hydroxyceramides are present on the extracellular surface of the CE, forming an external layer or coating of the CE, the 'lipid envelope', and some of them are covalently attached through ester linkages to the outer surface of the CE. Indeed, ω-hydroxyceramides are linked to involucrin through an ester bond by TG1, at least *in vitro*. Potentially, other CE proteins could also be covalently linked to ω-hydroxyceramides as the involucrin knockout mice show normal CE and stratum corneum.

Cell Adhesion and Desquamation

The epidermis is separated from the underlying dermis by the basement membrane composed of extracellular matrix proteins. The basal layer adheres to the basement membrane through integrin-containing hemidesmosomes (Watt, 2002). The importance of integrins in epidermal physiology is apparent from knockout studies since their deletion results in defects in organization of the basement membrane, impaired wound healing, blistering and abnormal keratinocyte proliferation/differentiation.

In the epidermis two types of junctions are responsible for intercellular adhesion and cohesion of the stratum corneum: (1) the adherence junctions (connecting the actin cytoskeleton of neighbouring cells) and (2) desmosomes (connecting the keratin filament cytoskeleton of adjacent cells). Certain proteins constituting the adhesion complexes are specifically expressed during keratinocyte differentiation, including desmoglein-1, desmocollin-1, envoplakin, periplakin, plakophilin-1, and corneodesmosin. The cohesion of the stratum corneum depends on modified desmosomes called corneodesmosomes. At the transition between the granular layer and the stratum corneum, profound changes are observed in desmosome morphology, namely the disappearance of the cytoplasmic plaque and formation of a homogeneous electron-dense plaque in the extracellular core (Serre *et al.*, 1991; Simon *et al.*, 2001). The two major components of the corneodesmosomes are desmoglein-1 and desmocollin-1, which are two glycoproteins belonging to the family of cadherins, calcium-dependent cell adhesion molecules. Another constituent of the corneodesmosomes is corneodesmosin located in the core of its extracellular compartment. Corneodesmosin is synthesized in the upper spinous and granular layer, and is a phosphorylated and glycosylated protein. It is a GS-rich protein, and these sequences are likely to form Ω-loop-related domains similar to loricrin and keratins as described above. The suggested function of these structural motifs is to interact with identical loops on the same or neighbouring proteins. Corneodesmosome degradation is of major importance during the desquamation process. In several pathological conditions (psoriasis, xerosis) a great increase in the number of persisting corneodesmosomes at the surface of the stratum corneum is observed. Several serine proteases, including the stratum corneum chymotryptic enzyme (SCCE) and the stratum corneum tryptic enzyme (SCTE) are thought to be involved in corneodesmosome proteolysis (see below) (Brattsand and Egelrud, 1999; Ekholm *et al.*, 2000). Deregulation of desmosome formation is observed in several degenerative cutaneous diseases: pemphigus, an autoimmune blistering disease with autoantibodies against components of the desmosome, and desmoglein-1 and desmoglein-3 in pemphigus foliaceous and pemphigus vulgaris, respectively. Mutations in plakophilin-1 are responsible for ectodermal dysplasia, while mutations in desmoplakin are found in patients affected by palmoplantar keratoderma. Calcium is also necessary for desmosomal integrity reflected in Hailey-Hailey disease and Darier's disease, both characterized by desmosomal dysfunction caused by mutations in the calcium pumps.

During the transition from the granular layer to the cornified layer, all keratinocytes simultaneously lose their nucleus and other cellular organelles, a process that requires a massive activation of epidermal proteases. Many skin proteases have been identified, but for most of them the precise role in epidermal differentiation remains elusive. Proteases are involved in at least three processes during keratinocyte differentiation: (1) certain CE precursors require proteolytic processing before they are linked by TGs into the CE; (2) the loss of nuclei and mitochondria requires proteolytic processing; (3) the dead cells (corneocytes) of the uppermost cornified layer are shed into the environment by a process called desquamation which requires proteolysis of the structural components of the corneodesmosomes.

Intracellular Proteases

During cornification several CE precursors are crosslinked by TGs to assemble the CE, a process that requires proteolytic activity since TGs themselves require proteolytic activation. A mutation in TG1 that prevents its processing has been identified as one of the causes of lamellar ichthyosis (Candi *et al.*, 1998). The cysteine protease m-calpain or furin may be the enzyme responsible for the activation of TG1 (Kim and Bae, 1998; Kim *et al.*, 1994; Egberts *et al.*, 2004). In addition TG substrates, such as profilaggrin and involucrin, are first cleaved into small fragments before crosslinking. During differentiation, filaggrin is the substrate of many proteases such as PEP1, calpain, furin, MT-SP1 and caspase-14. The exact order of the different proteolytic events is not completely resolved, but it is believed that dephosphorylation precedes filaggrin processing. Among the intracellular proteases are (1) lysosomal protease. The lysosomes contain a large number of hydrolases for the degradation of macromolecules. The signals for the release and activation of lysosomal enzymes during keratinocyte differentiation are unclear as is the mechanism by which they degrade the organelles. The large cathepsin family contains both exo- and endopeptidases and includes the lysosomal cysteine, aspartic and serine proteases. Different members of the cathepsin family (cathepsin A, B, C, D, E, H, L, L2 and S) are present in keratinocytes, and cathepsin activity has been measured in epidermal extracts. (2) Calcium-dependent proteases: In the epidermis an increasing calcium-gradient exists concomitant with terminal differentiation both in the intra- and extracellular compartments, suggesting the involvement of calcium-dependent proteases. μ-Calpain (calpain I) and m-calpain (calpain II) are calcium-dependent cysteine proteases that require μM or mM calcium, respectively, for their activity. Both calpain I and II are present in the suprabasal layers of the epidermis. Other calcium-dependent serine proteases are the subtilisin-like protein convertases (SPCs) involved in the secretory pathway. Several mammalian SPCs have been described and at least four of them are found in the epidermis at the mRNA level: furin (SPC1), SPC4, SPC6 and SPC7. (3) Cell death–related proteases: Different caspases are found in the skin, but the pro-apoptotic family members are present as pro-enzymes. No skin abnormalities have so far been described in pro-apoptotic caspase knockout mice. However, caspase-14 is mainly expressed in the suprabasal layers of the epidermis, is not involved in apoptosis, but is processed at the transition from the granular to the cornified layer during keratinocyte differentiation. The epidermis of caspase-14-deficient mice is shiny and lichenified, indicating an altered stratum corneum composition, and it is characterized by reduced skin-hydration levels and increased water loss (Denecker *et al.*, 2007; Nicotera and Melino, 2007).

Extracellular Proteases

Extracellular proteases, abundant in the stratum corneum, have been correlated with desquamation, thus suggesting that they may be involved in the degradation of the corneodesmosomes. Corneodesmosomal proteins that are degraded during desquamation are

Table 2 Common and distinctive features of cornification and apoptosis

Common features		
Intrinsic control mechanism		
Elimination of cells without tissue destruction, inflammation or scar formation		
Disintegration of DNA and nuclei		
Modulation by glucocorticoids and retinoids		
Distinctive features	**Cornification**	**Apoptosis**
Elimination of dead cells by	Desquamation	Phagocytosis
Cell organelles are	Lysed	Encapsulated
Nucleus	Enlargement	Chromatin condensation
	No DNA fragmentation	DNA fragmentation
	TUNEL negativity	TUNEL positivity
Plasma membrane	Fused with lamellae	Intact
Duration of the process	Slow (20 days)	Rapid (20 minutes)
Activation by Ca^{2+} ions	Obligatory	Facultative
Enzymes involved	Transglutaminases	Caspases, calpains

desmoglein 1, desmocollin 1, plakoglobin and corneodesmosin. The enzymes responsible for their processing and degradation are not yet identified, although possible candidates are SCCE, SCTE and stratum corneum cathepsin-L-like enzyme (SCCL). Other proteases identified in the cornified layer are cathepsin D and stratum corneum thiol protease (SCTP) or cathepsin L2. SCCE and SCTE are expressed in the granular layer and belong to the serine protease kallikrein family. The SCCE precursor is present in the lamellar bodies that extrude their contents at the transition between the granular and cornified layers into the extracellular space. The activation occurs by a tryptic-like cleavage, removing the *N*-terminus (Hansson *et al.*, 1994; Sondell *et al.*, 1995). SCTE is also confined to the differentiating suprabasal layers of the epidermis. Both SCCE and SCTE can degrade 52–56 kDa corneodesmosin *in vitro*.

Conclusions

Two types of epidermal cell death are found in the skin: apoptosis (occurring in the basal layer) versus cornification (occurring in the upper layers). In human skin, apoptosis occurs primarily in the basal layer of the epidermis in association with cell damage, for example in lichenoid disease, psoriasis, and after ultraviolet irradiation. Apoptosis is an active, energy-dependent, gene-directed biochemical pathway of cell self-destruction. Both keratinocyte terminal differentiation and apoptosis are molecular programmes resulting in cell death. Because of certain similarities between the two processes, several investigators have suggested that terminal cell differentiation of epidermal keratinocytes is a specialized form of apoptosis. **Table 2** summarizes common and distinctive features of cornification and apoptosis.

1. *Similarities*. For both apoptosis and keratinocyte differentiation, the interaction with the extracellular matrix through integrins plays an important role. Keratinocytes will start to differentiate when contact with the basal membrane is lost, and indeed keratinocytes can be induced to differentiate *in vitro* by forced suspension growth. Similarly, other cell types will die by apoptosis when they become detached from the substrate (anoikis). The transcription factor c-myc has also been shown to be involved in both processes. In certain cells the constitutive activation of c-myc results in apoptosis while overexpression of c-myc in keratinocytes induces terminal differentiation. The Bcl-2 family members, in particular Bcl-2 itself, are downregulated during terminal differentiation, whereas expression of pro-apoptotic Bax and Bak is still found in the suprabasal layers of the epidermis. Similarly, increases in calcium concentration induce both keratinocyte differentiation and apoptosis, although, for apoptosis, this does not appear to be obligatory.

2. *Differences*. Growing evidence strongly suggests that keratinocyte differentiation and apoptosis occur through different molecular pathways. Apoptosis is a rapid process, all the components for signal transduction are present in the cells and no de novo protein synthesis is required to execute the apoptotic pathway. In contrast,

keratinocyte differentiation is a slow, coordinated process in space and time, which requires the ordered production of typical differentiation-associated proteins. The plasma membrane during apoptotic body formation remains intact (membrane blebbing), whereas during keratinocyte differentiation the plasma membrane disappears after fusion with the lamellar bodies. The corneocytes, the final product of keratinocyte differentiation, fulfil an important physiological role before they desquamate into the environment, whereas the apoptotic cell has no obvious function and is eliminated by phagocytosis. Both apoptosis and keratinocyte differentiation involve the destruction of the nucleus, although the mechanisms are distinct. Nuclear degradation during apoptosis is marked by chromatin condensation and DNA laddering detected by TUNEL (terminal dUTP-biotin nick end labelling) staining. Although TUNEL positivity has been sporadically observed in differentiating keratinocytes, the nuclei are enlarged instead of condensed and neither chromatin condensation nor DNA laddering is detected. Caspases have a major role in apoptosis, where they cleave a large number of substrates. Different caspases are expressed in the lower layers of the epidermis. However, no abnormal skin formation has been reported in any pro-apoptotic caspase-deficient mice. On the other hand, caspase-14, which is not involved in apoptosis, is processed in the later stages of epidermal differentiation (Nicotera and Melino, 2007).

In conclusion, in human skin two distinct types of cell death occur via two completely unrelated mechanisms: (1) terminal keratinocyte differentiation, with a defined physiological function of the dead cells (corneocytes) constituting the skin barrier; and (2) apoptosis which occurs mainly in the basal layer as a defence–reaction to eliminate heavily damaged cells and prevent propagation of nuclear damage to daughter progeny.

References

Bouwstra JA, Honeywell-Nguyen PL, Gooris GS *et al.* (2003) Structure of the skin barrier and its modulation by vesicular formulations. *Progress in Lipid Research* **42**: 1–36.

Brattsand M and Egelrud T (1999) Purification, molecular cloning, and expression of a human stratum corneum trypsin-like serine protease with possible function in desquamation. *Journal of Biological Chemistry* **274**: 30033–30040.

Broome AM, Ryan D and Eckert RL (2003) S100 protein subcellular localization during epidermal differentiation and psoriasis. *Journal of Histochemistry and Cytochemistry* **51**: 675–685.

Candi E, Melino G, Lahm A *et al.* (1998) Transglutaminase 1 mutations in lamellar ichthyosis. Loss of activity due to failure of activation by proteolytic processing. *Journal of Biological Chemistry* **273**: 13693–13702.

Candi E, Melino G, Mei G *et al.* (1995) Biochemical, structural, and transglutaminase substrate properties of human loricrin, the major epidermal cornified cell envelope protein. *Journal of Biological Chemistry* **270**: 26382–26390.

Denecker G, Hoste E, Gilbert B *et al.* (2007) Caspase 14 protects against epidermal UVB photodamage and water loss. *Nature Cell Biology* **9**(6): 666–674.

Egberts F, Heinrich M, Jensen JM *et al.* (2004) Cathepsin D is involved in the regulation of transglutaminase 1 and epidermal differentiation. *Journal of Cell Science* **117**: 2295–2307.

Ekholm IE, Brattsand M and Egelrud T (2000) Stratum corneum tryptic enzyme in normal epidermis: a missing link in the desquamation process? *Journal of Investigation Dermatology* **114**: 56–63.

Fuchs E and Cleveland DW (1998) A structural scaffolding of intermediate filaments in health and disease. *Science* **23**: 514–519.

Hansson L, Strömqvist M, Bäckman A *et al.* (1994) Cloning, expression, and characterization of stratum corneum chymotryptic enzyme: a skin-specific human serine proteinase. *Journal of Biological Chemistry* **269**: 19420–19426.

Hohl D, Mehrel T, Lichti U *et al.* (1991) Characterization of human loricrin. Structure and function of a new class of epidermal cell envelope proteins. *Journal of Biological Chemistry* **266**: 6626–6636.

Kalinin A, Marekov LN and Steinert PM (2001) Assembly of the epidermal cornified cell envelope. *Journal of Cell Science* **114**: 3069–3070.

Kartasova T, Darwiche N, Kohno Y *et al.* (1996) Sequence and expression patterns of mouse SPR1: correlation of expression with epithelial function. *Journal of Investigation Dermatology* **106**: 294–304.

Kim SY and Bae CD (1998) Calpain inhibitors reduce the cornified cell envelope formation by inhibiting proteolytic processing of transglutaminase 1. *Experimental & Molecular Medicine* **30**: 257–262.

Kim IG, Lee SC, Lee JH *et al.* (1994) Structure and organization of the human transglutaminase 3 gene: evolutionary relationship to the transglutaminase family. *Journal of Investigation Dermatology* **103**: 137–142.

Lorand L and Graham RM (2003) Transglutaminases: crosslinking enzymes with pleiotropic functions. *Nature Reviews. Molecular Cell Biology* **4**: 140–156.

Madison KC (2003) Barrier function of the skin: 'la raison d'etre' of the epidermis. *Journal of Investigation Dermatology* **121**: 231–241.

Matsuki M, Yamashita F, Ishida-Yamamoto A *et al.* (1998) Defective stratum corneum and early neonatal death in mice lacking the gene for transglutaminase 1 (keratinocyte transglutaminase). *Proceedings of the National Academy of Sciences of the United States of America* **95**: 1044–1049.

Nicotera P and Melino G (2007) Caspase-14 and epidermis maturation. *Nature Cell Biology* **9**: 621–622.

Porter RM and Lane EB (2003) Phenotypes, genotypes and their contribution to understanding keratin function. *Trends in Genetics* **19**: 278–285.

Rice RH and Green H (1977) The cornified envelope of terminally differentiated human epidermal keratinocytes consists of cross-linked protein. *Cell* **11**: 417–422.

Serre G, Mils V, Haftek M *et al.* (1991) Identification of late differentiation antigens of human cornified epithelia, expressed in re-organized desmosomes and bound to cross-linked envelope. *Journal of Investigation Dermatology* **97**: 1061–1072.

Simon M and Green H (1988) The glutamine residues reactive in transglutaminase-catalyzed cross-linking of involucrin. *Journal of Biological Chemistry* **263**: 18093–18098.

Simon M, Jonca N, Guerrin M *et al.* (2001) Refined characterization of corneodesmosin proteolysis during terminal differentiation of human epidermis and its relationship to desquamation. *Journal of Biological Chemistry* **276**: 20292–20299.

Sondell B, Thornell LE and Egelrud T (1995) Evidence that stratum corneum chymotryptic enzyme is transported to the stratum corneum extracellular space via lamellar bodies. *Journal of Investigation Dermatology* **104**: 819–823.

Steinert PM, Cantieri JS, Teller DC *et al.* (1981) Characterization of a class of cationic proteins that specifically interact with intermediate filaments. *Proceedings of the National Academy of Sciences of the United States of America* **78**: 4097–4101.

Steinert PM and Marekov LN (1995) The proteins elafin, filaggrin, keratin intermediate filaments, loricrin, and small proline-rich proteins 1 and 2 are isodipeptide cross-linked components of the human epidermal cornified cell envelope. *Journal of Biological Chemistry* **270**: 17702–17711.

Steven AC and Steinert PM (1994) Protein composition of cornified cell envelopes of epidermal keratinocytes. *Journal of Cell Science* **107**: 693–700.

Strelkov SV, Hemmann H and Aebi U (2003) Molecular architecture of intermediate filaments. *BioEssays* **25**: 243–251.

Watt FM (2002) Role of integrins in regulating epidermal adhesion, growth and differentiation. *EMBO Journal* **21**: 3919–3926.

Further Reading

Aberdam D, Candi E, Knight RA and Melino G (2008) miRNAs, 'stemness' and skin. *Trends in Biochemical Sciences* **33**(12): 583–591.

Blanpain C and Fuchs E (2009) Epidermal homeostasis: a balancing act of stem cells in the skin. *Nature Reviews. Molecular Cell Biology* **10**(3): 207–217.

Candi E, Schmidt R and Melino G (2005) The cornified envelope: a model of cell death in the skin. *Nature Reviews. Molecular Cell Biology* **6**(4): 328–340. Review.

Elias PM (2005) Stratum corneum defensive functions: an integrated view. *Journal of Investigation Dermatology* **125**(2): 183–200. Review.

Fuchs E (2007) Scratching the surface of skin development. *Nature* **445**(7130): 834–842.

Getsios S, Huen AC and Green KJ (2004) Working out the strength and flexibility of desmosomes. *Nature Reviews. Molecular Cell Biology* **5**(4): 271–281. Review.

Godsel LM, Hobbs RP and Green KJ (2008) Intermediate filament assembly: dynamics to disease. *Trends in Cell Biology* **18**(1): 28–37. Review.

Melino G, De Laurenzi V, Catani MV *et al.* (1998) The cornified envelope: a model of cell death in the skin. *Results and Problems in Cell Differentiation* **24**: 175–212.

Nemes Z and Steinert PM (1999) Bricks and mortar of the epidermal barrier. *Experimental & Molecular Medicine* **31**(1): 5–19. Review.

Nemes Z and Steinert PM (1999) Bricks and mortar of the epidermal barrier. *Experimental & Molecular Medicine* **31**(1): 5–19.

Steinert PM, Steven AC and Roop DR (1985) The molecular biology of intermediate filaments. *Cell* **42**(2): 411–420. Review.

Apoptosis: Inherited Disorders

Helen C Su, *National Institutes of Health, Bethesda, Maryland, USA*

Michael J Lenardo, *National Institutes of Health, Bethesda, Maryland, USA*

This article is dedicated to the memory of Dr Peter M Steinert, in acknowledgement of his immense contribution to skin biology.

Advanced article

Article Contents

- Introduction
- Mechanisms of Apoptosis
- Apoptosis and Autoimmune Disease
- Apoptosis and Immunodeficiency
- Apoptosis and Malignancy
- Acknowledgement

Apoptosis is a mechanism of programmed cell death used in organ remodelling during development and for eliminating unnecessary or dangerous cells. In the immune system, apoptosis maintains homeostasis by limiting lymphocyte responses. Genetic defects in apoptosis leads to disturbed lymphocyte homeostasis and, potentially, autoimmunity or neoplasia. An important apoptosis disorder in humans involves the death receptor Fas. However, recently, disorders in the apoptosis pathways triggered by trophic factor withdrawal or deoxyribonucleic acid (DNA) damage have also been described. Dual roles of certain molecules for both apoptotic and normal nonapoptotic lymphocyte function can account for combined apoptosis and immunodeficiency disorders.

Introduction

It is easy to think about how cells proliferate in the course of growth, differentiation and response to the environment. A less obvious, but equally important, process is the elimination of cells. Cell removal must occur in many instances during the lifetime of multicellular organisms, such as in the formation of spaces between the fingers and toes during fetal development. Apoptosis, a form of programmed cell death, is the term given to an orderly activation of a cellular programme, usually involving specific gene products, that causes the cell to shrink and disperse into small fragments, which are then ingested by nearby phagocytic cells. All multicellular organisms have such death programmes. The gene products involved are homologous from simple worms to mammals, indicating their vital function in evolutionary survival. Apoptosis is distinct from necrosis, the type of cell death that follows infection, injury or damage. Necrosis spills cellular contents into the tissues, inciting an inflammatory response with recruitment of immune cells and mediators. By contrast, apoptosis is believed to induce no inflammatory tissue reaction. Defects in apoptosis in humans have been recognized in the immune system and in the development of cancer. It remains to be determined whether programmes of necrosis also have been a regulatory role. **See also**: Apoptosis: Regulatory Genes and Disease

Mechanisms of Apoptosis

What happens during apoptosis?

On receiving a signal to die by apoptosis, a cell activates an irreversible biochemical programme of suicide. In many cases, the proteins involved in the death programme are preformed so that no new gene expression or protein synthesis is required. The central event of this programme is the activation of one or more members of a special family of proteolytic enzymes called caspases. Active caspases then carry out cleavage of selected substrates to produce the several characteristic biochemical and morphological changes, some of which are shown in **Figure 1**. One of the first recognizable changes is in the structure and permeability of the cell membrane. At the cell surface, proteins and lipids which are normally hidden become exposed. The deoxyribonucleic acid (DNA) inside the cell nucleus becomes fragmented in a characteristic manner, resulting in a set of segments of incremental sizes, visualized as 'DNA ladders' when separated by gel electrophoresis. Mitochondria break down and the cell nucleus becomes fragmented. The cell surface develops blebs that contain cytoplasmic and nuclear material, and these blebs break off to form small cell remnants or apoptotic bodies. The containment of cytosolic contents until the apoptotic bodies have been ingested by local phagocytes prevents injurious inflammation that would occur if these contents were released.

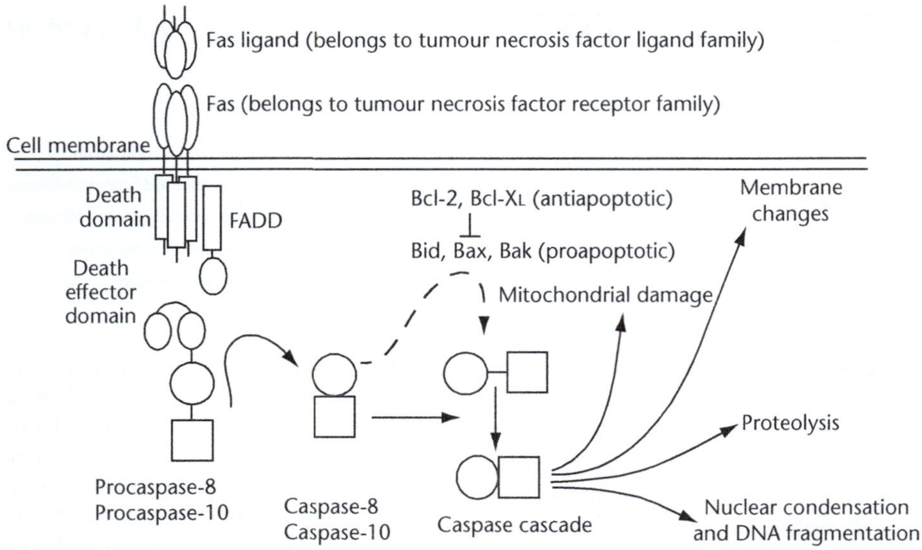

Figure 1 The Fas-mediated apoptosis pathway. Apoptosis is initiated by trimerized Fas ligand (FasL) binding to its extracellular docking site on trimerized Fas. Fas–FasL interaction initiates intracellular recruitment of FADD (Fas-associated protein with death domain) to the trimerized cytoplasmic death domains of Fas, which together with caspase-8 and -10 molecules form a death-inducing signalling complex. Activation of the initiator caspase-8 and -10 leads to a lethal proteolytic cascade, involving other effector caspases and ultimately causing mitochondrial damage, membrane changes, proteolysis, nuclear condensation and chromosomal degradation. Intracellular changes sensed by Bcl-2 family members can also lead to mitochondrial disturbances that activate caspases.

What triggers apoptosis?

Either withdrawal of growth factors or of other mitochondrial disturbances (intrinsic apoptosis) or specific interactions between cell surface apoptosis receptors and their ligands (extrinsic apoptosis) can initiate the cell death programme of apoptosis. For example, activated T lymphocytes require interleukin 2 (IL-2), a growth factor, to survive. Removal or depletion of IL-2 leads to apoptosis. Alternatively, engagement of the T-cell receptor (TCR) or the cell surface receptor Fas with its ligand induces apoptosis even in the presence of IL-2. These triggers are coordinately regulated to bring the immune system back to a certain set point. When there is increased antigen and IL-2 production, leading to a stronger proliferative response, there is also a correspondingly greater apoptotic response to bring it back down. This form of negative regulation in lymphocytes is termed 'propriocidal regulation' (Bidere *et al.*, 2006).

Apoptosis mediators

The complexity and variety of proteins involved in apoptosis reflect the exquisite regulation of this critical process. The Fas (tumour necrosis factor receptor (TNFR) superfamily, member 6 – TNFRSF6, also called CD95 or APO-1) apoptosis pathway of lymphocytes summarized in **Figure 1** represents a general framework to illustrate some of the essential principles of apoptosis. The maintenance of proper lymphocyte homeostasis requires that lymphocyte expansion, resulting from immune responses, be balanced by lymphocyte elimination or death. Lymphocyte apoptosis is an important mechanism for maintaining this balance. There are multiple pathways, both antigen-driven and intrinsic, that induce apoptosis in lymphocytes, many of which use common intracellular molecules including caspases and regulators of mitochondrial stability that comprise the Bcl-2-related gene family such as Bim. One major mediator of extrinsic apoptosis in lymphocytes is the cell surface receptor Fas, which on interaction with its specific ligand, Fas ligand (FasL), induces T-cell apoptosis. This receptor is a member of the TNFR gene superfamily which includes other similar apoptosis-inducing 'death receptors'. These death receptors share a similar cytoplasmic region called the 'death domain' that conveys the signal to the caspase activation machinery. The gene encoding Fas is called TNFRSF6 (also known as CD95, APO-1). Other key genes in this superfamily are TNFR1, also called TNFR superfamily, member 1A

(TNFRSF1A); translocating chain-association membrane protein (TRAMP), also called TNFR superfamily, member 25 (TNFRSF25); TRAIL-R1, also called tumor necrosis factor receptor superfamily, member 10a (TNFRSF10A) and TRAIL-R2, also called tumor necrosis factor receptor superfamily, member 10b (TNFRSF10B). All of these may be classified as death receptors because they harbour cytoplasmic death domains. There is also a corresponding gene family of specific ligands related to TNF and FasL, each of which interacts with one or more specific TNFRs. Many of the members of these ligand and receptor gene superfamilies mediate apoptosis, whereas others inhibit apoptosis and still others have unrelated immune cell regulatory functions. Receptor–ligand interactions for the death receptors can lead to apoptosis rapidly, without the need for new gene transcription or protein synthesis (Bidere *et al.*, 2006). Structurally, Fas is a 45-kDa transmembrane glycoprotein with extracellular, transmembrane and intracellular domains. Its extracellular domain consists of cysteine-rich domains (CRDs) characteristic of the TNFR superfamily that associate three Fas monomers together and allow them to interact with the trimeric ligand, FasL. FasL may be bound to the membrane of another cell or may be secreted. Engagement of Fas with FasL induces activation of the apoptosis programme, beginning with formation of a death-inducing signal complex of molecules that associate with an intracellular Fas domain called the 'death domain'. A component of the complex is the adapter molecule FADD (Fas-associated protein with death domain), the 'Fas-associated death domain' protein, which at one end interacts with Fas through its death domain. At its other end, FADD interacts with proteolytic proenzymes called caspase-8 or caspase-10 through death effector domains (DEDs) found on both FADD and caspase-8 or caspase-10. On association with Fas, FADD recruits caspase-8 and caspase-10, which become activated to cleave downstream caspases from proenzymes to their active forms. Other cytoplasmic caspases then become activated, ultimately causing cellular demise events by the apoptosis programme. In some instances, caspase-8 cleaves Bid, a member of the Bcl-2 family. Cleaved Bid cooperates with other pro-apoptotic Bcl-2 family members such as Bax to permeabilize the mitochondrial outer membrane. This step activates the intrinsic pathway of apoptosis, and in this way amplifies death signals initiated through the extrinsic pathway. The intrinsic pathway of apoptosis can also be initiated by intracellular signals, such as DNA damage or chemical toxicity, which are sensed by Bcl-2 family members, independent of the extrinsic pathway. **See also**: Death Receptors; Death Receptors at the Molecular Level: Therapeutic Implications; Death Receptor-Induced Necroptosis

The physiological function of the Fas receptor has been best elucidated in T lymphocytes (T cells). Resting T cells normally express a low level of surface Fas, which is increased following an encounter with an antigen that engages the T cell's antigen receptor and delivers a competency signal for cell death. During this initial phase of T-cell activation, the low level of Fas molecules expressed does not provide the functional capacity to activate the Fas apoptotic pathway. After activation and a number of cell divisions, T cells express high levels of surface Fas and become sensitive to Fas-induced apoptosis. The ability of activated T cells to proliferate and initiate an immune response is highly dependent on the autocrine interaction between IL-2 and its receptor; thus, IL-2 is an important permissive factor for T-cell apoptosis (Bidere *et al.*, 2006). After antigen re-engagement of an activated T cell, a competency signal from TCR together with engagement of Fas entrains the caspase-activating mechanism to the surface events thereby triggering apoptosis.

Apoptosis and Autoimmune Disease

Mouse studies of autoimmunity

A mouse strain with a tendency to develop enlarged lymph nodes and spleens along with autoimmune diseases such as immune complex glomerulonephritis was serendipitously discovered and extensively studied because of its resemblance to patients with the disease systemic lupus erythematosus (Cohen and Eisenberg, 1991). These mice, called *lpr* and *gld*, were found to have defects in Fas and FasL, respectively (Watanabe-Fukunaga *et al.*, 1992; Takahashi *et al.*, 1994). Although lymphocytes in these mice proliferated normally in response to environmental challenges, they were not expunged at the end of an immune response because of defective Fas-mediated apoptosis, and accumulated abnormally. This caused marked enlargement of the lymphoid tissues. Furthermore, some lymphocytes were able to react to a degree with the murine host tissues, and this self-recognition triggered development of autoimmune disease. Almost all mature T cells express either CD4 or CD8 cofactors along with CD3 receptors for antigen. A hallmark of these mice, however, was the presence of mature

Table 1 Selected mouse models of human autoimmunity

Mouse strain	Genetic/molecular defect	Clinical findings	Human correspondence
lpr/lpr, lpr^{cg}	Fas	Splenomegaly, lymphadenopathy, hypergammaglobulinemia, autoantibodies, DNT expansion and glomerulonephritis	ALPS type 1A
gld/gld	Fas ligand	Splenomegaly, lymphadenopathy, hypergammaglobulinemia, autoantibodies, DNT expansion and glomerulonephritis	SLE, ALPS type 1B
$bim^{-/-}$	Bim	Splenomegaly, lymphadenopathy, hypergammaglobulinemia, autoantibodies, no DNT expansion and glomerulonephritis	Unknown

Note: DNTs, double-negative T cells.

T lymphocytes that expressed CD3, but no longer expressed either CD4 or CD8. These cells became known as double-negative T cells (DNTs) and may contribute to disease manifestations.

Homologous recombination and transgenic technologies can be used in mice to manipulate individual genes in specific tissues. These technologies have facilitated the experimental examination of how Fas regulates the cells that contribute to autoimmunity. Selective ablation of the Fas/Fas ligand genes or interfering with Fas signalling in dendritic-, B- or T cells established that loss of Fas function in any cell subset promotes disease by perturbing normal cellular control mechanisms (Stranges *et al.*, 2007; Chen *et al.*, 2006; Hao *et al.*, 2008; Mabrouk *et al.*, 2008). For example, loss of Fas function in dendritic cells or B cells increases the pool of antigen presenting cells, which stimulate the T helper cells that in turn promote T and B cell proliferation. Loss of Fas function in T cells can also lead to the same end result. These technologies have additionally been used to study the physiological role of Bcl-2 family members, which can act on mitochondria to influence apoptosis. Of the many Bcl-2 family members, only Bim ablation resulted in abnormal T- and B-lymphocyte accumulation and autoimmunity in mice (Bouillet *et al.*, 1999; Fischer *et al.*, 2007). Recent studies using mice deficient in both Bim and Fas have shown that these key molecules for intrinsic and extrinsic pathways cooperatively regulate apoptosis induced by repeated antigenic stimulation of T cells (Snow *et al.*, 2008; Hutcheson *et al.*, 2008; Hughes *et al.*, 2008; Weant *et al.*, 2008). Thus, multiple death mechanisms exist to maintain immune homeostasis and prevent autoimmunity *in vivo*.

See **Table 1** for a summary of mouse models of human autoimmunity.

Humans with genetic defects in apoptosis

Sneller *et al.* (1992) reported two patients with enlarged lymph nodes and spleens, autoimmunity and DNTs, and suggested that the condition closely resembled the *lpr* and *gld* mice. In 1995, a series of patients with this syndrome were demonstrated to have defects in lymphocyte apoptosis due to deleterious mutations in the gene encoding Fas (Fisher *et al.*, 1995; Rieux-Laucat *et al.*, 1995). The human disease is known as autoimmune lymphoproliferative syndrome (ALPS) or, alternatively, as the Canale–Smith syndrome (Drappa *et al.*, 1996) or lymphoproliferative syndrome with autoimmunity. Since these original reports, hundreds of individuals with ALPS have been studied. Their clinical and laboratory features are presented in **Table 2**. To meet the definition of ALPS, a patient must have chronic adenopathy and/or splenomegaly in the early years of life, increased numbers of DNTs and defective lymphocyte apoptosis. Almost all such persons have autoantibodies or overt autoimmune disease at some point, a result of chronic persistence and activation of both T cells that stimulate B-cell maturation and the B cells themselves. The most common autoimmune complications of ALPS are autoimmune haemolytic anaemia, thrombocytopaenia and neutropaenia, cell deficits that are usually due to antibody-mediated destruction (**Table 1**). During infancy and childhood, the lymphoproliferation is often massive, with distortion of normal anatomical relationships (**Figure 2**). Both lymphoproliferation and autoimmunity become less severe through adolescence and adulthood and may regress completely. Treatment of autoimmune episodes with conventional agents such as corticosteroids or corticosteroid-sparing agents such as mycophenolate mofetil is generally effective, whereas

Table 2 Clinical and laboratory findings in patients with autoimmune lymphoproliferative syndrome

Sign	Details	Per cent of patients affected
Chronic accumulation of non-malignant lymphoid cells	Lymphadenopathy	100
	Splenomegaly	87, with 32 requiring splenectomy
	Hepatomegaly	45
Autoimmunity	Autoantibodies in the blood	>50
	Autoimmune haemolytic anaemia	53
	Autoimmune thrombocytopaenia	44
	Autoimmune neutropaenia	31
	Autoimmune diseases of kidney, liver, nervous system or eye	<0 Each
Lymphoma	Hodgkin or non-Hodgkin type	Risk at least 50-fold increased
Abnormally high levels of DNTs lacking CD4 or CD8	In blood or histological specimens	100
Defective apoptosis	Poor apoptosis after *in vitro* activation and exposure to apoptotic stimulus	100
Germline mutation in an apoptosis mediator (*TNFRSF6* encoding Fas)		65

Note: DNTs, double-negative T cells.

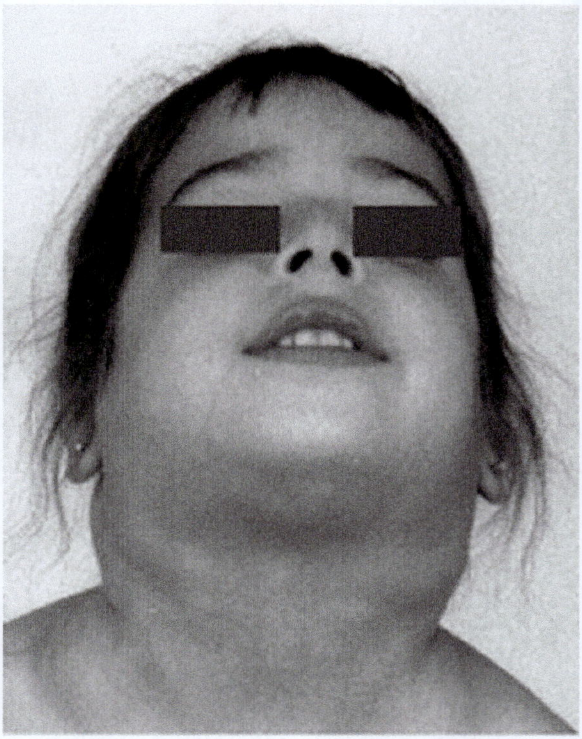

Figure 2 Girl with autoimmune lymphoproliferative syndrome (ALPS), showing massive cervical adenopathy.

treatment for benign lymph node enlargement is only transiently effective and thus not indicated.

In addition to immunological aberrations, the extended survival of lymphocytes due to defective Fas-mediated apoptosis may allow malignant transformation to occur. Fas and other members affecting apoptosis pathways, especially p53, play physiological roles as tumour suppressors. Now that many patients with ALPS and their families have been identified, it is clear that there is an elevated incidence of lymphomas in ALPS (Straus *et al.*, 2001). Individuals with inherited mutations in Fas have a significantly increased risk of Hodgkin (relative risk (RR) = 55) and non-Hodgkin (RR = 14) lymphoma. A variety of different lymphoid malignancies have been reported in patients with ALPS and their mutation-positive family members: T-cell rich B-cell lymphoma, Burkitt lymphoma, atypical lymphoma and nodular lymphocyte predominant Hodgkin disease. This implies that a general antineoplastic mechanism has been impaired within lymphocytes by genetic defects in the Fas pathway. Most lymphoma cases have occurred in families with mutations affecting the death domain of Fas. Interestingly, the propensity for lymphoid malignancy may hinge on the ability of Fas to simultaneously activate nuclear factor-kappa B (NF-κB) and mitogen-activated protein kinase (MAPK) signals for cell growth (Legembre *et al.*, 2004). In ALPS patients, Fas mutations can impair death signals to a greater degree than growth signals, and these effects can cooperatively lead to increased lymphoma risk.

The majority of ALPS patients have heterozygous mutations in the gene encoding Fas, located on chromosome 10q24.1. Some patients have somatic mutations of Fas within DNT cells (Holzelova *et al.*, 2004). A minority of individuals with ALPS does not have defects in Fas. A few have defects in caspase-10 or FasL (Wang *et al.*, 1999; Zhu *et al.*, 2006; Del-Rey *et al.*, 2006; Bi *et al.*, 2007). More than 60 unique Fas mutations have been found in over 70 families with ALPS. The majority of mutations are in exon 9, which encodes the intracellular death domain. Almost all mutations involve changes, deletions or additions of one or a small number of nucleotides, either in the coding regions or the splice sites of the gene (**Figure 3**). Although de novo mutations have been detected in a few families, genetic analysis has revealed autosomal dominant inheritance in most families, although a few patients with a severe, early onset clinical phenotype have mutations in both Fas alleles. Most Fas mutations exert a dominant negative effect. In these cases, the mutant allele leads to production of an abnormal protein that incorporates into Fas trimers at the cell surface (Siegel *et al.*, 2000; Vaishnaw *et al.*, 1999). Next, trimers of normal and abnormal Fas proteins disrupt the death-inducing signal complex, preventing transmission of a death signal. Nonetheless, within families, different individuals carrying the same Fas mutation may have highly variable expression of their disease, suggesting that additional factors besides defective Fas signalling are important in the pathogenesis of ALPS (Jackson *et al.*, 1999).

Nearly all ALPS patients studied have genetic defects that impair the extrinsic apoptosis pathway mediated by Fas. However, one patient was recently identified who

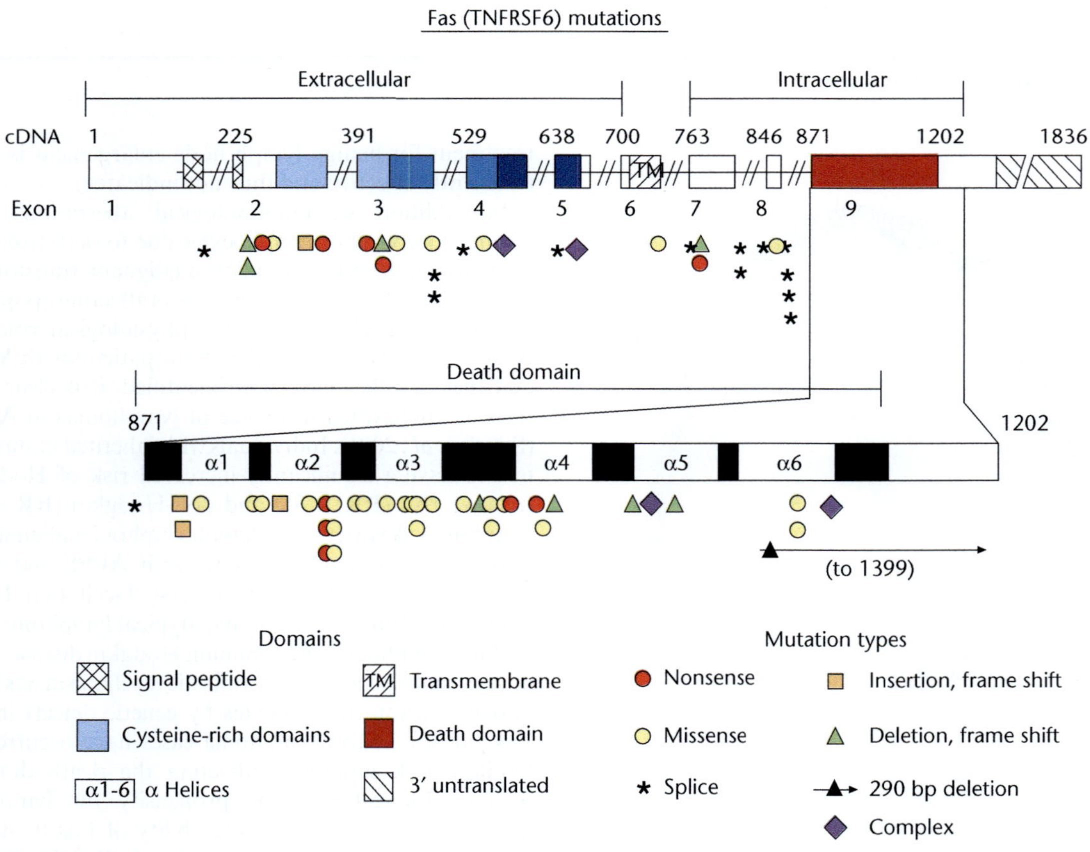

Figure 3 Schematic representation of Fas mutations found in ALPS patients. Most mutations are found in exon 9, within the intracellular death domain of Fas. This causes a dominant interfering effect in which most Fas receptors, which are composed of trimers, have at least one mutant chain. The defective complex is thus unable to recruit the other signalling components necessary for propagating signals for programmed cell death.

had defective cytokine-deprivation apoptosis with no Fas impairment (Oliveira *et al.*, 2007). T cells from the patient exhibited prolonged survival resulting from failure to upregulate the pro-apoptotic molecule Bim after IL-2 removal. Bim upregulation is required for mitochondrial depolarization to trigger intrinsic apoptosis. The Bim suppression occurred because of an upstream activating gain-of-function G13D germline mutation in the small guanosine triphosphatase (GTPase) neuroblastoma Ras (NRAS). Somatic mutations in this gene are associated with cancer, and intriguingly, this patient also had a history of childhood leukaemia and adult lymphoma. Similar defects in intrinsic apoptosis pathways could potentially account for other cases of ALPS with unidentified molecular pathogenesis.

In summary, although ALPS is a rare genetic disease that is not completely understood, the syndrome provides a model for investigating many more common forms of autoimmunity and unbalanced lymphocyte homeostasis in humans.

Apoptosis and Immunodeficiency

Although it may seem at first that immunodeficiency should not coexist with defective apoptosis, two disorders in humans have shown otherwise. The first, caspase-8 deficiency state (CEDS), was discovered in two siblings who had mildly enlarged lymph nodes and spleen, variable DNT cell numbers, autoimmunity and defective lymphocyte apoptosis (Chun *et al.*, 2002). However, their major problems were repeated infections – both sinus and lung infections, as well as herpes simplex virus outbreaks of the skin and mucous membranes – which have not been observed in ALPS. This immunodeficiency occurred because of defective lymphocyte activation stemming from failure to activate the gene transcription factor NF-κB (Su *et al.*, 2005). NF-κB turns on an extensive cellular programme of genes in immune cells for activation and proliferation. In the patients, NF-κB was not induced after stimulating T- or B lymphocytes through antigen receptor, or through certain innate receptors found on B cells and natural killer (NK) cells. This constellation of clinical and biochemical abnormalities resulted from a homozygous caspase-8 loss-of-function mutation. Because of its distinct phenotype, the caspase-8 deficiency moniker serves to differentiate it from classical ALPS.

Recently, a second immunodeficiency, X-linked lymphoproliferative (XLP) disease, was also discovered to have defective apoptosis. XLP, typically caused by mutations in SH2D1A (SAP), is characterized by hypogammaglobulinemia that results from poor T cell–B cell interactions with impaired germinal centre formation. These patients usually die from overwhelming uncontrollable *Epstein–Bar virus* (EBV) infections with concurrent expansion of CD8 T cells. Cell accumulation apparently results because loss of SAP impairs certain T cell receptor signals required for apoptosis that is induced on repeated antigenic stimulation (Snow *et al.*, 2008). This finding explains a key observation pertaining to the pathogenesis of this disease. Thus, like CEDS, XLP reveals that immunodeficiency featuring lymphoproliferation can result from mutations in molecules that normally participate in apoptosis and yet also possess nonapoptotic functions.

Apoptosis and Malignancy

The importance of apoptosis in maintaining the balance between cell survival and cell death is also relevant to the development of malignancy. Many genes encode products that normally act to inhibit cell growth or get rid of unwanted cells; loss-of-function of one or more of these gene products in a cell can lead to unregulated proliferation of that cell, or cancer. Many genes, including those involved in apoptosis, have been found mutated in malignancies, but their relative contribution at an early stage of carcinogenesis is less clear. As discussed earlier in the section on 'Humans with genetic defects in apoptosis', Fas mutations are associated with predisposition to lymphoid malignancy. Germline mutations in other genes, especially p53, but DAPK1 (death-associated protein kinase 1) and phosphatase and tensin homolog (PTEN), have also been associated with development of cancers.

The p53 transcription factor has been recognized as a central modulator of cell growth and activation. Among the genes activated by p53 are p21Waf1/Cip1, which is well known to cause cell cycle arrest, as well as those encoding Fas, Bax, Bid, Noxa and PUMA (p53 upregulated modulator of apoptosis). p53 is a checkpoint protein that sensitizes cells to apoptosis during genotoxic stress, such as DNA damage, when p53 accumulates. However, p53 regulates apoptosis not only by regulating gene transcription in the nucleus, but also by regulating the Bcl-2 family members Bax and Bcl-X_L in the cytoplasm and at the mitochondria (Chipuk *et al.*, 2005). p53 can activate Bax, which permeabilizes mitochondria, but this does not occur when p53 is bound in the cytoplasm to the antiapoptotic molecule Bcl-X_L. It is only after p53 induces PUMA transcription, when newly synthesized PUMA displaces p53 from Bcl-X_L, that the freed p53 becomes able to activate Bax to cause apoptosis. **See also**: BH3-Only Proteins; P53 and Cell Death

Given its role in inducing a large number of cellular regulatory pathways including apoptosis, it is not surprising that p53 is also a tumour-suppressor gene. Loss of the p53 gene (tumour protein p53, *TP53* (Li–Fraumeni syndrome)) has been found to be a characteristic of over half of human tumours, and many tumours also have lost genes whose products stabilize p53, such as ataxia telangiectasia mutated (includes complementation groups A, C and D) (*ATM*). The Fas gene is also missing, and activating NRAS mutations are found, in many tumours. In addition to this indirect evidence connecting loss of apoptosis and cell control with tumourigenesis, homozygous mice with a targeted disruption of the p53 gene all develop fatal tumours.

Humans with defects in p53 also are cancer-prone. The Li–Fraumeni syndrome is an autosomal dominant disorder associated with development of solid tumours such as sarcomas, breast cancer and brain tumours as well as lymphomas. Affected members of Li–Fraumeni syndrome families have a germline mutation in one allele of *TP53*. A somatic mutation in the remaining copy of the gene in a single cell of any tissue can lead to uncontrolled proliferation and malignant transformation of that cell. Individuals may suffer multiple, different malignancies, and the onset of malignancies is at a younger age, on average, than sporadic cases of similar tumours in the general population.

Germline mutations in other genes with well-documented effects on apoptosis have been found to confer cancer susceptibility states. In Cowden syndrome, autosomal dominant germline mutations in PTEN are associated with increased risk of developing various malignancies including breast and thyroid cancers. PTEN normally exerts its pro-apoptotic effects by antagonizing the phosphoinositide 3 kinase (PI3K)/Akt pathway that suppresses cell death. In addition, recently an autosomal dominant DAPK1 mutation that silences expression has been associated with chronic lymphocytic leukaemia (CLL) in a large kindred (Raval *et al.*, 2007). DAPK1 is a member of the calcium/calmodulin kinase family that exerts pro-apoptotic effects through its activation of p53.

Thus, genes for both p53 and Fas, as well as other genes important in maintaining stability in cell numbers through apoptosis, normally inhibit tumour growth in our bodies, and their loss, whether in the germline or occurring in a single somatic cell, is associated with increased and unregulated proliferation. **See also**: Apoptosis: Regulatory Genes and Disease

Acknowledgement

This work was supported by the Division of Intramural Research of NIAID, NIH.

References

Bi LL, Pan G, Atkinson TP *et al.* (2007) Dominant inhibition of Fas ligand-mediated apoptosis due to a heterozygous mutation associated with autoimmune lymphoproliferative syndrome (ALPS) type 1b. *BMC Medical Genetics* **8**: 41–55.

Bidere N, Su HC and Lenardo MJ (2006) Genetic disorders of programmed cell death in the immune system. *Annual Review of Immunology* **24**: 321–352.

Bouillet P, Metcalf D, Huang DC *et al.* (1999) Proapoptotic Bcl-2 relative Bim required for certain responses, leukocyte homeostasis, and to preclude autoimmunity. *Science* **286**: 1735–1738.

Chen M, Wang Y, Wang Y *et al.* (2006) Dendritic cell apoptosis in the maintenance of immune tolerance. *Science* **311**: 1160–1164.

Chipuk JE, Bouchier-Hayes L, Kuwana T *et al.* (2005) PUMA couples the nuclear and cytoplasmic proapoptotic function of p53. *Science* **309**: 1732–1735.

Chun HJ, Zheng L, Ahmad M *et al.* (2002) Pleiotropic lymphocyte activation defects due to caspase-8 mutation cause human immunodeficiency. *Nature* **419**: 395–399.

Cohen PL and Eisenberg RA (1991) *Lpr* and *gld*: single gene models of systemic autoimmunity and lymphoproliferative disease. *Annual Review of Immunology* **9**: 243–269.

Del-Rey M, Ruiz-Contreras J, Bosque A *et al.* (2006) A homozygous Fas ligand gene mutation in a patient causes a new type of autoimmune lymphoproliferative syndrome. *Blood* **108**: 1306–1312.

Drappa J, Vaishnaw AK, Sullivan KE, Chu J-L and Elkon KB (1996) Fas gene mutations in the Canale–Smith syndrome, an inherited lymphoproliferative disorder associated with autoimmunity. *New England Journal of Medicine* **335**: 1643–1649.

Fischer SF, Bouillet P, O'Donnell K *et al.* (2007) Proapoptotic BH3-only protein Bim is essential for developmentally programmed death of germinal center-derived memory B cells and antibody-forming cells. *Blood* **110**: 3978–3984.

Fisher GH, Rosenberg FJ, Straus SE *et al.* (1995) Dominant interfering Fas gene mutations impair apoptosis in a human autoimmune lymphoproliferative syndrome. *Cell* **81**: 935–946.

Hao Z, Duncan GS, Seagal J *et al.* (2008) Fas receptor expression in germinal-center B cells is essential for T and B lymphocyte homeostasis. *Immunity* **29**: 615–627.

Holzelova E, Vonarbourg C, Stolzenberg MC *et al.* (2004) Autoimmune lymphoproliferative syndrome with somatic Fas mutation. *New England Journal of Medicine* **351**: 1409–1418.

Hughes PD, Betz GT, Fortner KA *et al.* (2008) Apoptosis regulators Fas and Bim cooperate in shutdown of chronic immune responses and prevention of autoimmunity. *Immunity* **28**: 197–205.

Hutcheson J, Scatizzi JC, Siddiqui AM *et al.* (2008) Combined deficiency of proapoptotic regulators Bim and Fas results in the early onset of systemic autoimmunity. *Immunity* **28**: 206–217.

Jackson CE, Fischer RE, Hsu AP *et al.* (1999) Autoimmune lymphoproliferative syndrome with defective Fas: genotype influences penetrance. *American Journal of Human Genetics* **64**: 1002–1014.

Legembre P, Barnhart BC, Zheng L *et al.* (2004) Induction of apoptosis and activation of NF-kappaB by CD95 require different signalling thresholds. *EMBO Reports* **5**: 1084–1089.

Mabrouk I, Buart S, Hasmin M *et al.* (2008) Prevention of autoimmunity and control of recall response to exogenous antigen by Fas death receptor ligand expression on T cells. *Immunity* **29**: 1–12.

Oliveira JB, Bidere N, Niemela JE *et al.* (2007) NRAS mutation causes a human autoimmune lymphoproliferative syndrome. *Proceedings of the National Academy of Sciences of the USA* **194**: 8953–8958.

Raval A, Tanner SM, Byrd JC *et al.* (2007) Downregulation of death-associated protein kinase 1 (DAPK1) in chronic lymphocytic leukemia. *Cell* **129**: 879–890.

Rieux-Laucat F, Le Deist F, Hivbroz C *et al.* (1995) Mutations in Fas associated with human lymphoproliferative syndrome and autoimmunity. *Science* **268**: 1347–1349.

Siegel RM, Frederiksen JK, Zacharias DA *et al.* (2000) Fas preassociation required for apoptosis signaling and dominant inhibition by pathogenic mutations. *Science* **288**: 2354–2357.

Sneller MC, Straus SE, Jaffe ES *et al.* (1992) A novel lymphoproliferative/autoimmune syndrome resembling murine *lpr/gld* disease. *Journal of Clinical Investigation* **90**: 334–341.

Snow AL, Oliveira JB, Zheng L *et al.* (2008) Critical role for Bim in T cell receptor restimulation-induced death. *Biology Direct* **3**: 34–51.

Stranges PB, Watson J, Cooper CJ *et al.* (2007) Elimination of antigen-presenting cells and autoreactive T cells by Fas contributes to prevention of autoimmunity. *Immunity* **26**: 629–641.

Straus SE, Jaffe ES, Puck JM *et al.* (2001) The development of lymphomas in families with autoimmune lymphoproliferative syndrome with germ line Fas mutations and defective lymphocyte apoptosis. *Blood* **98**: 194–200.

Su HC, Bidere N, Zheng L *et al.* (2005) Requirement for caspase-8 in NF-kappaB activation by antigen receptor. *Science* **307**: 1465–1468.

Takahashi T, Tanaka M, Brannan CI *et al.* (1994) Generalized lymphoproliferative disease in mice, caused by a point mutation in the Fas ligand. *Cell* **76**: 969–976.

Vaishnaw AK, Orlinick JR, Chu J-L *et al.* (1999) The molecular basis for apoptotic defects in patients with CD95 (Fas/APO-1) mutations. *Journal of Clinical Investigation* **103**: 355–363.

Wang J, Zheng L, Lobito A *et al.* (1999) Inherited human caspase 10 mutations underlie defective lymphocyte and dendritic cell apoptosis in autoimmune lymphoproliferative syndrome type II. *Cell* **98**: 47–58.

Watanabe-Fukunaga R, Brannan CI, Copeland NG *et al.* (1992) Lymphoproliferative disorder in mice explained by defects in Fas antigen that mediates apoptosis. *Nature* **356**: 314–317.

Weant A, Michalek RD, Khan IU *et al.* (2008) Apoptosis regulators Bim and Fas function concurrently to control autoimmunity and CD8+ T cell contraction. *Immunity* **28**: 218–230.

Zhu S, Hsu AP, Vacek MM *et al.* (2006) Genetic alterations in caspase-10 may be causative or protective in autoimmune lymphoproliferative syndrome. *Human Genetics* **119**: 284–294.

Further Reading

Kroemer G, Gulluzzi L, Vandenabeele P *et al.* (2009) Classification of cell death: recommendations of the Nomenclature Committee on Cell Death 2009. *Cell Death and Differentiation* **16**: 3–11.

National Institute of Allergy and Infectious Diseases (2008) *Autoimmune Lymphoproliferative Syndrome* http://www3.niaid.nih.gov/topics/ALPS/.

OMIM *Ataxia telangiectasia Mutated (Includes Complementation Groups A, C and D) (ATM)*; MIM number: 208900. http://www.ncbi.nlm.nih.gov/entrez/dispomim.cgi?id=208900.

OMIM *Caspase 8 (CASP8)*; MIM number 601763. http://www.ncbi.nlm.nih.gov/entrez/dispomim.cgi?id=601763.

OMIM *Caspase 10 (CASP10)*; MIM number 601762. http://www.ncbi.nlm.nih.gov/entrez/dispomim.cgi?id=601762.

OMIM *Tumor Necrosis Factor Ligand Superfamily, Member 6 (TNFSF6)*; MIM number 134638. http://www.ncbi.nlm.nih.gov/entrez/dispomim.cgi?id=134638.

OMIM *Tumor Necrosis Factor Receptor Superfamily, Member 6 (TNFRSF6)*; MIM number: 134637. http://www.ncbi.nlm.nih.gov/entrez/dispomim.cgi?id=134637.

OMIM *Tumor protein p53 (Li-Fraumeni syndrome) (TP53)*; MIM number: 191170. http://www.ncbi.nlm.nih.gov/entrez/dispomim.cgi?id=191170.

From Reactive Oxygen and Nitrogen Species to Therapy

Scott R McKercher, *Burnham Institute for Medical Research, La Jolla, California, USA*

Tomohiro Nakamura, *Burnham Institute for Medical Research, La Jolla, California, USA*

Stuart A Lipton, *Burnham Institute for Medical Research, La Jolla, California, USA*

Advanced article

Article Contents

- Introduction
- Generation of ROS/RNS by Ca^{2+} Influx through NMDA Receptor Channels in Response to Glutamatergic Signalling
- Nitrosative Stress, Protein Misfolding, Synaptic Injury and Neuronal Cell Death
- Protein Misfolding and Aggregation in Neurodegenerative Diseases
- *S*-nitrosylation of Parkin and the UPS
- *S*-nitrosylation of PDI Mediates Protein Misfolding and Neurotoxicity in Cell Models of PD and AD
- Potential Treatments for Protecting Neurons from Free Radicals and Oxidative/Nitrosative Stress
- Conclusions
- Acknowledgements

Excessive free radicals, including reactive oxygen species (ROS) and nitric oxide (NO), contribute to pathological production of misfolded proteins, synaptic damage and apoptosis. Misfolded protein aggregates occur in several chronic neurodegenerative disorders, including Parkinson and Alzheimer diseases. In rare cases, the cause is a genetic mutation; the majority is sporadic and may be a response to environmental factors that generate free radicals. Overactivity of the *N*-methyl-D-aspartate (NMDA)-subtype of glutamate receptor can generate ROS and NO species that mediate protein misfolding, synaptic damage and apoptosis, and thus, neurodegenerative disease. We review current evidence that excessive ROS and NO contribute to protein misfolding by *S*-nitrosylation of the E3 ubiquitin ligase parkin and protein-disulfide isomerase. We also discuss NMDA receptor antagonists memantine and NitroMemantine, drugs that block excessive glutamate excitation, thereby limiting production of ROS and NO, and therapeutic electrophiles, drugs that protect cells from oxidative stress by activating the Nrf2 transcriptional pathway.

Introduction

In many neurodegenerative diseases, a prominent feature is excessive generation of reactive oxygen species (ROS) and reactive nitrogen species (RNS), including nitric oxide (NO), which can contribute to neuronal cell injury and death (Beal, 2001; Lipton, 2006; Lipton and Rosenberg, 1994). *N*-methyl-D-aspartate (NMDA)-type glutamate receptors have been linked to ROS and NO production in the nervous system. Overactivation of NMDA receptors causes excessive influx of Ca^{2+} ions, which generates ROS and activates neuronal NO synthase (nNOS) (Abu-Soud and Stuehr, 1993; Dawson *et al.*, 1991). Reaction of the NO group with critical cysteine thiols of target proteins results in the formation of an *S*-nitrosoprotein (SNO-P) and can thus regulate protein function (Lipton *et al.*, 1993). This process, called *S*-nitrosylation, was first described by Lipton and Stamler, and can mediate either protective or neurooxic effects depending on the action of the protein target. Additionally, free radicals, principally ROS, are formed during normal mitochondrial respiration as well as during excessive calcium influx. One species of ROS, superoxide anion ($O_2^{\bullet -}$), reacts rapidly with free radical NO to form the very toxic product peroxynitrite ($ONOO^-$) (Lipton *et al.*, 1993; **Figure 1**).

Misfolded proteins that can form aggregates are found in many neurodegenerative diseases, and soluble oligomers of these aberrantly folded proteins are thought to adversely affect cell function by interfering with normal cellular processes or initiating cell death signalling pathways. As examples, α-synuclein and synphilin-1 in Parkinson disease (PD), and amyloid-β (Aβ) and tau in Alzheimer disease (AD) form toxic aggregates that consist of oligomers of nonnative secondary structures. The formation of larger aggregates may be an attempt of the cell to wall off these toxic

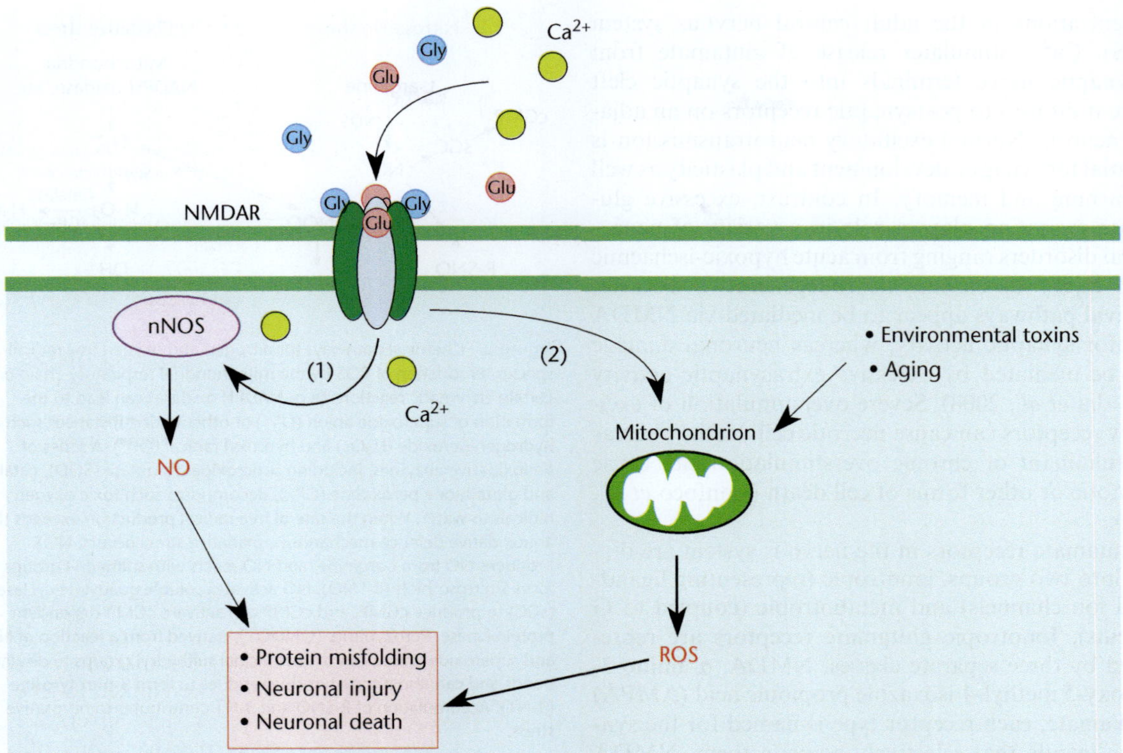

Figure 1 Excess activation of the NMDA receptor (NMDAR) by glutamate (Glu) and glycine (Gly) induces Ca^{2+} influx and activates excitotoxic pathways. NMDAR hyperactivation triggers (1) generation of NO by neuronal NO synthase (nNOS) and (2) production and release of ROS from mitochondria. Additionally, generation of mitochondrial ROS occurs on environmental toxin stimuli or during the normal ageing process. Excessively produced NO and ROS contribute to protein misfolding, neuronal cell injury and death.

proteins. Protein aggregation is also a signature of Huntington disease (a polyQ disorder), amyotrophic lateral sclerosis (ALS) and prion disease (Ciechanover and Brundin, 2003). These protein aggregation disorders are also termed 'conformational diseases' (Kopito and Ron, 2000).

Sporadic forms of neurodegenerative diseases, rather than single gene mutation, constitute the vast majority of cases and pathologic protein aggregation in these diseases may be the result of posttranslational changes to the protein engendered by nitrosative and/or oxidative stress which can mimic the more rare genetic variants of the disease (Yao *et al.*, 2004; Chung *et al.*, 2004). Here we focus on specific examples that show the critical roles of *S*-nitrosylation of ubiquitin E3 ligases, for example, parkin and endoplasmic reticulum (ER) chaperones, such as protein-disulfide isomerase (PDI), in accumulation of misfolded proteins in neurodegenerative diseases (Chung *et al.*, 2004; Uehara *et al.*, 2006; Yao *et al.*, 2004). We also discuss therapeutic applications of the NMDA receptor open-channel blockers memantine and its newer NO-derivatives for preventing excess ROS and NO production. Additionally, we discuss the use of a class of electrophiles for neuroprotection from neurodegenerative disorders (Lipton, 2006, 2007).

Generation of ROS/RNS by Ca^{2+} Influx through NMDA Receptor Channels in Response to Glutamatergic Signalling

The major excitatory neurotransmitter in the brain is the amino acid glutamate, which is present at high concentrations in the adult central nervous system (CNS). Ca^{2+} stimulates release of glutamate from presynaptic nerve terminals into the synaptic cleft where it diffuses to postsynaptic receptors on an adjacent neuron. Normal excitatory neurotransmission is essential for synaptic development and plasticity as well as learning and memory. In contrast, excessive glutamate excitation plays a role in a variety of neurological disorders ranging from acute hypoxic-ischaemic brain injury to chronic neurodegenerative diseases. Survival pathways appear to be mediated via NMDA receptor synaptic activity, whereas neuronal damage may be mediated by excessive extrasynaptic activity (Papadia *et al.*, 2008). Severe overstimulation of excitatory receptors can cause necrotic cell death, whereas less fulminant or chronic overstimulation can cause apoptotic or other forms of cell death (Bonfoco *et al.*, 1995).

Glutamate receptors in the nervous system are divided into two groups, ionotropic (representing ligand-gated ion channels) and metabotropic (coupled to G proteins). Ionotropic glutamate receptors are represented by three separate classes, NMDA, α-amino-3-hydroxy-5 methyl-4-isoxazole propionic acid (AMPA) and kainate; each receptor type is named for the synthetic ligands that selectively activate them. NMDA receptors, unlike most other types, are highly permeable to Ca^{2+}. Depolarization relieves blockade of NMDA receptor-coupled ion channels by Mg^{2+} (Mayer *et al.*, 1984). Ca^{2+} influx promotes many normal intracellular signalling pathways, but excessive influx causes pathological signalling, contributing to cell injury and death via production of free radicals such as ROS and NO as well as other enzymatic processes (Bonfoco *et al.*, 1995; Dawson *et al.*, 1991; Lipton *et al.*, 1993; Lipton and Rosenberg, 1994; **Figure 2**), although the exact mechanisms are not fully understood. 'Excitotoxicity' is defined as neuronal damage caused by excessive activation of glutamate receptors (Olney, 1969) and is at least partly mediated by excessive Ca^{2+} influx through the ion channels activated by NMDA-type receptors (Lipton, 2006; Lipton and Rosenberg, 1994).

A connection between ROS/RNS and mitochondrial dysfunction in neurodegenerative diseases, especially in PD, has recently been postulated (Beal, 2001; Betarbet *et al.*, 2000). Mitochondrial complex I (NADH:ubiquinone oxidoreductase) is particularly sensitive to inhibition from pesticides and other environmental toxins. This complex consists of 43 subunits and 8 iron–sulfur clusters; all electrons pass through the N2 iron–sulfur cluster, which is connected to ubiquinone by a single subunit. Potent complex I inhibitors, representing considerable structural diversity, bind at this point of the complex; thus, the binding site is very promiscuous. These facts may account for the high sensitivity of complex I to pesticide-induced malfunction. Consequent oxidative and nitrosative stress can contribute to aberrant protein accumulation (Chung *et al.*, 2004; Uehara *et al.*, 2006; Yao *et al.*, 2004). Many features of sporadic PD, such as degeneration of dopaminergic neurons, overproduction and aggregation of α-synuclein, accumulation of Lewy body-like intraneuronal inclusions, and impairment of behavioural functions can be reproduced in animal models by treatment with complex I inhibitors, such as 1-methyl-4-phenyl-1,2,3,6-tetrahydropyridine (MPTP), 6-hydroxydopamine, rotenone and paraquat, which result in overproduction of ROS/RNS (Beal, 2001; Betarbet *et al.*, 2000). Studies such as these strongly suggest a relationship between ROS/RNS and protein misfolding. Each may be the pathogenic trigger for the other in neurodegenerative diseases, but the mechanism has not yet been determined.

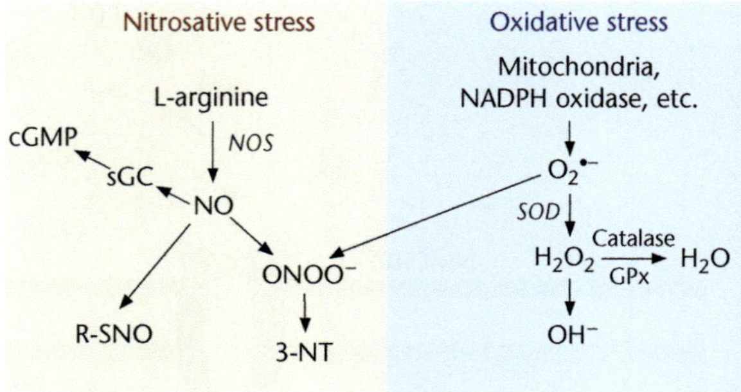

Figure 2 Chemical pathways for nitrogen and oxygen free radical species. Production of ROS by the mitochondrial respiratory chain or certain enzymatic reactions (e.g. NADPH oxidase) can lead to the formation of superoxide anion ($O_2^{\bullet-}$) or other toxic substances such as hydrogen peroxide (H_2O_2) and hydroxyl radical ($OH^{\bullet}$). A series of antioxidative enzymes, including superoxide dismutase (SOD), catalase and glutathione peroxidase (GPx), decomposes such toxic oxygen radicals to water. When the rate of free radical production exceeds the antioxidative defence mechanisms, oxidative stress occurs. NOS produces NO from L-arginine, and NO reacts with sulfhydryl groups to form *S*-nitrosothiols (R-SNO). NO activates soluble guanylate cyclase (sGC) to produce cGMP, and cGMP can activate cGMP-dependent protein kinase. Peroxynitrite ($ONOO^-$), derived from a reaction of NO and superoxide anion, can oxidize vicinal sulfhydryl groups to disulfide bonds and can also nitrate tyrosine residues to form 3-nitrotyrosine (3-NT). Accumulation of R-SNO and 3-NT contributes to nitrosative stress.

Alternatively, an accumulation of ROS can trigger caspase activation resulting in synaptic damage and apoptosis (Shibata *et al.*, 2006). This process can be exacerbated by the action of endogenous 'neurotoxic' electrophiles such as 15d-PGJ2 and catecholamine metabolites (including dopamine) (Spencer *et al.*, 1994). Such electrophiles compromise the reductive capacity of the cell by binding reduced cysteine residues, such as in glutathione (GSH), through a reaction called *S*-alkylation. **See also**: Apoptosis: Regulatory Genes and Disease

Nitrosative Stress, Protein Misfolding, Synaptic Injury and Neuronal Cell Death

NO participates in cellular signalling pathways that regulate broad aspects of brain function, including synaptic plasticity, normal development and neuronal cell death (Dawson *et al.*, 1991). These effects were thought to be largely achieved by activation of guanylate cyclase to form cyclic guanosine-3′,5′-monophosphate (cGMP), but emerging evidence suggests that a more prominent reaction of NO is *S*-nitros(yl)ation of regulatory protein thiol groups (Lipton *et al.*, 1993; Stamler *et al.*, 2001). *S*-nitrosylation is the covalent addition of an NO group to a cysteine thiol/sulfhydryl (RSH or, more properly, thiolate anion, RS^-) to form an *S*-nitrosothiol derivative (R-SNO). Such regulatory modifications are broadly found in mammalian, plant and microbial proteins. A consensus motif of nucleophilic residues (generally an acid and a base) surround a critical cysteine, increasing the susceptibility of the sulfhydryl to *S*-nitrosylation. This process is counterbalanced by denitrosylation by means of thioredoxin/thioredoxin reductase, protein-disulfide isomerase (PDI), intracellular glutathione and other mechanisms. We first identified the physiological relevance of the redox-based mechanisms by which NO and related RNS exert paradoxical effects (Lipton *et al.*, 1993). NO is neuroprotective through *S*-nitrosylation of NMDA receptors and caspases, yet is neurodestructive through formation of peroxynitrite or *S*-nitrosylation of matrix metalloproteinase (MMP)-9, glyceraldehydes-3-phosphate dehydrogenase (GAPDH), and other targets (Gu *et al.*, 2002; Hara *et al.*, 2005; Lipton *et al.*, 1993; **Figure 1** and **Figure 3**). For example, our recent work has shown that *S*-nitrosylation of dynamin-related protein 1 (Drp1) causes excessive mitochondrial fragmentation and occurs in Alzheimer brains. The mitochondrial fragmentation compromises the bioenergetics of electron transport, leading to ATP depletion and consequent synaptic damage in AD (Cho *et al.*, 2009).

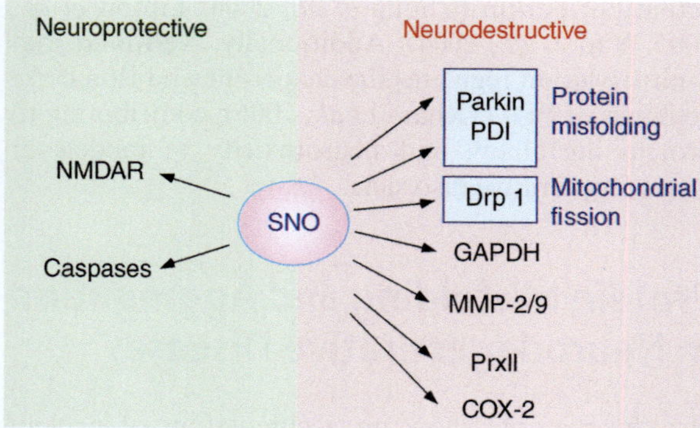

Figure 3 *S*-nitrosylation of neuronal proteins. Physiological levels of NO mediate neuroprotective effects, at least in part, by *S*-nitrosylating the NMDAR and caspases, thus inhibiting their activity. In contrast, overproduction of NO can be neurotoxic via *S*-nitrosylation of Parkin, PDI, GAPDH, MMP-2/9, PrxII and COX-2. *S*-nitrosylated parkin and PDI contribute to neuronal cell injury by triggering accumulation of misfolded proteins. *S*-nitrosylation of Drp1 causes excessive mitochondrial fragmentation in neurodegenerative conditions.

Emerging evidence suggests that *S*-nitrosylation is analogous to phosphorylation in regulating the biological activity of many proteins (Chung *et al.*, 2004; Gu *et al.*, 2002; Hara *et al.*, 2005; Lipton *et al.*, 1993; Stamler *et al.*, 2001; Uehara *et al.*, 2006; Yao *et al.*, 2004). However, the chemistry of NO is much more complex. NO is often a good 'leaving group', resulting in further oxidation of the thiol to a disulfide bond between neighbouring (vicinal) cysteine residues. Alternatively, as NO 'leaves' for another reaction partner, the resulting thiol can react with ROS to yield sulfenic (-SOH), sulfinic ($-SO_2H$) or sulfonic ($-SO_3H$) acid derivatives of the protein (Gu *et al.*, 2002; Uehara *et al.*, 2006; Yao *et al.*, 2004). *S*-nitrosylation may possibly produce a nitroxyl disulfide, in which the NO group is shared by close cysteine thiols (Houk *et al.*, 2003).

What is the consequence of these many oxidative/nitrosative reactions? Recent evidence suggests that NO-related species may play a significant role in the process of protein misfolding. Increased nitrosative and oxidative stress are associated with chaperone and proteasomal dysfunction, resulting in accumulation of misfolded aggregates (Isaacs *et al.*, 2006; Zhang and Kaufman, 2006). However, until recently little was known regarding the molecular and pathogenic mechanisms underlying contributions of NO to the formation of aggregates such as amyloid plaques in AD or Lewy bodies in PD. We and our colleagues recently presented physiological and chemical evidence that *S*-nitrosylation modulates the ubiquitin E3 ligase activity of parkin (Chung *et al.*, 2004; Lipton *et al.*, 2005; Yao *et al.*, 2004). Additionally, we found that *S*-nitrosylation regulates the chaperone and isomerase activities of PDI (Uehara *et al.*, 2006), contributing to protein misfolding and neurotoxicity in models of neurodegenerative disorders.

Protein Misfolding and Aggregation in Neurodegenerative Diseases

Healthy neurons show no accumulation of protein aggregates, indicating that the appearance of such structures is a response to pathological stresses. Considerable evidence suggests that misfolded or otherwise abnormal proteins are produced even in healthy cells. The difference can largely be accounted for by cellular mechanisms for quality control, such as molecular chaperones, the ubiquitin-proteasome system (UPS) and autophagy/lysosomal degradation. A reduction in molecular chaperone or proteasome activities can result in deposition and accumulation of aberrant proteins either within or outside of cells in the brain under pathological conditions. Several mutations in molecular chaperones or UPS-associated enzymes are known to contribute to neurodegeneration. For example, a reduction in proteasome activity was found in the substantia nigra of PD patients (McNaught *et al.*, 2004), and overexpression of the molecular chaperone HSP70 prevented neurodegeneration *in vivo* in models of PD (Auluck *et al.*, 2002). Aggregated proteins were first considered to be pathogenic. However, recent evidence suggests that macroscopic aggregates are an attempt by the cell to sequester aberrant proteins, whereas soluble (micro-) oligomers of such proteins are the most toxic forms (Arrasate *et al.*, 2004).

S-nitrosylation of Parkin and the UPS

Studies of rare mutations have revealed key components of the mechanism for protein aggregation and pathology in PD, including sporadic forms of the disease. Such studies revealed that mutated α-synuclein is a major constituent of Lewy bodies in PD patient brains, and that mutant forms of the ubiquitin E3 ligase parkin or the ubiquitin *C*-terminal hydrolase UCH-L1 (a deubiquinating enzyme) may result in UPS dysfunction and also result in hereditary forms of PD. Attachment of polyubiquitin chains to proteins in mammalian cells directs them to the UPS for degradation. E3 ubiquitin ligases are primarily responsible for specific substrate recognition (Ross and Pickart, 2004). The activating and conjugating enzymes, E1 and E2, respectively, catalyse attachment of the ubiquitin chain to the protein.

Parkin is a member of a large family of E3 ubiquitin ligases. Parkin contains a total of 35 cysteine residues, many of which coordinate a structurally important zinc atom involved in catalysis (Marin and Ferrus, 2002). Parkin recruits substrate proteins as well as an E2 enzyme (e.g. UbcH7, UbcH8 or UbcH13). Mutations in both alleles of the parkin gene will cause dysfunction in its activity, although not all mutations result in loss of parkin E3 ligase activity. Parkin can also mono-ubiquitinate Eps15, HSP70 and itself, possibly at multiple sites. These activities may explain why some parkin mutations result in the formation of Lewy bodies while others do not. Synphilin-1 (α-synuclein-interacting protein) is a substrate for parkin ubiquitination, and is found in Lewy body-like inclusions in cultured cells when coexpressed with α-synuclein. Accumulation of these proteins portends a poor prognosis for the survival of dopaminergic neurons in familial PD and possibly also in sporadic PD.

PD is the second most prevalent neurodegenerative disease and is characterized by the progressive loss of dopamine neurons in the substantia nigra pars compacta. Aberrant protein accumulation is observed in patients with genetically encoded mutant proteins, and recent evidence from our laboratory suggests that nitrosative and oxidative stress are potential causal factors for protein accumulation in the much more common sporadic form of PD. Nitrosative/oxidative stress can mimic PD by promoting protein misfolding in the absence of a genetic mutation (Chung *et al.*, 2004; Lipton *et al.*, 2005; Yao *et al.*, 2004). In fact, *S*-nitrosylation and further oxidation of parkin or UCH-L1 result in dysfunctional enzymes and disruption of UPS function (Chung *et al.*, 2004; Yao *et al.*, 2004; **Figure 4**). Nitrosative stress produces *S*-nitrosylation of parkin (forming SNO-parkin) in rodent models of PD and in brains of human patients with PD and the related α-synucleinopathy, DLBD (diffuse Lewy body disease). Initially, *S*-nitrosylation of parkin stimulates its ubiquitin E3 ligase activity, which can contribute to Lewy body formation. Subsequently, we found that the E3 ligase activity of SNO-parkin decreases, creating a futile cycle (Lipton *et al.*, 2005; Yao *et al.*, 2004). Such *S*-nitrosylation of parkin on critical cysteine residues also compromises its neuroprotective activity (Chung *et al.*, 2004; Yao *et al.*, 2004). It is likely that the enzymatic functions of similar ubiquitin E3 ligases are

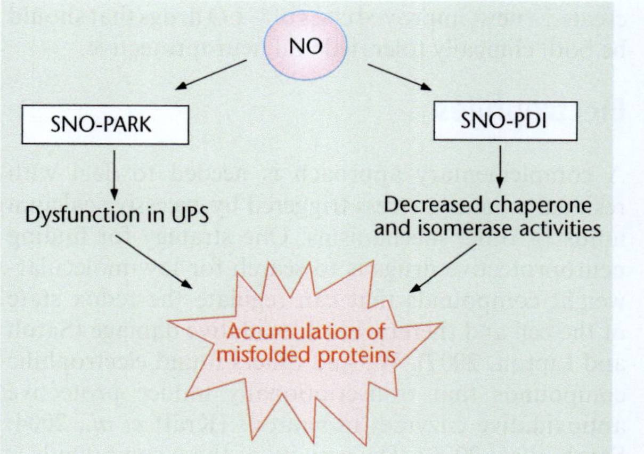

Figure 4 Possible mechanism whereby *S*-nitrosylated species contribute to the accumulation of aberrant proteins and neuronal damage. *S*-nitrosylation of parkin (forming SNO-PARK) and PDI (forming SNO-PDI) can contribute to neuronal cell injury in part by triggering accumulation of misfolded proteins. Specifically, *S*-nitrosylation of parkin disturbs its E3 ligase activity, thus inducing dysfunction in the ubiquitin proteasome system (UPS) (*left*). *S*-nitrosylated PDI decreases its molecular chaperone and disulfide isomerase activity (*right*).

S-nitrosylated, suggesting that this may be involved in a number of degenerative disorders.

S-nitrosylation of PDI Mediates Protein Misfolding and Neurotoxicity in Cell Models of PD and AD

During protein folding in the ER, PDI can introduce disulfide bonds into proteins (oxidation), break disulfide bonds (reduction) and catalyse thiol/disulfide exchange (isomerization), thus facilitating disulfide bond formation, rearrangement reactions and structural stability (Lyles and Gilbert, 1991). During oxidation of a target protein, oxidized PDI catalyses disulfide formation in the substrate protein, resulting in the reduction of PDI. In contrast, the reduced form of the active-site cysteines can initiate isomerization by attacking the disulfide of a substrate protein and forming a transient intermolecular disulfide bond. As a consequence, an intramolecular disulfide rearrangement occurs within the substrate itself, resulting in the generation of reduced PDI. Several mammalian PDI homologues, such as ERp57 and PDIp, also localize to the ER and may manifest similar functions (Conn *et al.*, 2004). Increased expression of PDIp in neuronal cells under conditions mimicking PD suggest the possible contribution of PDIp to neuronal survival (Conn *et al.*, 2004). In many neurodegenerative disorders and cerebral ischaemia, the accumulation of immature and denatured proteins results in ER dysfunction (Conn *et al.*, 2004), but upregulation of PDI represents an adaptive response promoting protein refolding and may offer neuronal cell protection (Conn *et al.*, 2004). We recently reported that the *S*-nitrosylation of PDI (to form SNO-PDI) disrupts this neuroprotective role (Uehara *et al.*, 2006).

The redox environment of the ER can influence the stability of protein *S*-nitrosylation and oxidation reactions (Forrester *et al.*, 2006). The ER normally has a more positive redox potential than that of the cytosol and mitochondria. Excessive NO can create ER stress by disruption of Ca^{2+} homeostasis, affecting the function of many chaperone proteins. One possible mechanism is the increased activity of the ER Ca^{2+} channel-ryanodine receptor through *S*-nitrosylation (Xu *et al.*, 1998). Excessive NO, as well as rotenone exposure, can *S*-nitrosylate the active-site thiols of PDI, inhibiting its isomerase and chaperone activities (Uehara *et al.*, 2006; **Figure 4**). *S*-nitrosylation of PDI prevented its attenuation of neuronal cell death triggered by ER stress, misfolded proteins or proteasome inhibition. Also, PDI was *S*-nitrosylated in the brains of virtually all cases we examined of sporadic AD and PD. These results suggest that NO may mediate cell death or injury via *S*-nitrosylation of PDI and consequent protein misfolding. Similarly, *S*-nitrosylation reactions on a variety of proteins can contribute to cell injury or death.

The activity of the UPS and molecular chaperones normally decline with age (Paz Gavilan *et al.*, 2006). Since we have not found detectable levels of SNO-parkin and SNO-PDI in normal aged brain but only in disease states (Chung *et al.*, 2004; Uehara *et al.*, 2006; Yao *et al.*, 2004), it is likely that *S*-nitrosylation of these and similar proteins is a key mechanism contributing to neurodegenerative conditions.

Potential Treatments for Protecting Neurons from Free Radicals and Oxidative/Nitrosative Stress

Memantine and derivatives

Neuronal cell injury and death induced by glutamate is mediated at least in part by oxidative and nitrosative stress, and several antioxidant molecules have been reported to protect neurons against such assaults

(Murphy *et al.*, 1990). Inhibition of NMDA receptors can block excessive Ca^{2+} influx to prevent ROS and RNS formation, but inhibiting physiological activity of these receptors causes unacceptable side effects. However, we have demonstrated that the adamantane derivative, memantine, preferentially blocks excessive (pathological) NMDA receptor activity while allowing normal (physiological) activity to proceed. Memantine effectively blocks only excessively open channels through an uncompetitive mechanism in conjunction with a relatively fast off-rate and a low affinity for the NMDA receptor. An uncompetitive antagonist acts allosterically at a site other than the agonist-binding site, and its action is contingent on prior activation of the receptor by the agonist. Thus, the same amount of memantine will block higher concentrations of agonist to a relatively greater degree than lower concentrations of agonist, providing greater protection when more glutamate is present. Thus, the critical features of memantine's mode of action are its *u*ncompetitive mechanism and *f*ast *o*ff-rate, or what we call a UFO drug – a drug that is present at its site of inhibitory action only when you need it and then quickly disappears (**Figure 5**). Many studies *in vitro* and in animal models of stroke and neurodegenerative disease showed that memantine protects neurons from NMDA receptor-mediated excitotoxic damage (Chen *et al.*, 1992). In fact, memantine has been approved for human use by the FDA for moderate-to-severe AD and is currently being studied for other neurodegenerative disorders, including human immunodeficiency virus (HIV)-associated dementia and Huntington disease.

Therapeutic results with memantine are meaningful, but as with most first generation drugs, improvements are possible through creation of derivatives. The reaction of NO with the sulfhydryl groups of critical cysteine residues of the NMDA receptor, especially C399 on the external domain of the NR2A subunit, downregulates (but does not completely shut off) receptor activity (Lipton, 2006, 2007). Therefore, we have developed combinatorial drugs called NitroMemantines that use memantine to target NO to the nitrosylation sites of the NMDA receptor to avoid the systemic side effects of NO, whereas still utilizing the uncompetitive channel blocking properties of memantine. Preliminary studies show NitroMemantines to be a more effective neuroprotectant *in vitro* and in animal models than memantine and at lower concentrations (Lipton, 2006). Though much research remains to be done on these second generation NitroMemantine drugs, the combination of memantine with NO has created a new, improved class of UFO drugs that should be both clinically tolerated and neuroprotective.

Electrophiles

A complementary approach is needed to deal with residual oxidative stress triggered by excessive calcium influx or other mechanisms. One strategy for finding neuroprotective drugs is to search for low-molecular-weight compounds that can regulate the redox state of the cell and thereby block oxidative damage (Satoh and Lipton, 2007). We and others found electrophilic compounds that transcriptionally induce protective antioxidative enzymes in neurons (Kraft *et al.*, 2004; Satoh *et al.*, 2006). The activity of these compounds is attained by inducing the transcription of so-called 'phase 2 enzymes', such as haeme oxygenase-1 (HO-1), that regulate the intracellular redox state of the cell (Itoh *et al.*, 2004; Satoh and Lipton, 2007). The Keap1/Nrf2 transcriptional pathway is key in regulating the activity of phase 2 enzymes. Keap1 facilitates ubiquitination of Nrf2 until electrophiles reacting with a critical cysteine on Keap1 block this activity, allowing Nrf2 to translocate to the nucleus where it binds to the antioxidant-responsive element (ARE) in upstream promoters of phase 2 enzymes to activate their transcription (Itoh *et al.*, 2004).

The specific nature of the electrophile is important. Chemically, enones, such as curcumin (Yazawa *et al.*, 2006) and neurite outgrowth-promoting prostaglandin (NEPP11) (Satoh *et al.*, 2006) are true electrophiles. Some enone-type neuroprotective electrophilic compounds that we have studied, such as NEPP11, accumulate in neurons, exerting a direct protective action through induction of HO-1 via the Keap1/Nrf2 pathway (Satoh *et al.*, 2006). In contrast, however, we advocate the use of catechols that only become electrophiles on oxidative conversion to quinones (Nakamura *et al.*, 2003). Thus, catechol-type compounds can function as prodrugs that exert their effects only under oxidative conditions (Lipton, 2004; Satoh and Lipton, 2007). In this manner, catechols, unlike enones, avoid reaction with other cysteine groups, for example, those on glutathione, that could paradoxically decrease the antioxidative capacity of the cell. Some catechol-type neuroprotective electrophilic compounds act preferentially in astrocytes, protecting neurons by a paracrine mechanism (Kraft *et al.*, 2004). Carnosic acid (CA) is a naturally occurring catechol-type polyphenolic diterpene obtained from Rosmarinus officinalis (the herb rosemary) (Kosaka and Yokoi, 2003). Prior work had suggested that CA could exert free radical-scavenging activity (Aruoma

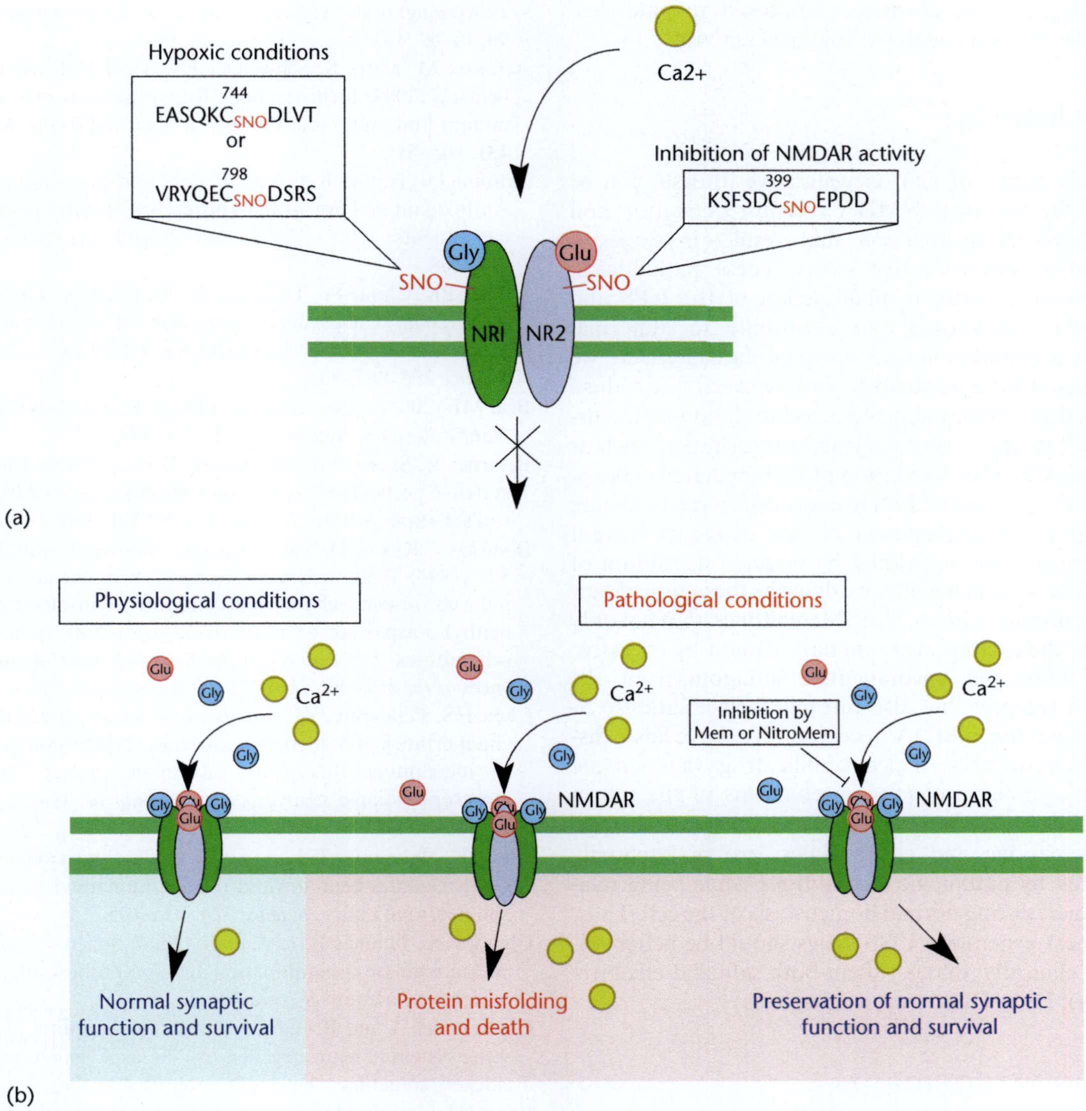

Figure 5 *S*-nitrosylation and Memantine/NitroMemantine regulate NMDA receptor activity. (a) NO-induced inhibition of NMDA receptor. Relatively hypoxic conditions (brain oxygen levels are normally low under physiological conditions) render the NMDA receptor exquisitely sensitive to *S*-nitrosylation of Cys744 and Cys798 on the NR1 subunit. NO modification of these two thiols in NR1 further enhances *S*-nitrosylation of the Cys399 site on the NR2A subunit, resulting in the inhibition of receptor activity by very low levels of NO. Under certain physiological conditions, NR2A can also be nitrosylated at Cys87 and Cys320 (not shown for clarity). (b) Memantine and NitroMemantine preferentially block excessive extrasynaptic NMDA receptor activity. (*Left*) Normal (physiological/synaptic) activity of the NMDAR is required for synaptic function and neuronal survival. (*Middle*) Excessive activation of the NMDAR, predominantly at extrasynaptic sites, is thought to induce neuronal cell injury and death, and is associated with the accumulation of misfolded proteins. (*Right*) Memantine (Mem) and the newer NitroMemantine drugs (NitroMem) preferentially block excessive (pathological) extrasynaptic NMDA receptor activity, whereas relatively sparing normal (physiological) synaptic activity.

et al., 1992), but we have recently found that CA exerts a primary action by activating the Keap1/Nrf2 pathway at sites of oxidative insult (Satoh *et al.*, 2008). In Parkinson disease, oxidative stress has not only a crucial role in disease progression (Jenner, 2007) but also can be used to activate such pro-electrophilic compounds to provide neuroprotection where it is needed. This approach represents a novel strategy against neurodegenerative disorders that could activate electrophilic drugs through pathological activity.

Conclusions

Sporadic forms of neurodegenerative diseases can be caused by excessive NMDA receptor activation and mitochondrial dysfunction that results in excessive nitrosative and oxidative stress. These pathological processes can result in malfunction of the UPS and molecular chaperones and contribute to abnormal protein accumulation and neuronal damage. Here we have described a mechanistic link between free radical production, abnormal protein accumulation and neuronal cell injury in neurodegenerative disorders such as PD and AD. The elucidation of NO-mediated *S*-nitrosylation of parkin and PDI in neurodegenerative disease may promote development of new drugs to prevent aberrant protein misfolding by targeted disruption of this process. Additionally, we describe the action of new memantine derivatives, NitroMemantines, that not only address the excitotoxicity damage caused by excessive Ca^{2+} influx via uncompetitive antagonism of the NMDA receptor, but also through their abilities to *S*-nitrosylate the NMDA receptor. We have also discovered a new class of electrophilic drugs that activate the endogenous protective mechanisms of the cell in response to oxidative stress. Both of these types of drugs, the memantine and electrophiles, are preferentially activated by pathological conditions while being relatively inert during normal homeostasis of the cell. Thus, these next generation CNS drugs should be better tolerated clinically, making them both safe and effective.
See also: Drug Discovery in Apoptosis

Acknowledgements

This work was supported in part by a JSPS Postdoctoral Fellowship for Research Abroad (to T.N.), NIH grants P01 HD29587, R01 EY05477, R01 EY09024, the American Parkinson's Disease Association, San Diego Chapter and an Ellison Senior Scholars Award in Aging (to S.A.L.). Additional support was provided by the NIH Blueprint Grant for La Jolla Interdisciplinary Neuroscience Center Cores P30 NS057096.

References

Abu-Soud HM and Stuehr DJ (1993) Nitric oxide synthases reveal a role for calmodulin in controlling electron transfer. *Proceedings of the National Academy of Sciences of the USA* **90**: 10769–10772.

Arrasate M, Mitra S, Schweitzer ES, Segal MR and Finkbeiner S (2004) Inclusion body formation reduces levels of mutant huntingtin and the risk of neuronal death. *Nature* **431**: 805–810.

Aruoma OI, Halliwell B, Aeschbach R and Loligers J (1992) Antioxidant and pro-oxidant properties of active rosemary constituents: carnosol and carnosic acid. *Xenobiotica* **22**: 257–268.

Auluck PK, Chan HY, Trojanowski JQ, Lee VM and Bonini NM (2002) Chaperone suppression of alpha-synuclein toxicity in a *Drosophila* model for Parkinson's disease. *Science* **295**: 865–868.

Beal MF (2001) Experimental models of Parkinson's disease. *Nature Reviews. Neuroscience* **2**: 325–334.

Betarbet R, Sherer TB, MacKenzie G *et al.* (2000) Chronic systemic pesticide exposure reproduces features of Parkinson's disease. *Nature Neuroscience* **3**: 1301–1306.

Bonfoco E, Krainc D, Ankarcrona M, Nicotera P and Lipton SA (1995) Apoptosis and necrosis: two distinct events induced, respectively, by mild and intense insults with *N*-methyl-D-aspartate or nitric oxide/superoxide in cortical cell cultures. *Proceedings of the National Academy of Sciences of the USA* **92**: 7162–7166.

Chen HS, Pellegrini JW, Aggarwal SK *et al.* (1992) Open-channel block of *N*-methyl-D-aspartate (NMDA) responses by memantine: therapeutic advantage against NMDA receptor-mediated neurotoxicity. *Journal of Neuroscience* **12**: 4427–4436.

Cho DH, Nakamura T, Fang J *et al.* (2009) S-nitrosylation of Drp1 mediates beta-amyloid-related mitochondrial fission and neuronal injury. *Science* **324**: 102–105.

Chung KK, Thomas B, Li X *et al.* (2004) S-nitrosylation of parkin regulates ubiquitination and compromises parkin's protective function. *Science* **304**: 1328–1331.

Ciechanover A and Brundin P (2003) The ubiquitin proteasome system in neurodegenerative diseases: sometimes the chicken, sometimes the egg. *Neuron* **40**: 427–446.

Conn KJ, Gao W, McKee A *et al.* (2004) Identification of the protein disulfide isomerase family member PDIp in experimental Parkinson's disease and Lewy body pathology. *Brain Research* **1022**: 164–172.

Dawson VL, Dawson TM, London ED, Bredt DS and Snyder SH (1991) Nitric oxide mediates glutamate neurotoxicity in primary cortical cultures. *Proceedings of the National Academy of Sciences of the USA* **88**: 6368–6371.

Forrester MT, Benhar M and Stamler JS (2006) Nitrosative stress in the ER: a new role for *S*-nitrosylation in neurodegenerative diseases. *ACS Chemical Biology* **1**: 355–358.

Gu Z, Kaul M, Yan B *et al.* (2002) S-nitrosylation of matrix metalloproteinases: signaling pathway to neuronal cell death. *Science* **297**: 1186–1190.

Hara MR, Agrawal N, Kim SF *et al.* (2005) S-nitrosylated GAPDH initiates apoptotic cell death by nuclear translocation following Siah1 binding. *Nature Cell Biology* **7**: 665–674.

Houk KN, Hietbrink BN, Bartberger MD *et al.* (2003) Nitroxyl disulfides, novel intermediates in transnitrosation reactions. *Journal of American Chemical Society* **125**: 6972–6976.

Isaacs AM, Senn DB, Yuan M, Shine JP and Yankner BA (2006) Acceleration of amyloid beta-peptide aggregation by physiological concentrations of calcium. *Journal of Biological Chemistry* **281**: 27916–27923.

Itoh K, Tong KI and Yamamoto M (2004) Molecular mechanism activating Nrf2-Keap1 pathway in regulation of adaptive response to electrophiles. *Free Radical Biology & Medicine* **36**: 1208–1213.

Jenner P (2007) Oxidative stress and Parkinson's disease. *Handbook of Clinical Neurology* **83**: 507–520.

Kopito RR and Ron D (2000) Conformational disease. *Nature Cell Biology* **2**: E207–209.

Kosaka K and Yokoi T (2003) Carnosic acid, a component of rosemary (*Rosmarinus officinalis* L.), promotes synthesis of nerve growth factor in T98G human glioblastoma cells. *Biological & Pharmaceutical Bulletin* **26**: 1620–1622.

Kraft AD, Johnson DA and Johnson JA (2004) Nuclear factor E2-related factor 2-dependent antioxidant response element activation by tert-butylhydroquinone and sulforaphane occurring preferentially in astrocytes conditions neurons against oxidative insult. *Journal Neuroscience* **24**: 1101–1112.

Lipton SA (2004) Turning down, but not off. *Nature* **428**: 473.

Lipton SA (2006) Paradigm shift in neuroprotection by NMDA receptor blockade: memantine and beyond. *Nature Reviews. Drug Discovery* **5**: 160–170.

Lipton SA (2007) Pathologically activated therapeutics for neuroprotection. *Nature Reviews. Neuroscience* **8**: 803–808.

Lipton SA, Choi YB, Pan ZH *et al.* (1993) A redox-based mechanism for the neuroprotective and neurodestructive effects of nitric oxide and related nitroso-compounds. *Nature* **364**: 626–632.

Lipton SA, Nakamura T, Yao D *et al.* (2005) Comment on "*S*-nitrosylation of parkin regulates ubiquitination and compromises parkin's protective function". *Science* **308**: 1870.

Lipton SA and Rosenberg PA (1994) Excitatory amino acids as a final common pathway for neurologic disorders. *New England Journal of Medicine* **330**: 613–622.

Lyles MM and Gilbert HF (1991) Catalysis of the oxidative folding of ribonuclease A by protein disulfide isomerase: dependence of the rate on the composition of the redox buffer. *Biochemistry* **30**: 613–619.

Marin I and Ferrus A (2002) Comparative genomics of the RBR family, including the Parkinson's disease-related gene parkin and the genes of the ariadne subfamily. *Molecular Biology and Evolution* **19**: 2039–2050.

Mayer ML, Westbrook GL and Guthrie PB (1984) Voltage-dependent block by Mg^{2+} of NMDA responses in spinal cord neurones. *Nature* **309**: 261–263.

McNaught KS, Perl DP, Brownell AL and Olanow CW (2004) Systemic exposure to proteasome inhibitors causes a progressive model of Parkinson's disease. *Annals of Neurology* **56**: 149–162.

Murphy TH, Schnaar RL and Coyle JT (1990) Immature cortical neurons are uniquely sensitive to glutamate toxicity by inhibition of cystine uptake. *FASEB Journal* **4**: 1624–1633.

Nakamura Y, Kumagai T, Yoshida C *et al.* (2003) Pivotal role of electrophilicity in glutathione *S*-transferase induction by tert-butylhydroquinone. *Biochemistry* **42**: 4300–4309.

Olney JW (1969) Brain lesions, obesity, and other disturbances in mice treated with monosodium glutamate. *Science* **164**: 719–721.

Papadia S, Soriano FX, Leveille F *et al.* (2008) Synaptic NMDA receptor activity boosts intrinsic antioxidant defenses. *Nature Neuroscience* **11**: 476–487.

Paz Gavilan M, Vela J, Castano A *et al.* (2006) Cellular environment facilitates protein accumulation in aged rat hippocampus. *Neurobiology of Aging* **27**: 973–982.

Ross CA and Pickart CM (2004) The ubiquitin-proteasome pathway in Parkinson's disease and other neurodegenerative diseases. *Trends in Cell Biology* **14**: 703–711.

Satoh T, Kosaka K, Itoh K *et al.* (2008) Carnosic acid, a catechol-type electrophilic compound, protects neurons both in vitro and in vivo through activation of the Keap1/Nrf2 pathway via *S*-alkylation of targeted cysteines on Keap1. *Journal of Neurochemistry* **104**: 1116–1131.

Satoh T and Lipton SA (2007) Redox regulation of neuronal survival mediated by electrophilic compounds. *Trends in Neuroscience* **30**: 37–45.

Satoh T, Okamoto SI, Cui J *et al.* (2006) Activation of the Keap1/Nrf2 pathway for neuroprotection by electrophilic [correction of electrophillic] phase II inducers. *Proceedings of the National Academy of Sciences of the USA* **103**: 768–773.

Shibata T, Iio K, Kawai Y *et al.* (2006) Identification of a lipid peroxidation product as a potential trigger of the p53 pathway. *Journal of Biological Chemistry* **281**: 1196–1204.

Spencer JP, Jenner A, Aruoma OI *et al.* (1994) Intense oxidative DNA damage promoted by L-dopa and its metabolites. Implications for neurodegenerative disease. *FEBS Letters* **24**: 246–250.

Stamler JS, Lamas S and Fang FC (2001) Nitrosylation. The prototypic redox-based signaling mechanism. *Cell* **106**: 675–683.

Uehara T, Nakamura T, Yao D *et al.* (2006) S-nitrosylated protein-disulphide isomerase links protein misfolding to neurodegeneration. *Nature* **441**: 513–517.

Xu L, Eu JP, Meissner G and Stamler JS (1998) Activation of the cardiac calcium release channel (ryanodine receptor) by poly *S*-nitrosylation. *Science* **279**: 234–237.

Yao D, Gu Z, Nakamura T *et al.* (2004) Nitrosative stress linked to sporadic Parkinson's disease: *S*-nitrosylation of parkin regulates its E3 ubiquitin ligase activity. *Proceedings of the National Academy of Sciences of the USA* **101**: 10810–10814.

Yazawa K, Kihara T, Shen H *et al.* (2006) Distinct mechanisms underlie distinct polyphenol-induced neuroprotection. *FEBS Letters* **580**: 6623–6628.

Zhang K and Kaufman RJ (2006) The unfolded protein response: a stress signaling pathway critical for health and disease. *Neurology* **66**: S102–109.

Microbial Inhibitors of Apoptosis

Georg Häcker, *Institute for Medical Microbiology, Technische Universität München, Munich, Germany*

Advanced article

Article Contents

- Introduction
- Apoptosis Induction and the Benefits of Apoptosis Inhibition for Microbes
- Inhibitors of Apoptosis Induction: Activation of Pro-survival Pathways
- Viral Inhibitors of Apoptosis
- Bacterial Inhibitors of Apoptosis
- Parasitic Protozoa and the Inhibition of Apoptosis
- Antiapoptotic Effects in the Immune Response
- Concluding Remarks

Infection with microbial agents (bacteria, fungi, viruses and parasites) is a challenge to a host cell. The host cell's response to infection is complex and entails changes in the activity of a number of intracellular signalling pathways, commonly including the apoptotic machinery. Microbes, if they have evolved the capacity to infect mammals or other complex hosts, have to be able to deal with such defence systems. The lifestyle of microbes varies substantially – viruses, for instance, depend on an intact cell, whereas many bacteria grow on surfaces and are not sensitive to host cell apoptosis. What we know about microbial interference with host cell apoptosis is in accordance with these lifestyles. Viruses often directly target and inhibit the apoptosis machinery with specific proteins whereas other microbes frequently also interfere with host apoptosis but more indirectly.

Introduction

Microbes are incredibly common. The vast majority of microbes do not live in association with humans, require growth conditions that are not found in the human body and can therefore not cause human disease. There are, however, still a very considerable number of microbial species that live on or in the human body, and a considerable number of microbial agents that can cause disease, from the groups of bacteria, fungi, viruses and parasites.

It is helpful in this context to divide an infection into two parts. Initially, the infectious agent will come into contact with immunologically nonspecialized tissue cells at the site of entry into the body or perhaps on distribution with the blood stream in a distant organ: *Staphylococcus aureus* faces a keratinocyte in the skin, *Hepatitis B virus* infects a hepatocyte. These first cells will respond to this infection in some way, and we will look at the question of apoptosis and apoptosis inhibition in this context later. At the second stage of infection, the agent is recognized by specialized cells from the immune system, which will undertake to combat and to destroy the infectious agent. Apoptosis also plays a significant role in shaping this immune response. I will briefly touch on the matter of apoptosis and apoptosis inhibition in immune cells through components of microbes later.

Why do microbes need apoptosis inhibitors? The most straightforward explanation is certainly that it is beneficial to the microbial agent to keep a host cell or a population of host cells alive. Whether, and to which extent this is important to the microbe will greatly depend on the replication cycle and the growth requirements of the infectious agent (**Figure 1**). Here the main factor is probably whether the microbe needs host cells to grow and to replicate. The case is simplest for viruses. Viruses are obligate intracellular parasites that cannot replicate without the host cell and that usually use the cell's machinery for the synthesis of their macromolecules. No virus can replicate outside a cell, and apoptosis can be an effective protection against viral replication in a host organism. If the first cell that is infected is able to die by apoptosis, the death of the cell will usually also kill the virus, which is probably caught in a state where it cannot infect other cells and is then further degraded by the katabolic enzymes active in a cell dying by apoptosis. Apoptosis is thus a defence system that the host uses to protect itself against the infection. The effectiveness of this system was first illustrated in the interaction of an insect virus and its host cells. Taking away one single gene from the virus led to the massive induction of apoptosis in the host cell during infection, and this gene, p35, was later characterized to be a potent inhibitor of host cell caspases (Clem *et al.*, 1991). In mammals, there is an additional factor that for the host makes apoptosis a desirable consequence of viral infection. If a virus-infected cell dies, it will be taken up by phagocytes, and there is good evidence that this uptake can make the viral antigen available to the immune system, cause the stimulation of lymphocytes and thereby initiate a potent immune response to viral antigen (Fonteneau *et al.*, 2002).

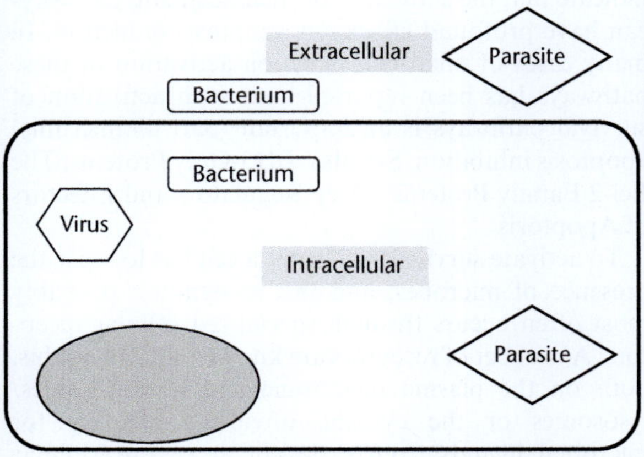

Figure 1 The interaction between microbes and human host cells is determined by the microbial lifestyle. Viruses are intracellular parasites. They need parts of the cellular machinery for their replication and are vulnerable to apoptosis by the host cell. In the groups of bacteria and parasites there are different lifestyles. Some bacteria can grow only within the host cell and undergo differentiation into specialized forms before they can be transmitted between host cells and between hosts (obligate intracellular bacteria, e.g. Chlamydiae). Other bacterial genera can grow or at least live both inside and outside host cells (*Listeria*, *Salmonella*, *Legionella* and others; facultative intracellular bacteria). Apoptosis of the host cell usually does not vitally affect these bacteria although it may have indirect consequences such as through its effect on the immune response. Some bacteria are at least mainly extracellular, for instance, *Streptococci* and *Escherichia coli*. Apoptosis inhibition for these bacteria is least likely but may occur through the activation of pro-survival pathways. Among the groups of protozoan parasites the lifestyles are similar to the ones discussed for bacteria.

Apoptosis Induction and the Benefits of Apoptosis Inhibition for Microbes

When we look at it from this angle, it becomes clear that the cell will probably 'try' to undergo apoptosis when it detects the infection. The virus thus can be seen to 'induce' apoptosis and, to ensure its own replication, it will require means to inhibit apoptosis. It should be pointed out here that the types of infections that pathogens cause in their natural host and that we see today have to be the products of evolution and therefore have found a certain balance where both, infectious agent and host can stay alive. A virus that kills its host quickly has poor chances of being transmitted; a virus that is eliminated before it can replicate will disappear. We can therefore not expect a clear-cut advantage in one or the other direction, and the same is probably also true when looking at individual functions such as apoptosis. Apoptosis will be induced but it will not act as a perfect sterilizing weapon, nor is apoptosis inhibition by the virus likely to make the virus undefeatable.

The situation is less clear for bacteria. Most bacteria do not require an intact host cell to replicate and they grow either extracellularly or can grow both outside and inside human cells (usually these bacteria can also be grown on agar plates). There is however a group of bacteria that are referred to as obligate intracellular organisms; the medically most prominent bacteria here belong to the genus *Chlamydia*/*Chlamydophila*. Like viruses, these bacteria require an intact host cell for replication, and host cell apoptosis can be an effective means of inhibiting growth (this has at least been shown *in vitro* for *Chlamydia trachomatis*; Ying *et al.*, 2008a, b). It should be pointed out that, unlike viruses, these bacteria are still self-sufficient in terms of replication although they require some macromolecules from the host cell for growth. However, even extracellular bacteria are likely to be affected by the extent of host cell death, as cell death will affect tissue structure and impact on inflammation and the antibacterial immune response. Bacterial products and components are efficiently recognized not only by the immune system but also by tissue resident cells, and while the cellular response ensuing on such recognition is more strikingly one of the activation and differentiation, the changes in the cell's susceptibility to apoptosis must not be overlooked (see later discussion).

Parasitism is extremely common in nature. Although, for instance, a virus could also be classified as a parasite (as it replicates at the host's expense), use of the medical term is usually confined to two groups of eukaryotes, protozoa (single-celled organisms) and metazoa (worms and arthropods). The interaction of worms and the host is in most cases not well understood, and the discussion of modulation of apoptosis by worms has at present to be limited to the general effects on the immune response (see later discussion). A number of protozoa have an intracellular lifestyle and are accordingly probably vulnerable to apoptosis. Antiapoptotic activities have been described for some of the better-researched.

Inhibitors of Apoptosis Induction: Activation of Pro-survival Pathways

One important point to be considered in the context of microbial apoptosis inhibitors is the interconnection of the apoptosis apparatus with other cellular signalling pathways. We know a lot about the molecular details of the actual apoptosis apparatus (most instances here concern mitochondrial apoptosis). Mitochondrial apoptosis is triggered typically when one or several BH3-only proteins are activated, which then go on to activate the effectors of cytochrome *c* release, Bax and/or Bak, whereas the antiapoptotic Bcl-2-like proteins prevent this activation (Youle and Strasser, 2008). The steps upstream of these events are, however, still not very clear. BH3-only proteins can be regulated by gene induction and probably also activated by posttranslational regulation. Likewise, the resistance to mitochondrial apoptosis can be determined by the levels of pro-survival Bcl-2 (B-cell lymphoma 2) proteins, and although the sum of these Bcl-2 proteins determines resistance to apoptosis, they are not functionally equivalent. The complexity of these processes makes it very challenging to work out how upstream pathways regulate the activation of apoptosis. A number of pathways have been reported to regulate apoptosis in various cell types. The most prominent ones are probably the nuclear factor-κB (NFκB) and the phospho-inositol-3 kinase (PI3K)/Akt pathways, but the pathways involving Jun *N*-terminal kinase (JNK) and extracellular signal-regulated kinase (ERK) have also been implicated. NFκB, PI3K/Akt and ERK pathways are usually considered antiapoptotic whereas JNK activity has been found to be pro-apoptotic. However, these effects are molecularly not very well defined. NFκB can induce antiapoptotic genes such as cFLIP (cellular FLICE-inhibitory protein), cIAP2 (inhibitor of apoptosis protein 2), Bcl-x_L and A1 (Naugler and Karin, 2008). However, it has not often been attempted actually to map the antiapoptotic activity to one or several NFκB-regulated gene products. Likewise, the PI3K/Akt-pathway has been proposed to regulate apoptosis via phosphorylation of the BH3-only protein Bad. However, this is unlikely fully to explain the antiapoptotic effect of Akt activity; indeed, in some cells the Akt effect was found to depend on antiapoptotic Mcl-1 (myeloid cell leukemia 1) (Hahnel *et al.*, 2008). Without going into further detail, this will suffice to indicate that the activation of such upstream pathways can have profound effects on apoptosis induction. In many cases of microbial infection activation of these pathways has been reported, and such activation of survival pathways is an important part of microbial apoptosis inhibition. **See also**: BH3-Only Proteins; The Bcl-2 Family Proteins – Key Regulators and Effectors of Apoptosis

To activate survival pathways, a cell has to sense the presence of microbes, and this recognition probably most often occurs through specialized cellular receptors. A number of receptors are known that can do this, both on the plasma membrane and in endosomes/lysosomes or the cytosol, obviously catering for microbial ligands from extracellular or intracellular pathogens. Since these receptors typically recognize ligands that vary between individual bacterial groups but that still conform to the same structural pattern (such as bacterial lipopolysaccharide that in one form or another is found in all Gram-negative bacteria), these receptors are often called pattern recognition receptors (PRR). The most prominent class of PRR known are the Toll-like receptors (TLR). TLR are potent activators of NFκB, ERK and PI3K pathways, and other PRR have similar activities (**Figure 2**; Ishii *et al.*, 2008). Since most or all human pathogens are recognized by PRR, activation of these survival pathways is the rule rather than the exception on the recognition of bacteria. Accordingly, a pro-survival effect

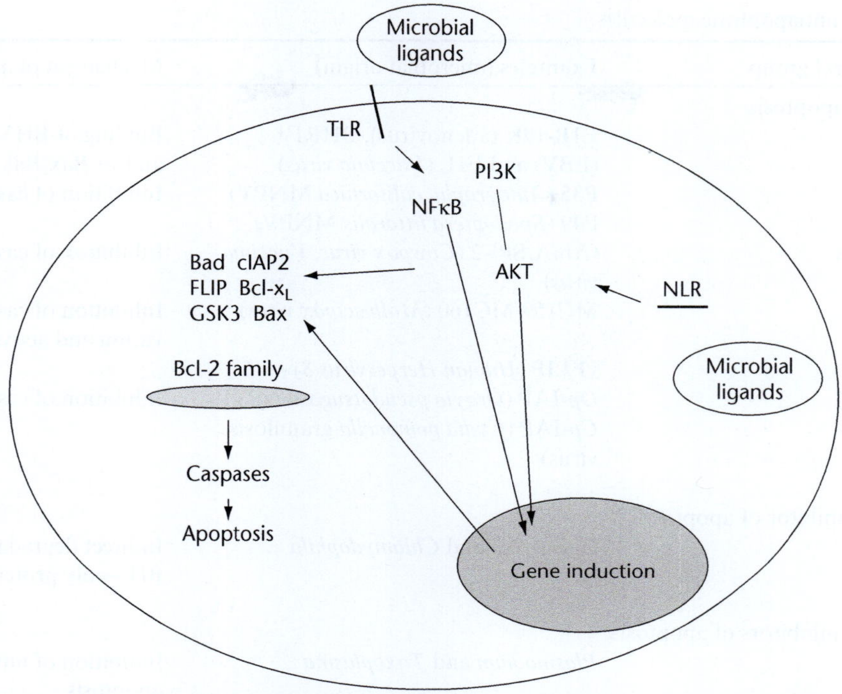

Figure 2 Activation of pro-survival pathways by microbial agents. Cellular receptors (shown are by way of example of two groups, the Toll-like receptors, TLR, and the nod-like receptors, NLR) can sense microbial components. These ligands are typically recognized by their structural composition but may vary considerably in actual molecular identity (e.g. TLR4 recognizes bacterial lipopolysaccharide from many different bacteria that vary in terms of fatty acid and polysaccharide structure). Signalling through these receptors causes among other events the activation of pro-survival pathways, such as the NFκB and PI3K pathways. In most cases, these pathways induce pro-survival genes that make the cell less susceptible to pro-apoptotic stimuli; modification of proteins such as bad phosphorylation has also been reported. The mechanism acts upstream of the initiation of mitochondrial apoptosis. The precise nature of these genes in a given situation is largely unknown but some candidates are shown.

has been described for numerous microbial organisms. Especially, a large number of bacteria probably inhibit apoptosis via such pathways (see later discussion). A number of the microbial strategies that microbes employ to inhibit host cell apoptosis are summarized in **Table 1**. **See also**: Caspases in Inflammation and Immunity

Viral Inhibitors of Apoptosis

Viral genes, in general, tend to be recognizable derivatives of genes of the cells they infect. Indeed, it is likely that the only way for a virus to acquire a gene is to hijack a cellular gene into the viral genome. In terms of antiapoptotic proteins, it is therefore not surprising that quite a number of viral antiapoptotic proteins are encoded by viral genes that have detectable resemblance to host cell antiapoptotic proteins. The largest family of viral antiapoptotic proteins is the group of proteins that resemble Bcl-2; this resemblance is usually but not always structural and often functional, although the latter has in many cases not been carefully explored. However, other viral strategies such as caspase inhibition have also been described (**Figure 3**).

Bcl-2-like viral proteins

A number of viral apoptosis inhibitors can be classified as Bcl-2 like, most often based on structural homology. The cellular group of Bcl-2-like proteins has a number of short domains known as Bcl-2-homology (BH) domains, normally four in the antiapoptotic, Bcl-2-like group (Cory and Adams, 2002). Viral Bcl-2-like proteins (vBcl-2s) are characterized by having preserved this domain structure although to a varying extent. Functionally, vBcl-2s seem to act like Bcl-2, at least in the sense that they inhibit the BH3-only

Table 1 Microbial antiapoptotic molecules

Functional/structural group	Examples (microbial origin)	Mechanism of action
Viral inhibitors of apoptosis		
Bcl-2-like	E1B-19K (adenovirus), BHRF1 (EBV) and F1L (*Vaccinia virus*)	Binding of BH3-only proteins and/or Bax/Bak
P35	P35 (*Autographa californica* MNPV) P49 (*Spodoptera littoralis* MNPV)	Inhibition of caspases
Serpin-like proteins	CrmA/SPI-2 (*Cowpox virus*, *Vaccinia virus*)	Inhibition of caspases
FLIP-like proteins	MC159/MC160 (*Molluscipox virus*)	Inhibition of caspase-8 activation and activation of NFκB
	vFLIP (*Human Herpesvirus 8*)	
IAP (insect viruses)	*Op*-IAP (*Orgyia pseudotsugata* NPV) *Cp*-IAP (*Cydia pomonella* granulosis virus)	Inhibition of caspase activation
Specific bacterial inhibitor of apoptosis		
CPAF	*Chlamydia* and *Chlamydophila*	Indirect degradation of BH3-only proteins
Specific protozoan inhibitors of apoptosis		
Unknown	*Plasmodium* and *Toxoplasma*	Inhibition of mitochondrial apoptosis
Various microbes		
Ligands of pattern recognition receptors	LPS (Gram-negative bacteria) and microbial nucleic acids (bacteria, viruses and protozoa)	Activation of pro-survival pathways

protein-dependent activation/activity of Bax/Bak. The precise molecular mechanism has, however, in many cases not been confirmed experimentally. **See also**: BH3-Only Proteins; Structure and Function of IAP and Bcl-2 Proteins; The Bcl-2 Family Proteins – Key Regulators and Effectors of Apoptosis

The first vBcl-2 to be identified and characterized was the 19 kDa protein encoded by the E1B gene of adenovirus (E1B-19K). E1B-19K can interact with Bax and Bak and block their function (Cuconati and White, 2002). Viral Bcl-2s have also been found in a number of herpesviruses. *Epstein–Barr virus* (EBV) has two genes whose products have sequence similarity to Bcl-2, termed BALF1 and BHRF1. BHRF1 is an undisputed apoptosis inhibitor (Henderson *et al.*, 1993) and while BALF1 has also been described as blocking apoptosis (Marshall *et al.*, 1999), it has also been suggested to be a Bcl-2 inhibitor (and thus to promote apoptosis) (Bellows *et al.*, 2002). A recent study found that in the context of the infection of B cells by EBV (its natural host), these vBcl-2s were required during the initial stages of infection; a mutant virus lacking these genes induced apoptosis and was unable to replicate. EBV typically establishes a latent infection in B cells but its vBcl-2s were not expressed during latency (Altmann and Hammerschmidt, 2005). This is a good example of the balance of apoptosis induction and inhibition during a viral infection. The infected B cell seems to be able to sense the infection and reacts to this by undergoing apoptosis. The virus counters this response by virtue of its apoptosis inhibitors.

Other herpesviruses also have vBcl-2s: human herpesvirus 8 (although this protein does not appear to interact with Bax or Bak), *Herpesvirus Saimiri*, cytomegalovirus (vMIA, viral mitochondrial-localized inhibitor of apoptosis) and others. Similarly, a number of poxviruses carry genes that resemble Bcl-2 (avipoxvirus, vaccinia virus, myxomavirus; Everett and McFadden, 2002), African swine fever virus, *Hepatitis C virus* (a *Flavivirus*) and even in a *Birnavirus* infecting

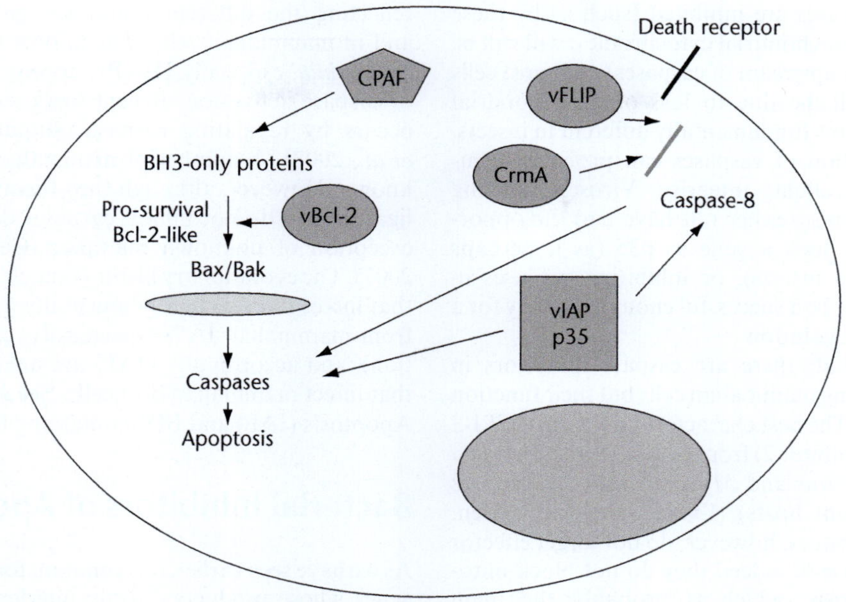

Figure 3 Direct antiapoptotic effects by microbial proteins. Direct attack of the host cell's apoptosis machinery by microbial inhibitors is largely the domain of viruses. The biggest group of known viral apoptosis inhibitors is the family of proteins with similarity to Bcl-2-like pro-survival proteins (vBcl-2s), which block the activation of Bax/Bak at or upstream of mitochondria. This strategy is used by a number of different viruses such as poxviruses, herpesviruses and adenoviruses. The viral proteins CrmA/SPI-2 and vFLIP can, at least in experimental conditions, block death receptor-induced apoptosis. Both may have different biological functions, as CrmA can also inhibit caspase-1 (which may be more important because this blocks the release of pro-inflammatory mediators) and vFLIP may interfere with NFκB activation. The inhibitors vIAP and p35/p49 are only known in insect-infecting baculoviruses where vIAPs probably prevent the activation of initiator caspases whereas p35 as a pseudosubstrate blocks active caspases. CPAF is the only bacterial protein whose autonomous antiapoptotic activity is well-documented. CPAF indirectly induces the degradation of BH3-only proteins in *Chlamydia*-infected cells and makes them resistant to apoptosis.

fish (Hong *et al.*, 2002). There are, however, intriguing aspects to these proteins. A number of them have very limited sequence similarity to human Bcl-2 but a very similar 3-dimensional structure, for instance, poxviral M11L and F1L (Kvansakul *et al.*, 2008). It is also uncertain whether they indeed have the same molecular function as Bcl-2s (which is the binding of BH3-only proteins and perhaps of Bax/Bak). F1L can bind Bak and Bim but not Bax or other BH3-only proteins. vMIA has a number of effects other than apoptosis inhibition, such as a strong effect on mitochondrial morphology (McCormick *et al.*, 2003). Finally, some poxviral proteins have been reported to assume a Bcl-2-like structural fold without overt antiapoptotic function (Graham *et al.*, 2008). Apoptosis inhibition of the Bcl-2 way is very potent, and the wide prevalence of this strategy in different viruses suggests that it is very useful to a virus. The book is, however, not closed on this aspect of viral biology and we may expect to learn more from these proteins not only about viruses but also about the way human Bcl-2-like proteins work.

Viral caspase inhibitors: serpin-like proteins and vFLIPs

A different strategy is also employed by baculoviruses, namely the direct inhibition of caspases. Two proteins are known to have this molecular function, p35 and p49 (from different viruses; Clem *et al.*, 1991; Du *et al.*, 1999). A baculovirus lacking p35 induces massive cell death and can reproduce only to a very limited extent (Clem *et al.*, 1991).

These proteins have slightly different specificities but can generally inhibit most caspases by serving as caspase substrates that on cleavage remain tightly associated with the caspase and prevent its further activity (this was first described for a class of cellular serine protease inhibitors, or serpins) (Bump *et al.*, 1995). No similar inhibitor is known in viruses infecting mammalian cells, nor is the cellular gene known from which these inhibitors are derived. It is possible that this confinement to insect viruses reflects the different apoptosis machinery in insects and mammals. In mammals, if caspases are inhibited (such as by these viral proteins), mitochondrial cytochrome *c* will still be released (as this is upstream of caspases) and most cells will probably still die due to loss of mitochondrial integrity. This seems fundamentally different in insects, where the inhibition of caspases can probably completely preserve cellular integrity. Viruses infecting mammalian cells may either not have had the opportunity to acquire such a gene as p35 (as it perhaps is present only in insects), or inhibiting caspases in mammals may not be a successful enough strategy for a virus to allow its evolution.

Having said that, there are caspase inhibitors in poxviruses infecting mammalian cells but their function may be different. The best characterized is CrmA/SPI-2 (serin protease inhibitor 2) from *Cowpox virus*, *Vaccinia virus*, *Rabbitpox virus* and *Myxoma virus* (poxviruses that infect different hosts) (Everett and McFadden, 2002). These inhibitors, however, do not target effector caspases or caspase-9; indeed they do not block mitochondrial apoptosis, which is probably the main apoptosis pathway *in vivo*. These proteins are able to block caspase-8 (activated downstream of death receptors) and caspase-1 (responsible for the secretion of pro-inflammatory cytokines but not apoptosis). It may well be speculated that their activity is more important in blocking the inflammatory immune response rather than apoptosis. A last instance of an again functionally distinct group are vFLIPs (viral FLICE-inhibitory proteins), which are known in herpesviruses as well as in poxviruses (Everett and McFadden, 2002). vFLIPs may inhibit the activation of caspase-8 by interfering with its recruitment into the activation scaffold known as death-inducing signalling complex (DISC) following death receptor ligation. However, vFLIPs may also have functions in NFκB signalling (Bagneris *et al.*, 2008), and it is unclear what their relevance for the infection is.

Viral caspase inhibitors: inhibitor of apoptosis proteins

A class of proteins now collectively known as IAPs was first discovered in a baculovirus (a virus infecting insects) and termed OpIAP. Later, cellular homologues (from which the viral protein was presumably derived) were discovered, both in insects (e.g. DIAP1 in *Drosophila*) and in mammalian cells. Viral and insect IAPs seem to be able to block caspases. However, rather than blocking the active enzyme they function very likely in regulating and inhibiting its activation. Like p35, vIAPs have only been found in baculovirus, probably again reflecting the different apoptosis systems in insects and in mammalian cells. The known cellular IAPs in *Drosophila*, especially DIAP1, appear to be inhibitors of caspase activation. Recent work suggests that this occurs by regulating caspase ubiquitination (Ditzel *et al.*, 2008). A number of mammalian IAPs are also known. However, although they function as ubiquitin ligases, they do not inhibit caspases directly with one exception of unknown relevance (Callus and Vaux, 2007). The evolutionary history is unclear but it appears that insect IAPs have fundamentally distinct functions from mammalian IAPs (specifically, caspase inactivation), and accordingly, vIAPs are unknown in viruses that infect mammalian host cells. **See also**: Inhibitor of Apoptosis (IAP) and BIR-containing Proteins

Bacterial Inhibitors of Apoptosis

As we have seen earlier, it is common for viruses to carry genes whose products directly interfere with the host cell's apoptosis apparatus. This is different for bacteria, where no such direct attack is known, possibly with the exception of Chlamydiae. The reason for this may be 2-fold. First, and probably more importantly, bacteria lack a virus' ability to incorporate a host cell's genes into its own genome, at least on a large scale, and cannot therefore simply adopt the host's strategy to modulate the apoptosis machinery. Secondly, with a few exceptions, bacteria do not depend as stringently on the host cell for replication and therefore may not have been under the same pressure to evolve antiapoptotic mechanisms. It is also possible that the host cell does not use apoptosis as a defence against bacteria to the same extent as it does against viruses. In most cases, cell death will not prevent bacterial spread and will therefore not be a helpful strategy. Perhaps this is different in cases of obligate intracellular bacteria such as Chlamydiae.

Chlamydiae are a group of bacteria that can only reproduce inside host cells (the newer nomenclature also recognizes the very similar genus *Chlamydophila*). Some chlamydial species infect animals, and two species are important human pathogens: *C. trachomatis*, common cause of preventable blindness in the world and the most common bacterial agent of sexually transmitted disease and *Chlamydia pneumoniae*, which causes airway infections and may contribute to atherosclerosis. Chlamydiae replicate in a specialized inclusion within the cytosol and affect the host cell in many ways (Fields and Hackstadt, 2002; Wyrick, 2000). In terms of host cell apoptosis, the infection with *Chlamydia* is the best characterized of bacterial infections. *Chlamydia* can inhibit apoptosis in host cells by probably more than one mechanism and actually induces a form of cell death at later stages of infection that appears to be nonapoptotic (Ying and Hacker, 2007). Two different mechanisms of apoptosis inhibition by Chlamydiae have been proposed. On the one hand, it has been suggested that Chlamydiae induce cellular antiapoptotic proteins via the activation of the cellular pro-survival signalling pathways discussed earlier, NFκB and PI3K (the ERK-pathway has also been proposed). This mechanism may operate during the early stages of the infection, and it has been suggested that IAPs as well as the Bcl-2-like protein Mcl-1 are involved (Rajalingam *et al.*, 2008). At later stages, however, the cells are profoundly protected against experimentally induced apoptosis without the requirement for IAPs or Mcl-1 (Ying *et al.*, 2008a). The protection against apoptosis during this stage is very likely due to the degradation of cellular BH3-only proteins, the essential initiators of mitochondrial apoptosis (Fischer *et al.*, 2004). The exact mechanism of this degradation is unknown but it appears to be an indirect effect of a chlamydial protease, CPAF (chlamydial protease activity factor) being secreted by the bacteria into the host cytosol (Paschen *et al.*, 2008). It is in fact not clear whether *Chlamydia* indeed depends on the inhibition of apoptosis for its replication. Perhaps the cell can detect the infection and react with apoptosis which is at the same time inhibited by *Chlamydia*. The net result may then be the observed nonapoptotic cell death. This is, however, speculation, and apoptosis inhibition by *Chlamydia* may be purely a byproduct of the infection. Intriguingly, the protease CPAF is already found in *Chlamydia*-related bacteria, *Protochlamydia amoebophila*, which parasitizes (single-celled) amoeba where apoptosis very likely plays no role (Horn *et al.*, 2004). Although, CPAF from *P. amoebophila* can also induce degradation of BH3-only proteins is unknown; the mechanism in human-pathogenic Chlamydiae may therefore be an adaptation of an existing protease or indeed be simply a side effect that does not benefit but also not harm the bacteria.

There is one more example where direct targeting of mitochondria by bacteria has been reported, namely bacterial agents of sexually transmitted disease and meningitis, *Neisseria gonorrhoeae* and *Neisseria meningitidis*. The agent these bacteria use for this purpose is a protein from the class of porins, transporters across the bacterial membrane. Porins are related to a structural class of mitochondrial membrane proteins known as β-barrel proteins. Porins have been suggested to localize to mitochondria in infected cells, and different groups of authors have proposed that porins either induce or inhibit apoptosis (see, e.g. Massari *et al.*, 2000; Muller *et al.*, 2000). In either case, porins are bacterial proteins that serve independent functions in bacteria (molecule transport over membranes), and any interference with the host cell's apoptosis system may therefore be coincidental.

As discussed earlier, bacteria are often recognized by cellular receptors, which then activate pro-survival pathways, and many examples have been reported where this is the case. Examples include the activation of NFκB by *Ehrlichia*, by surface proteins of *Helicobacter pylori* and fimbriae of *Porphyromonas gingivalis* or infection with the intracellular bacterium *Rickettsia rickettsii* as well as the activation of the PI3K/Akt-pathway by *Po. gingivalis* and by *Salmonella*.

Although molecular players and mechanisms have in some cases been suggested, it is basically unclear how this activation of survival pathways is achieved. There must be cellular receptors that recognize bacterial molecules and that then activate these cellular pathways (see **Figure 2**). Although we have some understanding of the receptor families involved, the precise pairing of receptor and ligand in a bacterial infection is mostly uncertain. It should also be pointed out that it is difficult to assess what the role of bacterial apoptosis inhibition is in the context of the infection. I will use one example to illustrate this complexity. It has been suggested that *Mycobacterium tuberculosis*, the main agent of human tuberculosis has to prevent macrophage apoptosis for successful infection (Porcelli and Jacobs, 2008). However, others have concluded the exact opposite: apoptosis is required for bacterial virulence (Briken and Miller, 2008). This illustrates the difficulty in assessing the beneficial effect of cell death during an infection, where the consequences of local cell death are overlaid with the effects of the local and systemic immune response and both will affect the outcome.

In a few instances the direct interference of bacterial activities with caspases has been suggested but further experimentation will be necessary to confirm this. Thus, *Shigella flexneri* has been reported to inhibit caspase-9 but it is unclear how this could be achieved and whether this could be important (Clark and Maurelli, 2007).

Parasitic Protozoa and the Inhibition of Apoptosis

Some protozoa replicate exclusively inside their host cell and may therefore be vulnerable to the apoptotic defence of the host. Parasites again lack the capacity to acquire host cell genes into their genome, and protozoan apoptosis inhibitors would have to be of their own invention. Protozoan diseases are hugely important world wide, and some of the most prevalent agents are relatively well researched. No dedicated apoptosis inhibitor in protozoa has been identified but the function of apoptosis inhibition during infection is well documented and has in some cases been mapped carefully on a functional level.

Plasmodium, the agent of malaria, appears to have some antiapoptotic activities. The clinically apparent stages of malaria have their correlate in the parasites multiplying in erythrocytes, which probably cannot undergo apoptosis (death of erythrocytes has been termed eryptosis but this is likely to follow different molecular mechanisms; Lang *et al.*, 2008). When first infecting the mammalian host, however, *Plasmodium* requires a cycle of differentiation and replication in the liver, and it has been reported that infected hepatocytes are resistant to apoptosis. During the early stages of infection, this has been demonstrated to be the result of the activation of PI3K (Leiriao *et al.*, 2005). It has further been found that hepatocytes infected with the rodent pathogen *Plasmodium berghei* are protected against experimental apoptosis (van de Sand *et al.*, 2005) through a molecularly unclear mechanism. *Toxoplasma gondii* is a very common parasite that causes only an inapparent acute infection in most immunocompetent individuals but that very commonly establishes latent infections. *To. gondii* resides within a parasitophoric vacuole in the host cell cytoplasm, from where it is able to inhibit apoptosis induced by a great number of experimental stimuli (Goebel *et al.*, 2001; Nash *et al.*, 1998) suggesting a target at the central signal transduction machinery of apoptosis (or a variety of targets). The molecular identity of the postulated toxoplasmal inhibitor of apoptosis is still uncertain but there appears to be a mechanism that disrupts especially mitochondrial apoptosis (Hippe *et al.*, 2008). *Theileria parva*, a protozoan parasite of cattle has also been found to inhibit apoptosis in infected host cells by modulation of a number of signalling pathways (Heussler and Stanway, 2008). It will be very interesting to see the molecular nature of these functionally described protozoan apoptosis inhibitors.

Antiapoptotic Effects in the Immune Response

The immune response to microbial pathogens is strongly regulated through apoptosis. Lymphocytes expand during the immune reaction, and the majority dies by apoptosis once the pathogen has been cleared away. In the innate immune system, apoptosis is particularly prominent in neutrophil granulocytes. Neutrophils are very short-lived, produced in the bone marrow and physiologically die by apoptosis after probably less than a day in the peripheral blood. Neutrophils phagocytose and destroy especially bacteria, and for this purpose they enter tissues. In inflamed tissues, the lifespan of neutrophils appears to be considerably prolonged, which may be necessary for their effective antibacterial activity. Experimentally, the activation of TLR with bacterial components can keep neutrophils alive, very likely through the activation of the earlier survival pathways. This can thus be seen as a mechanism by which the host exploits bacterial antiapoptotic activities in the defence against the same microbial agents. There is no simple answer to the question whether apoptosis inhibition in the immune system has benefits for the host or the infectious agents. However, modulation of apoptosis in the responding immune system is likely to affect the outcome of the infection in many cases.

Concluding Remarks

Microbial agents are amazingly diverse, in terms of lifestyle and growth requirements, of their ability to establish an infection and in terms of stimulating and inhibiting host cell signalling pathways. With regard to apoptosis inhibition, the major theme in host–microbe interaction is clearly the inhibition of apoptosis by viral gene products. This ability has alerted researchers to the ability of cells to use apoptosis in self-defence, and apoptosis induction is, alongside defence actions such as interferon secretion, now a recognized response of host cells to viral infection. In line with the advancement of our molecular understanding of apoptosis and its inhibition by viruses, blockade of viral apoptosis inhibitors is a promising future strategy of antiviral chemotherapy. In particular, the first inhibitors of cellular Bcl-2s are being tested at present, and vBcl-2s may be similarly amenable to pharmaceutical inhibition. Other parasitic microbes may or may not require inhibition of apoptosis, very likely depending on their replication requirements. Like in many other ways, through their interference with the apoptotic apparatus microbes teach us about their approaches to infecting a host, and an understanding of this interaction will help us appreciating coevolution and its various forces.

References

Altmann M and Hammerschmidt W (2005) *Epstein–Barr virus* provides a new paradigm: a requirement for the immediate inhibition of apoptosis. *PLoS Biology* **3**: e404.

Bagneris C, Ageichik AV, Cronin N *et al.* (2008) Crystal structure of a vFlip-IKKgamma complex: insights into viral activation of the IKK signalosome. *Molecular Cell* **30**: 620–631.

Bellows DS, Howell M, Pearson C, Hazlewood SA and Hardwick JM (2002) *Epstein–Barr virus* BALF1 is a BCL-2-like antagonist of the herpesvirus antiapoptotic BCL-2 proteins. *Journal of Virology* **76**: 2469–2479.

Briken V and Miller JL (2008) Living on the edge: inhibition of host cell apoptosis by *Mycobacterium tuberculosis*. *Future Microbiology* **3**: 415–422.

Bump NJ, Hackett M, Hugunin M *et al.* (1995) Inhibition of ICE family proteases by antiapoptotic protein p35. *Science* **269**: 1885–1888.

Callus BA and Vaux DL (2007) Caspase inhibitors: viral, cellular and chemical. *Cell Death and Differentiation* **14**: 73–78.

Clark CS and Maurelli AT (2007) *Shigella flexneri* inhibits staurosporine-induced apoptosis in epithelial cells. *Infection and Immunity* **75**: 2531–2539.

Clem RJ, Fechheimer M and Miller LK (1991) Prevention of apoptosis by a baculovirus gene during infection of insect cells. *Science* **254**: 1388–1390.

Cory S and Adams JM (2002) The Bcl2 family: regulators of the cellular life-or-death switch. *Nature Reviews. Cancer* **2**: 647–656.

Cuconati A and White E (2002) Viral homologs of BCL-2: role of apoptosis in the regulation of virus infection. *Genes & Development* **16**: 2465–2478.

Ditzel M, Broemer M, Tenev T *et al.* (2008) Inactivation of effector caspases through nondegradative polyubiquitylation. *Molecular Cell* **32**: 540–553.

Du Q, Lehavi D, Faktor O, Qi Y and Chejanovsky N (1999) Isolation of an apoptosis suppressor gene of the *Spodoptera littoralis* nucleopolyhedrovirus. *Journal Virology* **73**: 1278–1285.

Everett H and McFadden G (2002) Poxviruses and apoptosis: a time to die. *Current Opinion on Microbiology* **5**: 395–402.

Fields KA and Hackstadt T (2002) The chlamydial inclusion: escape from the endocytic pathway. *Annual Review of Cell and Developmental Biology* **18**: 221–245.

Fischer SF, Vier J, Kirschnek S *et al.* (2004) *Chlamydia* inhibit host cell apoptosis by degradation of proapoptotic BH3-only proteins. *Journal of Experimental Medicine* **200**: 905–916.

Fonteneau JF, Larsson M and Bhardwaj N (2002) Interactions between dead cells and dendritic cells in the induction of antiviral CTL responses. *Current Opinion in Immunology* **14**: 471–477.

Goebel S, Gross U and Luder CG (2001) Inhibition of host cell apoptosis by *Toxoplasma gondii* is accompanied by reduced activation of the caspase cascade and alterations of poly (ADP-ribose) polymerase expression. *Journal of Cell Science* **114**: 3495–3505.

Graham SC, Bahar MW, Cooray S *et al.* (2008) Vaccinia virus proteins A52 and B14 Share a Bcl-2-like fold but have evolved to inhibit NF-kappaB rather than apoptosis. *PLoS Pathogens* **4**: e1000128.

Hahnel PS, Thaler S, Antunes E *et al.* (2008) Targeting AKT signaling sensitizes cancer to cellular immunotherapy. *Cancer Research* **68**: 3899–3906.

Henderson S, Huen D, Rowe M *et al.* (1993) *Epstein–Barr virus*-coded BHRF1 protein, a viral homologue of Bcl-2, protects human B cells from programmed cell death. *Proceedings of the National Academy of Sciences of the USA* **90**: 8479–8483.

Heussler VT and Stanway RR (2008) Cellular and molecular interactions between the apicomplexan parasites *Plasmodium* and *Theileria* and their host cells. *Parasite* **15**: 211–218.

Hippe D, Lytovchenko O, Schmitz I and Luder CG (2008) Fas/CD95-mediated apoptosis of type II cells is blocked by *Toxoplasma gondii* primarily via interference with the mitochondrial amplification loop. *Infection and Immunity* **76**: 2905–2912.

Hong JR, Gong HY and Wu JL (2002) IPNV VP5, a novel anti-apoptosis gene of the Bcl-2 family, regulates Mcl-1 and viral protein expression. *Virology* **295**: 217–229.

Horn M, Collingro A, Schmitz-Esser S *et al.* (2004) Illuminating the evolutionary history of Chlamydiae. *Science* **304**: 728–730.

Ishii KJ, Koyama S, Nakagawa A, Coban C and Akira S (2008) Host innate immune receptors and beyond: making sense of microbial infections. *Cell Host & Microbe* **3**: 352–363.

Kvansakul M, Yang H, Fairlie WD *et al.* (2008) Vaccinia virus anti-apoptotic F1L is a novel Bcl-2-like domain-swapped dimer that binds a highly selective subset of BH3-containing death ligands. *Cell Death and Differentiation* **15**: 1564–1571.

Lang F, Gulbins E, Lerche H *et al.* (2008) Eryptosis, a window to systemic disease. *Cellular Physiology and Biochemistry* **22**: 373–380.

Leiriao P, Albuquerque SS, Corso S *et al.* (2005) HGF/MET signalling protects *Plasmodium*-infected host cells from apoptosis. *Cellular Microbiology* **7**: 603–609.

Marshall WL, Yim C, Gustafson E *et al.* (1999) *Epstein–Barr virus* encodes a novel homolog of the bcl-2 oncogene that inhibits apoptosis and associates with Bax and Bak. *Journal of Virology* **73**: 5181–5185.

Massari P, Ho Y and Wetzler LM (2000) *Neisseria meningitidis* porin PorB interacts with mitochondria and protects cells from apoptosis. *Proceedings of the National Academy of Sciences of the USA* **97**: 9070–9075.

McCormick AL, Smith VL, Chow D and Mocarski ES (2003) Disruption of mitochondrial networks by the human cytomegalovirus UL37 gene product viral mitochondrion-localized inhibitor of apoptosis. *Journal of Virology* **77**: 631–641.

Muller A, Gunther D, Brinkmann V *et al.* (2000) Targeting of the pro-apoptotic VDAC-like porin (PorB) of *Neisseria gonorrhoeae* to mitochondria of infected cells. *EMBO Journal* **19**: 5332–5343.

Nash PB, Purner MB, Leon RP *et al.* (1998) *Toxoplasma gondii*-infected cells are resistant to multiple inducers of apoptosis. *Journal of Immunology* **160**: 1824–1830.

Naugler WE and Karin M (2008) NF-kappaB and cancer-identifying targets and mechanisms. *Current Opinion in Genetic Development* **18**: 19–26.

Paschen SA, Christian JG, Vier J *et al.* (2008) Cytopathicity of *Chlamydia* is largely reproduced by expression of a single chlamydial protease. *Journal of Cell Biology* **182**: 117–127.

Porcelli SA and Jacobs WR Jr (2008) Tuberculosis: unsealing the apoptotic envelope. *Nature Immunology* **9**: 1101–1102.

Rajalingam K, Sharma M, Lohmann C *et al.* (2008) Mcl-1 is a key regulator of apoptosis resistance in *Chlamydia trachomatis*-infected cells. *PLoS ONE* **3**: e3102.

van de Sand C, Horstmann S, Schmidt A *et al.* (2005) The liver stage of *Plasmodium berghei* inhibits host cell apoptosis. *Molecular Microbiology* **58**: 731–742.

Wyrick PB (2000) Intracellular survival by *Chlamydia*. *Cellular Microbiology* **2**: 275–282.

Ying S, Christian JG, Paschen SA and Hacker G (2008a) *Chlamydia trachomatis* can protect host cells against apoptosis in the absence of cellular inhibitor of apoptosis proteins and Mcl-1. *Microbes and Infection* **10**: 97–101.

Ying S and Hacker G (2007) Apoptosis induced by direct triggering of mitochondrial apoptosis proceeds in the near-absence of some apoptotic markers. *Apoptosis* **12**: 2003–2011.

Ying S, Pettengill M, Latham ER *et al.* (2008b) Premature apoptosis of *Chlamydia*-infected cells disrupts chlamydial development. *Journal of Infectious Disease* **198**: 1536–1544.

Youle RJ and Strasser A (2008) The BCL-2 protein family: opposing activities that mediate cell death. *Nature Review of Molecular and Cellular Biology* **9**: 47–59.

Further Reading

Faherty CS and Maurelli AT (2008) Staying alive: bacterial inhibition of apoptosis during infection. *Trends in Microbiology* **16**: 173–180.

Galluzzi L, Brenner C, Morselli E, Touat Z and Kroemer G (2008) Viral control of mitochondrial apoptosis. *PLoS Pathogens* **4**: e1000018.

Ishii KJ, Koyama S, Nakagawa A, Coban C and Akira S (2008) Host innate immune receptors and beyond: making sense of microbial infections. *Cell Host & Microbe* **3**: 352–363.

Luder CG, Stanway RR, Chaussepied M, Langsley G and Heussler VT (2009) Intracellular survival of apicomplexan parasites and host cell modification. *International Journal of Parasitology* **39**: 163–173.

Luo JL, Kamata H and Karin M (2005) The anti-death machinery in IKK/NF-kappaB signaling. *Journal of Clinical Immunology* **25**: 541–550.

Strasser A (2005) The role of BH3-only proteins in the immune system. *Nature Review of Immunology* **5**: 189–200.

Drug Discovery in Apoptosis

Tom O'Brien, *Genentech Inc., South San Francisco, California, USA*

Vishva M Dixit, *Genentech Inc., South San Francisco, California, USA*

Advanced article

Article Contents

- Introduction
- Role of Apoptosis in Disease
- Discovery of Small Molecule Caspase Inhibitors
- Caspase Inhibitors in Clinical Development
- Alternative Approaches to Caspase Inhibitor Discovery
- Noncaspase Targets for Apoptosis Drug Discovery
- Summary

Regulation of cell death (apoptosis) is a crucial process that has to be precisely modulated during normal cell growth, and inappropriate regulation of this process has been implicated in a large number of ailments ranging from cancer to neurodegenerative diseases. Given that apoptosis can be either induced (e.g. following tissue injury or ischaemia) or attenuated (e.g. in cancer), there are a large number of potential

intervention strategies that can be taken. To date, there are no approved therapeutics whose primary mode of action involves modulation of a critical component of the apoptotic pathway. However, there is an increasing number of therapeutics that are actively being tested in clinical trials, some of which are likely to become approved therapeutics.

Introduction

Both the apoptotic and inflammatory caspase pathways have been implicated in a wide range of indications. For instance, apoptotic caspase family members are normally responsible for the timely induction of apoptosis; however, there are a growing number of indications where this tight regulation has been lost. Similarly, the inflammatory caspase pathway, which is mediated predominantly by caspase-1, is responsible for the proteolytic maturation of the pro-interleukin 1β (IL-1β) and pro-IL-18 cytokines to their mature forms (IL-1β and IL-18) (Kersse *et al.*, 2008). The main function of IL-1β is to promote the infiltration of leukocytes to sites of injury or sites of infection and to initiate a local inflammatory response to an infection, whereas the main function of IL-18, which is a member of the IL-1 family, is the stimulation of additional cytokines such as interferon γ. Thus, inhibiting caspase-1 or the apoptotic caspases should help alleviate symptoms associated with inflammatory or apoptotic pathologies, respectively. **See also**: Apoptosis: Inherited Disorders; Apoptosis: Regulatory Genes and Disease; The Siren's Song: This Death that Makes Life Live

In contrast to the discovery of compounds that inhibit the inflammatory and apoptotic caspase-dependent pathways, there is a growing realization that activating the apoptotic pathway may be beneficial in settings where it is attenuated, for example, in tumour cell growth. Clearly there are a number of potential intervention points other than caspases that can be utilized for this purpose, and these will be discussed.

Role of Apoptosis in Disease

There is ample evidence, and even an approved therapeutic (Kineret, which is an IL-Rα antagonist), showing that inhibition of the IL-1β cytokine pathway can modulate the severity of rheumatoid arthritis (RA) (Fleischmann *et al.*, 2004). As would be expected given its role in IL-1β maturation, caspase-1 knockout mice or peptide-based caspase-1 inhibitors have implicated a role of caspase-1 in rodent inflammation-based models (e.g. a caspase-1 peptide inhibitor reduces the severity of a type II collagen-induced arthritis model in mice) (Kersse *et al.*, 2008). Given that an RA therapeutic would have to be administered chronically, it is imperative that prolonged inhibition of caspase-1 does not have any detrimental side effects. Fortunately, caspase-1 knockout mice are viable and do not appear to have any developmental defects (Li *et al.*, 1995), suggesting that chronic inhibition of this target may be clinically acceptable (this will be discussed in more detail in the later section). Thus, both genetic and small molecule inhibitor approaches support the important role of caspase-1 in the development of RA.

The inflammatory caspase pathway has also been implicated in additional autoinflammatory disorders. Osteoarthritis, a degenerative disease of the joint which is induced by excessive levels of the IL-1β and IL-18 cytokine levels in the synovium fluid, leads to degradation of the cartilage matrix (Kersse *et al.*, 2008). An advanced caspase-1 inhibitor (Pralnacasan, VX-745, discussed in section on Caspase inhibitors in clinical development) was shown to reduce joint damage in a murine model of knee osteoarthritis (Rudolphi *et al.*, 2003), suggesting that caspase inhibition could be useful in this clinical setting.

Sepsis is a life threatening condition that is triggered by an infectious agent and results in a systemic inflammatory response. In this case, not only is there an enhanced cytokine response (predominantly IL-1β, IL-6, IL-8, IL-18 and tumour necrosis factor α (TNFα)), but there is an associated high level of apoptosis. Mice that are deficient for caspase-1 are protected in a murine model of sepsis (endotoxin challenged mice); however, mice that contain an IL-1β/IL-18 double knockout are not (Sarkar *et al.*, 2006). Thus, the protection offered by the caspase-1 knockout may be due to its indirect role in the apoptotic response. In support of this, a pan-caspase inhibitor or a caspase-3-specific inhibitor were able to enhance viability in this model (Hotchkiss *et al.*, 2000). Thus, even though there is an inflammatory component to this disease, it appears that inhibiting the apoptotic caspase response rather than the inflammatory caspase response may be more biologically relevant.

The apoptotic pathway is implicated in a number of acute indications that result in tissue injury. One of the key indications is cardiac ischaemia, which according to the latest data published by the American Heart Association, results in the death of 2400 people in America each day (equivalent to 1 death every 37 s) from cardiac disease (Lloyd-Jones *et al.*, 2008). Though it is encouraging that the death rate has declined 26.4% between 1995 and 2005, it is clear that this still remains an unmet medical need. There is a large body of evidence implicating enhanced levels of apoptosis in cardiac tissue following an ischaemic response, such as increased levels of activated caspase-3 and apoptosis markers in cardiomyocytes (Matsui, 2008). Tissue-specific overexpression of caspase-3 in cardiomyocytes results in an increase in infarct size associated with a decrease in cardiac function (Condorelli *et al.*, 2001). Conversely, inhibiting caspase activity through the early infusion of peptide-based inhibitors following hypoxic injury reduced infarct size by 20–30% in a rat model (Mocanu *et al.*, 2000). Thus, cardiac ischaemia is one therapeutic area where a caspase inhibitor could beneficially affect clinical outcome.

Additional indications include stroke, which, like cardiac ischaemia, has a strong apoptotic response associated with tissue damage and is ranked the number three killer among all causes of death in the USA. There are a number of examples demonstrating the utility of caspase inhibitors in this setting using animal models (Matsui, 2008); however, even though there is data demonstrating that inhibition of caspase activity can reduce infarct size, the correlation between the reduction of infarct size and cognitive outcome is still ambiguous. Also, there may be a limited time window following a stroke when administration of a caspase inhibitor may be beneficial. Thus, given the complexities of these type of clinical trials, and the issues associated with timing of administration, this is not likely to be an indication readily entered into for a caspase inhibitor proof of concept clinical trial.

Early observations demonstrating that caspase inhibitors could protect mice from liver failure following administration of an anti-Fas antibody (Hoglen *et al.*, 2001) suggested that inhibition of caspase activity could be a useful clinical option for various types of liver diseases. Liver diseases with an apoptotic component include Hepatitis B and C virus (HBV and HCV) infections, alcoholic liver disease, nonalcoholic steatohepatitis (NASH) and cholestatic liver disease, many of which eventually lead to liver fibrosis. In most cases the initiating apoptotic event is the death of liver hepatocytes, and thus it was proposed that inhibiting hepatocyte apoptosis could minimize or reduce the extent of the disease. The first apoptotic caspase inhibitor to enter clinical trials (Emricasan) was a pan-caspase inhibitor and was initially tested in patients with hepatic dysfunction in Phase I and II clinical trials (discussed in section Discovery of small molecule caspase inhibitors) (Linton *et al.*, 2005).

There are a number of additional indications where increased apoptosis has been documented, and where inhibition of caspase activity may be beneficial. These include neurodegenerative diseases such as Alzheimer, Huntington and Parkinson diseases. **See also**: Apoptosis: Inherited Disorders; From Reactive Oxygen and Nitrogen Species to Therapy

Discovery of Small Molecule Caspase Inhibitors

Caspases have an extremely well-defined active site that cleaves substrates adjacent to an aspartic acid with the ideal recognition sequence being four amino acids long, although caspase-2 appears to have a preference for a substrate sequence that contains five amino acid residues. Early selective inhibitors were derived from the known preferred substrate sequences, and these peptide-based inhibitors were used to help validate the role of caspase activity in a range of animal models. Even though these inhibitors were generally very potent and displayed reasonable selectivity, they were unlikely to become approved agents given the liabilities associated with peptide-based inhibitors (e.g. lack of cell permeability and poor metabolic stability). Thus, the search for molecules that could be used clinically focused on the discovery of nonpeptidic inhibitors. **See also**: Caspases and Cell Death

There are a few general features that are commonly found in caspase inhibitors: (a) a warhead, (b) a P1-aspartic acid residue and (c) the main body of the inhibitor that binds to the S2–S4 pocket on the enzyme and is the key determinant of selectivity (see **Figure 1a**). Nearly every active site caspase inhibitor contains a warhead and the nature of the electrophile determines whether the inhibitor binds reversibly or irreversibly (see **Figure 1b**). An aldehyde-based warhead, which binds reversibly, was routinely used in peptide inhibitors and was a valuable tool in proof-of-concept studies. However, a disadvantage with an aldehyde warhead (and reversible binding warheads in general) is the reduced cellular activity that is often observed compared to an inhibitor containing an irreversible binding warhead. This is thought to be due to the constant need for high levels of sustained caspase inhibition, given that the pathway may only need a small percentage of caspases to be active to drive apoptosis (Methot *et al.*, 2004). Despite this potential disadvantage, two

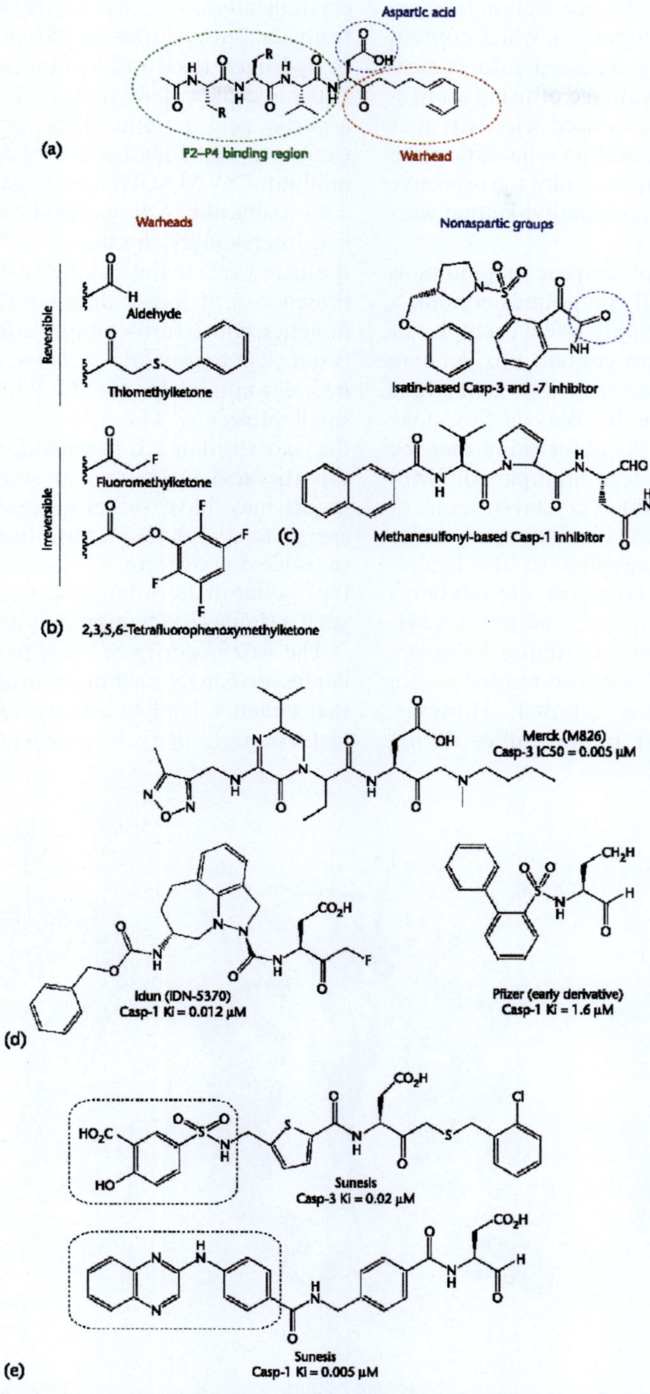

Figure 1 Anatomy of an active site caspase inhibitor. (a) Generalized structure with the three key regions highlighted. (b) Examples of different warheads. (c) Two examples of inhibitors that lack a P1 aspartic functionality. (d) Examples of caspase inhibitors with associated inhibitory activity. (e) Two caspase inhibitors that were derived from a fragment-based discovery approach. The initial fragments that were identified from screening are highlighted with a dashed line, and these fragments were subsequently linked to an aspartic acid and warhead to generate a fully elaborated molecule.

compounds (VX-740 and VX-765, see section on Caspase inhibitors in clinical development), which contain an aldehyde warhead, have progressed into clinical studies. Conversely, the disadvantage of using an irreversible binding warhead is the reduced selectivity that is routinely observed, which could be a liability when used for chronic administration (for a nice retrospective discussion on the history of fluoromethyl-ketone warheads; see Van Noorden, 2001).

Another defining feature of caspase inhibitors is the observation that nearly all compounds contain a P1-aspartic moiety. There are only a few examples of reported inhibitors that do not contain this defining group (Lee *et al.*, 2000; Okamoto *et al.*, 1999), and these examples are shown in **Figure 1c**. Why is this? One explanation could stem from the observation that the active site of caspase-1 can adopt multiple conformations (Romanowski *et al.*, 2004). A crystal structure of ligand-free caspase-1 revealed that the active site appeared very disordered compared to the ligand-bound form (e.g. containing an active site inhibitor; **Figure 2**). In this case one of the key residues (Arg341, contained within Loop 3) that contributes hydrogen bond interactions with the P1 aspartic residue within the oxyanion hole is solvent exposed. However, when malonate was included in the buffer during crystallization, the critical Arg341 residue is relocated from the surface into the S1 pocket and is involved in hydrogen bond interactions with the bound malonate molecule, interactions that are similar to a bound aspartic acid residue. This conformation was very similar to that found when a fully elaborated peptide inhibitor (YVAD-CHO, the tetrapeptide inhibitor with a reversible aldehyde warhead) was bound in the active site. Interestingly, this indicates that the transition from the ligand-free to the ligand-bound conformation in the presence of malonate does not require P2–P4 binding moieties, which further implies that the P1 aspartic acid is sufficient to generate an active site conformation that now accommodates the P2–P4 binding groups of the small molecule. This observation may help explain the near absolute requirement for inhibitors to have an aspartic acid in their P1 position. The aspartic acid group may have sufficient recognition and binding energy to stimulate the transition from the disordered to ordered active site, or alternatively, could help trap the enzyme in the ordered active site form. It is not clear yet if a similar mechanism may apply to other caspases.

The vast majority of small molecule drug discovery has focused on the identification of compound properties that enhance binding affinity while at the same time increasing selectivity. Examples of some of the inhibitors

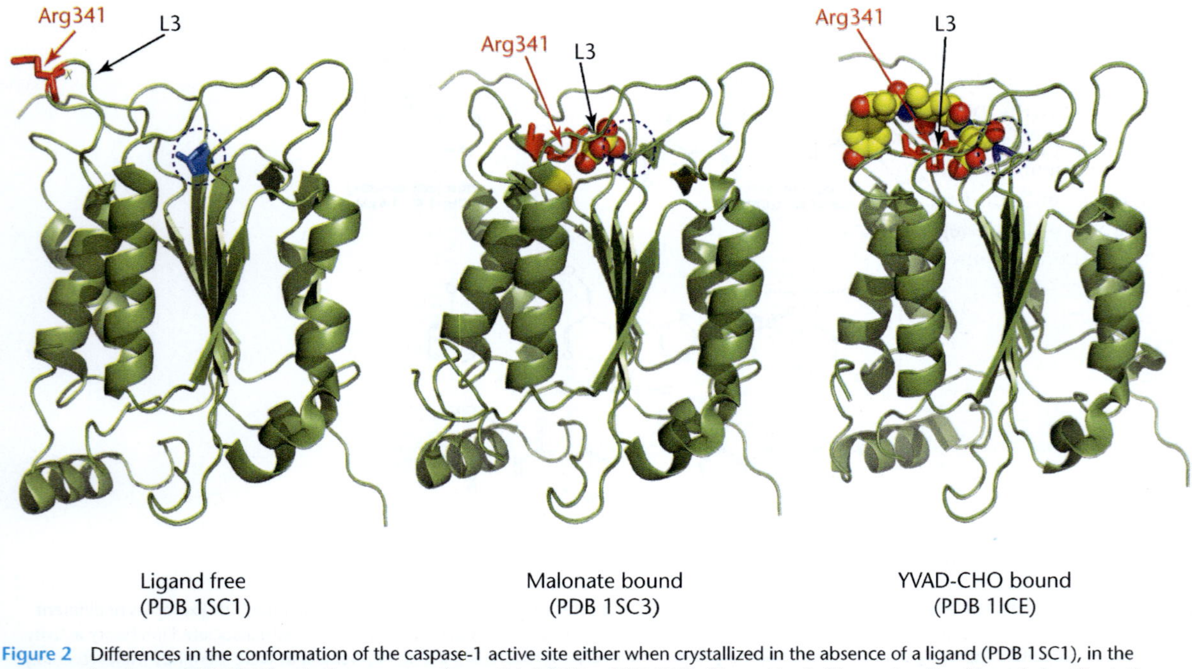

Figure 2 Differences in the conformation of the caspase-1 active site either when crystallized in the absence of a ligand (PDB 1SC1), in the presence of malonate (PDB 1SC3) or in the presence of a peptide inhibitor (PDB 1ICE). The residue in blue and outlined by a dashed line is the active site cysteine residue, and the critical Arg341 (in red) and loop 3 are indicated.

that have been identified are shown in **Figure 1d**. In each case the compounds contain a P1 aspartic acid functionality, a warhead to interact with the active site cysteine reside and a P2–P4 binding moiety that contributes binding affinity and selectivity. These compounds span an extremely wide range of diversity, and in most cases were discovered by traditional high-throughput screening followed by structure-guided optimization. However, fragment-based discovery approaches have also been undertaken, and examples of compounds derived from one of these approaches are shown in **Figure 1e**. In both of these cases an initial binding fragment was discovered using mass spectrometry and then rapidly elaborated into larger molecules with high binding affinity (Erlanson *et al.*, 2003; Fahr *et al.*, 2006).

Caspase Inhibitors in Clinical Development

Only four caspase inhibitors have entered clinical trials (**Figure 3**), and of these, only one is still actively being pursued. The first caspase inhibitor to enter Phase I clinical trials was VX-740, a reversible binding caspase-1 inhibitor. VX-740 progressed into Phase II clinical trials for RA and osteoarthritis, and was delivered orally as an acetyl-prodrug that is rapidly hydrolysed to the active form which contains an aldehyde warhead (Rudolphi *et al.*, 2003; Vertex Pharmaceuticals, 2006). In RA patients a dose-dependent decrease in IL-1β levels along with positive anti-inflammatory effects were observed. This is in contrast to a trial in osteoarthritis patients where no significant differences in the primary endpoint were observed between placebo and Pralnacasan-treated groups, even though urine and serum markers from patients suggested that bone and cartilage turnover were modulated. Further clinical development of this compound was discontinued in 2003 due to the appearance of liver fibrosis in a 9-month nonclinical toxicology study in a single species. Unfortunately the exact nature of the toxicity has not been disclosed, and it is difficult to determine if this is compound- (possibly due to the reactive nature of the aldehyde warhead?) or target-related. The latter is considered unlikely, given that caspase-1 knockout mice are phenotypically normal (Li *et al.*, 1995).

Vertex progressed a second caspase-1 inhibitor (VX-765) into clinical trials with the same aldehyde-based warhead as VX-740, which would suggest that the toxicity observed with VX-740 is not target related. VX-765 is also delivered as an acetyl-prodrug, further suggesting that the toxicity of VX-740 was not due to the reactive nature of the aldehyde warhead (Wannamaker *et al.*, 2007). This second compound progressed into a four-week Phase IIa trial in 68 patients with psoriasis. At the completion of the trial Vertex indicated that the results warrant further development of the compound; however, they have indicated that they are unlikely to progress the compound any further in the absence of a corporate partner (Vertex Pharmaceuticals, 2006).

A third compound that entered clinical trials is IDN-6556/PF-3491390 (Emricasan). This compound, an irreversible pan-caspase inhibitor, was developed by Idun Pharmaceuticals who had initiated Phases I and II clinical trials to examine its utility in hepatic injury induced by HCV infection and also to assess if it can improve outcome in liver transplantation (Linton *et al.*, 2005). In 2005, Pfizer acquired Idun and took over responsibility for clinical development. Surprisingly, in early 2008 Pfizer announced that they were discontinuing Phase II clinical trials with this compound, which they claimed was due to strategic reasons and shifting portfolio decisions.

The fourth compound, LB84451/GS-9450, is also an irreversible pan-caspase inhibitor, and is being pursued in indications focused on hepatic injury (hepatitis C and fibrosis), which is a similar strategy as employed by Idun/Pfizer. Little information on this compound has been disclosed, however, in late 2007 Gilead Inc. licensed this compound from LG Life Sciences Ltd. and they are currently pursuing its clinical development (Gilead, 2007). Given that this is now the only compound in clinical development, it will be very interesting to see how it progresses as its success could signal a revival in caspase inhibitor discovery.

Alternative Approaches to Caspase Inhibitor Discovery

As described earlier, all inhibitors that have so far been described target the active site, and thus contain many of the same liabilities associated with having a reactive warhead and an aspartic residue. However, another interesting avenue that could be considered was recently disclosed. A fragment-based approach used to discover novel small molecule caspase inhibitors serendipitously discovered an allosteric binding site in caspase-3 and -7 (Hardy *et al.*, 2004). In this case, the authors used mass spectrometry to identify cysteine containing small fragments that would bind to the active site cysteine of caspase-3 via disulfide linkage. Only two fragments were identified in the screen that

VX-740
(Pralnacasan)

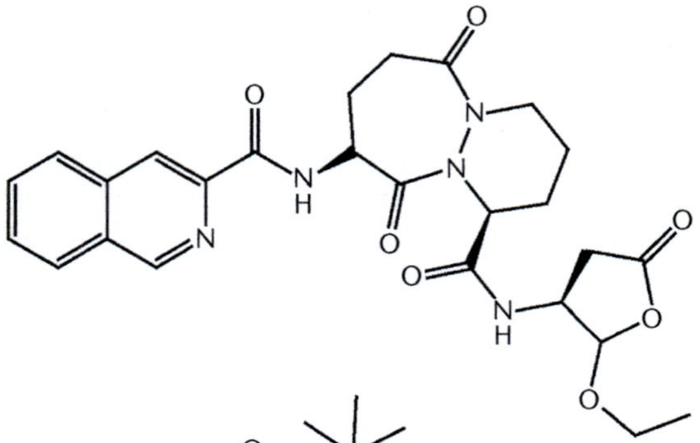

VX-765

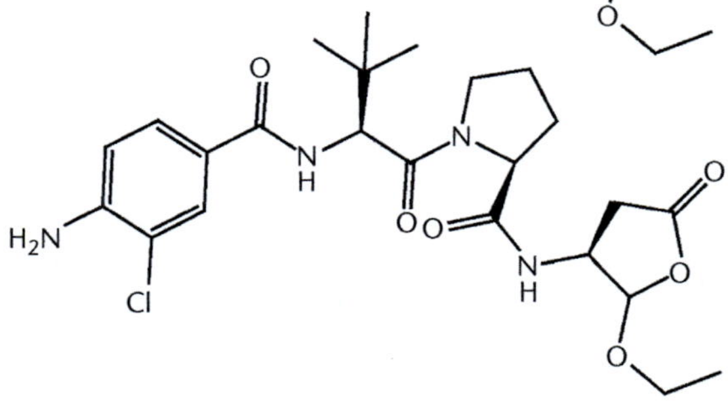

IDN-6556/PF-03491390
(Emricisan)

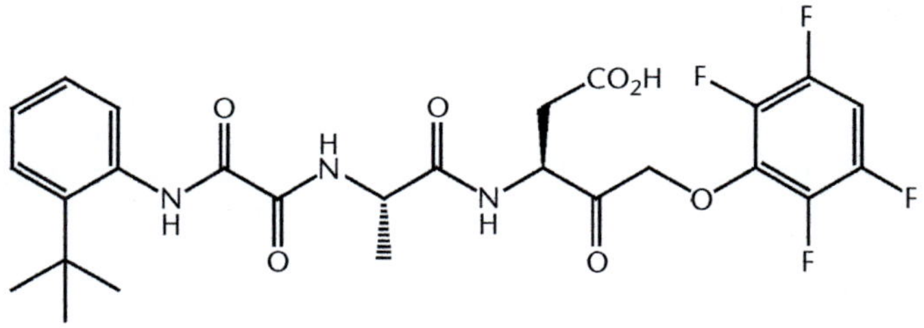

LB-84451/GS-9450

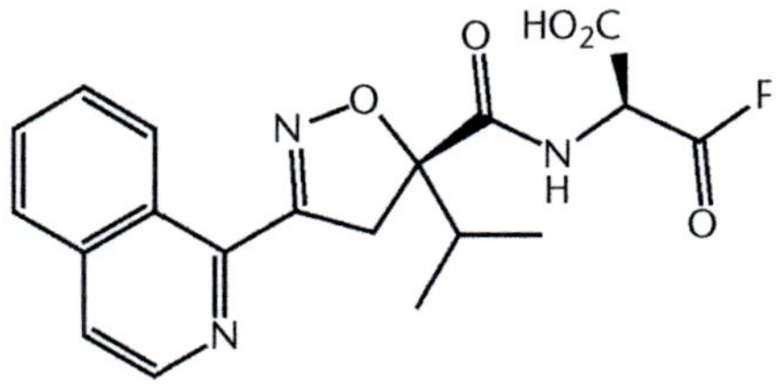

Figure 3 Caspase inhibitors that have entered clinical trials.

bound to caspase-3, and surprisingly both of these bound to a single cysteine residue on the small subunit and not the large subunit (which contains the active site cysteine residue). The small subunit only contains three cysteine residues, and by peptide mapping and generating Cys to Ala point mutations, Cys264 was identified as the binding residue. This cysteine residue resides within a small pocket that is located between the two dimers of the active heterotetramer.

Interestingly, when either of these fragments was covalently bound to Cys264, caspase-3 enzymatic activity was inhibited, suggesting that binding of the fragments to this site allosterically blocked the active site. Conversely, when the active site is prebound with an inhibitor (e.g. DEVD-CHO), the allosteric site was no longer accessible to these cysteine-containing fragments. The mechanism underlying this allosteric regulation became apparent once a cocrystal structure was solved. In this case, a cocrystal structure of caspase-7 with one of these fragments (DICA, 2-(2,4-dichlorophenoxy)-N-(2-mercapto-ethyl)-acetamide) verified that it was indeed bound to the cysteine residue at the interface between the two heterodimers (the cysteine residue in caspase-7 is Cys290, which is equivalent to Cys264 in caspase-3), and explained why the enzyme is catalytically inactive (**Figure 4**). When the fragments are bound within the allosteric site, the critical residues necessary for forming the S1-binding pocket (the active site Cys186 and Arg187) are displaced from the S1 pocket and prevented from forming a functional complex with a substrate. Conversely, when the active site is occupied with bound inhibitor, Arg187 re-orientates into the allosteric pocket and sterically hinders binding of the DICA compound.

What is interesting about this discovery is that all the caspases with a published structure appear to have a similar putative allosteric site. For example, a recent report has demonstrated the existence of a similar allosteric mechanism in caspase-1 (Scheer *et al.*, 2006). Moreover, the protein sequence within these sites is very divergent, possibly indicating that high selectivity could be obtained with elaborated compounds. As yet, there are no reports of noncovalent-binding compounds that bind to the allosteric site and modulate caspase activity, but it is nevertheless a very intriguing approach that could be taken to discover novel caspase inhibitors.

Noncaspase Targets for Apoptosis Drug Discovery

So far we have discussed strategies for inhibiting apoptosis. However, there are many clinical advantages and uses for compounds that can induce apoptosis, most of which are oncology focused. One way is to induce the apoptotic signalling pathway through the use of a molecule that mimics the agonist activity the Trail ligand, which binds to the DR4 and DR5 receptors and activates the extrinsic caspase pathway. As malignant cancer cells grow with few cell cycle restrictions and tend to harbour genetic instability, it is thought that the apoptotic pathway within these cells is primed for activation. Thus, a recombinant ligand or antibody that binds to the apoptotic receptors and functions agonistically could tip the balance in favour of apoptosis. These approaches have been utilized very successfully *in vitro* and *in vivo* and there are now at least seven different molecules in Phase I or II clinical trials, with preliminary data indicating that these have various levels of clinical responses (Ashkenazi, 2008).

An alternative approach is to utilize small molecule inhibitors that target the pro-apoptotic protein family Bcl-2. One example of this is ABT-263, which is in Phase I clinical trials and is co-developed by Abbott Laboratories and Genentech Inc. (Tse *et al.*, 2008). Normally,

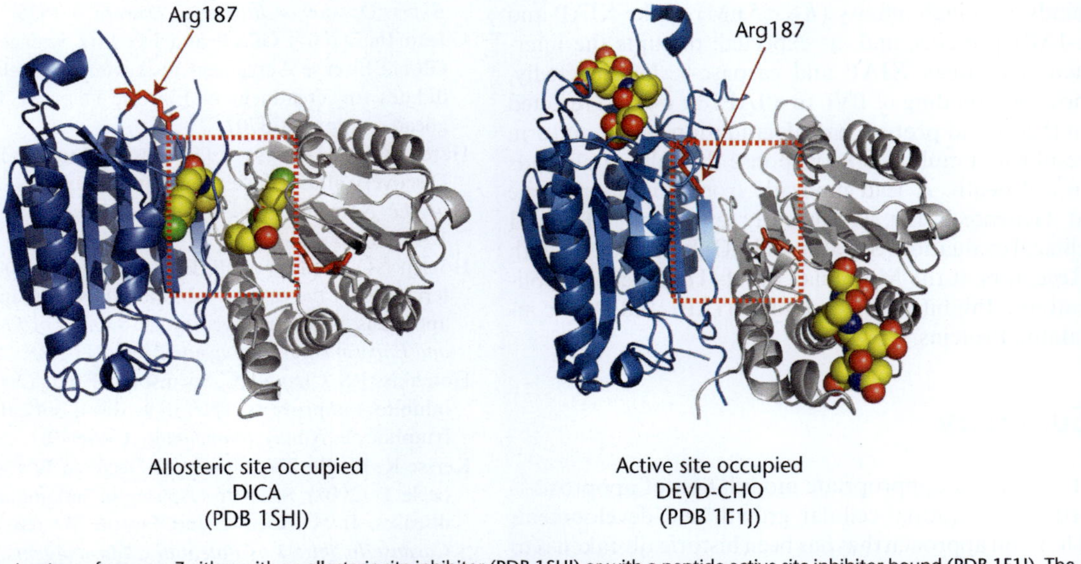

Figure 4 The structure of caspase-7 either with an allosteric site inhibitor (PDB 1SHJ) or with a peptide active site inhibitor bound (PDB 1F1J). The dashed box contains the allosteric binding site. The critical Arg187 that forms part of the S1 pocket is shown in red, and is displaced from the allosteric site when the DICA compound is bound (left panel). This residue occupies the allosteric pocket when the active site is occupied with a tetrapeptide inhibitor (DEVD-CHO) (right panel), thus preventing binding of DICA.

pro-apoptotic proteins such as Bad bind to pro-survival proteins to induce apoptosis. However, malignant cells tend to over express many of the pro-survival family members, for example Bcl-2, Bcl-x_L and Mcl-1, which contribute to the ability of cancer cells to evade apoptosis. A proof-of-concept molecule, ABT-737, binds to Bcl-x_L, Bcl-2 and Bcl-w with high affinity and induces apoptosis in cells and in xenograft models. ABT-263, an orally bioavailable compound with a similar binding profile to ABT-737, is currently undergoing Phase I clinical evaluation. In cells this compound disrupts the interaction between Bcl-x_L and Bim (pro-death protein) and in xenograft models was shown to significantly reduce tumour growth in multiple models.

The inhibitor of apoptosis (IAP) protein family, such as XIAP (X-linked IAP), c-IAP1 and c-IAP2 (cellular-IAP1 and 2), modulates apoptosis by inhibiting caspase activity either directly or indirectly. Initiation of the apoptotic cascade results in the release of the Smac (second mitochondria derived activator of caspase)/DIABLO protein complex from mitochondria, which binds directly to XIAP and prevents its interaction with caspases. The c-IAPs are thought to function by binding directly to Smac, thus preventing their ability to relieve XIAP-mediated caspase inhibition. Only four amino acids located at the *N*-terminal region of Smac are required for binding to the IAPs, thus small molecule compounds that mimic Smac binding to the IAPs could act to antagonize IAP activity. One example of an IAP antagonist is BV6 (Varfolomeev *et al*., 2007) which binds with high affinity ($K_D < 5$ nM) to the XIAP and c-IAP1 proteins, and, as expected, disrupts the interaction between XIAP and caspase-9. Unexpectedly, however, binding of BV6 to c-IAP1 or c-IAP2 resulted in their rapid proteasomal degradation which, in turn, results in a multifaceted response that ultimately ends in cell death. A lead molecule from this programme at Genentech Inc. is currently undergoing Phase I clinical evaluation. **See also**: BH3-Only Proteins; Death Receptors at the Molecular Level: Therapeutic Implications; Inhibitor of Apoptosis (IAP) and BIR-containing Proteins.

Summary

It is clear that appropriate modulation of apoptosis is critical for normal cellular growth and development. The main approach that has been historically taken is to directly inhibit caspase activity using active site inhibitors. Success in this area has been mixed, and it is clear that enthusiasm for caspase inhibitors has been reduced in recent years. The success, or failure, of the last remaining caspase inhibitor undergoing clinical evaluation could either solidify this viewpoint or re-ignite interest. In contrast, a new plethora of approaches are being taken to induce apoptosis as a way to target tumourigenic cells. In particular, a large number of molecules are undergoing clinical evaluation to determine which of these approaches may be clinically relevant. The next couple of years will be extremely interesting and it is likely that additional approaches may be discovered that have not yet been explored.

References

Ashkenazi A (2008) Directing cancer cells to self-destruct with pro-apoptotic receptor agonists. *Nature Reviews of Drug Discovery* **7**: 1001–1012.

Condorelli G, Roncarati R, Ross J *et al*. (2001) Heart-targeted overexpression of caspase3 in mice increases infarct size and depresses cardiac function. *Proceedings of the National Academy of Sciences of the USA* **98**: 9977–9982.

Erlanson DA, Lam JW, Wiesmann C *et al*. (2003) *In situ* assembly of enzyme inhibitors using extended tethering. *Nature Biotechnology* **21**: 308–314.

Fahr BT, O'Brien T, Pham P *et al*. (2006) Tethering identifies fragment that yields potent inhibitors of human caspase-1. *Bioorganic & Medicinal Chemistry Letters* **16**: 559–562.

Fleischmann R, Stern R and Iqbal I (2004) Anakinra: an inhibitor of IL-1 for the treatment of rheumatoid arthritis. *Expert Opinion on Biological Therapy* **4**: 1333.

Gilead Inc. (2007) Gilead and LG Life Sciences announce Global License Agreement to Advance Novel Drug Candidates for Treatment of Fibrotic Diseases. http://www.gilead.com/pr_1073707.

Hardy JA, Lam J, Nguyen JT, O'Brien T and Wells JA (2004) Discovery of an allosteric site in the caspases. *Proceedings of the National Academy of Sciences of the USA* **101**: 12461–12466.

Hoglen NC, Hirakawa BP, Fisher CD *et al*. (2001) Characterization of the caspase inhibitor IDN-1965 in a model of apoptosis-associated liver injury. *Journal of Pharmacology and Experimental Therapeutics* **297**: 811–818.

Hotchkiss RS, Chang KC, Swanson PE *et al*. (2000) Caspase inhibitors improve survival in sepsis: a critical role of the lymphocyte. *Nature Immunology* **1**: 496–501.

Kersse K, Berghe TV, Lippens S, Declercq W and Vandenabeele P (2008) Role of caspases in inflammation-driven diseases. In: O'Brien T and Linton SD (eds) *Design of Caspase Inhibitors as Potential Clinical Agents*, pp. 19–58. Boca Raton, FL: CRC Press.

Lee D, Long SA, Adams JL *et al*. (2000) Potent and selective nonpeptide inhibitors of caspases 3 and 7 inhibit apoptosis and maintain cell functionality. *Journal of Biological Chemistry* **275**: 16007–16014.

Li P, Allen H, Banerjee S *et al.* (1995) Mice deficient in IL-1b-converting enzyme are defective in production of mature IL-1[beta] and resistant to endotoxic shock. *Cell* **80**: 401–411.

Linton SD, Aja T, Armstrong RA *et al.* (2005) First-in-class pan caspase inhibitor developed for the treatment of liver disease. *Journal of Medicinal Chemistry* **48**: 6779–6782.

Lloyd-Jones D, Adams R, Carnethon M *et al.* (2008) Heart Disease and Stroke Statistics – 2009 update. A Report from the American Heart Association Statistics Committee and Stroke Statistics Subcommittee. Circulation, CIRCULATIONAHA.108.191261.

Matsui T (2008) Role of caspases in apoptotic driven indications. In: O'Brien T and Linton SD (eds) *Design of Caspase Inhibitors as Potential Clinical Agents*, pp. 75–92. Boca Raton, FL: CRC Press.

Methot N, Vaillancourt JP, Huang J *et al.* (2004) A caspase active site probe reveals high fractional inhibition needed to block DNA fragmentation. *Journal of Biological Chemistry* **279**: 27905–27914.

Mocanu M, Baxter GF and Yellon DM (2000) Caspase inhibition and limitation of myocardial infarct size: protection against lethal reperfusion injury. *British Journal of Pharmacology* **130**: 197–200.

Okamoto Y, Anan H, Nakai E *et al.* (1999) Peptide based interleukin-1 beta converting enzyme (ICE) inhibitors: synthesis, structure activity relationships and crystallographic study of the ICE-inhibitor complex. *Chemical & Pharmaceutical Bulletin (Tokyo)* **47**: 11–21.

Romanowski MJ, Scheer JM, O'Brien T and McDowell RS (2004) Crystal structures of a ligand-free and malonate-bound human caspase-1: implications for the mechanism of substrate binding. *Structure* **12**: 1361–1371.

Rudolphi K, Gerwin N, Verzijl N, van der Kraan P and van den Berg W (2003) Pralnacasan, an inhibitor of interleukin-1b converting enzyme, reduces joint damage in two murine models of osteoarthritis. *Osteoarthritis and Cartilage* **11**: 738–746.

Sarkar A, Hall MW, Exline M *et al.* (2006) Caspase-1 regulates *Escherichia coli* sepsis and splenic B cell apoptosis independently of interleukin-1b and interleukin-18. *American Journal of Respiratory and Critical Care Medicine* **174**: 1003–1010.

Scheer JM, Romanowski MJ and Wells JA (2006) A common allosteric site and mechanism in caspases. *Proceedings of the National Academy of Sciences of the USA* **103**: 7595–7600.

Tse C, Shoemaker AR, Adickes J *et al.* (2008) ABT-263: a potent and orally bioavailable Bcl-2 family inhibitor. *Cancer Research* **68**: 3421–3428.

Van Noorden CJF (2001) The history of Z-VAD-FMK, a tool for understanding the significance of caspase inhibition. *Acta Histochemica* **103**: 241–251.

Varfolomeev E, Blankenship JW, Wayson SM *et al.* (2007) IAP antagonists induce autoubiquitination of c-IAPs, NF-κB activation, and TNFa-dependent apoptosis. *Cell* **131**: 669–681.

Vertex Pharmaceuticals I (2006) Annual report. http://investors.vrtx.com/sec. cfm?DocType = Annual&Year = .

Wannamaker W, Davies R, Namchuk M *et al.* (2007) (S)-1-((S)-2-{[1-(4-amino-3-chloro-phenyl)-methanoyl]-amino}-3,3-dimethyl-butanoyl)-pyrrolidine-2-carboxylic acid ((2R,3S)-2-ethoxy-5-oxo-tetrahydro-furan-3-yl)-amide (VX-765), an orally available selective interleukin (IL)-converting enzyme/caspase-1 inhibitor, exhibits potent anti-inflammatory activities by inhibiting the release of IL-1b and IL-18. *Journal of Pharmacology and Experimental Therapeutics* **321**: 509–516.

Further Reading

LaCasse EC, Mahoney DJ, Cheung HH *et al.* (2008) IAP-targeted therapies for cancer. *Oncogene* **27**: 6252–6275.

Linton SD (2005) Caspase inhibitors: a pharmaceutical perspective. *Current Topics in Medicinal Chemistry* **5**: 1697–1717.

O'Brien T and Linton SD (2009) *Design of Caspase Inhibitors as Potential Clinical Agents*. CRC Enzyme Inhibitors Series. Boca Raton, FL: CRC Press.

Subject Index

Notes

Page numbers in **bold** indicate major discussions on the subject indexed.
Page numbers suffixed with *T* indicate tables: page numbers suffixed with *F* indicate figures.
To save space in the index, the following abbreviations have been used:
IAPs – inhibitors of apoptosis proteins
MOMP – mitochondrial outer membrane permeabilization

A

A1 70
expression regulation 73
induction 73
structure 78
A20 129
ABT-263 74, 86, 289–290
clinical trials 85*T*
ABT-737 163, 206, 290
Acinus, nuclear degradation 65
acquired immunodeficiency syndrome (AIDS) 11–12
activation-induced cell death (AICD) 220–221
acute hypoxic-ischaemic brain injury 264
acute renal failure (ARF), caspase-1 219
adalimumab 210
addiction molecules, bacteria 15, 16
adenine nucleotide translocase (ANT), MOMP 95
adenovirus E1B-19K 276
adenylic nucleotides, inner mitochondrial membrane (IMM) 98
adherence junctions, skin cornification 248
Advexin 236
ageing 18
autophagy *see* autophagy
necroptosis 134
AIDS (acquired immunodeficiency syndrome) 11–12
AIF (apoptosis-inducing factor), nuclear degradation 65
alarmins 213
ALDH4 gene activation, p53 234
ALFY (autophagy-linked FYVE protein) 194
ALPS *see* autoimmune lymphoproliferative syndrome (ALPS)
ALS (amyotrophic lateral sclerosis), protein misfolding 263
altruistic cell death, *Caenorhabditis elegans* 14
Alzheimer disease
amyloid plaques, nitric oxide 265–266
apoptosis dysfunction 48
autophagy 181
caspases 36
cell models 267
dynamin-related protein 1 (Drp1) *S*-nitrosylation 265
necroptosis 134
protein misfolding 262
AMBRA1, autophagy regulation 190
Ameisen, Jean Claude 9
AMG655 118*T*
clinical trials 124
AMPA receptors 264
AMP-dependent protein kinase (AMPK), autophagy regulation 180
AMPK (AMP-dependent protein kinase), autophagy regulation 180
amplification, apoptosis 33
AMPs (antimicrobial peptides), TOLL pathway 214–215
β-amyloid, misfolding 262
amyloid plaques, Alzheimer disease 265–266
amyotrophic lateral sclerosis (ALS), protein misfolding 263
anakinra 220
androgen receptor, autophagy in neurodegenerative disease 194
animal models
autoimmune disease *see* autoimmune diseases/disorders
Parkinson disease 264
ankylosing spondylitis, TNF receptors 209
ANT (adenylic nucleotides), inner mitochondrial membrane (IMM) 98
antiapoptotic Bcl-2 family proteins *see* Bcl-2 family proteins
anti-FasL antibodies 221
antigen-presenting cells (APCs) 213
anti-inflammatory cytokines, apoptotic cell engulfment 172
antimicrobial peptides (AMPs), TOLL pathway 214–215
Apaf-1 (apoptosis-activating factor-1) 38–39
apoptosome 38, 39
caspase-9 activation 225
cytochrome *c* association 24
in disease 205*T*
isoforms 39
knockout animals 48
development 47*F*
oligomerization, intrinsic apoptosis pathway 54, 91
procaspase-9 complex 152, 153*F*
regulation, transcription/translation 43
structure 38–39, 41*F*
CARD 38–39, 149–150, 151*F*
NB-ARC (nucleotide-binding Apaf-1/Rgene/CED4) domain 38–39
NBD (nucleotide α/β binding domain) 39
NOD 38–39, 150
Apaf-1 (apoptosis-activating factor-1) (*continued*)
WD-40 region 38
WD repeat domain 150
WHD (winged-helix domain) 39
see also Dark protein
Apaf-1-interacting protein 43
APO-1 *see* Fas/CD95
APO-2 *see* TRAIL-R1 (DR3)
Apo2L.DR5-8 122
APO-3 *see* TRAIL-R1 (DR3)
apomab 118*T*, 123–124
clinical trials 124
structure 124*F*
apoptosis
amplification 33
autophagy *vs.* 185*F*, 185*T*
cell dismantling *see* demolition phase of apoptosis
cell microscopy 128*F*
definition 6, 128
demolition 33
determination 33
in disease 283–284
see also specific diseases/disorders
initiation 33
innate immunity *vs.* 217*F*
mechanisms 253
molecular events 8*F*
see also extrinsic apoptosis pathway; granzyme B apoptosis pathway; intrinsic apoptosis pathway; *specific processes*
apoptosis-activating factor-1 *see* Apaf-1 (apoptosis-activating factor-1)
apoptosis- and splicing-associated protein (ASAP), nuclear degradation 65
apoptosis antigen-1 (APO-1) *see* Fas/CD95
apoptosis-inducing factor (AIF), nuclear degradation 65
apoptosomes **37–50**, 61, 91
assembly 39
Apaf-1 38, 39
ATP 39
cytochrome *c* 39
extrinsic apoptosis pathway 200
C. elegans 43, 45, 46*F*
caspases 33
caspase-3 38
caspase-7 38
caspase-9 38, 40–41, 149–150
cytochrome *c* 38, 41–42
D. melanogaster 43, 45–47, 46*F*
Buffy 46
cytochrome *c* 45
apoptosomes (*continued*)
Dark protein 45
Dcp-1 45
Debel 46
Drice 45
Drp1 46–47
definition 38
evolution 43, 45–47
formation modulation 42–43
mammalian development 47–48
structure 38–42, 41*F*, 45*F*
three-dimensional 39
see also specific components
arsenic trioxide (ATO), Bid effects 82
ASAP (apoptosis- and splicing-associated protein), nuclear degradation 65
asymmetric DD:DD interactions 153*F*, 154
asymmetric division
ageing 18
programmed cell death 15
Saccharomyces cerevisiae 18
ataxia telangiectasia 260
Atg genes/proteins, autophagy 176
Atg1, autophagy 178
atg1 gene, autophagy regulation 190
Atg3, autophagy regulation 190
Atg5
Atg12 conjugation system 176
autophagy regulation 190
in cancer 182
Atg6, autophagy regulation 190
Atg7, autophagy 176, 190
Atg8
autophagy regulation 190
LC3–PE complex 176
atg8 gene mutation, *Drosophila melanogaster* 194
Atg10, autophagy regulation 190
Atg12
Atg5 conjugation system 176
autophagy regulation 190
Atg14, autophagy 178
Atg16L complex, autophagy 176–177
Atg17, autophagy 178
ATP
apoptosome assembly 39
autophagy regulation 180
synthesis, mitochondria 97
ATP-dependent aminophospholipid translocases 167
autoantibodies, FasL 221
Autographa californica, p53 caspase inhibition 34

autoimmune diseases/disorders 203, 255–259
animal models 255–256, 256*T*
cancers 257
extrinsic apoptosis pathway 221
lymphocyte apoptosis 256
MFG-E8 169–170, 170*F*
see also specific diseases/disorders
autoimmune lymphoproliferative syndrome (ALPS) 148
aetiology 256
appearance 257*F*
caspase-8 209, 221, 221*F*
caspase-10 209, 221, 221*F*
clinical findings 257*T*
Fas–FasL pathway 208
Fas mutations 258*F*
autophagy 67–68, **175–188, 189–197**
ageing 194–195
apoptosis *vs.* 185*F*, 185*T*
Atg genes/proteins 176, 178, 190
C. elegans
neuronal cell death 193
nutrient restriction 192
cancer 181–182, 182*F*
anticancer therapy 182, 183*F*
cell death 184–185
D. melanogaster 189, 192–193
ageing 194
atg8 gene mutation 194
autophagy-linked FYVE protein (ALFY) 194
blue cheese (bchs) gene 194
caspases 193
cell death 192–193
development 192
nutrient restriction 192
definition 128, 189
diseases/disorders 180–184
cancer *see above*
cardiac disease 181
infections *see below*
inflammatory disease 183–184
models 193–195
myodegenerative disease 181, 184
neurodegenerative disease 181, 193–194
functions 191–193
infections 183–184, 193
phagocytosis 184*F*
kinases 178
3-methyladenine (3-MA) inhibition 184
molecular components 177*F*
nutrient restriction 192
oxidative stress 192
programmed cell death type II 176
Rabs 178–179
regulation 179–180, 180*F*, 189–191, 191*F*
AMBRA1 190
AMP-dependent protein kinase (AMPK) 180
Atg proteins 190
ATP levels 180
BECLIN1 190
Beclin-1 179
cytoplasm to vacuole-targeting (CVT) pathway 190
eukaryotic initiation factor 4E (eIF4E) 179
eukaryotic initiation factor 4E-binding protein-1 (4E-BP1) 179
insulin 180
PI3K 179, 190
protein phosphatase 2A (PP2A) 180
S cerevisiae 189–190
autophagy *(continued)*
Tap42 180
TOR 179, 180, 190
VPS15 190
VPS34 190
vps34 mutants 190–191
SNAREs 178–179
ubiquitin-like conjugation systems 176–178, 190
Atg7 176
Atg8/LC3–PE complex 176
Atg12–Atg5 conjugation system 176
Atg16L complex 176–177
phagophore 176
autophagy-linked FYVE protein (ALFY) 194
axon pruning, apoptotic cell engulfment 173

B

Bacillus subtilis
programmed cell death 14
asymmetric division 15
SpoA sporulation factor 15
bacteria
addiction molecules 15, 16
apoptosis induction 273
apoptosis inhibitors 278–279
host cell interactions 273*F*
paracrine killers 15
programmed cell death 14
see also individual species
baculoviruses, caspase inhibitors 277
Bad 80*T*, 84
Bcl family member binding 71
chronic myeloid leukaemia (CML) 83
BAI1 (brain-specific angiogenesis inhibitor 1), apoptotic cell engulfment 172
Bak 101*F*
activation 72, 93*F*
BH3-only proteins 79*F*
granzyme B pathway 202
tBid (truncated Bid) 62
Bax double knockouts 104
Bcl family member binding 71, 72*F*
in disease 205*T*
intrinsic apoptosis pathway 54, 274
knockout animals 73
membrane location 160
MOMP 92, 105
regulation 198
structure 78
BALF1 276
Bax 101*F*
activation 72, 93*F*
BH3-only proteins 79*F*
granzyme B pathway 202
tBid (truncated Bid) 62
Bak double knockouts 104
Bcl family member binding 71, 72*F*
in disease 204*T*
cancer 259
colorectal cancer 74
intrinsic apoptosis pathway 54, 274
intrinsic apoptotic pathway 233–234
knockout animals 73
membrane location 160
MOMP 92, 105
outer mitochondrial membrane 92
regulation 198
structure 78, 159*F*
B cell(s)
autoimmune disease 255–256
survival, Bcl-2 206
B-cell lymphomas
Bik/Nbk/Blk 83
Bim 83
granzyme B 207
Bcl-2
B-cell survival 206
blockable apoptosis *see* intrinsic apoptosis pathway
demolition phase of apoptosis 67
in disease 204*T*
autophagy in cancer 182
carcinogenesis 78
follicular lymphoma *see* follicular lymphomas
haematopoietic cancers 205
lymphoid cancers 205
identification 4, 70
knockout animals 73
overexpression 70
structure 78
T-cell survival 206
viral homologues 73
Bcl-2 family proteins 24, **69–75**, 77*F*, 91*F*
antagonists 74, 118, 162–163
BH3-protein mimetics 163
antiapoptotic (survival) 70*F*, 77*F*, 92, 158
binding characteristics 71
C-terminus 70
location 160
proapoptotic balance 198
BH3-only *see* BH3-only proteins
in disease 205–206
cancer 74, 78–79
evolution 72
functions **156–165**
apoptosis modulation 115
caspase regulation 157*F*
intrinsic apoptosis pathway 54, 233
MOMP 91–92, 92–93
interactions 71–72, 93*F*
in vivo roles 73
membrane interactions 160
mitochondria network *see* mitochondria network
outer mitochondrial membrane (OMM) association 98
proapoptotic 70*F*, 77*F*, 158
antiapoptotic balance 198
signal integration/regulation 72–73
structure 70–71, 70*F*, **156–165**, 159*F*, 161*F*
BIR domain 160, 161*F*
domain organization 158*F*
RING domain 160–161, 161*F*
structure–function relationship 159, 160–161
see also intrinsic apoptosis pathway; *specific proteins*
Bcl-2 homology (BH) domains 70
Bcl2L11 *see* Bim
Bcl-2-like viral proteins (vBcl-2) 275–277
adenovirus E1B-19K 276
BALF1 276
BHRF1 276
cytomegalovirus 276
Herpesvirus Saimiri 276
human herpesvirus 8 276
poxviruses 276–277
Bcl-B *see* Bcl-w (Bcl-B)
Bcl-w (Bcl-B) 70, 101*F*
lymphoid cells 206
structure 78, 159*F*
Bcl-x 70
expression regulation 73
knockout animals 73
Bcl-x_L 101*F*
autophagy 182
demolition phase of apoptosis 67
in disease 206
cancer 78, 182
knockout animals 104
lymphoid cells 206
mitochondria network 104
overexpression 104
structure 78
Beclin 1
autophagy 178, 179, 190
in cancer 182
deficiency 182
BH3-only proteins 70*F*, 71, **75–90**, 77*F*
Bax/Bak activation 79*F*
Bcl family member binding 71
function 80–81*T*, 159–160
intrinsic apoptosis pathway 274
mimetic drugs 86, 163
see also specific drugs
MOMP 93
pro-survival protein interaction 160
regulation 80–81*T*, 198
signalling 78, 78*F*
structure 71, 78, 158–159
BHRF1 276
Bid 71, 80*T*, 82
activation 62
caspase-8 200
granzyme B 225
see also tBid (truncated Bid)
arsenic trioxide (ATO) effects 82
CD95 signalling pathways 113
cisplatin effects 82
granzyme B pathway 202, 225
histone deactylase inhibition (HDACi) effects 82
knockout animals 82
MOMP 94, 95
proteasome inhibitor effects 82
Bik 80*T*, 84
B-cell lymphomas 83
Bcl family member binding 71
deficiency 83
prostate cancer 83
T-lymphoma cells 83
Bim 80*T*, 82–84
ablation, autoimmune disease 256
anticancer drug efficacy 83
B-cell lymphomas 83
Burkitt lymphoma 83
DNA damage-induced diagnosis 83
homeostasis regulation 83
induction 83
isoforms 82
in lymphoma 83
MOMP 94, 95
posttranslational regulation 83
systemic lupus erythematosus-like disease 83
T-cell development 220
as tumour suppressor 83
BIR (baculovirus IAP repeat) 138
caspase inhibitors 34
BIRC1 *see* NAIP (BIRC1)
BIRC2 *see* cIAP1
BIRC3 *see* cIAP2
BIRC4 *see* XIAP (X-linked inhibitor of apoptosis)
BIRC5 *see* Survivin
BIRC6 *see* BRUCE
BIRC7 142
see also ML-IAP
BIR-containing proteins **138–147**
BIR-fold structure 138, 140
see also inhibitors of apoptosis proteins (IAPs); *specific proteins*
BIR domains 138, 140
Bcl-2 family proteins 160, 161*F*

blebbing, demolition phase of apoptosis 55
Blk *see* Bik
blue cheese (bchs) gene, *Drosophila melanogaster* autophagy 194
Bmf 81*T*, 84, 86
Bcl family member binding 71
knockout animals 83, 86
necroptosis pathways 134
splice variants 86
Bod *see* Bim
Bok 70
MOMP 92
structure 78
brain-specific angiogenesis inhibitor 1 (BAI1), apoptotic cell engulfment 172
breast cancer, p53, therapy effects 236
Brenner, Sydney 3
bridging molecules, apoptotic cell engulfment *see* engulfment of apoptotic cells
BRUCE 138, 140*T*, 144–145
autophagy 193
caspase inhibition 35*T*
structure 139*F*
Buffy, *Drosophila melanogaster* apoptosome 46
Burkitt lymphoma, Bim 83

C

C1q-deficient animals, apoptotic cell engulfment 171
CA (carnosic acid) 268–269
CAD (caspase activated dnase)
DNA condensation/degradation 56
nuclear degradation 65
Caenorhabditis elegans 17, **21–29**
apoptosis
cell engulfment 24–26, 25*F*, 167
germline apoptosis 26
induction 25*F*
apoptosome 43, 45, 46*F*
autophagy *see* autophagy
C. elegans p53-like 1 (CEP-1) 17–18
caspase homologues 51
CED-3 *see* CED-3
CED-4 *see* CED-4
CED-9 *see* CED-9
cell death
altruistic 14
multiple pathways 17–18
regulation 24
cell lineage 21–22, 22*F*
conserved genes 23–24, 24*F*
see also specific genes
'death genes' 14
development 21, 22*F*
DNA damage-induced germline apoptosis 26–27
Egl-1 18
embryogenesis 21
historical aspects 21
Nomarski (DIC) optics 21, 22*F*
p53 232
programmed cell death evolution 13–14
second larval stage 21
Caenorhabditis elegans p53-like 1 (CEP-1) 17–18, 27
calcium-dependent necroptosis pathways 133
calcium-independent phospholipase A (iPLA) 67
calpains, skin cornification 249
Canale–Smith syndrome *see* autoimmune lymphoproliferative syndrome (ALPS)
cancer 11–12, 259–260
autoimmune disease 257
autophagy *see* autophagy
Bax 259
Bcl-2 family proteins 74
caspase-8 209
cell killing, TRAIL 207–208
cIAPs 210
DAPK1 (death-associated protein kinase 1) 259
definition 230
p53 203
PTEN (phosphatase and tensin homolog) 259
PUMA 259–260
therapy
autophagy 182, 183*F*
Bim efficacy effects 83
TRAIL receptors 124–125
TRAIL-Rs, differential expression 119
carcinogenesis 90, 203
Bcl-2 78
Bcl-2 family proteins 78–79
Bcl-X_L 78
p53 inactivation 162–163
CARD (caspase-recruitment domain) **147–156**
Apaf-1 38–39, 149–150
CARD interactions 152, 153*F*
homotypic interactions 219
inflammasome 219
caspases 51, 61
cIAP1 142
cIAP2 142
functions 148–150
group II (initiator) caspases 61
sequence homology 152
structure 149*F*, 150
cardiac disease, autophagy 181
cardiac ischaemia 283–284
CARDINAL/TUCAN, apoptosome formation 43
carnosic acid (CA) 268–269
caspase(s) **30–37**, **50–60**, 51*F*, 61–66, 91
activation 52–54, 52*F*, 62, 62*F*, 91, 111–112
apoptosome 33
caspase-3 54
DISC (death initiation signalling complex) 31–32, 114
FLIP 149
proteolytic processing 51–52
reactive oxygen species 264–265
see also extrinsic apoptosis pathway; granzyme B apoptosis pathway; intrinsic apoptosis pathway
active site 31–32
adaptive immune response 220–221
autophagy, *D. melanogaster* 193
ced-3 gene 23
classification 30–31, 31*F*, 51–52
see also specific classes
death receptor signalling 111–112
definition 4, 61
group I (inflammation) 30, 51, 197, **212–223**, 218*F*
diseases/disorders 219–220
group II (initiator) 30, 51, 61, 91
death effector domain (DED) 61
knockout animals 61*T*
group III (effector) 30–31, 51, 61, 91
activation 31
knockout animals 61*T*
prodomain 31
history 30–31
inhibitors *see* caspase inhibitors
innate immunity **212–223**, 214*F*
caspase(s) (*continued*)
knockout animals 34, 36*T*, 61, 61*T*
mechanism of action 32–33, 32*F*
molecular actions 197
necroptosis 135
pathology 35–36
PRR binding 215–219
regulation
Bcl-2 family proteins 157*F*
IAPs 157*F*, 161–162
XIAP 162
sequence of action 33, 33*F*
structure 31–32, 31*F*, 51–52, 62*F*, 111
caspase recruitment domain (CARD) 51
caspase recruitment domain (DED) 61
death effector domain (DED) 51, 61
dimers 31
substrates 33, 58
recognition 32–33
specificity 32, 34*F*
therapeutic implications 36
see also specific types
caspase-1 30
activation 51
inflammasome 150
active form 218, 218*F*
autophagy 193
binding site, inhibitor design 289
in disease
acute renal failure 219
cutaneous T-cell lymphoma (CTCL) 219
inflammation 217
metastatic melanoma 219
multiple sclerosis 219
pyroptosis 217
septic shock 219
systemic inflammatory response syndrome (SIRS) 219
inhibitors 35*T*, 217–218
knockout animals 36*T*
drug discovery 283
sepsis, protection from 283
caspase-2 30
activation 31, 33, 61
PIDDosome 150
inhibitors 35*T*
knockout animals 36*T*, 61*T*
small molecule inhibitors 284
caspase-3 30, 158
actions 33
activation, caspase-9, by 33
apoptosis demolition phase 54
apoptosome 38
binding site, inhibitor design 287–288
cardiac ischaemia 284
caspases, activation of 54
caspase-9 42
deficiency 62
granzyme B apoptosis pathway 225
inhibitors 35*T*
knockout animals 36*T*, 61*T*
development 47–48
mechanism of action 33
caspase-4 30
inflammation 217
inhibitors 35*T*
caspase-5 30
inflammation 217
inhibitors 35*T*
caspase-6 30
activation 33
apoptosis demolition phase 54
caspase-6 (*continued*)
inhibitors 35*T*
knockout animals 36*T*, 61*T*
caspase-7 30, 158
actions 33
activation, caspase-9, by 33
apoptosis demolition phase 54
apoptosome 38
binding site, inhibitor design 287–289, 289*F*
deficiency 62
granzyme B apoptosis pathway 225
inhibitors 35*T*
knockout animals 36*T*, 61*T*
caspase-8 30
activation 31–32, 148–149
dimerization 52
DISC (death-inducing complex) 61
TNFR1 201
Bid activation 200
CD95 signal transduction 30
in disease 204*T*
autoimmune lymphoproliferative syndrome 221, 221*F*
inhibitors 35*T*
knockout animals 36*T*, 61*T*
mutations 209
in regulation 158
RIP1, cleavage of 131
RIP3, cleavage of 131
structure 255
caspase-8 deficiency state (CEDS) 259
caspase-9 30
activation 30–31, 31–32, 33, 40, 61
Apaf-1 225
apoptosome 149–150
caspase-3 by 33, 42
dimerization 52
apoptosome 38, 40–41
in disease 205*T*
functions 33
in regulation 158
inhibitors 35*T*
knockout animals 36*T*, 61*T*
development 47, 47*F*
mammalian development 47–48
regulation 43
structure 40–41, 41*F*
see also procaspase-9
caspase-10 30
activation 148–149
CD95 signal transduction 30
in disease 204*T*
autoimmune lymphoproliferative syndrome 209, 221, 221*F*
inhibitors 35*T*
structure 255
caspase-11, knockout animals 36*T*
caspase-12 30
inflammation 217
knockout animals 36*T*
caspase-14 31
caspase activated dnase *see* CAD (caspase activated dnase)
caspase-independent cell death 67–68
see also autophagy
caspase inhibitors 33–34, 35*T*, 284, 286–287
alternative approaches 287–289
baculoviruses 277
BIR (baculovirus IAP repeat) 34
clinical development 287
general features 284, 286
natural 33–34
pharmacological 33–34
poxviruses 278
Rabbitpox virus 278
structure 285*F*, 286*F*

caspase inhibitors (*continued*)
viruses 277–278
'warhead' 284, 286
XIAP 199
caspase-recruitment domain *see* CARD (caspase-recruitment domain)
catenins, cell detachment 55
cathepsin(s)
necroptosis pathways 133
skin cornification 249
cathepsin D, skin cornification 249
CD95 *see* Fas/CD95
CD95L *see* FasL/CD95L
CD120a *see* tumour necrosis factor receptor 1 (TNF-R1)
CED-1, apoptotic cell engulfment 24–25, 168, 168*T*
CED-2, apoptotic cell engulfment 25, 168, 168*T*
CED-3 18, 23
germline apoptosis 26
loss-of-function mutants 17
see also caspase-3
ced-3 gene 3, 23–24
mutants 23
ionizing radiation 27
CED-4 18, 23
CED-9 association 45
germline apoptosis 26
see also Apaf-1 (apoptosis-activating factor-1)
ced-4 gene 3, 23–24
loss-of-function mutants 17
mutants 23
ionizing radiation 27
CED-4L 23
CED-5
apoptotic cell engulfment 168, 168*T*
dying cell engulfment 25
CED-6, apoptotic cell engulfment 24–25, 168*T*
CED-7, apoptotic cell engulfment 24–25, 167, 168, 168*T*
CED-9 18, 23
CED-4 association 45
EGL-1 association 45
germline apoptosis 26
loss-of-function mutants 17
mitochondria network 103–104
ced-9 gene 3, 23–24
mutants, ionizing radiation 27
CED-10, apoptotic cell engulfment 25, 168, 168*T*
CED-11, apoptotic cell engulfment 168*T*
CED-12, apoptotic cell engulfment 25, 168, 168*T*
cell adhesion, skin cornification *see* skin cornification
cell–cell connectivity, cell detachment 55
cell fate 1
mapping of 3
cell lineage, *Caenorhabditis elegans* 21–22, 22*F*
cell models, Alzheimer disease 267
CEP-1 (Caenorhabditis elegans p53-like 1) 17–18, 27
c-FLIP proteins 149
apoptosis modulation 114–115, 115*F*
pro-survival signalling 201
chemoattractants, apoptotic cell engulfment 166
chemotherapy
induced apoptosis 236
resistance 79
Chlamydia, apoptosis inhibitors 273–274
chlamydial protease activity factor (CPAF) 279
Chlamydia pneumoniae, apoptosis inhibitors 278
Chlamydia trachomatis, apoptosis inhibitors 278
chronic lymphocytic leukaemia (CLL)
DAPK1 mutations 260
PUMA upregulation 81
sensitization to anti-TRAIL receptor antibodies 122
TRAIL resistance 125
chronic myeloid leukaemia (CML), Bad 83
cIAP(s)
in disease 210
drug design 290
cIAP1 138, 140*T*, 142–143, 158
caspase inhibition 35*T*
in disease 204*T*, 210
multiple myelomas 143
knockout animals 142
pro-survival signalling 201
signalling 142–143
TNF signalling 142–143, 143*F*
structure 139*F*
caspase recruitment domain (CARD) 142
TRAF1/TRAF2 binding 142, 162
cIAP2 138, 140*T*, 142–143, 158
caspase inhibition 35*T*
in disease 204*T*, 210
multiple myelomas 143
knockout animals 142
signalling 142–143
pro-survival signalling 201
structure 139*F*
caspase recruitment domain (CARD) 142
TRAF1/TRAF2 binding 142, 162
cisplatin, Bid effects 82
clinical trials
R-roscovitine 203, 205
TRAIL receptor agonists 123–124, 124–125
CLK-2RAD-5, DNA damage-response pathways 27
CLL *see* chronic lymphocytic leukaemia (CLL)
CML (chronic myeloid leukaemia), Bad 83
CO-31398 (styrylquinazoline) 237
coevolution, apoptosis and multicellularity 9
colorectal cancer, Bax mutations 74
complement, apoptotic cell engulfment 171
corneodesmosin, skin cornification 248–249
Cowden syndrome 260
Cowpox virus, caspase inhibitors 278
CPAF (chlamydial protease activity factor) 279
cristae remodelling, mitochondria 106–107
CrkII, apoptotic cell engulfment 168–169
CrmA (cytokine response modifier A), caspase inhibition 34, 35*T*
Crohn disease, *Nod2* mutations 220
crosslinking, small proline-rich proteins 246–247
CS-1008 *see* TRA-8 (CS-1008)
CTCL (cutaneous T-cell lymphoma), caspase-1 219
C-terminus, antiapoptotic Bcl-2 family proteins 70
Currie, Alistair 3
cutaneous T-cell lymphoma (CTCL), caspase-1 219
CVT (cytoplasm to vacuole-targeting) pathway, autophagy regulation 190
CX3CL1 (fractalkine)
apoptotic cell engulfment 167
phagocytosis signal 57
CYLD (cylindromatosis) 129
cylindromatosis (CYLD) 129
cysteine aspartic acid-specific proteases *see* caspase(s)
cytochrome c
Apaf-1 association 24
apoptosome 38, 39, 41–42
D. melanogaster 45
CD95 signalling pathways 113
deficiency 48
heme-binding 42
intrinsic apoptosis pathway 54, 91
knockout animals 47*F*
release 42, 62
structure 41*F*
cytokine response modifier A (CrmA), caspase inhibition 34, 35*T*
cytokines, inflammatory 209–210
cytomegalovirus, Bcl-2-like viral proteins (vBcl-2) 276
cytoplasm to vacuole-targeting (CVT) pathway, autophagy regulation 190
cytoskeleton
cell rounding 55
demolition phase of apoptosis 64, 68
nuclear fragmentation 56
cytotoxic T-cells 202, 223–224
cell killing 225*F*
extrinsic apoptosis pathway 53
FasL 208

D

DAMPs (danger-associated molecular patterns) 213
danger-associated molecular patterns (DAMPs) 213
DAPK1 (death-associated protein kinase 1)
autophagy 182
cancer 182, 259
chronic lymphocytic leukaemia (CLL) 260
mutations 260
Dark protein 45
Dcp-1, *Drosophila melanogaster* apoptosome 45
DcR1 *see* TRAIL-R3
DcR2 *see* TRAIL-R4
DcR3 110
structure 111*F*
DD (death domain) **147–156**
functions 148–150
interactions 152, 153*F*, 154
sequence homology 152
structure 149*F*, 151*F*
see also CARD (caspase-recruitment domain); DED (death effector domain); PYD (pyrin domain)
death-associated protein kinase 1 *see* DAPK1 (death-associated protein kinase 1)
death domain *see* DD (death domain)
death effector domain *see* DED (death effector domain)
'death genes,' *Caenorhabditis elegans* 14
death ligands (DLs) 110–111
see also specific ligands
death receptors **110–116, 117–127**
apoptosis modulation 114–115
c-FLIP proteins 114–115, 115*F*
definition 110
induced necrosis *see* necroptosis
death receptors (*continued*)
ligands 110–111
pro-survival signalling 201–202
signalling 200*F*
caspases 111–112
initiation 110–111
pro-survival 201–202
structure 110, 111*F*
as therapeutic targets 117–119, 207–208
see also extrinsic apoptosis pathway; *specific receptors*
Debel, *Drosophila melanogaster* apoptosome 46
DED (death effector domain) **147–156**, 255
caspases 51, 61
DED interactions 152, 153*F*, 154
functions 148–150
group II (initiator) caspases 61
sequence homology 152
structure 149*F*, 150
DEF40 *see* CAD (caspase activated dnase)
demolition phase of apoptosis 33, 54–56, **60–69**
blebbing 55
caspases *see* caspase(s)
cell physiology 56
cytoskeletal events 64, 68
detachment 55, 63
DNA condensation/degradation 56
effectors 54
gain-of-function substrate cleavage 62–63, 64*F*
Golgi apparatus 66
hallmark events 54, 55*F*
loss-of-function substrate cleavage 62–63, 64*F*
membrane blebbing 64
morphological changes 64–66, 65*F*, 90–91
nuclear events 64–66
Acinus 65
AIF (apoptosis-inducing factor) 65
ASAP (apoptosis- and splicing-associated protein) 65
CAD (DNA fragmentation factor 40/DEF40 65
endonuclease G 65
inhibitor of CAD (ICAD) 65
nuclear lamina proteins 65
nucleoporins 65
PARP-1 (poly(adenosine diphosphate (ADP)-ribose)polymerase) 65
SAF-A (scaffolding attachment factor) 65
nuclear fragmentation 56
phagocytosis signals 67
rounding 55
survival pathway targeting 67
transcriptional machinery 66–67
pro-apoptotic gene activation 65
RNA splicing 65
survival gene inactivation 65
translational machinery 66–67
initiation factors 65
see also engulfment of apoptotic cells
dendritic cells, autophagy in infections 183–184
desmocollin-1, skin cornification 248
desmoglein-1, skin cornification 248
desmosomes, skin cornification 248
desquamation, skin cornification 248–249
detachment, demolition phase of apoptosis *see* demolition phase of apoptosis

determination, apoptosis 33
DIABLO *see* Smac/DIABLO
DIAP1, *Drosophila melanogaster* 141–142
dimerization
caspase-8 activation 52
caspase-9 activation 52
direct activator model
Bak activation 93*F*
Bax activation 93*F*
MOMP induction 93–94, 93*F*
DISC (death initiation signalling complex)
caspase activation 31–32, 114
caspase-8 61
extrinsic apoptosis pathway 199
formation 112, 232
structure 112, 112*F*
disease models, autophagy 193–195
displacement model
Bak activation 93*F*
Bax activation 93*F*
MOMP induction 93*F*, 94
disulfide bonds, PDI 267
DNA
condensation/degradation *see* demolition phase of apoptosis
damage-induced apoptosis
Bim 83
NOXA 79, 81
PUMA 79, 81
damage-response pathways 27
DNA fragmentation factor 40 *see* CAD (caspase activated dnase)
Dnm1p, mitochondrial fission 102
Dock180, apoptotic cell engulfment 168–169
downstream signalling, pro-survival signalling 202
DP5 *see* Hrk/DP5
DR1 *see* tumour necrosis factor receptor 1 (TNF-R1)
DR2 *see* Fas/CD95
DR3 *see* TRAIL-R1 (DR3)
DR4 *see* TRAIL-R2
DR5, p53 extrinsic apoptosis pathway 232
DR6 110
signalling 114
structure 111*F*
DREDD 213–215, 214
IMD pathway 213, 214–215
TOLL pathway 213–214
Drice, *Drosophila melanogaster* apoptosome 45
Drosophila melanogaster
apoptosome *see* apoptosomes
autophagy *see* autophagy
caspase homologues 51
DIAP1 141–142
p53 232
Drp1 (dynamin-related protein 1) 101*F*
Alzheimer disease 265
D. melanogaster apoptosome 46–47
knockout animals 101–102, 103
mitochondrial fission 94, 101
mitochondrial fusion 94
negative mutants 105
S-nitrosylation 265
trafficking, mitochondria network 105
drug discovery **282–291**
caspase inhibitors *see* caspase inhibitors
noncaspase inhibitors 289–290
see also specific drugs
dying cell engulfment, *Caenorhabditis elegans* *see Caenorhabditis elegans*
dynamin-related GTPases 94
dynamin-related protein 1 *see* Drp1 (dynamin-related protein 1)

E

"eat-me" signals, apoptotic cell engulfment *see* engulfment of apoptotic cells
EBV see Epstein–Barr virus (EBV) infection
20-ecdysone, autophagy 192
EDF (extranuclear death factor), *Escherichia coli* 16
EDN (eosinophil-derived neurotoxin) 213
EGg Laying defective *see* EGL-1 (EGg Laying defective)
EGL-1 (EGg Laying defective) 23
CED-9 association 45
egl-1 gene 23–24
C. elegans 18
transcription regulation 24
eIF4E (eukaryotic initiation factor 4E), autophagy regulation 179
electrophiles, oxidative/nitrosative stress protection 268–270
ELMO1, apoptotic cell engulfment 168–169
embryogenesis, *Caenorhabditis elegans* 21
Emricasan 284, 287
structure 288*F*
endonuclease G, nuclear degradation 65
endoplasmic reticulum (ER)
demolition phase of apoptosis 56
redox environment 267
stress in, nitric oxide (NO) 267
endosomal sorting complex required for transport (ESCRT machinery), autophagy 179
engulfment of apoptotic cells **165–175**
bridging molecules 169–171
complement 171
growth arrest-specific 6 (Gas6) 170
MFG-E8 169–170
protein S 170
TAM receptors 170–171
"eat-me" signals 166, 167–168
ATP-dependent aminophospholipid translocases 167
C. elegans studies 167
phosphatidylserine 167
phospholipid scramblases 167
"find-me" signals 166–167
chemoattractants 166
CX3CL1 (fractalkine) 167
lysophosphatidylcholine (LPC) 166
phospholipase A2 166–167
future work 172–173
immunosuppressive role 172
phosphatidylserine receptors 171–172
signalling pathways 166*F*, 168–169, 168*T*
see also demolition phase of apoptosis
eosinophil-derived neurotoxin (EDN) 213
epidermal differentiation *see* skin cornification
epidermal proteases, skin cornification 249
Epstein–Barr virus (EBV) infection
adenovirus E1B-19K 276
granzyme B 207
XIAP 206
Epstein–Barr virus (EBV) infection (*continued*)
X-linked lymphoproliferative (XLP) syndrome 259
ERK *see* extracellular signal-regulated kinase (ERK)
erythropoiesis, apoptotic cell engulfment 172
Escherichia coli 9–10
extranuclear death factor (EDF) 16
mazE/mazF addiction molecule 16
toxin pleiotropy 17
ESCRT machinery (endosomal sorting complex required for transport), autophagy 179
etanercept 210
eukaryotic initiation factor 4E (eIF4E), autophagy regulation 179
eukaryotic initiation factor 4E-binding protein-1 (4E-BP1), autophagy regulation 179
evolution (of cell death) 5, 9–11, 10*F*
mitochondrial damage 11
'original sin' hypothesis 16–17
stability 15–16
symbioses 10–11
see also Red Queen metaphor
extracellular matrix, detachment 63
extracellular proteases, skin cornification *see* skin cornification
extracellular signal-regulated kinase (ERK)
intrinsic apoptosis pathway 274
procaspase-9 phosphorylation 43
extranuclear death factor (EDF), *Escherichia coli* 16
extrinsic apoptosis pathway 38, 52, 76, 77–78, 117, 157, 198, 254
apoptosome formation 200
cytotoxic T lymphocytes 53
definition 91
DISC 199
in disease 204*T*, 207–210
autoimmune disease 221
immunodeficiencies 221
FADD (Fas-associated death domain) 53, 199
Fas 53, 199
intrinsic pathway cross-talk 200
network 38*F*, 53*F*, 63*F*, 76*F*, 200*F*, 254*F*
p53 *see* p53
regulatory genes 199–201
as therapeutic target 118
TNFR1 199
TNF receptor super family (TNFRSF) 199
TRAIL (TNF-related apoptosis-inducing ligand) 199
TRAIL (TNF-related apoptosis-inducing ligand) receptor 53
see also death receptors

F

FADD (Fas-associated death domain) 129, 148–149
in disease 204*T*
extrinsic apoptosis pathway 53, 199
Fas complex 153*F*, 154
structure, DED 150, 151*F*
familial cold urticaria (FCU), NALP3 219–220
familial haemophagocytic lymphohistiocytosis (FHL), perforin 207
Fas-associated death domain *see* FADD (Fas-associated death domain)
Fas/CD95 110, 117, 148, 254
discovery 7, 118–119
Fas/CD95 (*continued*)
in disease 204*T*, 208–209
extrinsic apoptosis pathway 53, 199, 232
FADD complex 153*F*, 154
knockout animals, neurotoxin-induced Parkinson disease 202
ligand *see* FasL/CD95L
mutations, autoimmune lymphoproliferative syndrome (ALPS) 258*F*
pro-survival signalling 201–202
Pseudomonas aeruginosa infection 208
signalling pathways 112–114, 113*F*, 257–258
Bid 113
caspase-8 30
caspase-10 30
CD95L 112–113
cytochrome *c* release 113
tBid 113
structure 111*F*, 255
DD 150, 151*F*
T cells 255
see also DISC (death initiation signalling complex)
Fas–FasL pathway, autoimmune lymphoproliferative syndrome (ALPS) 208
FasL/CD95L 110, 254
cytotoxic T lymphocytes 208
in disease 204*T*
autoantibodies 221
hypergammaglobulinaemia 221
lymphomas 221
extrinsic apoptosis pathway 232
natural killer cells 208
rituximab combination 119
signalling 112–113
FCU (familial cold urticaria), NALP3 219–220
ferredoxin reductase, upregulation by 5-fluorouracil 234
FHL (familial haemophagocytic lymphohistiocytosis), perforin 207
filaggrin 242, 247
keratin intermediate filaments 247–248
"find-me" signals, apoptotic cell engulfment *see* engulfment of apoptotic cells
Fis1 101*F*
mitochondrial fission 102
FLIPs
caspase activation 149
cellular *see* c-FLIP proteins
vFLIPs (viral FLICE-inhibitory proteins) 278
fluorescence resonance energy transfer (FRET), Rac1 169
5-fluorouracil, ferredoxin reductase upregulation 234
follicular lymphomas 73, 78
translocations 74
fractalkine *see* CX3CL1 (fractalkine)
FRET (fluorescence resonance energy transfer), Rac1 169

G

gamma irradiation, p53 apoptosis 81
Gas6 (growth arrest-specific 6), apoptotic cell engulfment 170
Gdap1 101*F*
GDC-0152 144
Gendicine 236
gene duplication, Bcl-2 family protein evolution 72

germinal centre expression, MFG-E8 169
germline apoptosis, *Caenorhabditis elegans* 26
giantin, demolition phase of apoptosis 56
β-glucocerebroside, skin cornification 248
GLUD1 (glutamate dehydrogenase 1), RIP3 interaction 131
GLUL (glutamate-ammonia ligase), RIP3 interaction 131
glutamate-ammonia ligase (GLUL), RIP3 interaction 131
glutamate dehydrogenase 1 (GLUD1), RIP3 interaction 131
glutamate receptors 264
Golgi apparatus, demolition phase of apoptosis 56, 66
gout, NALP3 220
granule exocytosis-induced death 224
granule-mediated caspase-independent cell death 26, 226*F*
granulocyte-macrophage colony stimulating factor (GM-CSF), A1 induction 73
granzyme(s) **223–229**
  cell death 226–228
  substrates 227*F*
  see also specific granzymes
granzyme A 226
  actions 227
  substrates 227, 227*F*
granzyme B
  characteristics 225
  in disease 207
  functions 202, 225*F*
    Bid cleavage 225
  granzyme B pathway 53
  knockout animals 226–227
  substrates 227*F*
granzyme B apoptosis pathway 52, 53–54, 53*F*, 198, 202–203, 224–226
  caspase-3 225
  caspase-7 225
  in disease 205*T*
  phagocytosis 224
  reactive oxygen species 202–203
granzyme C 226, 228
  substrates 227*F*
granzyme H 226, 228
  substrates 227*F*
granzyme K 226, 228
  substrates 227*F*
granzyme M 226, 228
  substrates 227*F*
growth arrest-specific 6 (Gas6), apoptotic cell engulfment 170
growth factors, demolition phase of apoptosis 67
GS-9450 *see* LB84451/GS-9450

H

haematopoietic cancers, Bcl-2 205
HARE (hyaluronic acid receptor for endocytosis), apoptotic cell engulfment 172
Hax1 (CLS1 associated protein), MOMP 95
HCLS1 associated protein (Hax1), MOMP 95
HD *see* Huntington disease (HD)
HDAC6 (histone deactylase 6), autophagy in neurodegenerative disease 194
HDACi (histone deactylase inhibition), Bid effects 82
head and neck squamous cell carcinoma (HNSCC), NOXA 82
heat shock protein(s), apoptosome formation 43
heat shock protein 70 (Hsp70), phagocytosis signal 57
Heidegger, Martin 2
heme, cytochrome *c* binding 42
hepatitis B 284
hepatitis C 284
herpesviruses, vFLIPs (viral FLICE-inhibitory proteins) 278
Herpesvirus Saimiri, Bcl-2-like viral proteins (vBcl-2) 276
hFis1 (human fission 1), mitochondrial fission 94
HGS1029, IAP antagonists 144
HGS-ETR1 (mapatumumab) 118*T*
  clinical trials 124
HGS-ETR2 (lexatumumab) 118*T*
  clinical trials 124
histone deactylase 6 (HDAC6), autophagy in neurodegenerative disease 194
histone deactylase inhibition (HDACi), Bid effects 82
historical aspects (apoptosis) 1–3, 7*F*
HMGB1 57
HNSCC (head and neck squamous cell carcinoma), NOXA 82
Hodgkin disease 257
holocytochrome, structure 41*F*
homeostasis 3*F*
  Bim 83
homotypic interactions, DD (death domain) 154
Horvitz, H Robert 3
Hrk/DP5 81*T*, 86
  nerve growth factor deprivation 86
Hsp70 (heat shock protein 70), phagocytosis signal 57
HtrA2/Omi
  apoptosome formation 42
  MOMP 95
HTT (HUNTINGTON protein), autophagy in neurodegenerative disease 194
human fission 1 (hFis1), mitochondrial fission 94
human herpesvirus 8, Bcl-2-like viral proteins (vBcl-2) 276
Huntington disease (HD)
  apoptosis dysfunction 48
  caspase 36
  necroptosis 134
  protein misfolding 263
HUNTINGTON protein (HTT), autophagy in neurodegenerative disease 194
HUS-1, DNA damage-response pathways 27
hyaluronic acid receptor for endocytosis (HARE), apoptotic cell engulfment 172
hyperactivation, autophagy in cancer 181–182
hypergammaglobulinaemia, FasL 221
hyperglycosylation, TRAIL-Rs 119
hypogammaglobulinaemia, XIAP 206
hypoxic-ischaemic brain injury, acute 264

I

IAPs *see* inhibitors of apoptosis proteins (IAPs)
ICAD (inhibitor of CAD), nuclear degradation 65
IDN-6556 *see* Emricasan
ILP-2, caspase inhibition 35*T*
IMD pathway, DREDD 213, 214–215
IMM *see* inner mitochondrial membrane (IMM)
immune clearance, apoptotic cells 56–57
immune response
  anti-apoptotic effects 280
  see also innate immunity
immune surveillance hypothesis 207
immune tolerance 56–57
immunodeficiencies 259
  extrinsic apoptosis pathway 221
  see also specific diseases/disorders
immunosuppression, apoptotic cell engulfment *see* engulfment of apoptotic cells
'induced conformation' model, caspase-9 activation 40
'induced proximity' model, caspase-9 activation 40
infections
  autophagy *see* autophagy
  route of entry 272
  see also specific infections
inflammasome 219
  caspase-1 activation 150
inflammation, caspase-1 217
inflammatory cytokines 209–210
inflammatory disease, autophagy 183–184
infliximab 210
inherited disorders of apoptosis **253–261**
  see also specific diseases/disorders
inhibitor of CAD (ICAD), nuclear degradation 65
inhibitors of apoptosis proteins (IAPs) **138–147**
  antagonists 118, 144, 162–163
    peptidomimetic antagonists 144, 163, 163*F*
  binding, apoptosome formation 42
  in cancer 143–144
  discovery 138
  drug design 290
  function **156–165**
    caspase regulation 34, 157*F*, 161–162
    RING E3-ligase activity 162
  historical aspects 139
  identification 158
  overexpression 225–256
  structure 139*F*, **156–165**
    RING domains 141
    UBA domains 141
  see also BIR-containing proteins; cIAP(s); *specific proteins;* XIAP (X-linked inhibitor of apoptosis)
innate immunity
  apoptosis *vs.* 217*F*
  autophagy in infections 183
  caspases **212–223**, 214*F*
  definition 212
inner mitochondrial membrane (IMM) 98
  mitochondrial fusion regulation 99–100
  MOMP 95
insulin, autophagy regulation 180
integrins, skin cornification 248
interest in apoptosis 1–2, 6
  scientific papers 6, 6*F*
interleukin-1β conversion enzyme (ICE), knockout animals 36*T*
interleukin-1β pathway inhibition, rheumatoid arthritis 283
interleukin-2 (IL-2) 254
intracellular proteases, skin cornification *see* skin cornification
intracellular signals 10
intrinsic apoptosis pathway 38, 52, 76, 77–78, 91–92, 157–158, 198, 254
  Apaf-1 oligomerization 54, 91
  Bak 54, 274
  Bax 54, 274
  Bcl-2 protein family 54
  BH3-only proteins 274
  caspase activation 62
  cytochrome *c* release 54, 91
  damaging agents 198
  definition 91
  in disease 203–206, 204–205*T*
    protein folding diseases 198
  extracellular signal-regulated kinase (ERK) 274
  extrinsic apoptosis pathway cross-talk 200
  Jun *N*-terminal kinase (JNK) 274
  mitochondrial damage 198–199
  network 38*F*, 53*F*, 63*F*, 76*F*, 199*F*
  NFκB 274
  p53 *see* p53
  regulatory genes 198–199
involucrin 245
ionizing radiation
  C. elegans 26–27
  ced-3 gene mutants 27
  ced-4 gene mutants 27
  ced-9 gene mutants 27
  DNA damage-induced germline apoptosis 26–27
ionotropic glutamate receptors 264
IPAF, inflammasome 219
iPLA (calcium-independent phospholipase A) 67

J

Jacob, François 8–9
Jun *N*-terminal kinase (JNK) 274

K

kainate receptors 264
keratin intermediate filaments (KIFs) 247
  filaggrin 247–248
keratin K5 241–242
keratin K14 241–242
keratinocyte differentiation 240–241
Kerr, John F 3
KIAP *see* ML-IAP
KIFs *see* keratin intermediate filaments (KIFs)
KILLER *see* TRAIL-R2
kinases
  autophagy 178
  see also specific kinases
knockout animals
  Apaf-1 (apoptosis-activating factor-1)
    see Apaf-1 (apoptosis-activating factor-1)
  apoptotic cell engulfment 171
  Bak 73
  Bax 73
  Bcl-2 73
  BclX 73
  Bid 82
  Bmf 83
  caspases 34, 36*T*, 61, 61*T*
  cIAP1 142
  cIAP2 142
  Drp1 101–102, 103
  granzyme B 226–227
  MFG-E8 169–170
  Mfn (mitofusin) 1 100, 103
  Mfn (mitofusin) 2 100, 103

knockout animals (*continued*)
Opa1 103
phosphatidylserine receptors 171
phospholipid scramblases (PLSCRs) 167
RIP1 (receptor-interacting protein 1) 129
Survivin 144–145
TNF receptor-associated death domain (TRADD) 129
TRAIL 208
transglutaminase 1 (TG1) 242
transglutaminase 2 (TG2) 244–245
XIAP 142, 206
Kuhn, Thomas 2

L

lactic acid production, necroptosis pathways 133
LAG-2 (Lin-12 And Glp-1) growth factor 26
lamellar granules (Odland bodies), skin cornification 248
lamins, nuclear fragmentation 56
LAMP-1 (lysosomal-associated membrane protein-1), autophagy 179
LAMP-2 (lysosomal-associated membrane protein-2), autophagy 179
Langerhans cell expression, MFG-E8 170
LARD (lymphocyte-associated receptor death) *see* TRAIL-R1 (DR3)
LB84451/GS-9450 287
structure 288*F*
LBY135 118*T*
clinical trials 124
Lewy bodies *see* Parkinson disease
lexatumumab *see* HGS-ETR2 (lexatumumab)
Li–Fraumeni syndrome 260
p53 203, 231
Lin-12 And Glp-1 (LAG-2) growth factor 26
lipid envelope, skin cornification *see* skin cornification
Listeria monocytogenes, autophagy 193
Livin *see* ML-IAP
LMP (lysosomal membrane permeabilization), necroptosis pathways 133
Lockshin, Richard 2
loricrin 245–246, 246*F*
LPC *see* lysophosphatidylcholine (LPC)
lung adenocarcinomas, p73 235
lymphocyte apoptosis 254, 280
autoimmune disease 256
lymphocyte-associated receptor death (LARD) *see* TRAIL-R1 (DR3)
lymphoid cancers, Bcl-2 205
lymphomas
Bim 83
cIAPs 210
FasL 221
NOXA 82
p53 205
XIAP 206
lysophosphatidylcholine (LPC)
apoptotic cell engulfment 166
phagocytosis signals 57, 67
lysosomal-associated membrane protein-1 (LAMP-1), autophagy 179
lysosomal-associated membrane protein-2 (LAMP-2), autophagy 179
lysosomal membrane permeabilization (LMP), necroptosis pathways 133

M

macrophages, autophagy in infections 183
MALT (mucosa-associated lymphoid tissue) lymphomas
cIAPs 210
IAPs 143
mammals
development 47–48
programmed cell death 17–18
mapatumumab *see* HGS-ETR1 (mapatumumab)
MARCH5 (membrane-associated RING-CH 5) 101*F*
mitochondrial fission 102
mazE/mazF addiction molecule 16
MC159 v-FLIP, DED:DED interactions 152, 153*F*, 154
MCL-1 70
expression regulation 73
lymphoid cells 206
mitochondria network 105
structure 78, 159*F*
T-cell development 220
MDM2 (murine double minute 2)
p53 activation 231, 232*F*
p53 regulation 203
Mdm4, p53 activation 231
MDP (muramyl dipeptide) 217
mechanisms (of cell death) 4*F*
mediators (of apoptosis) 254–255
see also specific mediators
melanoma
apoptosis dysfunction 48
caspase-1 219
PUMA downregulation 82
memantine 267–268
NMDA receptor 269*F*
membrane-associated RING-CH 5 *see* MARCH5 (membrane-associated RING-CH 5)
membrane blebbing, demolition phase of apoptosis 64
metabotropic glutamate receptors 264
metastases, caspase-1 219
3-methyladenine (3-MA) inhibition, autophagy 184
Mff (mitochondrial fission factor) 101*F*
mitochondrial fission 102
MFG-E8
apoptotic cell engulfment 169–170
autoimmune disease development 169–170, 170*F*
germinal centre expression 169
knockout animals 169–170
Langerhans cell expression 170
microglia expression 170
systemic lupus erythematosus 169, 170
Mfn 1 (mitofusin 1) 101*F*
knockout animals 100, 103
mitochondrial fusion regulation 100
overexpression, apoptosis resistance 106
Mfn 2 (mitofusin 2) 101*F*
knockout animals 100, 103
mitochondrial fusion regulation 100
overexpression, apoptosis resistance 106
Mfn-binding protein (MiB) 101*F*
mitochondrial fusion regulation 100
MiB *see* Mfn-binding protein (MiB)
microbial inhibitors **272–282**
benefits of 273–274
direct effects 277*F*
microbial inhibitors (*continued*)
pro-survival pathway activation 274–275, 275*F*
see also bacteria; viruses; *specific inhibitors*
microglia, MFG-E8 170
micro ribonucleic acids (microRNAs), Bcl-2 family proteins expression 73
MIG-2, *C. elegans* dying cell engulfment 25
MIRA (propoxy-methyl meleimide) 237–238
mitochondria
ATP synthesis 97
cristae remodelling 106–107
damage
evolution (of cell death) 11
intrinsic apoptosis pathway 198–199
fission *see* mitochondrial fission
fusion *see* mitochondrial fusion
intrinsic apoptotic pathway 24
morphogenesis proteins 98–99
outer membrane permeabilization *see* mitochondrial outer membrane permeabilization (MOMP)
ultrastructure 97–98
inner mitochondrial membrane (IMM) *see* inner mitochondrial membrane (IMM)
outer mitochondrial membrane (OMM) *see* outer mitochondrial membrane (OMM)
mitochondria-associated phospholipase D (mito-PLD) 100, 101*F*
mitochondrial fission **97–109**
Dnm1p 102
Drp1 101
Fis1 102
MARCH5 (membrane-associated RING-CH5) 102
Mff (mitochondrial fission factor) 102
MOMP *see* mitochondrial outer membrane permeabilization (MOMP)
regulation 101–102, 101*F*
signalling pathways 102
SUMO 102
mitochondrial fission factor *see* Mff (mitochondrial fission factor)
mitochondrial fusion **97–109**
MOMP *see* mitochondrial outer membrane permeabilization (MOMP)
regulation 99–100, 101*F*
mitochondrial outer membrane permeabilization (MOMP) **90–96**, 105, 232
Bcl-2 family 91–92, 92–93
BH3-only proteins 93
bioenergetics 94–95
Bok 92
cytochrome *c* release 42
granzyme B actions 225
induction 93–94
direct activator model 93–94, 93*F*
displacement model 93*F*, 94
internal control 95
mitochondrial fission 94
dynamin-related GTPases 94
dynamin-related protein 1 (Drp1) 94
human fission 1 (hFis1) 94
mitochondrial fusion 94
regulation 107
Smac/DIABLO 91–92
mitochondrial permeability transition (MPT), MOMP 95
mitochondria-mediated apoptosis *see* apoptosomes
mitochondria network 98–99, 98*F*, 99*F*
Bcl-2 protein family 103–105
Bax/Bak double knockouts 104
Bcl-X_L knockouts 104
Bcl-X_L overexpression 104
CED-9 cells 103–104
Drp1 trafficking 105
Mcl-1 105
proapoptotic proteins 104
cellular role 102–103
diseases/disorders 102
mitochondrial morphogenesis proteins 98–99
regulation 98*F*
mitofusin 1 *see* Mfn 1 (mitofusin 1)
mitofusin 2 *see* Mfn 2 (mitofusin 2)
mito-PLD (mitochondria-associated phospholipase D) 100, 101*F*
ML-IAP 138, 140*T*, 142
caspase inhibition 35*T*
structure 139*F*
monocytes, autophagy in infections 183
Monod, Jacques 8–9
MPT (mitochondrial permeability transition), MOMP 95
mPTP, necroptosis pathways 133
MRT-2, DNA damage-response pathways 27
Mst1, DNA condensation/degradation 56
Mtd *see* Bok
MTP18 101*F*
Muckle–Wells syndrome, NALP3 220
multicellularity, apoptosis coevolution 9
multiple death pathways, *Caenorhabditis elegans* 17–18
multiple myelomas 143
multiple sclerosis, caspase-1 219
multivesicular bodies (MVBs), autophagy 179
muramyl dipeptide (MDP) 217
murine double minute 2 *see* MDM2 (murine double minute 2)
MVBs (multivesicular bodies), autophagy 179
myc-induced B-cell lymphoma development, PUMA deficiency 81
Mycobacterium tuberculosis
apoptosis inhibitors 278
autophagy 193
myelomas, multiple 143
myodegenerative disease, autophagy *see* autophagy
Myxobacteria 16
programmed cell death 14
Myxoma virus, caspase inhibitors 278

N

NADH dehydrogenase (ubiquinone) FE-S protein 1 (NDFUS1), MOMP 94–95
NADH dehydrogenase ubiquinone flavoprotein (NDUFS3)
granzyme A cleavage 227–228
reactive oxygen species generation 228
NAIP (BIRC1) 138, 140*T*, 145
caspase inhibition 35*T*
inflammasome 219
structure 139*F*
NALP3, inflammatory disease 219–220
NALPs, inflammasome 219

NASH (nonalcoholic steatohepatitis) 284
natural killer (NK) cells 202, 223–224
actions 226*F*
FasL 208
granzyme B pathway 53
NBD (nucleotide α/β binding domain), Apaf-1 39
Nbk *see* Bik
NDFUS1 (NADH dehydrogenase (ubiquinone) FE-S protein 1), MOMP 94–95
NDUFS3 *see* NADH dehydrogenase ubiquinone flavoprotein (NDUFS3)
Nec-1 (necrostatin-1), RIP1 inhibition 129
necroptosis **127–137**, 201
in vitro assays 130*T*
mechanisms 128
NADPH oxidase 128
necrosome 133*F*
pathways 133–134
phospholipase A 128
physiological role 134–135
lysosomal membrane permeabilization 128
respiration 131–132
reactive oxygen species (ROS) 128, 131
RIP1 (receptor-interacting protein 1) 128, 129, 131
RIP3 (receptor-interacting protein 3) 128
signalling complexes 129–131, 134*F*
necrosis 90
cell microscopy 128*F*
definition 128
necrosome, necroptosis 133*F*
necrostatin(s) 134–135
necrostatin-1 (Nec-1), RIP1 inhibition 129
negative mutants, Drp1 105
Neisseria gonorrhoeae, apoptosis inhibitors 278
Neisseria meningitidis, apoptosis inhibitors 278
neonatal-onset multisystem inflammatory disease (NOMID), NALP3 220
NEP11 (neurite outgrowth-promoting prostaglandin) 268
nerve growth factor deprivation, Hrk/DP5 86
neural tube development, Apaf-1 knockout animals 48
neurite outgrowth-promoting prostaglandin (NEP11) 268
neurodegenerative diseases 90
autophagy *see* autophagy
p53 230
protein misfolding 266
reactive nitrogen species 262
reactive oxygen species 262
see also specific diseases/disorders
neuronal cell death, autophagy 193
neuronal progenitor cells, Apaf-1 knockout animals 48
neurotoxin-induced Parkinson disease, Fas knockout animals 202
neutrophils, apoptosis 280
NF-κB *see* nuclear factor-κB (NF-κB)
nitric oxide (NO) 262
Alzheimer disease 265–266
endoplasmic reticulum, stress in 267
Lewy bodies 265–266
Parkinson disease 265–266
protein *S*-nitrosylation 265, 265*F*
NitroMemantine 268
NMDA receptor 269*F*
nitrosative stress 265–266
S-nitrosylation 262
NMDA receptor 269*F*
parkin 266–267
PDI 267
NLRP1, structure 150, 151*F*
NLRs (nod-like receptors) 215
NMDA receptors 264
excess activation 263*F*
memantine 269*F*
NitroMemantine 269*F*
S-nitrosylation 269*F*
reactive nitrogen/oxygen species 262, 263–265
acute hypoxic-ischaemic brain injury 264
NO *see* nitric oxide (NO)
NOD (nucleotide-binding and oligomerization domain), Apaf-1 38–39, 150
Nod2, Crohn disease 220
nod-like receptors (NLRs) 215
Nod signalosome 215, 217
Nomarski (DIC) optics, *Caenorhabditis elegans* 21, 22*F*
nomenclature problems 5
NOMID (neonatal-onset multisystem inflammatory disease), NALP3 220
nonalcoholic steatohepatitis (NASH) 284
non-Hodgkin lymphoma 257
non-small cell lung carcinoma (NSCLC), NOXA 82
NOXA 79, 80*T*, 81–82
activation 72
in disease 205
head and neck squamous cell carcinoma (HNSCC) 82
lymphomas 82
non-small cell lung carcinoma (NSCLC) 82
DNA damage-induced apoptosis 79, 81
intrinsic apoptotic pathway 233
NSCLC (non-small cell lung carcinoma), NOXA 82
nuclear factor-κB (NF-κB)
activation, TNF-R1 118
demolition phase of apoptosis 67
intrinsic apoptosis pathway 274
pro-survival signalling 201
signalling pathway, tumour necrosis factors 114
TOLL pathway 214
nucleoporins, nuclear degradation 65
nucleotide α/β binding domain (NBD), Apaf-1 39
nucleotide-binding and oligomerization domain (NOD), Apaf-1 38–39, 150
nucleotide-binding Apaf-1/Rgene/CED4 (NB-ARC) domain, Apaf-1 38–39
nucleus
degradation, CAD (DNA fragmentation factor 40/DEF40) 65
envelope fragmentation 56
fragmentation *see* demolition phase of apoptosis
lamina proteins 65
'nurse cell model,' *Caenorhabditis elegans* germline apoptosis 26
Nutlins 237
nutrient restriction, autophagy 192

O

Odland bodies (lamellar granules), skin cornification 248
Omi/HtrA2, apoptosome formation 42
OMM *see* outer mitochondrial membrane (OMM)
Opa1 (optic atrophy 1) 101*F*
downregulation, apoptosis increase 106
knockout animals 103
mitochondrial fusion regulation 100
OPG *see* osteoprotegerin (OPG)
OpIAP, caspase inhibition 35*T*
optic atrophy 1 *see* Opa1 (optic atrophy 1)
'original sin' hypothesis, evolution (of cell death) 16–17
osteoarthritis 283
VX-740 287
osteoprotegerin (OPG) 110
structure 111*F*
outer mitochondrial membrane (OMM) 91, 98
Bak 105
Bax 92, 105
Bcl-2 protein family association 98
mitochondrial fusion regulation 99–100
permeabilization *see* mitochondrial outer membrane permeabilization (MOMP)
transporter of the outer membrane (TOM) 98
voltage-dependent anionic channel (VDAC) 98
oxidative stress
autophagy 192
Parkinson disease 269–270
see also reactive oxygen species (ROS)

P

P1-aspartic moiety, caspase inhibitors 286
P35, caspase inhibition 35*T*
p53 79, **230–240**
activation 231, 231*F*
Mdm2 231, 232*F*
stress signals 230
C. elegans 232
cancer therapy 236–238, 237*F*
breast cancer 236
mutation effects 79
D. melanogaster 232
in disease 203, 204*T*, 205
cancer 203, 230–231
Li–Fraumeni syndrome 203, 231
lymphoma 205
neurodegenerative disease 230
extrinsic apoptosis pathway 232–233, 233*F*
family members 235–236
chemotherapy-induced apoptosis 236
see also specific members
future work 238
inactivation
carcinogenesis 162–163
tumour cell chemotherapy resistance 118
intrinsic apoptotic pathway 232, 233–234, 233*F*
mutations, chemotherapy resistance 79
pro-survival signalling inhibition 234–235
reactivation 236, 237*F*
redox signalling 234
regulation, MDM2 (murine double minute 2) 203
roles/functions 232–235
structure 235*F*
transactivation-independent activities 234
as transcription factor 230
p53 *(continued)*
transcription-defective mutants 234
tumour suppression 236
p55 *see* tumour necrosis factor receptor 1 (TNF-R1)
p60 *see* tumour necrosis factor receptor 1 (TNF-R1)
p63 235
structure 235*F*
p73 235
structure 235*F*
p115 56
PAMP *see* pathogen-associated molecular patterns (PAMP)
paracrine killers, bacteria 15
parasites 274
apoptosis inhibitors 279–280
host cell interactions 273*F*
parkin
definition 266
S-nitrosylation 266–267
Lewy bodies 266, 267*F*
Parkinson disease 266, 267*F*
Parkinson disease 266–267
animal models 264
apoptosis dysfunction 48
cell models 267
Lewy bodies
nitric oxide 265–266
parkin *S*-nitrosylation 266, 267*F*
necroptosis 134
oxidative stress 269–270
protein misfolding 262
'sporadic' 264
PARL (presenilin-associated rhomboid-like protein) 106–107
PARP-1 (poly(adenosine diphosphate (ADP)-ribose)polymerase) 65
pathogen-associated molecular patterns (PAMP) 213
IMD pathway 214
pattern recognition receptors (PRRs) 274–275
caspase binding 215–219
definition 212
PDB 1ICE, structure 286*F*
PDB 1SC1, structure 286*F*
PDB 1SC3, structure 286*F*
Pelle:Tube complex, DD:DD interactions 152, 153*F*
peptidoglycans (PGNs) 214
peptidomimetic antagonists, IAPs 144, 163, 163*F*
perforin 202–203, 224
in disease 205*T*, 207
familial haemophagocytic lymphohistiocytosis (FHL) 207
granzyme B pathway 53
periodic fever syndrome, TNFR1 knockout animals 209
peroxynitrite (ONOO-) 262
PF-3491390 *see* Emricasan
PGN recognition proteins (PGRPs), IMD pathway 214
phage-display cloning 171
phagocytosis
apoptotic cells 57
autophagy in infections 184*F*
granzyme B apoptosis pathway 224
see also demolition phase of apoptosis; engulfment of apoptotic cells
phagophore, autophagy 176
phosphatase and tensin homolog *see PTEN* (phosphatase and tensin homolog)
phosphatidylinositol 3-kinase (PI3K)
autophagy 178, 179, 190
intrinsic apoptosis pathway 274
p53 234–235

phosphatidylserine (PS)
apoptotic cell engulfment 167, 172
dead cell recognition 26
phagocytosis 57, 67
receptors *see* engulfment of apoptotic cells
phospholipase A2, apoptotic cell engulfment 166–167
phospholipid scramblases (PLSCRs) 167
PI3K *see* phosphatidylinositol 3-kinase (PI3K)
PIDD, RAIDD complex 153*F*, 154
PIDDosome, caspase-2 activation 150
PKB-Akt, procaspase-9 phosphorylation 42–43
plakophilin-1 mutations, skin cornification 249
Plasmodium, apoptosis inhibitors 280
PLSCRs (phospholipid scramblases) 167
polycystic kidney disease, TNF receptors 209
poly(adenosine diphosphate (ADP)-ribose)polymerase (PARP-1) 65
porins, apoptosis inhibitors 278
posttranslational regulation, Bim 83
poxviruses
Bcl-2-like viral proteins (vBcl-2) 276–277
caspase inhibitors 278
vFLIPs (viral FLICE-inhibitory proteins) 278
PP2A (protein phosphatase 2A), autophagy regulation 180
presenilin, MOMP 95
presenilin-associated rhomboid-like protein (PARL) 106–107
PRF1 *see* perforin
PRIMA (p53 reactivation and induction of massive apoptosis) 237–238
prion diseases, protein misfolding 263
pro-apoptotic Bcl-2 family proteins 70*F*
pro-apoptotic gene activation, demolition phase of apoptosis 65
pro-apoptotic proteins, mitochondria network 104
procaspase-9 24
activation 40–41
Apaf-1 complex 152, 153*F*
differential localization 43
isoforms 41
phosphorylation, apoptosome formation 42–43
see also caspase-9
profilaggrin 242, 247–248
programmed cell death
bacteria 14
definition 6
evolution **13–20**
mammals 17–18
type II 176
unicellular organisms 14
propoxy-methyl meleimide (MIRA) 237–238
prostate cancer, Bik/Nbk/Blk 83
proteasome inhibitors, Bid effects 82
protein misfolding 262, 265–266
Huntington disease 263
intrinsic apoptosis pathway 198
prion diseases 263
protein *S*-nitrosylation, nitric oxide (NO) 265, 265*F*
protein phosphatase 2A (PP2A), autophagy regulation 180
protein S, apoptotic cell engulfment 170
proteolytic processing, caspase activation 51–52
'proximity-driven dimerization' model, caspase-9 activation 40
PRRs *see* pattern recognition receptors (PRRs)
PS *see* phosphatidylserine (PS)
Pseudomonas aeruginosa infection, Fas-induced apoptosis 208
psoriasis
skin cornification 249
TNF receptors 209
Ptainacasan *see* VX-740
PTEN (phosphatase and tensin homolog)
cancer 259
p53 234
PUMA 79, 80*T*, 81–82
activation 72
cancer 259–260
deficiency, *myc*-induced B-cell lymphoma development 81
in disease 205
chronic lymphocytic leukaemia 81
melanoma 82
DNA damage-induced apoptosis 79, 81
downregulation, melanoma 82
intrinsic apoptotic pathway 233
upregulation, chronic lymphocytic leukaemia 81
putative HLA-DR-associated protein-1 *(PHAP1)*, apoptosome formation 43
PYD (pyrin domain) **147–156**
functions 148–150
structure 149*F*, 150, 152
pyrin domain *see* PYD (pyrin domain)
pyroptosis, caspase-1 217

R

Rab(s), autophagy *see* autophagy
Rab5, apoptotic cell engulfment 169
Rab11, autophagy 179
Rab24, autophagy 179
Rabbitpox virus, caspase inhibitors 278
Rac1
apoptotic cell engulfment 168–169
fluorescence resonance energy transfer (FRET) 169
RacGEF, *Caenorhabditis elegans* dying cell engulfment 25
RAD-5/CLK-2, DNA damage-response pathways 27
RAIDD (RIP-associated ICH-1homologous protein with a death domain) 150
ras/MAPK pathway, *Caenorhabditis elegans* germline apoptosis 26
reactivation of transcriptional reporter activity (RETRA) 238
reactive nitrogen species (RNS) **262–272**
neurodegenerative disease 262
NMDA receptor channels *see* NMDA receptors
protection from 267–270
see also specific treatments
see also individual species
reactive oxygen species (ROS) **262–272**
caspase activation 264–265
generation, NADH dehydrogenase ubiquinone flavoprotein (NDUFS3) 228
granzyme B pathway 202–203
necroptosis 131
neurodegenerative disease 262
NMDA receptor channels *see* NMDA receptors
reactive oxygen species (ROS) *(continued)*
protection from 267–270
see also specific treatments
see also individual species
receptor-interacting protein *see* RIP1 (receptor-interacting protein 1)
receptor-interacting protein 2 (RIP2), Nod signalosome 215
receptor-interacting protein 3 (RIP3) 131
redox environment , endoplasmic reticulum 267
redox signalling, p53 234
Red Queen metaphor 14–15
REF(2)P, autophagy in neurodegenerative disease 194
regulatory genes **197–211**
RELISH 214–215
RETRA (reactivation of transcriptional reporter activity) 238
rhApo2L/TRAIL 118*T*
clinical trials 124, 125
structure 120, 121*F*
rheumatoid arthritis
interleukin-1β pathway inhibition 283
TNF receptors 209
tumour necrosis factor-α overexpression 209
VX-740 287
RING domains
Bcl-2 family proteins 160–161, 161*F*
IAPs 141
RING E3-ligase activity, IAPs 162
RIP1 (receptor-interacting protein 1)
caspase-8, cleavage by 131
knockout animals 129
necroptosis 129, 131
necrostatin-1 (Nec-1) 129
pro-survival signalling 201
RIP2 (receptor-interacting protein 2), Nod signalosome 215
RIP3 (receptor-interacting protein 3) 131
RIP-associated ICH-1homologous protein with a death domain (RAIDD) 150
RITA (reactivation of p53 and induction of tumour cell apoptosis) 237
rituximab, CD95L combination 119
RNA splicing, demolition phase of apoptosis 65
RNS *see* reactive nitrogen species (RNS)
ROCK1 (Rho-associated kinase)
cell rounding 55
gain-of-function substrate cleavage 63
ROS *see* reactive oxygen species (ROS)
R-roscovitine, clinical trials 203, 205
rounding, demolition phase of apoptosis *see* demolition phase of apoptosis

S

Saccharomyces cerevisiae
ageing 18
asymmetric division 18
autophagy regulation 189–190
SAF-A (scaffolding attachment factor), nuclear degradation 65
Saunders, John 3
SBMA (spinobulbar muscular atrophy) 194
scaffolding attachment factor (SAF-A), nuclear degradation 65
SCCE (stratum corneum chymotryptic enzyme) 249
SCCL (stratum corneum cathepsin-L-like enzyme) 249
scientific papers, interest in apoptosis 6, 6*F*
SCTE (stratum corneum tryptic enzyme) 249
SCTP (stratum corneum thiol protease) 249
Sec16, autophagy 179
Sec17, autophagy 179
self–non-self recognition 212
sepsis 283
septic shock
caspase-1 219
tumour necrosis factor-α 209
SH2D1A mutation, X-linked lymphoproliferative (XLP) syndrome 259
Shigella flexneri, apoptosis inhibitors 278
signalling pathways
BH3-only proteins 78, 78*F*
Fas 257–258
mitochondrial fission 102
see also specific pathways
SIRS (systemic inflammatory response syndrome), caspase-1 219
SKD1 (suppressor-of-potassium-transport-growth-defect-1) 179
skin cornification **240–252**, 243*F*, 244*F*
cell adhesion 248–249
desquamation 248–249
diseases/disorders 242
epidermal differentiation 240–242, 241*F*
apoptosis 241
keratin K5 241–242
keratin K14 241–242
keratinocyte differentiation 240–241
extracellular proteases 249–250
intracellular proteases 249
lipid envelope 248
skin transglutaminases 242, 244–245, 244*F*
substrates 245–248
see also specific substrates
see also specific enzymes
transglutaminases 242
skin transglutaminases *see* skin cornification
SLE *see* systemic lupus erythematosus (SLE)
Smac/DIABLO
apoptosome formation 42
MOMP 91–92
XIAP binding 141*F*
small proline-rich proteins 242, 246–247, 247*T*
SNAREs *see* autophagy
SPÄTZLE, TOLL pathway 214
SPCs (subtilisin-like protein convertases) 249
spinobulbar muscular atrophy (SBMA) 194
SpoA sporulation factor, *Bacillus subtilis* 15
'sporadic' Parkinson disease 264
stabilin-2, phosphatidylserine receptors 172
Stent, Gunther 2
stomatin-like protein 2 (Stoml2), mitochondrial fusion regulation 100
Stoml2 (stomatin-like protein 2), mitochondrial fusion regulation 100
stratum corneum, structure 242
stratum corneum cathepsin-L-like enzyme (SCCL) 249

stratum corneum chymotryptic enzyme (SCCE) 249
stratum corneum thiol protease (SCTP) 249
stratum corneum tryptic enzyme (SCTE) 249
Streptomyces, programmed cell death 14
stress signals, p53 activation 230
stroke 284
styrylquinazoline (CO-31398) 237
subtilisin-like protein convertases (SPCs) 249
Sulston, John 3
SUMO, mitochondrial fission 102
suppressor-of-potassium-transport-growth-defect-1 (SKD1) 179
survival gene inactivation, demolition phase of apoptosis 65
survivin 138, 140*T*, 144–145
  caspase inhibition 35*T*
  p53 234
  structure 139*F*, 144
symbioses, evolution (of cell death) 10–11
synphilin-1, misfolding 262
syntaxin 5, demolition phase of apoptosis 56
α-synuclein, misfolding 262
systemic inflammatory response syndrome (SIRS), caspase-1 219
systemic lupus erythematosus (SLE) 255–256
  MFG-E8 169, 170
systemic lupus erythematosus-like disease, Bim 83

T

TAM receptors, apoptotic cell engulfment 170–171
Tap42, autophagy regulation 180
TAT-1, apoptotic cell engulfment 167, 168*T*
tau, misfolding 262
tBid (truncated Bid) 82
  CD95 signalling pathways 113
  mitochondrial permeabilization 62
T cells
  autoimmune disease 255–256
  cytotoxic *see* cytotoxic T-cells
  development 220
  Fas 255
  survival, Bcl-2 206
  T-helper cells 212–213
TG1 *see* transglutaminase 1 (TG1)
TG2 *see* transglutaminase 2 (TG2)
TG3 (transglutaminase 3) 243*F*, 245
TG4 (transglutaminase 4) 243*F*
TG5 (transglutaminase 5) 243*F*, 245
TG6 (transglutaminase 6) 243*F*
TG7 (transglutaminase 7) 243*F*
Theileria parva, apoptosis inhibitors 280
T-helper cells 212–213
TIGAR gene activation, p53 234
Tim-4 expression, phosphatidylserine receptors 171–172
TIR (TOLL/interleukin 1 (IL-1) receptor) domain, TOLL pathway 214
TL1A 110
TLRs *see* Toll-like receptors (TLRs)
T-lymphoma cells, Bik/Nbk/Blk 83
TNFR-associated membrane protein *see* TRAIL-R1 (DR3)
TNF receptor-associated death domain *see* TRADD (TNF receptor-associated death domain)
TNF-related apoptosis-inducing ligand *see* TRAIL (TNF-related apoptosis-inducing ligand)
TNF-related apoptosis-inducing ligand receptor 1 *see* TRAIL-R1 (DR3)
TNF-related apoptosis-inducing ligand receptor 2 *see* TRAIL-R2
TNFRSF6 *see* Fas/CD95
TNFRSF10A *see* TRAIL-R1 (DR3)
TNFSF6 *see* FasL/CD95L
TNFSF10 *see* TRAIL (TNF-related apoptosis-inducing ligand)
TOLL/interleukin 1 (IL-1) receptor (TIR) domain, TOLL pathway 214
Toll-like receptors (TLRs) 215, 274–275
  autophagy in infections 183
  ligands 215
TOLL pathway, DREDD 213–214
TOR
  ageing 194
  autophagy regulation 179, 180, 190
toxin pleiotropy, *Escherichia coli* 17
Toxoplasma gondii, apoptosis inhibitors 280
TP53 *see* p53
TRA-8 (CS-1008) 118*T*
  clinical trials 124
TRADD (TNF receptor-associated death domain) 129
  pro-survival signalling 201
TRAF1, cIAP binding 142, 162
TRAF2
  cIAP binding 142, 162
  multiple myelomas 143
  pro-survival signalling 201
TRAF3, multiple myelomas 143
TRAIL (TNF-related apoptosis-inducing ligand) 110, 119–124
  CMV clearance 208
  extrinsic apoptosis pathway 199
  in disease 204*T*
  in drug design 289
  knockout animals 208
  p53 in extrinsic apoptosis pathway 232
  receptors 114
    see also specific receptors
  structure 119
  TRAIL receptor complex 122–123, 122*F*
  tumour cell killing 207–208
TRAIL receptor(s) (TNF-related apoptosis-inducing ligand receptor) 119–124
  agonists 118*T*
    clinical trials 123–124, 124–125
    design 122–123
    tumour cell selectivity 125
  cancer therapy 124–125
  extrinsic apoptosis pathway 53
  pro-survival signalling 201
  therapeutic approaches 120*F*
  TRAIL complex 122–123, 122*F*
  see also specific receptors
TRAIL-R1 (DR3) 110, 117, 255
  apoptosis contribution 121–122
  cancer cell expression 207
  hyperglycosylation 119
  in disease 204*T*
  signalling 114
  structure 111*F*
TRAIL-R2 110, 117, 255
  apoptosis contribution 121–122
  cancer cell expression 207
  hyperglycosylation 119
  signalling 114
  structure 111*F*
TRAIL-R3 110
TRAIL-R3 110 *(continued)*
  structure 111*F*
  tumour differential expression 119
TRAIL-R4 110
  structure 111*F*
  tumour differential expression 119
TRAMP *see* TRAIL-R1 (DR3)
transcription
  Apaf-1 regulation 43
  caspase-9 regulation 43
transglutaminase(s), small proline-rich proteins 246
transglutaminase 1 (TG1) 242, 243*F*
  knockout animals 242
transglutaminase 2 (TG2) 243*F*, 244–245
  knockout animals 244–245
transglutaminase 3 (TG3) 243*F*, 245
transglutaminase 4 (TG4) 243*F*
transglutaminase 5 (TG5) 243*F*, 245
transglutaminase 6 (TG6) 243*F*
transglutaminase 7 (TG7) 243*F*
translation
  Apaf-1 regulation 43
  caspase-9 regulation 43
transporter of the inner membrane (TIM) 98
transporter of the outer membrane (TOM) 98
TRICK2 *see* TRAIL-R2
truncated Bid *see* tBid (truncated Bid)
TUCAN/CARDINAL, apoptosome formation 43
tumourigenesis *see* carcinogenesis
tumour necrosis factor(s) 110
  complex I 114
  NF-κB signalling pathway 114
  signalling 114
    cIAP1 142–143, 143*F*
tumour necrosis factor α (TNF-α)
  A1 induction 73
  cancer therapy 207
  in disease 204*T*
    rheumatoid arthritis 209
  inflammatory cytokine expression 209–210
  overexpression 209
  septic shock 209
  signalling 114
tumour necrosis factor receptor(s)
  in disease 209–210
  extrinsic apoptosis pathway 199
tumour necrosis factor receptor 1 (TNF-R1) 110, 117, 254–255
  caspase-8 activation 201
  extrinsic apoptosis pathway 199
  knockout animals 201
    periodic fever syndrome 209
  NFκB activation 118
  signalling 129
  structure 111*F*
tumour necrosis factor-related apoptosis-inducing ligand *see* TRAIL (TNF-related apoptosis-inducing ligand)
tumours *see* cancer
tumour up-regulated CARD-containing antagonist of caspase nine *(TUCAN/CARDINAL)*, apoptosome formation 43

U

UBA domains, IAPs 141
ubiquitin-like conjugation systems, autophagy *see* autophagy
ubiquitin-proteasome system (UPS) 266
UFO drugs 268
ULK (UNC-51-like kinases), autophagy 178
ultraviolet irradiation, p53 apoptosis 81
ultraviolet irradiation resistance-associated gene (UVRAG) 178
UNC-51-like kinases (ULK), autophagy 178
UNC-73, *Caenorhabditis elegans* dying cell engulfment 25
unicellular organisms 9
  programmed cell death 14
  see also specific organisms
upstream inducers, programmed cell death 17

V

Vaccinia virus, caspase inhibitors 278
Vam3, autophagy 179
Vam7, autophagy 179
vBcl-2 *see* Bcl-2-like viral proteins (vBcl-2)
VDAC *see* voltage-dependent anion channel (VDAC)
vFLIPs (viral FLICE-inhibitory proteins) 278
viruses
  apoptosis induction 273
  apoptosis inhibitors 275–278
    caspase inhibitors 277–278
    necroptosis 135
    see also specific inhibitors
  Bcl-2 homologues 73
  cell death blocking 223
  definition 273
  host cell interactions 273*F*
viral FLICE-inhibitory proteins (vFLIPs) 278
viral mitochondrial-associated inhibitor of apoptosis (vMIA) 106
voltage-dependent anion channel (VDAC)
  necroptosis pathways 134
  outer mitochondrial membrane (OMM) 98
VPS15
  autophagy regulation 190
  deletion, neurodegenerative disease 193–194
VPS34
  autophagy regulation 190
  mutants 190–191
Vti1, autophagy 179
VX-740 286, 287
  structure 288*F*
VX-765 286, 287
  structure 288*F*

W

'warhead,' caspase inhibitors 284, 286
WD-40 region, Apaf-1 38
WD repeat domain, Apaf-1 150
WHD (winged-helix domain), Apaf-1 39
winged-helix domain (WHD), Apaf-1 39
WSL1 *see* TRAIL-R1 (DR3)
Wyllie, Andrew H 3

X

XIAP (X-linked inhibitor of apoptosis) 138, 140*T*, 141–142, 158
- apoptosis modulation 115
- apoptosome formation 42
- caspase regulation 35*T*, 162, 199
- in disease 205*T*, 206
  - *Epstein–Barr virus* infection 206
  - hypogammaglobulinaemia 206
  - lymphomas 206
  - X-linked lymphoproliferative (XLP) syndrome 206
- drug design 290
- *in vitro* assays 142
- knockout animals 142, 206
- Smac/DIABLO dimer binding 141*F*
- structure 139*F*

X-linked lymphoproliferative (XLP) syndrome 259
- XIAP 206

Y

Ykt6, autophagy 179

Z

zVAD
- necroptosis 135
- TNF-induced apoptosis 129

Printed and bound in the UK by
CPI Antony Rowe, Eastbourne

Printed and bound by CPI Group (UK) Ltd, Croydon, CR0 4YY
16/09/2026
14963384-0002